# Environmental Chemistry
## A Global Perspective

# Environmental Chemistry
## A Global Perspective

### THIRD EDITION

### Gary W. vanLoon

Department of Chemistry and School of Environmental Studies,
Queen's University, Kingston, Ontario

and

### Stephen J. Duffy

Department of Chemistry and Biochemistry, Mount Allison University,
Sackville, New Brunswick

# OXFORD
### UNIVERSITY PRESS

Great Clarendon Street, Oxford OX2 6DP

Oxford University Press is a department of the University of Oxford.
It furthers the University's objective of excellence in research, scholarship,
and education by publishing worldwide in

Oxford New York

Auckland  Cape Town  Dar es Salaam  Hong Kong  Karachi
Kuala Lumpur  Madrid  Melbourne  Mexico City  Nairobi
New Delhi  Shanghai  Taipei  Toronto

With offices in

Argentina  Austria  Brazil  Chile  Czech Republic  France  Greece
Guatemala  Hungary  Italy  Japan  South Korea  Poland  Portugal
Singapore  Switzerland  Thailand  Turkey  Ukraine  Vietnam

Published in the United States
by Oxford University Press Inc., New York

First edition published 2000
Second edition published 2005

British Library Cataloguing in Publication Data

Data available

Library of Congress Cataloging in Publication Data

Data available

ISBN 978–0–19–922886–7

1 3 5 7 9 10 8 6 4 2

Typeset by MPS Limited, a Macmillan Company
Printed in the UK by
CPI William Clowes, Beccles, NR34 7TL

# Preface to First Edition

During the past decade, environmental chemistry has come into its own as a respected subdiscipline in the field of chemical science—a subject that occupies an important place in both teaching and research activities of many academic institutions.[*] In its early phases, environmental chemistry was essentially a catalogue or description of chemical properties of the natural world and of concentrations of contaminants. As the subject matured, however, it has come to encompass challenging studies of highly complex systems, and not just in a static way. Current research focuses on the processes that operate within and between various environmental compartments and the ways in which human activities interact with the natural processes. Studies in environmental chemistry use information from all the traditional subdisciplines, building on this and contributing new knowledge in highly specific ways. Beyond the specialized research, environmental chemistry also involves attempts to integrate the particular ideas into a comprehensive picture of how the natural environment functions and responds to stresses. In this book, we have tried to introduce the basic concepts of the subject and to capture its vitality and relevance to key issues of global concern.

In preparing *Environmental Chemistry—A Global Perspective*, we have kept several ideas in mind.

- The book deals with *chemical principles* operating in the natural and altered environment.
- It builds on the fundamentals of physical, organic and inorganic chemistry. As such, it is directed toward students at the second- or third-year level in an undergraduate chemistry program.
- It presents a descriptive approach to the most important topics within the overall subject of environmental chemistry; at the same time, we give an introduction to some basic quantitative calculations.
- The subject is considered in a global context. Examples are chosen from all the continents, emphasizing the world-wide interconnectedness of all environmental issues.

The writing of the book builds on many years of research and teaching (for both of us) within the broad areas of environmental chemistry. We are indebted to students who have asked questions and to students who have found answers through their involvement in research projects. We also acknowledge support from colleagues in the Department of Chemistry and the School of Environmental Studies at Queen's University, as well as at Brock University. Finally, and most important, we wish to offer thanks to our family members for their support of our work over many years.

It is hoped that, in a small way, this book may contribute to understanding that is needed to maintain and restore this good Earth.

'. . . and God saw everything that he had made, and it was very good . . .'

Genesis 1.31a

December 1999

Gary W. vanLoon
Stephen J. Duffy

## Solutions manual

The complete worked solutions to the end of chapter problems are available for lecturers as PDF files on the book's companion web site: http://www.oxfordtextbooks.co.uk/orc/vanloon3e/

[*] Glaze, W. H., Environmental Chemistry Comes of Age, *Environ. Sci. Technol.*, **28**(4) (1994), 169A.

# Preface to Third Edition

Environmental Chemistry is a subject that presents chemists with a unique set of practical and academic challenges. Our laboratory is the Earth, with its mixture of air, water and soil phases, combined in homogeneous and heterogeneous systems made up of living and non-living components. A challenging subject that demands the most serious study!

And Environmental Chemistry is an evolving area of study with new data and new insights being continuously generated through the research efforts of scientists around the world. Throughout the third edition of *Environmental Chemistry – a Global Perspective*, we have tried to incorporate some of this new material into the flow of subjects, by updating information in fast-moving areas like atmospheric chemistry. In a new feature called *Literature Links*, which are occasionally included, we briefly summarize papers from diverse laboratories that describe current research related to the various themes of the book.

More and more, science practitioners including students are using the Web as a source of latest information. While recognizing the challenges of sifting through the seemingly endless supply of information to find reliable material, we have included references to websites of front line organizations such as the United Nations Environment Program, the World Resources Institute and the World Water Council. Some of the websites of such groups provide searchable databases and even realtime information, and we use these data in particular examples and problems.

We have maintained an overview that takes into account the global nature of Environmental Chemistry, by emphasizing the interconnectedness of natural processes and cycles with agriculture, industry, and personal human activities that take place in every society around the world.

At the time of writing we have been again and again reminded of the ultimate importance of understanding and respecting the Earth. Human life is at peril if we fail to work within the limits that nature has set. The massive oil release in the Gulf of Mexico, the impact of volcanic dust spewing from an eruption in Iceland, the ongoing increases in energy use around the world (and meagre controls set by the Copenhagen deliberations) were all news in late 2009 and early 2010. These, and many other less publicized issues, point to an ongoing need for all of us, including very importantly chemists, to work for human welfare within a limited environment. We hope that this book contributes to the understanding required for this endeavour.

As before, we acknowledge good ideas from users of the book in many countries, continuing support from colleagues at Queen's and Mount Allison Universities, and from our families.

Gary W. vanLoon
Stephen J. Duffy

# Contents

# Abbreviations

| | |
|---|---|
| AcE | acetylcholinesterase enzyme |
| ANC | acid neutralizing capacity |
| AQI | air quality index |
| AROTEL | airborne Raman, ozone, temperature and aerosol Lidar (see also Lidar) |
| ASP | activated sludge process |
| BCF | bioconcentration factor |
| BLM | biotic ligand model |
| BOD | biological (or biochemical) oxygen demand |
| BOQ | Bay of Quinte |
| CB | chlorobenzenes |
| CCF | chemical concentration factor |
| CEC | cation exchange capacity |
| CFC | chlorinated fluorocarbon |
| COD | chemical oxygen demand |
| COH | coefficient of haze |
| CP | chlorophenols |
| DDD | dichlorodiphenyldichloroethane |
| DDE | dichlorodiphenyldichloroethene |
| DDT | dichlorodiphenyltrichloroethane |
| DIAL | differential absorption Lidar (see also Lidar) |
| DOC | dissolved organic carbon |
| DOM | dissolved organic matter |
| DTPA | diethylenetriaminepentaacetate |
| DU | Dobson unit |
| EC | electrical conductivity |
| EDTA | ethylenediaminetetraacetic acid |
| en | electronegativity |
| EPA | (US) Environmental Protection Agency |
| ESP | exchangeable sodium percentage |
| FA | fulvic acid |
| FBC | fluidized-bed combustion |
| GAC | granular activated charcoal |
| GUS | groundwater ubiquity score |
| GWP | global warming potential |
| HA | humic acid |
| HCFC | hydrochlorofluorocarbon |
| HFC | hydrofluorocarbon |
| HM | humic material |
| HNLC | high nitrate, low chlorophyll |
| Hu | humin |
| IP | inhalable particulates |
| IPC | isopropyl-N-phenylcarbamate |
| IR | infra-red |
| $LC_{50}$ | lethal concentration$_{50}$ (concentration that would be lethal to 50% of an infinitely large population of the test organism over a given time period) |

| | |
|---|---|
| $LD_{50}$ | lethal dose$_{50}$ (dose that would be lethal to 50% of an infinitely large population of the test organism) |
| lidar | light detection and ranging |
| LPG | liquefied petroleum gas |
| MAMC | Metropolitan Area of Mexico City |
| MAS | magic angle spinning (in NMR) |
| MIC | methyl isocyanate |
| MMT | methylcyclopentadienyl manganese tricarbonyl |
| MSW | municipal solid waste |
| MTBE | methyl tertiary butyl ether |
| NMHC | non-methane hydrocarbon |
| NMR | nuclear magnetic resonance |
| NOAEL | no observable adverse effects levels (also called $LD_0$ or $LC_0$) |
| NOM | natural organic matter |
| NTA | nitrilotriacetic acid |
| ODP | ozone depletion potential |
| OM | organic matter |
| OMI | ozone monitoring instrument |
| PAH | polyaromatic aromatic hydrocarbon |
| PAN | peroxyacetic nitric anhydride |
| PAR | photosynthetically active radiation |
| PBDE | polybrominated diphenyl ethers |
| PCB | polychlorinated biphenyls |
| PCDD | polychlorinated dibenzo-$p$-dioxin |
| PCDF | polychlorinated dibenzofurans |
| PCP | pentachlorophenol |
| PM | particulate matter ($PM_{2.5}$, fraction of fine particles smaller than 2.5 µm) |
| POC | particulate organic carbon |
| POM | particulate organic matter |
| ppbv | parts per billion by volume |
| ppm | parts per million |
| ppmv | parts per million by volume |
| PPN | peroxypropionic nitric anhydride |
| PSC | polar stratospheric clouds |
| PTFE | polytetrafluoroethene |
| PVC | polyvinyl chloride |
| pzc | point of zero charge ($pH_0$) |
| RIRF | relative instantaneous radiative forcing (index) |
| RPP | reductive pentose phosphate (cycle) |
| RSP | respirable suspended particulates |
| RVP | Reid vapour pressure |
| SAR | sodium absorption ratio |
| SOLVE | Sage III ozone loss and validation experiment |
| SOF | soluble organic fraction (of diesel fuel) |
| SPF | sun protection factor |
| SPM | solid particulate matter |
| SS | suspended solids |
| TCE | trichloroethylene |
| TDF | total dustfall |
| TN | total nitrogen |

| | |
|---|---|
| TOC | total organic carbon |
| TOMS | total ozone mapping system |
| TP | total phosphorus |
| TS | total solids |
| TSP | total suspended particulate |
| TSS | total suspended solids |
| UF | urea formaldehyde (polymer) |
| UV | ultraviolet |
| VOC | volatile organic compound |
| zpc | zero point of charge ($pH_0$) |

# Environmental Chemistry
## A Global Perspective

### THIRD EDITION

We have to visualize the Earth as a small, rather crowded spaceship, destination unknown, in which humans have to find a slender thread of a way of life in the midst of a continually repeatable cycle of material transformations. In a spaceship, there can be no inputs or outputs. The water must circulate through the kidneys and the algae, the food likewise, the air likewise .... In a spaceship there can be no sewers and no imports.

Up to now the human population has been small enough so that we have not had to regard the Earth as a spaceship. We have been able to regard the atmosphere and the oceans and even the soil as inexhaustible reservoirs, from which we can draw at will and which we can pollute at will. There is writing on the wall, however .... Even now we may be doing irreversible damage to this precious little spaceship.

*K.E. Boulding*, 1966

# Chapter 1
## Environmental chemistry

The history of the Earth can be said to have begun more than 4.6 billion years ago.

For largely unexplained reasons, a cloud of molecular particles—mostly hydrogen—rotating through the galaxy began to contract and spin with increasing velocity. As the gravitational energy increased, contraction continued to accelerate and massive amounts of heat were generated. Initially, the heat was radiated out into space, but eventually it became trapped within the confines of the central body—the *protostar*—and its core became extremely dense and hot. The massive energy release caused hydrogen within the hottest regions to become ionized. The hydrogen nuclei became fuel for self-sustaining thermonuclear fusion reactions that maintained an interior temperature far in excess of 1 000 000 K.

The luminous sphere of gas that formed in this way could have been any typical star; this particular one is known to us as the Sun. The rapidly rotating core of matter that had contracted to form the Sun left on its periphery other matter that took the shape of a disc, known as the *solar nebula*. As nebula particles remote from the Sun cooled, gases in that part of the solar system began to interact to form compounds. Some atoms and molecules condensed to form more particles, and collisions amongst them, over time, gradually drew them together into solid bodies known as planetesimals. Eventually, with further coalescence, the small planetesimals grew to such a size, now planets, that they could retain an atmosphere. Reactions occurred within and between the atmosphere and the solid / liquid phases of the young planets. The elements that were present, and the changing affinities between these elements as the system cooled, determined the molecular species that were created.

One of these planets was the Earth.

### Early Earth history

In this earliest period of the Earth's life, the solid materials present in its core consisted of iron and alloys of iron, while the mantle and crust of the Earth were in large part made up of oxides and silicates of metals. The major gases in the primeval atmosphere were dihydrogen, dinitrogen, carbon monoxide, and carbon dioxide. Over time, much of the atmosphere was lost into space, whereas continued volcanism brought gases to the surface where they reacted to form other new gas species. Oxygen was abundant but there was no free dioxygen gas. In its entirety, this element was present in combined form—associated with metals or in the atmosphere as carbon dioxide.

Very early in the Earth's history, water was formed most likely by reactions such as

$$4H_2 + CO_2 \rightarrow CH_4 + 2H_2O \tag{1.1}$$

$$H_2 + CO_2 \rightarrow CO + H_2O \tag{1.2}$$

To occur to any significant extent, the two reactions require the presence of catalysts and these were available in the form of metal oxides on the surface of the primordial Earth.

Water making up the early seas may have been acidic due in part to dissolved carbon dioxide as well as to hydrochloric acid and sulphur species that were trace components of the early atmosphere. The acids, thought to be concentrated enough to generate an aqueous pH of about 2, and the warm temperatures of the early oceans were sufficient to cause significant dissolution of components in the associated rocks. Dissolution is a neutralizing process and the pH of the seas rose to a value (pH ~ 8) near that of the present-day oceans. At the same time, concentrations of metals in the water increased, sometimes exceeding the solubility products of secondary minerals. For example, the presence of dissolved aqueous carbonate species led to the formation of early sedimentary deposits of calcite ($CaCO_3$) and other carbonate minerals.

$$Ca^{2+} + CO_3^{2-} \rightarrow CaCO_3 \tag{1.3}$$

Also significantly affecting oceanic chemistry were continued underwater releases of gases and volcanic eruptions.

Volcanic activity, folding and uplift of rocks under pressure from the movement of tectonic plates, chemical and physical erosion, and sedimentation all changed the nature of the crust of the Earth over long periods of the Earth's early history.

## The beginnings of life

Because there was no free oxygen in the atmosphere, no ozone could be formed. The atmosphere was then transparent to a broad flux of solar radiation, including a large input of ultraviolet (UV) light. This highly energetic radiation and the presence of catalysts made it possible for simple organic compounds like methanol and formaldehyde to be synthesized.

$$CO + 2H_2 \rightarrow CH_3OH \tag{1.4}$$

$$CO + H_2 \rightarrow HCHO \tag{1.5}$$

Very early in the Earth's history these and other species—including HCN, $NH_3$, $H_2S$, and many others were formed. Some of the small molecules reacted further to produce more complex compounds, even including amino acids and simple peptides.

Very primitive forms of life are known to have developed as early as 4 billion years ago. The first cells used simple inorganic molecules as starting material for their synthesis and they, of course, lived in an environment devoid of free oxygen. With increasing complexity, around 3.5 billion years ago, some cells developed an ability to carry out photosynthesis—a reaction that released oxygen into the atmosphere.

$$CO_2 + H_2O \rightarrow \{CH_2O\} + O_2 \tag{1.6}$$

At first, as quickly as it formed, the free oxygen was removed by reaction with terrestrial materials. As the amount of aquatic plant life increased, however, free oxygen began to build up and by about 2 billion years ago the environment at the Earth's surface could be described as essentially oxidizing. Carbon dioxide gradually became a minor gas in the atmosphere. The presence of free oxygen led to the synthesis of ozone, which acted to partially shield the highly energetic components of solar radiation from reaching the Earth's surface. This opened the possibility for terrestrial life to emerge.

## The recent Earth

It was the development of life and an oxidizing atmosphere that dominated the change from the primitive to the present environment, and in the past billion years many features of the Earth's composition have remained relatively constant. Yet we should not leave the impression that geological and life processes were static during that period. On the contrary, the Earth is a dynamic system where processes such as volcanism, movement of tectonic plates, weathering,

**Table 1.1** Some important physical properties of the present-day Earth[a].

| Mass / kg | $5.98 \times 10^{24}$ | | |
| --- | --- | --- | --- |
| Radius / m | $6.38 \times 10^{6}$ | | |
| Density / kg m$^{-3}$ | 5520 | | |
| Distance from sun / km | $1.5 \times 10^{8}$ | | |
| Surface temperature / K | 290 | | |
| | Atmosphere | Oceans | Land |
| Mass / kg | $5.27 \times 10^{18}$ | $1.37 \times 10^{21}$ | |
| Surface area / m$^2$ | | $3.61 \times 10^{14}$ | $1.48 \times 10^{14}$ |
| Approximate density / kg m$^{-3}$ | 1.3 (at Earth's surface, 0 °C) | 1030 | 2700 (surface rocks) |
| Major components | $N_2$, $O_2$, $H_2O$, Ar | $H_2O$, dissolved species $Na^+$, $Cl^-$, $SO_4^{2-}$, $Mg^{2+}$ | Si, O, Al, Fe, Ca (as silicates, oxides, carbonates, etc.) |

[a] Additional data regarding the nature of the Earth are provided in Appendices A.1–A.3.

erosion, sedimentation, and the continuing evolution of life interact to provide the environment in which we now live. As changes occur in one compartment—through interactions and feedback changes occur over the Earth as a whole.

Nevertheless, from about 1 billion years ago to the present, the average composition of the atmosphere, the oceans, and the land, in their major components, has remained relatively constant. Table 1.1 lists some important physical features of our present-day Earth.

## 1.1 Environmental chemistry

### Systems and surroundings in environmental chemistry

In the terminology of thermodynamics, we define the *universe* as consisting of a *system* and its *surroundings*. The system is that portion of the universe under direct investigation, while the surroundings comprise everything beyond the system. We can apply this concept, for example, to consider an industrial chemical process such as the production of the wood preservative pentachlorophenol (PCP). The system—essentially what goes on in the factory reactor—is subject to investigation by chemists who develop and optimize appropriate synthetic reactions and by engineers who design and set up the manufacturing facility itself. Historically, much effort has been expended on examining the properties of systems but, increasingly, there is concern about surroundings.

In our example, when scientists move outside the factory and focus attention on the impact of the industrial process on its surroundings—such as release of the product PCP or by-products like dioxins—then the subject becomes one of environmental chemistry.

We may continue our example by moving away from manufacturing to the other end of the scientific spectrum where we examine mechanisms by which chemicals such as dioxins enter a living organism, the biochemical transformations they undergo, their molecular mode of action, and their elimination. The target organism has then become the system and is a subject for study by biologists, biochemists, and toxicologists.

What goes on in the surroundings outside the organism—transport of dioxins, associative reactions with soil and water, degradation process, etc.—is the subject matter of environmental chemistry. To express it simply, environmental chemistry is the chemistry of surroundings—the universe minus the system.

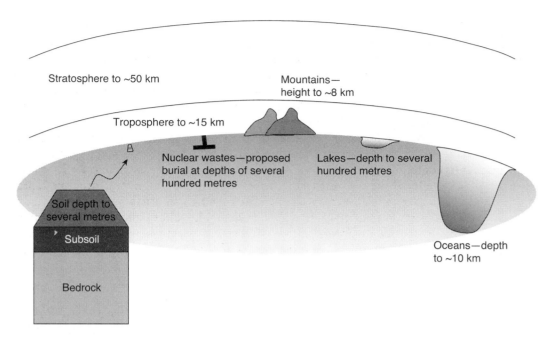

**Fig. 1.1** The environment near the surface of the Earth (not to scale).

Given this very broad concept of environmental chemistry, we must be clear about the physical extent of these surroundings. Beginning where humans live, we can move inward toward the centre of the Earth. We will find that, below a relatively thin layer of the Earth's crust, few chemical processes affect the environment, at least over a time-scale of years or even thousands of years. The layer may be as thin as a metre when considering many soil processes, ranging to tens of metres for lakes or a few kilometres for oceans and when considering the disposal of nuclear wastes. Even at its greatest, the layer is only a very small fraction of the 6380 km radius of the Earth (Fig. 1.1).

Moving outward into the atmosphere above the Earth's surface, there occur complex processes supported by and supporting Earth-bound reactions. Many of these processes take place at low altitudes but we know that chemical reactions at heights of 30 km or higher are critical to maintaining the Earth as it is.

Therefore, out of the total surroundings of the vast universe, a thin shell, perhaps 50 km in thickness, on and above the Earth's surface is the subject for most of what we say on the theme of environmental chemistry. Even these specified surroundings are very large and very complex—especially compared with the small, controlled systems with which chemists are usually involved in a laboratory. We will therefore frequently zero in on a more limited, defined portion such as the atmosphere of a building, a small lake, or a particular layer of soil, but when we do this we are really turning that part of the surroundings into a new system. We should never lose sight of the fact that each small part of the environment is connected with other parts and together they make an interrelated whole.

## What this book is about

This book has been written to provide the chemical basis for understanding our surroundings, the global environment. Having said that, we must point out that there are several important subjects that the book will *not* cover in detail.

The book is not about *environmental analysis*, though almost every environmental topic discussed is based on information obtained by analysis. Often, the quality of the discussion depends on the quality of the available data. For this reason, analytical chemistry is central to our understanding of the environment. And analysis is by no means straightforward. Determining ozone or nitrogen oxides in the stratosphere at an altitude of 25 km requires sophisticated instrumentation and interpretation. Finding the concentration of mercury that is available for uptake by fish in a body of water requires careful measurement and elegant reasoning related to the significance of the results. Environmental analytical chemistry is a large subject in its own right; it will be mentioned in examples and is the focus of some of our *literature links* (see below), but not as a separate subject in this book.

The book is not about *environmental toxicology*. One of the reasons—but by no means the only one—why twenty-first-century inhabitants of the Earth are so concerned about the environment is that its contamination and degradation inevitably have effects on human beings. The direct toxicological effects, the mechanisms by which they occur, and the quantitative aspects defining conditions under which organisms, particularly the human organism, are influenced by chemicals are the subjects of toxicology. In general, this book will be concerned with chemical behaviour external to specific organisms, although in several places we provide a brief discussion of basic toxicological principles, including factors affecting uptake and methods of assessing toxicity in the biosphere.

The book is not about *environmental control*—about detailed technologies for preventing or eliminating pollution, including important initiatives in 'green chemistry',[1] or about standards and laws that set guidelines and limits on levels of contaminants. These, too, are essential subjects for consideration if environmental science is to be more than a theoretical study. Where appropriate, we will document some important environmental guidelines and discuss technologies to achieve them. But for the most part, we will be concerned with the chemistry that underlies regulatory decisions and engineering design for control of pollution.

And, finally, the book is not about *environmental science* in its broadest sense. Books on environmental science attempt to cover the whole range of a topic, and they call on a large body of material from the sciences of climatology, geology, biology, and so on. To put any subject in context, we will open up the geographical, biological and historical perspectives but we will not dwell on them, important though they are.

Now, we can restate the purpose of our present exploration of environmental chemistry. The book has been written to provide the chemical basis for understanding our surroundings, the global environment. Emphasis will be on the composition of the natural environment, the processes that take place within it, and the kinds of changes that come about as a result of human activities. Many examples will be used to illustrate the principles being described and these will be chosen from situations throughout the world, because environmental chemistry is indeed a global subject. Nevertheless, there are many important specific types of problems that, inevitably, will not be mentioned. It is not the goal of this book to provide a complete collection of all environmental issues, but our objective is to give a chemical background so that one can have a basis for understanding such issues.

## 1.2 Environmental composition

In learning about the chemistry of a particular element or compound in the environment, we usually begin by observing where it is found and measuring the concentrations that are present. Several questions must be considered. With what materials is the chemical associated? Are its concentrations normal or unusually high or low? Is it toxic? Then, if there are places where there are environmental problems due to high concentrations, we investigate the source of these elevated levels. Thallium provides an interesting example.

---

[1] Green chemistry is an important branch of chemistry and engineering that considers design of industrial chemical processes in the context of their environmental impacts. As such, it includes subjects like the use of benign materials in production, minimizing energy consumption, finding uses for byproducts, and integrating processes, often in a cyclical manner. The goal of green chemistry is to carry out industrial activities in a way that reduces, even eliminates, adverse impacts of production on the environment.

In the last decade, some considerable interest has developed around the environmental chemistry of thallium. This element has been described as enigmatic because of its somewhat surprising 'chemical personality'. While it is the heaviest element in group 13 of the periodic table, its most common oxidation state is thallium (I) and in many ways its chemistry is similar to that of the alkali metals. The environmental interest arises in part because it is toxic to aquatic organisms, humans, and other mammals. Being similar in behaviour to the alkali metals, it substitutes for potassium and causes metabolic disturbances by inhibiting the activity of important enzymes and coenzymes. Actually, in part because of its toxicity and also because it is colourless and tasteless, it has been considered to be 'the perfect poison'.

All this means that it is important to know about the occurrence of thallium, *what are normal levels in air, water, and soil, and where and why is it present in greater abundance in certain locations and situations?*

Thallium is a surprisingly common element, being widely and relatively evenly distributed in the natural environment with a global average in the Earth's crust of approximately 0.7 mg kg$^{-1}$. In the aqueous environment, the mobile thallium (I) is present in varying concentrations, but in uncontaminated sites the values are typically around 10 ng L$^{-1}$ or less. With these 'background concentrations', areas that are subject to elevated levels can be identified. And this is what has been done in some instances.[2]

- Near coal-mining operations in Nova Scotia, Canada, water levels of 20 μg L$^{-1}$ and higher have been observed.

- Tailings from an abandoned gold mine in the same area have thallium concentrations up to 3.5 mg kg$^{-1}$.

- Soil concentrations averaging 20 mg kg$^{-1}$ have been measured in the south-west of Guizhou Province, China in areas of thallium-rich minerals.

Measuring concentrations such as these is only the beginning of an environmental chemistry study, but it is often an essential beginning—one that forms the basis for deeper studies of processes and effects.

## Species distribution

In many cases, it is helpful to go beyond measuring the total concentration of a particular element or compound in an environmental sample. Most substances can exist in more than one form, and a description of the species distribution for such substances is an important aspect of describing composition. In some instances, sophisticated analytical methods are used to distinguish between species in environmental samples. It is also possible to make use of distribution diagrams, of which there is a large variety of types, as an aid to evaluating species distribution. Figures 1.2 and 1.3 give examples of such diagrams for carbonate species and mercury in water. In the case of the carbonate system, the distribution is plotted as a function of pH of the water, while for mercury its composition is plotted against the chloride ion concentration. The diagrams are therefore limited by the variables that are chosen for their calculation.

Distribution diagrams are based in part on analytical data, both for concentrations of particular species in environmental samples and also for thermodynamic equilibrium constants of a variety of types. The carbonate diagram, for example, depends on knowing the association constant for aqueous carbon dioxide and water, as well as the acid dissociation constants for carbonic acid. The validity of a distribution diagram therefore depends on the availability and

[2] Cheam, V., Thallium contamination of water in Canada. *Water Qual. Res. J. Canada* **36** (2001), 851–77; Wong, H.K.T., A. Gauthier, and J.O. Niriagu, Dispersion and toxicity of metals from abandoned gold mine tailings at Goldenville, Nova Scotia, Canada. *Sci. Total Environ.*, **228** (1999), 35–47; Zhang, Z., B. Zhang, J. Long, X. Zhang, and G. Chen, Thallium pollution associated with mining of thallium deposits. *Sci. China* (Series D) **41** (1998), 75–81.

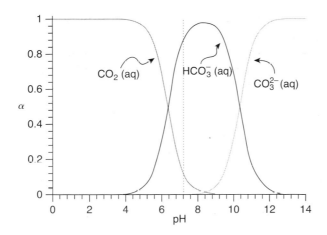

**Fig. 1.2** Distribution of carbonate species as a function of pH. The $\alpha$ value is the fraction of a particular species, $\alpha = \dfrac{\text{[individual species]}}{\text{[sum of all species]}}$.

quality of appropriate thermodynamic data, and also on the very important (and problematic) assumption that a particular natural system is at thermodynamic equilibrium.

We can use these diagrams to estimate the fraction of individual species under particular conditions such as at a given pH. On Fig. 1.2 the vertical line drawn at pH 7 intersects the carbon dioxide curve at $\alpha - 0.16$ and the hydrogen carbonate curve at $\alpha = 0.84$. At this pH, therefore, aqueous carbon dioxide makes up about 16% of all carbonate species, hydrogen carbonate about 84%, while carbonate is a negligible fraction.

The vertical line (a) on Fig. 1.3 refers to the oceans, which have a total chloride concentration of 0.56 mol L$^{-1}$. The $HgCl_4^{2-}$ is seen to make up 28% of mercury species, with $HgCl_3^-$ at 56%

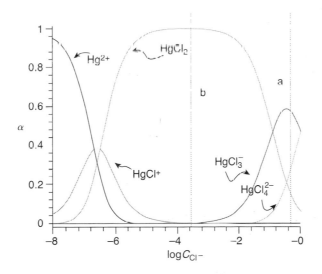

**Fig. 1.3** Distribution of mercury chloro species in water as a function of chloride ion concentration, $C_{Cl^-}$ (mol L$^{-1}$). The $\alpha$ value is the fraction of mercury in the form of a particular complex.

and $HgCl_2$ at 16%. Line (b) corresponds to well water with a chloride ion concentration of 9.5 ppm, where the only significant species of mercury is $HgCl_2$.

The species distribution of an element controls its behaviour in the environment and may be a major factor affecting biological availability. With thallium, in water its most common species is the soluble and mobile thallium (I) ion but, if it is oxidized to thallium (III), it is found mostly as insoluble colloidal species from which a portion may settle into the sediment.

In the case of mercury, in the aqueous environment it exists as inorganic species in the 0, +1, and +2 oxidation states depending on redox and other conditions. However, the major species found in fish is partially methylated mercury, $CH_3Hg^+$, which is produced in sediments by a variety of microbiological processes. This species is toxic both to the fish and to other animals (including humans) that may consume the fish. It is clearly important, then, to be aware not only of how much mercury is present in a sample, but also of the distribution of forms of the element in that sample.

## 1.3  Chemical processes

Knowledge of the composition of a particular compartment of the environment is a starting point for a description of environmental chemistry. However, if we stop there, we imply that the system is static, which, of course, is not true. There are many processes—physical, chemical, and biological—that operate within and connect the various components. The processes may be completely 'natural' and, in fact, over the geological time-scale, it is such processes that have contributed to making the Earth the way it is. Therefore, a second phase of developing understanding of environmental chemistry is to learn about the chemical reactions that are a part of the environmental processes.

In order to summarize the broad features of reactions involving environmental species, we will find it useful to think of the environment in terms of four principal compartments— the atmosphere (the gaseous environment), the hydrosphere (the liquid, essentially aqueous environment), the terrestrial (solid) environment, and the biosphere (the living environment). At first glance, these categories appear to be quite clear-cut; however, it will become evident that there are many overlapping areas. For example, we usually think of soil as part of the terrestrial environment, but the chemical behaviour of soil solutions and soil gases plays a major role in determining the environmental characteristics of soil itself. Concerning the compartments, we can then describe chemical processes within each, and also reactions that bring about transitions from one to the other.

Figure 1.4 shows a simple diagrammatic representation[3] of the global water system and its relation to the various compartments. Strictly speaking, most of the processes shown in this example are physical in that they involve phase changes, not chemical reactions, but the form of the diagram is similar to those constructed to show environmental chemical relationships.

The figure allows us to obtain an overview of the relations between the various forms of the substance. It also shows the cyclic nature of many natural processes. In the steady state, the parts of the cycle are balanced so that concentrations remain constant. This enables us to calculate residence times. In the water cycle, the total mass of water at any time in the atmosphere is approximately $1.3 \times 10^{16}$ kg. The inward flux is $4.23 \times 10^{17}$ and $7.29 \times 10^{16}$ kg y$^{-1}$ by evaporation from oceans and land, respectively. This is balanced by outward fluxes of $3.86 \times 10^{17}$ and $1.10 \times 10^{17}$ kg y$^{-1}$ precipitation onto the ocean and land. Therefore the total inward and outward flux is $4.96 \times 10^{17}$ kg y$^{-1}$ and the residence time of water in the atmosphere is determined as shown in Example 1.1.

---

[3] The form of this diagram will be used for other systems. For an example see Fig. 14.11 (The phosphorous cycle).

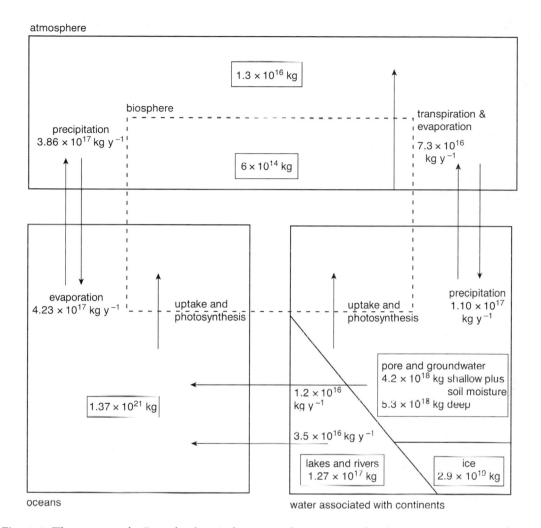

**Fig. 1.4** The water cycle. Boxed values in kg are total amounts in the given compartment. Values in kg y$^{-1}$ are fluxes or movement from one compartment to another. (Values taken from a number of sources and reported in Berner, E.K. and R.A. Berner, *The global water cycle*, Prentice Hall, Inc., New Jersey; 1987.)

---

### Example 1.1 **Calculation of residence time**

$$\tau = \text{residence time} = \frac{\text{steady state amount in the atmosphere}}{\text{flux (in or out)}}$$

$$= \frac{1.3 \times 10^{16}\ \text{kg}}{4.96 \times 10^{17}\ \text{kg y}^{-1}} = 0.0262\ \text{y} = 9.6\ \text{days}$$

This is the average time that a molecule of water spends in the atmosphere.

---

The box model cycle illustrated in Fig. 1.4 provides an overview of the major processes interconnecting compartments of the global water ecosystem, and may also allow for identification of important reactions within a single compartment. But there are many details involving specific reactions that are essential to completing the picture. As an example, a comprehensive description of water resource distribution in India is provided in Fig. 1.5. This shows details of the fate of water supplied by precipitation each year in that part of the subcontinent.

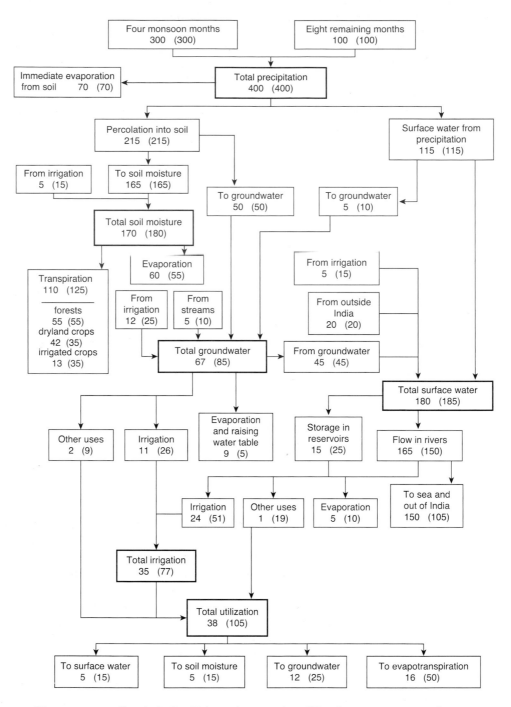

**Fig. 1.5** Water resource flux in India. Values given are in million hectare metres—a hectare metre is the volume of water required to cover 1 ha (10 000 m²) to a depth of 1 m. This is a volume of $10^4$ m³ or a mass of approximately $10^7$ kg. The first numerical figure of each pair is a value for 1974; the second, in parentheses, is an estimate for 2025. (Centre for Science and Environment, *The state of India's environment: a citizen's report*, India; 1978.)

The information contained in such a diagram along with appropriate chemical data could be used, for example, in developing a complete picture of water used for irrigation. We shall see later that irrigation is the major consumer of water on a global scale. The combined qualitative and quantitative data allow for estimating fluxes of individual chemicals—a requirement for evaluating sustainability of the water resource in agriculture. With reference to Fig. 1.5, there are two terms, *source* and *sink*, that are important and relevant to this and other environmental discussions. The *source* of all water in India is rainfall, as is shown at the top of the diagram. There are several *sinks*, that is the 'final' fate of the water after it has gone through a variety of processes. In this case, the sinks are identified as surface water, groundwater, soil moisture, and evapotranspiration.

Generating a valid description of environmental processes involves moving back and forth between the general, broad picture with its global context, and the particular detailed delineation of specific chemical reactions. Studies in both these areas contribute to our knowledge of the subject. One of the challenges of environmental chemistry is to bring together the general and the particular.

## 1.4 **Anthropogenic effects**

A third aspect of a study of environmental chemistry is to examine the effects of human (anthropogenic) activities on the natural processes which are always on-going. The anthropogenic effects may be catastrophic, usually in a localized area, and are then referred to as disasters. A tragic example was the uncontrolled release of toxic methyl isocyanate gas near Bhopal, in central India in 1984.

In the early morning of 3rd December 1984 just outside the city of Bhopal in central India, there was a massive release of four tonnes of methyl isocyanate ($CH_3-N=C=O$) gas from a storage tank at the Union Carbide plant where the chemical had been manufactured since 1980. The dense gas floated across the surrounding landscape killing more than 3800 persons and leaving at least 300 000 affected by exposure—some severely.

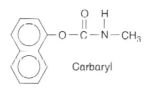

Carbaryl

Methyl isocyanate (MIC), is a starting material for the manufacture of the widely used carbamate pesticide, carbaryl, and was stored under refrigeration in an underground tank at the Bhopal plant site. A combination of technological problems—failure of the cooling unit, leakage of water into the tank, loss of nitrogen pressure above the MIC, and failure of several safety devices—was the immediate cause of the accident. But there were political, organizational, and human factors that provided a setting in which the multiple failures could occur simultaneously.[4] This tragic event is not unique but fortunately no spills or releases with such massive and deadly consequences have occurred in more recent years. In part this is due to adoption of voluntary programmes of *responsible care* by some members of the chemical industry.

Other environmental perturbations are more gradual and the effects may show up only in the medium or long term. This does not imply that the consequences are any less serious. For example, the possibility of global warming due to build-up of greenhouse gases in the troposphere

[4] A detailed account of the story of this, the worst industrial crisis in history, is given in Shrivastava, P., *Bhopal, anatomy of a crisis*, Ballinger Publishing Co., Cambridge, Massachusetts; 1987.

is of great concern with respect to life throughout the planet in the twenty-first century. Should adverse consequences due to global warming become a reality, its reversal, even if possible, could take as long a period of time as its advent.

In considering the potential effects of anthropogenic inputs into the environment, we make use of our knowledge of both the composition and the process data.

An example comes from consideration of the possible problems associated with the use of sewage sludge from municipal wastewater-treatment facilities as an amendment / fertilizer on soils used for agricultural purposes. This is a widely followed practice that has beneficial aspects as it supplies organic matter and small amounts of major and minor nutrients to the soil. One of the concerns regarding the practice is that potentially toxic concentrations of certain metals that are present in the sludge can be taken up by plants and incorporated in the food chain. Cadmium is one element of interest in this regard. Soils have a range of natural cadmium concentrations with a typical value being approximately 0.8 $\mu$g g$^{-1}$. Of the cadmium, it is thought that most is present as inorganic mineral species associated with the clay mineral phase of the soil. Sewage sludge is a variable and heterogeneous material but a typical cadmium concentration is 80 $\mu$g g$^{-1}$, some one hundred times greater than that of the soil itself. In considering whether additions of sludge to soil initiate a cadmium toxicity problem, the amount of sludge added becomes an issue as it will define the final concentration. Consideration of the form of cadmium in the sludge—it will probably be bound with the organic matrix—is also important. Finally, the geochemical and biochemical reactions that the added cadmium must undergo will be considered. These will include interconversions between various inorganic and organic species of the element in the soil, leaching of soluble forms, immobilization by ion exchange or other adsorption processes on the solid soil phases, and biological uptake by micro- or macroorganisms (Fig. 1.6).

The ultimate fate of the added cadmium depends on the extent of the individual reactions, how each process affects the others, and the ways in which a changing environment can alter the balance.

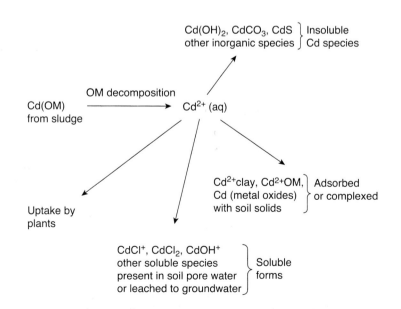

**Fig. 1.6** Cadmium biogeochemical reactions in soil after addition as a trace component in sewage sludge. OM, Organic matter.

## Fermi questions

Throughout the book, in some of the chapters you will encounter what is called a Fermi question.[5] These are questions that are unlike the calculations in the examples or the problems given at the end of each chapter, which are ones that usually require very specific and well-known procedures, information or data in order to determine a result. A Fermi question requires a different approach. For Fermi questions, 'back-of-the-envelope' type calculations make use of rough estimates of background information along with intuition and reasoning to come up with a very approximate but reasonable suggested conclusion. There is no 'correct answer' and a discussion is always necessary in order to make clear one's thinking and strategy. Below is our first Fermi question. As an example, we provide a suggested answer in this initial case.

---

### Fermi question

What volume of fuel would be consumed during the morning 'rush hour' of a typical large-sized North American city? What impacts would this have on the local environment? How would this situation compare to that in a city of similar size in Africa?

- We begin by choosing a reasonable 'large size'—say 2 million. Of this total population, about 40% could be either very young or older and retired and therefore unlikely to be driving during rush hour.
- We could estimate that half of the other 60% might be driving during rush hour—to work, transporting children to school, or for other purposes. In addition, there will be trucks and other commercial vehicles—say 100 000. Therefore, the total number of vehicles on the road = 2 × 0.6 × 0.5 = 0.6 million cars + 0.1 million others.
- A reasonable, but perhaps low, estimate of a one-way commute would be 15 km. Total distance driven = 0.7 × 15 km = 10.5 million km.
- An average fuel mileage (mixture of car sizes) might be 10 km per L. Total fuel = 1 100 000 L, around one million litres.

Emission of particulates and several gases from the exhaust of all vehicles contributes to poor air quality in the local environment. These will be described in detail in Chapter 4.

In Africa, in most cases the number of vehicles and distances travelled are much smaller, but there will be major differences between cities. For example, Cairo has a large population of cars while the number in Kinshasa is much smaller. Other features that will affect air quality are the climate and the type of vehicles, engines, and fuel. All of these are significantly different in most parts of Africa from the situation in the USA.

---

### Literature link  Environmental chemistry topics

Included in some chapters of the book are brief descriptions of recent research investigations in environmental chemistry that are taking place at different locations around the world. Each description, which we call a *literature link* is based on a paper that has been published in readily accessible research journals since 2003. The subject of environmental chemistry is a very active one and we encourage you to delve into the original papers; this will give you a sense of the exciting challenges that are encountered in this field of science—challenges in obtaining data, in interpreting it in the context of the specific area studied, and in fitting the information and interpretation into the global picture of how the Earth's environment functions.

---

[5] Fermi questions receive their name from Enrico Fermi, an Italian physicist who was instrumental in the development of quantum theory. Fermi questions make use of fundamental general knowledge, estimation and quantitative reasoning, communicating in simple mathematics, and questioning skills. Fermi questions encourage the use of different, sometimes non-traditional, approaches and emphasize the problem-solving process rather than the answer.

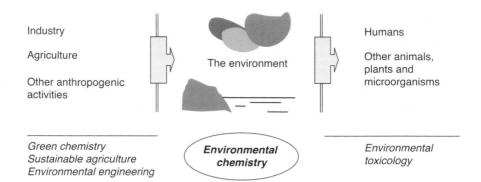

Industry

Agriculture

Other anthropogenic
activities

The environment

Humans

Other animals,
plants and
microorganisms

Green chemistry
Sustainable agriculture
Environmental engineering

*Environmental
chemistry*

*Environmental
toxicology*

**Fig. 1.7** Environmental chemistry, in the centre of the diagram, is concerned with the chemical nature, processes, and anthropogenic influences that take place in the atmosphere, the hydrosphere, and the terrestrial environment. Other branches of environmental science are directed more specifically toward the control and regulation of potentially harmful agents or to understanding how these agents affect living organisms.

## In summary

The subject matter in this book covers the chemistry of surroundings, the environment on and near the Earth's surface. These surroundings include air (the atmosphere), water (the hydrosphere), and land (the terrestrial environment), and we will deal with them in that order.

In considering particular environmental subjects, we focus on the basic chemistry and to do this it is necessary to examine some or all of the three factors—composition, chemical processes, and perturbations caused by natural or anthropogenic activities—that we have discussed above (Fig. 1.7). Although a systematic approach requires that individual topics be studied in isolation, the overlap, the connections, and the interdependence must always be kept in mind. Perhaps most important, as citizens of the Earth, while we are studying this vast subject, we should remember that we share together a single global environment.

---

## ADDITIONAL RESOURCES

1. Baird, C. and M. Cann, *Environmental chemistry*, 4th edn, W.H. Freeman and Co., Salt Lake City, Utah; 2008.

2. Eby, G.N., *Principles of environmental geochemistry*, Brooks / Cole-Thomson Learning, Pacific Grove, California; 2004.

3. Manahan, S.E., *Environmental chemistry*, 8th edn, Lewis Publishers, Inc., Chelsea, Michigan; 2004.

4. Spiro, T.G. and W.M. Stigliani, *Chemistry of the environment*, 2nd edn, Prentice Hall, Upper Saddle River, New Jersey; 2003.

5. Sawyer, C.N., P.L. McCarty, and G.F. Parkin, *Chemistry for environmental engineering and science*, 5th edn, McGraw Hill Companies, New York; 2002.

6. The United Nations Environment Program, *UNEP Yearbook 2009*, EARTHPRINT Ltd., Stevenage, Hertfordshire, UK, 2009.

7. The United Nations Environment Program, *Global Environment Outlook 4 (GEO-4): Environment for Development (available in English, French and Spanish)*, EARTHPRINT Ltd., Stevenage, Hertfordshire, UK, 2009.

The following two resources are online materials (course notes, videos, etc.) provided for courses on environmental chemistry given at two American universities.

8. Chemistry Department, *Environmental Chemistry CH390*, Oregon State University, http://oregonstate.edu/instruct/ch390/, accessed October 2009.

9. Foust, R. *Environmental chemistry*, Northern Arizona University, 2000 http://wlh.webhost.utexas.edu/, Search: Environmental Chemistry, accessed October 2009.

# Part A
# The Earth's atmosphere

When you understand all about the Sun
and all about the atmosphere
and all about the rotation of the Earth
you may still miss the radiance of the sunset.

*A. N. Whitehead, 1926*

# Chapter 2
## The Earth's atmosphere

The focus of this chapter is on the fundamental concepts needed to understand chemical processes that occur in the atmosphere—processes that will be examined in more detail over the next several chapters. We will accomplish this by:

- Describing the physical nature of the atmosphere, structure and composition, and the important role played by solar radiation in defining these features
- Providing basic concepts that will be used in later chapters to describe important chemical processes related to ozone, smog, acid precipitation chemistry, atmospheric aerosols, urban and indoor air, and global climate
- Reviewing fundamental physical chemistry properties of gases
- Introducing gas-phase reactions along with related thermodynamic and kinetic calculations, and including photochemical reactions as well as the extremely important category of free-radical reactions.

## 2.1 The Earth's atmosphere—the air we breathe

Of the planets in the solar system Mercury, nearest the Sun, has almost no atmosphere. The next three planets, Venus, Earth, and Mars, lost whatever gases were present at their formation and the atmospheres they now possess arise from gases released from their interiors and reactions that these have undergone. The atmosphere of each planet is unique. The outer planets have extremely dense atmospheres made up mostly of hydrogen and helium, little changed from the original composition at the time of formation.

The Earth's atmosphere is a thin shell of gases surrounding the globe. The thinness becomes clear when we realize that the part of the atmosphere in which all human activity takes place adds less than 0.3% to the radius of the Earth. Its unique chemistry, including compounds such as oxygen (molecular oxygen or dioxygen) and carbon dioxide that support the processes upon which all forms of life depend, distinguishes this atmosphere from that of other planets in the solar system. Table 2.1 shows the relative amounts of the four most abundant gases in the dry atmosphere.

The mixing ratios (concentrations, see Section 2.3) of the major gases remain relatively constant up to an altitude of about 80 km. The constancy arises because the kinetic energy of these molecules is sufficient to overcome any gravitational forces that would lead to settling. Because the mixing ratios of the major components are constant, the average molar mass, $\overline{M}_a$, of the atmosphere can be calculated.

**Table 2.1** Major components of the atmosphere near the surface of the Earth[a].

| Component | Mixing ratio (%) |
|---|---|
| Nitrogen | 78.08 |
| Oxygen | 20.95 |
| Argon | 0.93 |
| Carbon dioxide | 0.0387 |

[a] Concentrations (mixing ratios, a term that is defined below) are calculated on a dry atmosphere basis. The water content is a fifth major component, but its concentration is variable, ranging from <0.5 to 3.5%. A more complete set of data is given in Table 8.1.

---

Example 2.1 **The average molar mass of air in the lower atmosphere**

$$\overline{M}_a = M_{N_2} \times f_{N_2} + M_{O_2} \times f_{O_2} + M_{Ar} \times f_{Ar} + M_{CO_2} \times f_{CO_2} \tag{2.1}$$

where $M$ and $f$ refer to the molar mass and fractional abundance, respectively, of each component. Applied to the regions below 80 km of the Earth's dry atmosphere,

$$\overline{M}_a = 28.01 \text{ g mol}^{-1} \times 0.7808 + 32.00 \text{ g mol}^{-1} \times 0.2095 + 39.95 \text{ g mol}^{-1}$$
$$\times 0.0093 + 44.01 \text{ g mol}^{-1} \times 0.000387$$
$$= 28.96 \text{ g mol}^{-1}$$

It should not be surprising that the molar mass is close to, but somewhat greater than, that of pure $N_2$.

---

Above about 80 km altitude, the concentrations of the major species do begin to change significantly due to photochemical processes that cause dinitrogen and especially dioxygen to dissociate. These processes will be discussed later in this chapter.

In contrast to the major species of gases in the troposphere, the concentrations of some trace gases vary considerably. Reactive gases such as $NO_2$ and $SO_2$, the precursor compounds for acid precipitation (Chapter 5) persist for only hours to days and their tropospheric mixing ratios are not constant as they are dependent on local sources. Other trace gases such as volatile organic compounds (VOCs), or volatile / semi-volatile persistent organic pollutants (POPs) (Chapter 4) vary within a hemisphere due to their sources, distribution by air currents and oxidative removal processes. Finally, trace gases such as $CH_4$, $N_2O$ and chlorofluorcarbons (CFCs) have long atmospheric lifetimes and therefore show relatively constant compositions around the globe. More will be discussed about these gases in Chapter 8 along with carbon dioxide.

## Regions of the atmosphere

The atmosphere can be conveniently divided into four sections based on the direction of temperature change as one proceeds from lower to higher altitudes. The pattern is clearly shown as the zigzag line on Fig. 2.1. Beginning at the Earth's surface where the temporally and spatially averaged temperature is approximately 14°C (287 K), the atmospheric temperature falls steadily to a value of about −60°C (213 K) at an altitude of approximately 15 km. This part of the atmosphere closest to the Earth's surface, in which humans live and most biological activity occurs, is called the troposphere. The troposphere, a region of intense convective mixing, contains approximately 85% by mass of the entire atmosphere. The upper boundary of the troposphere, the tropopause, marks the altitude at which the direction of temperature change reverses. Above this, the next region is called the stratosphere where increasing altitude brings increasing temperature up to approximately −2°C (271 K) at 50 km, the stratopause; such a temperature profile is called an

### Regions of the Earth's atmosphere

- Troposphere ~ 0 to 15 km. Temperature decreases with altitude
- Stratosphere ~ 15 to 50 km. Temperature **increases** with altitude
- Mesosphere ~ 50 to 85 km. Temperature decreases with altitude
- Thermosphere ~ 85 to 500 km. Temperature **increases** with altitude

inversion. Due to the rising temperature and decreasing density with increasing altitude, there is little convective mixing and the stratosphere is a relatively stable region. At the stratopause, there is a second reversal in temperature and in the mesosphere, the region above, temperature decreases with altitude to −90°C (183 K) at 85 km, the mesopause. Continuing above this height, the temperature once more begins to increase in a region called the thermosphere, reaching a value of about 1200°C (1473 K) at 500 km. These and other features of the atmosphere are shown in Fig. 2.1.

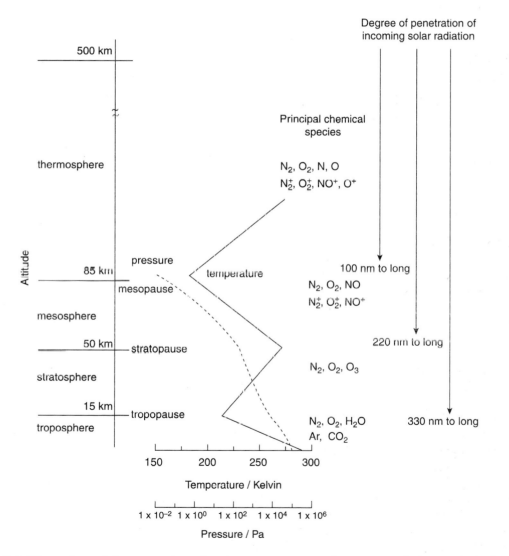

**Fig. 2.1** The regions of the atmosphere showing temperature and pressure variations, principal chemical species, and penetration of incoming solar radiation. Solid line, temperature; dashed line, pressure. 0°C = 273 K.

In rarefied atmospheres it is important to be aware that we are referring to *thermodynamic* temperatures, which are measures of the kinetic energy of molecules in the atmosphere. At high altitudes, if we were to use a conventional mercury thermometer to measure temperature, the reading would be much lower than 1200°C (1473 K) as the number of (highly energetic) collisions of gas molecules with the thermometer would be very small under the near-vacuum conditions.

## Atmospheric pressure

Atmospheric pressure, $P°$, is measured by the force of gravity acting on the atmosphere, divided by the total area of the Earth's surface, as calculated by eqn 2.2,

$$P° = \frac{M_{atm}\, g}{4\pi r^2} \tag{2.2}$$

where $P°$ is the pressure at the Earth's surface (sea level) = 101 325 Pa, $M_{atm}$ is the mass (in kg) of the atmosphere, $g$ is the acceleration due to gravity = 9.81 m s$^{-2}$ (this is approximately constant since most of the atmosphere is near the Earth's surface), and $r$ is the radius of the Earth = $6.37 \times 10^6$ m.

This equation can then be rearranged to calculate the total mass of the atmosphere. Solving for $M_{atm}$ gives a mass of $5.27 \times 10^{18}$ kg. We will make frequent use of this value.

> **Fermi question**
> What is the total mass of air in the room where you are now located?

Like temperature, pressure also changes with altitude but it undergoes a nearly steady decrease described by the equation

$$P_h = P°e^{-\overline{M}_a gh / RT} \tag{2.3}$$

where, in this equation, $P_h$ = pressure at given altitude / Pa, $P°$ = pressure at sea level = 101 325 Pa, h = altitude / m, $g$ = acceleration due to gravity = 9.81 m s$^{-2}$, $\overline{M}_a$ = average molar mass of atmospheric molecules = 0.0290 kg mol$^{-1}$, $R$ = gas constant = 8.314 J mol$^{-1}$K$^{-1}$, and $T$ = temperature / K. Note that an identical numerical result would be obtained if, simultaneously, altitude is expressed in km and molar mass in g mol$^{-1}$.

Because temperature appears in the denominator of the exponential term, there are small changes in slope of the pressure versus altitude curve within the different regions of the atmosphere and this is shown graphically in Fig. 2.1.

> **Main point 2.1** On the basis of the direction of temperature change with increasing altitude, the atmosphere is divided conceptually into four regions—the troposphere, stratosphere, mesosphere, and thermosphere. The relative amounts of the major gases remain constant for the first 80 km but, as indicated by the decreasing pressure with increasing altitude, the absolute amount of each gas decreases.

## 2.2 Solar influence on the chemical composition of the atmosphere

In order to understand the chemical composition of regions in the atmosphere it is convenient to begin with the thermosphere and proceed in a direction toward the Earth. At high altitudes, there is a near vacuum: at an altitude of 100 km (in the lower thermosphere), the pressure is approximately 0.025 Pa. This means that the concentration (molecules per unit volume) of all chemical species is only about one four-millionth of that at the Earth's surface (101 325 Pa / 0.025 Pa). As this region is on the outer fringe of the atmosphere, atoms and molecules are exposed to a full solar spectrum, including radiation in the ultraviolet (UV) region. This is high-energy radiation; for example, radiation of wavelength 100 nm, has associated energies calculated as follows.

---

**Example 2.2 Energy of 100-nm electromagnetic radiation**

For one photon,

$$E = h\upsilon = \frac{hc}{\lambda} = \frac{6.6 \times 10^{-34}\,\text{J s} \times 3.0 \times 10^{8}\,\text{m s}^{-1}}{100 \times 10^{-9}\,\text{m}} \tag{2.4}$$

$$= 2.0 \times 10^{-18}\,\text{J per photon}$$

and thus, for 1 mol of photons,

$$E = 2.0 \times 10^{-18} \times 6.0 \times 10^{23} - 1200\,\text{kJ mol}^{-1}$$

1200 kJ mol$^{-1}$ of energy is associated with radiation having a 100-nm wavelength.

Note: the notation $h\upsilon$ is used to represent energy ($E$) (at a specified wavelength ($\lambda$)) associated with a single photon in a photochemical reaction, (Planck's constant ($h$) multiplied by frequency ($\upsilon$)). When multiplied by Avogadro's number as shown in the calculation above and used below in various reactions, $h\upsilon$, then represents one mole of photons).

---

Such high energies are capable of bringing about dissociation of dinitrogen and dioxygen into their constituent atoms, and this occurs to a large extent in the thermosphere.

$$N_2 + h\upsilon(\lambda \leq 126\,\text{nm}) \rightarrow 2N \qquad \Delta H° \geq 945\,\text{kJ mol}^{-1} \tag{2.5}$$

$$O_2 + h\upsilon(\lambda \leq 240\,\text{nm}) \rightarrow 2O \qquad \Delta H° \geq 498\,\text{kJ mol}^{-1} \tag{2.6}$$

As shown, the energy required for dissociation in these two cases corresponds to electromagnetic radiation with $\lambda \leq 126$ nm and $\lambda \leq 240$ nm, respectively. Some dissociated atoms remain in the atomic form while others recombine, and new species such as NO are also produced in the thermosphere. The relative proportion of atoms to molecules increases with altitude and at 120 km the concentration of oxygen atoms approximately equals that of dioxygen molecules. A substantial portion of nitrogen, but less than that of oxygen, is also present as atoms. Because much of the oxygen and nitrogen gases is in the atomic form, the average molar mass, $\overline{M}_a$ becomes smaller than the value of 28.96 g mol$^{-1}$ that obtains in the lower atmosphere.

Besides breaking bonds, solar energy is also capable of ionizing both molecules and atoms (reactions 2.7 and 2.8). For this reason, the region above the mesopause is alternatively referred to as the ionosphere.

$$N_2 + h\upsilon(\lambda \leq 80\,\text{nm}) \rightarrow N_2^+ + e^- \qquad \Delta H° \geq 1500\,\text{kJ mol}^{-1} \tag{2.7}$$

$$O + h\upsilon(\lambda \leq 91\,\text{nm}) \rightarrow O^+ + e^- \qquad \Delta H° \geq 1310\,\text{kJ mol}^{-1} \tag{2.8}$$

Note the use of standard enthalpy rather than free energy to describe the energetics of atmospheric gas-phase reactions. This is convenient because temperature and pressure have less influence on enthalpy compared with free-energy changes. Also, it allows for description of the energetics of bond breaking and formation aside from statistical considerations of the bulk system.

In regions of the atmosphere where energy is absorbed causing ionization, the reverse exothermic electron-capture reactions also occur to some extent, releasing energy as kinetic energy. This is the reason for the high thermodynamic temperatures in the thermosphere. Proceeding closer to the Earth's surface but still in this region, there is a smaller flux of less highly energetic radiation to be absorbed and so the temperature and the population of atoms and ions decrease.

In the mesosphere, solar radiation begins to encounter new types of chemical species. Ozone ($O_3$), which has a particularly high concentration in the stratosphere, is also present to some extent above the stratopause and is capable of absorbing solar radiation of lower energy than that required for the dissociation and ionization of more stable species. This absorption of longer-wavelength radiation leads to an increase in temperature when moving to lower altitudes in the mesosphere.

The absorption of radiation by ozone causes it to dissociate, producing an oxygen molecule and atom, which are both in an excited state as indicated by the asterisk shown below, for absorption of radiation in the UV

$$O_3 + h\upsilon(\lambda \leq 325 \text{ nm}) \rightarrow O_2^* + O^* \qquad (2.9)$$

The temperature decline, as one moves down through the stratosphere, is explained in the same way as the decline in the thermosphere. Radiation within the energy range of ozone absorption has already been removed in the higher regions, producing elevated temperatures there; it therefore cannot penetrate further.

Little additional radiation is absorbed as solar photons pass through the troposphere. However, when the remaining spectrum of radiation strikes the Earth's surface it is partly absorbed by land and water and then re-emitted as lower energy infra-red (IR) radiation. Some of the IR radiation is absorbed by certain gases in the troposphere—the two principal ones being water vapour and carbon dioxide. Absorption causes heating near the Earth's surface, an effect that declines with increasing altitude as there is less radiation remaining to be absorbed and the concentration of absorbing gases is also declining. The heating of the Earth's lower atmosphere in this manner is the well-known 'greenhouse effect'. It is a key factor in supporting life as we know it on Earth, but anthropogenic perturbations could seriously alter the present balance. Much more will be said about this topic in Chapter 8.

It is important to be aware that the chemical composition of Earth's atmosphere is not an equilibrium system. If that were so, then all of the oxygen would exist combined with other elements, and none would be free to support life. Ultimately, the Sun provides energy for many otherwise energetically unfavourable processes, including photosynthesis that allows plants to make use of carbon dioxide and produce oxygen.

The constant supply of energy from the Sun is either absorbed, stored by chemical reactions, or reflected back into space by the Earth. The combination of these factors maintains the energy relations of the Earth in delicate balance.

## The troposphere

Most of the material concerning atmospheric chemistry discussed in this book relates to the troposphere, a region of intense air movement, so that gases of both natural and anthropogenic origin are constantly being mixed in vertical as well as lateral dimensions.

When the atmosphere becomes heated near the Earth's surface, the lighter, warmer air rises into regions of lower pressure. As the pressure declines, the upward-moving air parcel expands, a process that requires the gas molecules to do work. Work done means a loss of kinetic energy and a lower temperature (which is the average kinetic energy of the molecules in that portion of the air). This process by which there is a decrease in temperature without any transfer of heat outside the air parcel is called *adiabatic cooling*. The rate at which temperature decreases is called the *adiabatic lapse rate* which for dry air is approximately 10°C for every 1000 m gain in altitude. For moist air, the rate is somewhat reduced by processes of condensation of water vapour into clouds in the cooler atmosphere. The combination of factors leads to the temperature of −60°C at an altitude of 15 km at the tropopause.

**Table 2.2** Comparison of tropospheric atmospheres from various regions.

| Location | Atmospheric characteristic |
|---|---|
| Oceans | Sea-salt aerosol (sodium, calcium, magnesium, chloride, sulphate, and minor elements) |
| Land (dry) | Airborne dust (soil-related, plant pollens, etc.) |
| Urban | High levels of pollutants may be present (smoke, dust, primary and secondary smog chemicals) |
| Arid tropics | Low humidity, intense solar radiation |
| Humid tropics | High humidity, natural volatile organics, intense solar radiation |
| Arctic | Sunlight period variable on a yearly cycle, Arctic haze (see Chapter 6) is present (containing sulphate aerosols, soot, and metals) |

The upward movement of air is also one cause among several others that leads to convective mixing within the troposphere. A molecule released at the Earth's surface would typically be swept up to the top of the troposphere in one or two days. Lateral mixing also occurs, but at a slower rate. Within each hemisphere, good mixing on the spatial scale requires approximately one month. Therefore, any gaseous matter that has a long residence time is well mixed, making the overall composition of the troposphere homogeneous with respect to these species. This includes all the major gases listed in Table 2.1. Local variations in concentration are characteristic of physically or chemically reactive species—water vapour and many trace species being good examples. Because of their reactivity, such gases are not stable for a sufficiently long time to become well mixed and homogeneous in composition.

Other examples of how the components of tropospheric air depend on location including over the open ocean, over large continental areas, urban areas, the tropics, and the Arctic are given in Table 2.2.

---

**Main point 2.2** Temperature and compositional changes in the regions of the atmosphere are related to reactions that are determined by the energy of solar radiation at various altitudes. In the troposphere, the temperature profile is such that the atmosphere is usually well mixed and the chemical composition is relatively constant throughout, except for short-lived species.

---

## 2.3 Reactions and calculations in atmospheric chemistry

### Measures of atmospheric concentration

Some of the methods used to measure concentration of gases in the atmosphere may not be familiar to many people. For gases that make up a large proportion of the atmosphere, fractional or percentage concentrations are used but, in contrast to solids and liquids, these are usually expressed on a molar, not mass, basis. The Avogadro relationship then defines that molar ratios would also be equivalent to a volume ratio or a ratio of partial pressures. A general term used to express atmospheric concentration is 'mixing ratio' (see Table 2.1). As an example, the mixing ratio for nitrogen in the dry troposphere, expressed as a fraction, is 0.7808 or 78.08%. This means that, for every 100 mol of all gases except water, 78.08 mol are nitrogen. It also means that, at atmospheric pressure ($P° = 101\,325$ Pa), the partial pressure of nitrogen can be calculated as shown

$$0.7808 \times 101\,325\,\text{Pa} = 7.911 \times 10^4\,\text{Pa}$$

For gases present at low concentrations, the *parts per . . .* family of units is frequently used to express mixing ratios. As with the fractional and percentage concentrations, equivalent calculations are done on a molar, pressure, or volume basis. The term used to express concentrations is, for example, parts per million by volume for which the symbol is ppmv. The mixing ratio of methane in the dry troposphere is approximately 1.77 ppmv. This means that there are 1.77 μmol of methane for every 1.0 mol of the components of air. It also means that a fraction, 1.77 millionths, of the total pressure is due to methane. Note that ppmv is again different from the meaning of ppm applied to solids or liquids. We can, of course, do the conversion. Using an average molar mass of air of 29 g mol$^{-1}$, a 1.77 ppmv concentration is the same as 0.98 ppm by mass. The latter concentration unit is rarely used in atmospheric studies.

Units of mass per volume, moles per volume, or molecules per volume are other frequently used ways of expressing concentration of gases and particles in the atmosphere. The following calculation illustrates a simple conversion between mixing ratio and concentration for oxygen.

---

**Example 2.3 Conversion of units used to measure atmospheric concentrations of gases**
What is the concentration of oxygen in the atmosphere (0°C and 1 atm pressure) using units of g L$^{-1}$?

At 0°C and 101.3 kPa, the molar volume of a gas is 22.4 L. Therefore, 1 L of air contains $1/22.4 = 0.0446$ mol of all gases, of which only 20.95% is oxygen. The concentration of $O_2$ is then

$$0.0446 \text{ mol L}^{-1} \times 20.95/100 = 9.35 \times 10^{-3} \text{ mol L}^{-1}$$

In mass terms, this is equivalent to $32.0 \text{ g mol}^{-1} \times 9.35 \times 10^{-3} \text{ mol L}^{-1} = 0.299 \text{ g L}^{-1}$, since 1 mol of oxygen has a mass of 32.0 g.

---

A related set of units giving molecules per volume is commonly used for some important atmospheric species that are present in very small concentrations. For atmospheric aerosols, solid or liquid particles in air, it is necessary to use mass per volume units to measure concentration. This is because there is no single molar mass for heterogeneous particles like dust or smoke.

## Types of atmospheric reactions

A more detailed look at the chemistry involved in the troposphere and stratosphere will follow in subsequent chapters. Before beginning, we will examine some common features of atmospheric chemical reactions and introduce the types of calculations used in a quantitative description of the reactions. The discussion will make use of specific examples in order to illustrate important general principles, but the principles apply to all types of atmospheric reactions including 'natural' processes such as ozone creation and destruction in the stratosphere and also anthropogenic ones where a particular pollutant generated by human activity interacts with other atmospheric components. Some reactions involve major gases and occur on a global scale, while others deal with trace species in a localized setting. In discussing the composition of the troposphere, we made note of the variability in mixing ratios of reactive minor components. While these gases are present in very small concentrations, many of them have short residence times, which is a reflection of their high reactivity. Frequently, a study of atmospheric chemistry then centres on the reactions in which the trace species are involved.

To illustrate types of reactions and basic calculations used, we have selected examples of production and degradation of some oxides of nitrogen, frequently referred to as $NO_x$. Later chapters will present more systematic discussions of these and other important reactions in the atmosphere.

## Atmospheric reactions where thermodynamic calculations are appropriate

In some situations, thermodynamic calculations are appropriate for estimating concentrations of species produced in gas-phase reactions. For example, nitric oxide (NO) is generated during every combustion reaction including in forest fires, lightning discharges, industrial and domestic heating and internal combustion engines, as a result of the combination of nitrogen and oxygen from the air.

$$N_2 \text{ (g)} + O_2 \text{ (g)} \rightleftharpoons 2NO \text{ (g)} \tag{2.10}$$

This reaction has a *positive free energy* of formation at 25°C that can readily be calculated, as shown below in the example. The positive free energy indicates that the equilibrium of reaction 2.10 lies to the left.

---

**Example 2.4 The calculation of free energy of formation for NO (g) at 25°C**

From reaction 2.10,

$$\Delta G^\circ_{\text{rxn}} = 2G^\circ_f(NO) - \Delta G^\circ_f(N_2) - \Delta G^\circ_f(O_2)$$
$$= 2 \text{ mol} \times 86.55 \text{ kJ mol}^{-1} - 0 - 0$$
$$= +173.1 \text{ kJ}$$

The $\Delta G^\circ_f$ values for NO, $N_2$, and $O_2$ were obtained from Appendix B.2.

---

At ambient temperatures, therefore, nitric oxide is not formed to a significant extent. However, the reaction also has a positive entropy change as seen in the calculation below. As the temperature increases, therefore, the positive $\Delta G^\circ$ value becomes smaller. At the very high temperatures associated with combustion, it is sufficiently small that significant quantities of nitric oxide can be produced.

For example, in a cylinder of an internal combustion engine at the time of ignition the temperature may reach approximately 2500°C (= 2773 K). In order to calculate the free-energy change at the elevated temperature, we begin with data for the standard enthalpies of formation and absolute entropies, $\Delta H^\circ_f$ and $S^\circ$, respectively (consult Appendix B.2). The standard values are obtained at 25°C (= 298 K) and we assume that the $\Delta H^\circ_f$ values are unaffected by temperature. The values of $\Delta H^\circ_f$ for nitrogen and oxygen in the standard state are, by definition, zero.

---

**Example 2.5 The free energy and equilibrium constant for formation of NO (g) at 2773 K for reaction 2.10**

$\Delta G^\circ_{2773}$ is determined by

$$\Delta G^\circ_{2773} = 2\Delta H^\circ_{f(NO)} - T(2S^\circ_{NO} - S^\circ_{N_2} - S^\circ_{O_2})$$
$$= 2 \times 90.25 \text{ kJ mol}^{-1} - 2773 \text{ K} \times (2 \times 0.211 - 0.192 - 0.205) \text{ kJ mol}^{-1} \text{ K}^{-1}$$
$$= 111.2 \text{ kJ mol}^{-1}$$

From this final value of 111.2 kJ mol$^{-1}$, the equilibrium constant is then readily calculated.

$$\Delta G^\circ_T = -RT \ln K_p$$

$$\ln(K_p) = \frac{-\Delta G^\circ_T}{RT}$$

$$\ln K_p = \frac{-111\,200 \text{ J mol}^{-1}}{8.314 \text{ J mol}^{-1} \text{ K}^{-1} \times 2773 \text{ K}} = -4.82$$

$$K_p = e^{-4.82} = 0.0081$$

---

The partial pressure of nitric oxide in the exhaust gases produced under these conditions is evaluated assuming the equilibrium condition

$$K_p = \frac{(P_{NO}/P^\circ)^2}{(P_{N_2}/P^\circ)(P_{O_2}/P^\circ)}$$

In this equation, note that all pressures are relative to $P^\circ$, 101 325 Pa.

While the equilibrium constant is still small, it is now large enough at this elevated temperature that some NO is produced, as we shall see in Example 2.6.

---

### Example 2.6 **The mixing ratio of NO produced during combustion in the cylinder of an internal combustion engine**

Consider a situation where most of the oxygen has reacted with the fuel and the pressure of the residual oxygen is reduced to 1.0 kPa. Nitrogen does not take part in the combustion and its pressure in the compressed cylinder is 650 kPa. The temperature is 2773 K.

The residual $O_2$ reacts with $N_2$ to produce NO. After reaction, assume each has lost partial pressure of $x$ kPa; According to reaction 2.10, $P_{NO}$ will then be $2x$ kPa.

Using the equation for $K_p$ above and the conditions described,

$$\frac{(2x/101.3)^2}{((650 - x)/101.3)((1.0 - x)/101.3)} = 0.0081$$

Because the pressures in this equation are relative to the standard pressure $P^\circ$, all the fractions are dimensionless.

$$\frac{4x^2}{(650 - x)(1.0 - x)} = 0.0081$$

Assume $x \ll 650$ kPa and thus

$$\frac{4x^2}{650(1.0 - x)} = 0.0081$$

giving the quadratic

$$4x^2 + 5.27x - 5.27 = 0$$

$$x = \frac{-5.27 \pm \sqrt{(5.27)^2 - 4(4)(-5.27)}}{2(4)} = 0.66$$

The assumption that $x \ll 650$ kPa is valid and

$$P_{NO} = 2x = 1.32 \text{ kPa}$$

The nitric oxide mixing ratio in the hot compressed cylinder is then

$$1.32 \text{ kPa} \div 650 \text{ kPa} \times 10^6 \text{ ppmv} = 2030 \text{ ppmv}$$

---

Assuming no catalytic decomposition of nitric oxide in the exhaust system, the nitric oxide would be released to the atmosphere along with the other combustion products.

Concentrations of this order[1] have been observed in the exhaust of older automobiles built without emission control. For example, measurements on the exhaust of a 1966 Valiant (Chrysler Corporation) V8 midsize car running at 2000 rpm showed 1200 ppmv nitric oxide under no-load conditions and 2500 ppmv when the engine was operating with a 50-horsepower load. Fortunately, most cars that are built at present have efficient 'catalytic converters' that facilitate the conversion of nitric oxide and other undesirable combustion products into harmless gases. The chemistry of these processes will be described in Chapter 4.

---

[1] R.F. Gould (ed.), *Catalysts for the control of automotive pollutants*, Advances in Chemistry Symposium Series, no. 143, Chairman, J.E. McEvoy, American Chemical Society, Washington DC; 1975.

We can attempt to go further with calculations concerning nitric oxide by determining what its concentration would be when the gases are cooled and diluted after exiting the exhaust system. The positive $\Delta G$ for reaction 2.10 suggests that at equilibrium, most of the NO will revert back to $N_2$ and $O_2$. The question we are asking then is 'How much of the nitric oxide remains in the atmosphere?' To answer this question, we begin with another equilibrium calculation.

---

**Example 2.7 Equilibrium concentration of NO in the atmosphere at 25°C**

Suppose nitric oxide present in the exhaust gases at an approximate mixing ratio of 2030 ppmv becomes diluted in the open air by a factor of 20 000, giving an atmospheric concentration of 0.100 ppmv or 100 ppbv. Assume that the ambient temperature is 25°C. Using a similar method (Example 2.5) of calculation for $K_p$ we obtain a new equilibrium concentration of nitric oxide for the lower temperature

$$\ln(K_p) = \frac{-173\,100}{8.314 \times 298}$$

$$K_p = 4.54 \times 10^{-31}$$

Again, we use the $K_p$ expression

$$K_p = \frac{(P_{NO}/P^\circ)^2}{(P_{N_2}/P^\circ)(P_{O_2}/P^\circ)}$$

The atmospheric pressures of nitrogen and oxygen are extremely large compared to the 100 ppbv NO and even with reaction 2.10 going from right to left, this will not significantly alter the partial pressures of 79 and 21 kPa, respectively. In the calculation below, let $x$ be the partial pressure of nitrogen (and oxygen) produced as reaction 2.10 goes from right to left

$$4.54 \times 10^{-31} = \frac{((1.01 \times 10^{-5} - 2x)/101.3)^2}{((79+x)/101.3)((21+x)/101.3)}$$

Making assumptions that $x \lll 79$ and $\lll 21$ for $N_2$ and $O_2$, respectively, and therefore ignoring $x$ in the denominator, the equation is readily solved.

$$x = 5.05 \times 10^{-6}$$

Clearly, the assumptions of nearly constant partial pressure values for $N_2$ and $O_2$ were valid. Using these assumptions and returning to the original $K_p$ expression, the very small equilibrium partial pressure expected for NO can be calculated.

$$4.54 \times 10^{-31} = \frac{(P_{NO}/101.3)^2}{(79/101.3) \times (21/101.3)}$$

$$P_{NO} = 2.7 \times 10^{-14} \text{ kPa}$$

Converting the calculated partial pressure into a mixing ratio, we would estimate the ambient atmospheric concentration of nitric oxide near a combustion source to be extremely small.

$$(2.7 \times 10^{-14} \text{ kPa}/101.3 \text{ kPa}) \times 10^9 = 3 \times 10^{-7} \text{ ppbv}$$

---

In summary, the mixing ratios of nitric oxide that we have calculated are as follows:

- in the exhaust gases                                    ~2030 ppmv;
- in the atmosphere after reaction to equilibrium        ~$3 \times 10^{-7}$ ppbv.

Clearly, our calculation has indicated that an almost negligible concentration of nitric oxide should remain in the atmosphere. In fact, this is not what is observed. Small but significant amounts are frequently found (for example, ~100 ppbv as in Fig. 4.2(b)) near combustion sources. The actual mixing ratios are then about one billion times higher than the calculated equilibrium values.

The reason for the huge difference is from the use of thermodynamics, which assumes an equilibrium situation. In the engine cylinder, the temperature is high, reactions are very fast, and equilibrium is quickly attained, meaning that thermodynamic calculations are valid. In the ambient atmosphere, however, under moderate temperature conditions, reaction rates are slower and an equilibrium situation is never reached. As a consequence, the thermodynamic calculation gives a completely erroneous result for the concentration of nitric oxide at ambient temperatures.

## Kinetic calculations

The major discrepancy is therefore due to the fact that the reverse of reaction 2.10 is *extremely slow* at ambient temperatures. During combustion, there is sufficient thermal energy that equilibrium is rapidly established but, when cooling occurs, reactants and products are retained near levels corresponding to the high-temperature equilibrium. Sometimes, the term 'frozen' is used to describe a situation where thermodynamically unstable products do not react due to slow kinetics. The following calculation clearly shows this.

The second-order rate constant for the reaction (reverse of reaction 2.10)

$$2NO\,(g) \rightarrow N_2\,(g) + O_2\,(g) \tag{2.11}$$

is given by

$$k_2 = 2.6 \times 10^6 e^{-(3.21 \times 10^4)/T}\,m^3\,mol^{-1}\,s^{-1}$$

where $T$ = temperature / K. At 25°C, the value of $k_2$ is therefore $4.3 \times 10^{-41}\,m^3\,mol^{-1}\,s^{-1}$. Using these kinetic data allows us to calculate the rate of decomposition of nitric oxide by the simple reaction.

---

**Example 2.8  Rate of conversion of nitric oxide to nitrogen and oxygen**
The first step is to convert the mixing ratio of 100 ppbv estimated above for nitric oxide to a concentration $[NO]_i$ in mol m$^{-3}$

$$[NO]_i = \frac{n}{V} = \frac{P}{R \times T} = \frac{100 \times 10^{-9} \times 101\,325\,Pa}{8.314\,J\,mol^{-1}\,K^{-1} \times 298\,K}$$

$$= 4.1 \times 10^{-6}\,mol\,m^{-3}$$

The rate of decomposition can now be estimated

$$Rate = k_2[NO]_i^2$$
$$= 7.2 \times 10^{-52}\,mol\,m^{-3}\,s^{-1}$$

The reaction is clearly very slow and the half-life is given by[2]

$$t_{1/2} = \frac{1}{k[NO]_i}$$

$$t_{1/2} = \frac{1}{4.3 \times 10^{-41}\,m^3\,mol^{-1}\,s^{-1} \times 4.1 \times 10^{-6}\,mol\,m^{-3}}$$

$$= 5.7 \times 10^{45}\,s$$
$$= 1.8 \times 10^{38}\,y$$

---

The value of $1.8 \times 10^{38}$ y is more than $10^{28}$ times as long as the age of the Earth. It is obvious that the nitric oxide generated during combustion is *thermodynamically unstable*, but it is *kinetically extremely inert* at 25°C with respect to reaction 2.11.

---

[2] The half-life for first-order reactions is given by $t_{1/2} = \ln 2\,/\,k$ and for second order reactions, $t_{1/2} = 1\,/\,k[reactant]_{initial}$.

In this example of nitric oxide discharged into the atmosphere after being produced in an engine, thermodynamics predicted complete reversion of nitric oxide to dinitrogen and dioxygen. A kinetic evaluation for the same reaction predicted essentially the opposite result—the nitric oxide remains completely stable with respect to these two products. Experimental observations indicate that actual atmospheric nitric oxide concentrations fall between these two extremes (Table 5.2).

The observations imply that other reactions must be involved in controlling the atmospheric concentration of this gas and we must consider these simultaneously with reaction 2.11. The most important is the oxidation of nitric oxide to produce nitrogen dioxide. With dioxygen as oxidant (reaction 2.12), once again it can be shown that thermodynamics fails us, as a simple calculation shows that the ratio $P_{NO_2} / P_{NO}$ should always be greater than $10^6$.

$$2NO\,(g) + O_2\,(g) \rightarrow 2NO_2\,(g) \tag{2.12}$$

This is much greater than ratios normally found in the natural atmosphere. Once again, kinetics predicts that the oxidation should proceed slowly.

Consider the situation in a polluted urban atmosphere where the nitric oxide concentration in the morning might be approximately 150 ppbv. (0.15 ppmv = $1.5 \times 10^{-5}$ kPa, or $6.1 \times 10^{-6}$ mol m$^{-3}$ calculated in the same way as shown in the first step of example 2.8; also see Fig. 4.2(b)). Assuming reaction 2.12 to be an elementary process, the rate equals $k_3\,(P_{NO})^2\,(P_{O_2})$. The third-order rate constant $k_3$ has been evaluated to be $1.4 \times 10^{-2}$ (mol m$^{-3}$)$^{-2}$ s$^{-1}$ (or $2.4 \times 10^{-3}$ kPa$^{-2}$ s$^{-1}$) at 25°C. In Example 2.9 shown below, the partial pressures of the gases are provided in the familiar units of kPa. In doing the calculation, then, the uncommon kPa$^{-2}$ s$^{-1}$ unit for the rate constant is convenient. Using this information, we can calculate the initial rate of oxidation of nitric oxide.

---

**Example 2.9 Calculation of the rate of oxidation of NO by O$_2$**
Assume that oxygen has its normal partial pressure of approximately 21 kPa.

$$\text{Rate} = 2.4 \times 10^{-3}\ \text{kPa}^{-2}\ \text{s}^{-1}(1.5 \times 10^{-5}\ \text{kPa})^2\,(21\ \text{kPa})$$

$$= 1.1 \times 10^{-11}\ \text{kPa s}^{-1}$$

$$= \frac{1.1 \times 10^{-11} \times 10^9}{101.3} \times 3600 \times 24$$

$$= 9.7\ \text{ppbv day}^{-1}$$

---

This initial rate of oxidation is slow compared with the known concentration and cannot explain the rapid build-up of nitrogen dioxide observed during a smog event (see Figs. 4.2(a) and (b)). We shall see in Chapter 4 that dioxygen is not the most important oxidant in this reaction. Rather it is a combination of ozone (O$_3$) and peroxy radicals (ROO•)—also present in a tropospheric smog—that has the major effect on oxidizing nitric oxide. A general form of the reaction with peroxy radical species is

$$ROO\bullet + NO \rightarrow RO\bullet + NO_2 \tag{2.13}$$

Unfortunately, detailed calculations for this general reaction are difficult because of the variety of peroxy and oxy radical species and their variable and frequently unknown atmospheric concentrations. Free radical chemistry will be introduced later in this chapter.

---

**Main point 2.3** For any given atmospheric chemical process, either thermodynamic or kinetic control is possible, with the latter being frequently the case. Therefore, thermodynamic calculations alone often lead to erroneous predictions about concentrations, at least in the short term.

---

## Photochemical reactions

If we now consider the means by which nitrogen dioxide is decomposed and reverts to nitric oxide we encounter another feature characteristic of atmospheric reactions—the fact that many of these reactions have a photochemical component. By this we mean that the absorption of (usually solar) electromagnetic energy by species in the atmosphere is required to stimulate certain reactions. The reactant is excited to a higher energy state, thus enhancing bond-breaking or bond-making processes. The first step in a photochemical reaction is

$$XY + h\nu \rightarrow XY^* \quad \text{Absorption} \tag{2.14}$$

The asterisk here and elsewhere is used to denote an excited-state species. Absorption of energy is followed by further reactions that may take a variety of pathways such as

$$XY^* \rightarrow X + Y \quad \text{Decomposition} \tag{2.15}$$

or

$$XY^* + \text{other reactant(s)} \rightarrow \text{products} \tag{2.16}$$

The magnitude of the rate constant for a photochemical reaction, symbolized as $f$ to distinguish it from the thermal rate constant $k$, is determined by several factors (eqn 2.17).

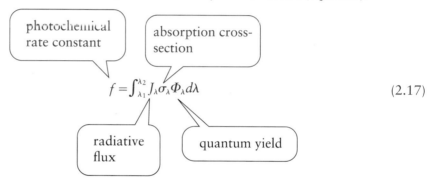

$$f = \int_{\lambda_1}^{\lambda_2} J_\lambda \sigma_\lambda \Phi_\lambda d\lambda \tag{2.17}$$

In this expression, $\lambda_1$ and $\lambda_2$ are the wavelength 'limits' of the solar radiation being considered. For example, at the Earth's surface the values would be approximately 300 and 800 nm, respectively, while the range would extend to shorter wavelengths in the stratosphere. The other terms in the expression all depend on the range of wavelengths under consideration. $J$ is the radiative flux, $\sigma$ is the absorption cross-section, which is a measure of the ability of the molecule of interest to absorb radiation of the type being considered, and $\Phi$ is the quantum yield, which is the ratio of the number of molecules undergoing the specific reaction of types such as reactions 2.15 and 2.16 to number of quanta of radiation absorbed.

When every absorption event leads to the particular reaction of interest, $\Phi$ has a value of unity. However, what happens far more frequently is that the excited species is deactivated by (i) 'quenching' through collisions with other gases (M) (reaction 2.18) or (ii) by transferring energy to other molecules leading to their excitation (reaction 2.19), and $\Phi$ is very small.

$$XY^* + M \rightarrow XY + M + \text{kinetic energy} \quad \text{Quenching} \tag{2.18}$$

$$XY^* + AB \rightarrow XY + AB^* \quad \text{Intermolecular transfer} \tag{2.19}$$

As another extreme, the quantum yield is greater than one when a single activated molecule sets off a chain of events leading to a sequence of reactions.

As an example of a photochemical reaction we can again consider a case from tropospheric $NO_x$ chemistry—the decomposition of nitrogen dioxide:

$$NO_2 + h\nu \rightarrow NO + O \tag{2.20}$$

The atomic oxygen produced in this reaction is in the ground state ($O(^3P)$) and, as an aside, it can be mentioned that $O(^3P)$ reacts with molecular dioxygen to generate ozone—the only significant pathway to produce ozone in the lower troposphere. The photolytic quantum yield for reaction 2.20 depends on the wavelength of the electromagnetic radiation and is near 1 for $\lambda < 360$ nm (near the high-energy end of the visible region of the spectrum) but falls off to 0 at about $\lambda > 440$ nm. It is commonly said that the minimum energy required to effect the process is associated with radiation of 400 nm. The rate of the reaction is given by

$$\text{Rate} = f_1[NO_2]$$

As usual, the photochemical first-order rate constant, $f_1$, depends on the intensity and wavelength of the incident radiation. For nitrogen dioxide photolysis, values of $f_1$ range from about $5.6 \times 10^{-3}$ s$^{-1}$ in intense sunlight to zero at night.

---

Example 2.10 **Half-life of nitrogen dioxide in intense sunlight**
For the first-order reaction,

$$t_{1/2} = \ln 2 \, / \, f_1$$
$$t_{1/2} = 0.693 \, / \, 5.6 \times 10^{-3} \text{ s}^{-1}$$
$$t_{1/2} = 124 \text{ s}$$

In sunlight, therefore, nitrogen dioxide would have a half-life of $\sim$ 120 s, indicating that the photolysis is rapid.

---

Decomposition of nitrogen dioxide is offset by synthesis as shown above in reaction 2.13, also a kinetically rapid process. The steady-state situation established between decomposition and synthesis is known as a photostationary state and is described by equating the rates of the production and consumption reactions

production (reaction 2.13) = consumption (reaction 2.20)

$$k_2[ROO\bullet][NO] = f_1[NO_2]$$

Therefore,

$$\frac{[NO_2]}{[NO]} = \frac{k_2[ROO\bullet]}{f_1}$$

(Note that ROO• is a composite of several species including ozone and $k_2$ is a composite rate constant.) In any case, the final equation determines the ratio of nitrogen dioxide to nitric oxide, a ratio that is established quickly and may vary from a small value in bright sunlight when $f_1$ is large, to a higher night-time value.

The example we have used to illustrate the importance of photochemistry in atmospheric reactions is one that occurs in the troposphere. There are many other reactions and, as we move upward in altitude to the stratosphere and above, we find that photochemistry becomes even more important. Obviously, this is because of the availability of previously unabsorbed higher energy radiation in the rarified upper atmosphere. There, photolysis of high binding energy molecules like dioxygen and dinitrogen becomes possible, as well as ionization of molecules and atoms that are 'normally' stable.

As we have noted, where kinetic properties determine the behaviour of chemical species in the atmosphere, we frequently use the term half-life, $t_{1/2}$, to describe the progress of the reaction. A similar, alternative description is given by the term residence time, $\tau$ (tau, also called atmospheric lifetime or residence time), which we defined in Chapter 1. Whereas $t_{1/2}$ represents the time taken for half of the reaction to occur, $\tau$ (the average lifetime) is the time during which the original concentration decreases to 37% of its original value. For a first-order or pseudo-first-order reaction, $t_{1/2}$ is equal to $\ln 2 / k_1$ (or $\ln 2 / f_1$), while $\tau$ is equal to $1 / k_1$ (or $1 / f_1$).

> **Main point 2.4** In addition to standard 'thermal reactions' there are important atmospheric processes that are, partly or completely, photochemical processes. The energy of solar radiation increases with increasing altitude so that photochemistry becomes even more important in the upper atmosphere.

## Free-radical reactions

A second key feature regarding chemical processes that occur in the atmosphere is the way in which free radicals[3] take part in many reactions. We will encounter a variety of radical species, but none is more important than the hydroxyl free radical (•OH), a species that will appear again and again in our study of atmospheric chemistry reactions. Neutral *hydroxyl radical* (not to be confused with the negatively charged *hydroxide ion*, $OH^-$, that is so pivotal in aqueous solution chemistry) is formed in the troposphere by a variety of means but the most important is a four-step process (including two photochemical steps).

$$NO_2 + h\nu \rightarrow NO + O \qquad (2.20)$$

$$O + O_2 + M \rightarrow O_3 + M \qquad (2.21)$$

In the second step, the combining of O and $O_2$ takes place as a result of an additional collision and an instantaneous energy-transfer process to another atmospheric molecule, a so-called 'third body', represented as M in reaction 2.21. Since dinitrogen and dioxygen are the most common species throughout the Earth's atmosphere, the third body is usually one of these molecules.

$$O_3 + h\nu \rightarrow O_2^* + O^* \qquad (2.22)$$

$$O^* + H_2O \rightarrow 2\cdot OH \qquad (2.23)$$

The atomic oxygen produced in reaction 2.20 is ground-state or triplet $(O(^3P))$ oxygen, while that in reaction 2.22 is in the excited state or singlet $(O(^1D))$ oxygen. For the former reaction, radiation in the visible region supplies sufficient energy to give a large value for the photo-chemical rate constant, $f_1$, while for the latter higher-energy $(\lambda < 325 \text{ nm})$ ultraviolet radiation is required. Excited-state oxygen atoms are essential in order to form hydroxyl by reaction with atmospheric (gaseous) water according to reaction 2.23. In competition with this reaction are quenching processes of the type

$$O^* + M \rightarrow O + M + \text{kinetic energy} \qquad (2.24)$$

Because of the availability of many M species ($O_2$, $N_2$, etc.) in air, reaction 2.24 proceeds readily and this results in a low quantum yield, $\Phi$, for the combined reactions 2.22 and 2.23.

Because of its low concentration and highly reactive nature, it is extremely difficult to measure[4] concentrations of hydroxyl radicals but, in many situations, levels have been estimated to be between $1 \times 10^4$ and $2.5 \times 10^6$ molecules per $cm^3$ in the troposphere, with values being higher at low altitudes and latitudes, in daytime, and in areas of heavy pollution. The global

---

[3] Free radicals are atoms or molecules that have one or more unpaired electrons. They are usually highly reactive species. In most cases, we will denote radicals using a dot (•) beside or above the radical species. While some compounds containing nitrogen (like $NO_2$) or chlorine (like ClO) atoms have an unpaired electron, we make an exception and will not usually show the unpaired electron in such compounds, unless some other radical species is involved in the reaction.

[4] F.L. Elsele and J.K. Bradshaw, The elusive hydroxyl radical: measuring OH in the atmosphere. *Analyt. Chem.* 65 (21), (1993) 927A–39A.

average tropospheric concentration has been estimated to be $8.1 \times 10^5$ molecules per cm$^3$, while in the tropics, an average concentration of $2 \times 10^6$ molecules per cm$^3$ may be a reasonable estimate. The factors determining actual concentrations of the hydroxyl radical relate to reactions 2.20 to 2.24.

We will later encounter a number of varied atmospheric reactions in which hydroxyl radicals take part. However, in a quantitative sense, the two most important are

$$\bullet OH + CO \rightarrow \underset{\text{hydrogen radical}}{\bullet H} + CO_2 \tag{2.25}$$

$$\bullet OH + CH_4 \rightarrow \underset{\text{methyl radical}}{\bullet CH_3} + H_2O \tag{2.26}$$

Reactions with carbon monoxide and methane account for approximately 70% and 30%, respectively, of all reactions involving hydroxyl radical in an unpolluted atmosphere. In the course of both reactions, other highly reactive radical species (the hydrogen and methyl radicals) are formed and these, in turn, react further:

$$\bullet H + O_2 + M \rightarrow \underset{\text{hydroperoxyl radical}}{HOO\bullet} + M \tag{2.27}$$

$$\bullet CH_3 + O_2 + M \rightarrow \underset{\text{peroxymethyl radical}}{CH_3OO\bullet} + M \tag{2.28}$$

The peroxy radical products are important oxidants as we have seen in the case of oxidation of nitric oxide (reaction 2.13).

In regions affected by either natural or anthropogenic gaseous emissions, other processes frequently contribute significantly to the loss of the hydroxyl radical. Most of the reactions may be classified into two categories. The first category is hydrogen abstraction as in the reactions with toluene (reaction 2.29) and formaldehyde (reaction 2.30).

$$\bullet OH + HCHO \longrightarrow H\overset{\bullet}{C}{=}O + H_2O \tag{2.30}$$

Similarly to the reaction with methane, in reactions 2.29 and 2.30 new highly reactive radical species are produced and they too react further with oxygen as in reactions 2.31 and 2.32, forming peroxy radical species.

$$H\overset{\bullet}{C}{=}O + O_2 \longrightarrow HOO\bullet + CO \tag{2.32}$$

The second category of reactions is usually an addition across a multiple bond as illustrated by the reactions with ethene (reaction 2.33) and benzene (reaction 2.35). DMS is an exception to this, reacting through both an abstraction and addition route. Again, new highly reactive species are produced and these typically begin a further reaction sequence by adding oxygen (reactions 2.34 and 2.36).

$$\bullet OH + H_2C{=}CH_2 \longrightarrow H_2\overset{\bullet}{C}{-}CH_2OH \tag{2.33}$$

$$H_2\overset{\bullet}{C}{-}CH_2OH + O_2 \xrightarrow{M} \underset{}{H_2C{-}O{-}\overset{\bullet}{O}} \tag{2.34}$$

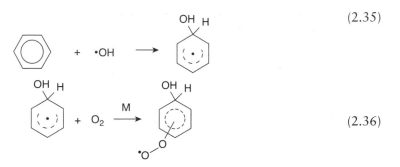

(2.35)

(2.36)

There are many organic species present in the atmosphere at low concentrations. These include ones of biogenic origin emitted by living or decomposing organisms and the multitude of chemicals released to the atmosphere from combustion of fossil fuels and from many industrial processes. Even present in buildings, organic species are given off by the diverse natural and synthetic materials used in construction. The possible chemical reactions with which these chemicals are involved are also limitless, but a surprising number are initiated and propagated by hydroxyl or other radical species. In many cases, the first step of these reactions is well understood, often involving principles such as those just shown; rate constants are also available so that the rate of destruction can be calculated. Frequently, the subsequent reaction steps are less clear and the mechanisms leading to ultimate destruction are not known.

> **Main point 2.5** Free radicals play a dominant role in many atmospheric reactions and the hydroxyl radical is especially prominent in this regard.

### Literature link  The hydroxyl radical factor

We cannot overestimate the importance of the role played by free radical species, especially the hydroxyl radical, in atmospheric chemistry. This will become evident as we proceed through later chapters of this book. A recent paper, *The cleansing capacity of the atmosphere,* by R.G. Prinn (Annu. Rev. Environ. Resour. 2003. 28:29–57), describes a wide range of processes by which trace gases in the atmosphere, many of human origin, are oxidized via radical-mediated processes. The ability of the atmosphere to destroy such compounds, thus cleansing itself, is referred to as its oxidation capacity.

The author points out important atmospheric conditions that affect the concentration of hydroxyl in the troposphere, and therefore the atmosphere's oxidation capacity. The concentration is lower at night-time, in winter, under cloudy conditions, in areas where nitric oxide emissions are reduced, in dry atmospheres and where the stratospheric ozone layer is especially thick. From what has been said in this chapter, you should be able to understand the reasons for each of these effects.

The most important anthropogenic factors affecting the concentration of hydroxyl are the emissions of carbon monoxide, hydrocarbons and nitrogen oxides. As we have seen, elevated emission rates of the former two compounds accelerate the consumption of hydroxyl and, if this were significant, would lead to a reduction in its abundance. On the other hand, increased amounts of nitrogen oxides will favour greater production of additional hydroxyl. It is interesting that carbon monoxide and nitrogen oxides are both products of combustion, so one is left with the question: 'Over time, will hydroxyl levels increase or decrease?' Most models described in the paper referenced above indicate that a decrease in the range of 10 to 20% is likely over time.

The word *sink* is used frequently to indicate a route by which chemical species are removed from the atmosphere. Radical reactions are an important sink for destroying undesirable species like volatile organic compounds, sulphur dioxide and nitrogen dioxide. But there are other sinks as well. For example, another removal mechanism for sulphur dioxide and nitrogen dioxide is by having these dissolve in water droplets in clouds, and then being rained out of the atmosphere.

## ADDITIONAL RESOURCES

1. Atkins, P. and J. de Paula, *Physical chemistry*, 8th edn, Oxford University Press, Oxford; 2006.
2. Finlayson-Pitts, B.J. and J.N. Pitts Jr., *Chemistry of the upper and lower atmosphere*, Academic Press, New York; 2000.
3. Wayne, R.P., *Chemistry of atmospheres, an introduction to the chemistry of the atmospheres of Earth, the planets, and their satellites*, 3rd edn, Clarendon Press, Oxford; 2000.
4. Chamberlain, J.W. and D.M. Hunten, *Theory of planetary atmospheres, an introduction to their physics and chemistry*, 2nd edn, Academic Press, Inc., London; 1987.

## PROBLEMS

1. The mixing ratio of oxygen in the atmosphere is 20.95%. Calculate the concentration in mol $L^{-1}$ and in g $m^{-3}$ at $P^{\circ}$ (101 325 Pa, 1.00 atm) and 25°C.

2. Calculate the atmospheric pressure at the stratopause. What are the concentrations (mol $m^{-3}$) of dioxygen and dinitrogen at this altitude? How do these concentrations compare with the corresponding values at sea level?

3. What is the total mass of the stratosphere (the region between 15 and 50 km above the Earth's surface)? What mass fraction of the atmosphere does this make up?

4. If the mixing ratio of ozone in a polluted urban atmosphere is 50 ppbv, calculate its concentration: (a) in mg $m^{-3}$; and (b) in molecules $cm^{-3}$.

5. The gases from a wood-burning stove are found to contain 1.8% carbon monoxide at a temperature of 65°C. Express the concentration in units of g $m^{-3}$.

6. Using Fig. 8.2 compare the mixing ratio (as %) of water in the atmosphere for a situation in a tropical rain forest in Kinshasa, DRC ($T = 36$°C, relative humidity = 92%) with that in Denver, USA ($T = 8$°C, relative humidity = 24%).

7. Calculate the maximum wavelength of radiation that could have sufficient energy to effect the dissociation of nitric oxide (NO). In what regions of the atmosphere would such radiation be available? Use data from Appendix B.2.

8. The average distance a gas molecule travels before colliding with another (mean free path, $S_{mfp}$) is given by the relation

$$S_{mfp} = \frac{kT}{\sqrt{2}P\sigma_c}$$

where $T$ and $P$ are the temperature (K) and pressure (Pa), respectively, $k$ is Boltzmann's constant = $1.38 \times 10^{-23}$ J $K^{-1}$, and $\sigma_c$ is the collision cross-section of the molecule ($m^2$). Calculate the mean free path of a dinitrogen molecule at the Earth's surface ($P^{\circ}$ and $T = 25$°C), and at the stratopause. The value of $\sigma_c$ for dinitrogen is 0.43 $nm^2$. What does this indicate regarding gas-phase reaction rates in these two locations?

9. Calculate the maximum wavelength of radiation required to bring about dissociation of: (a) a dinitrogen molecule; (b) a dioxygen molecule. Account qualitatively for the difference.

10. Carbon dioxide in the troposphere is a major greenhouse gas. It absorbs infra-red radiation, which causes changes in the frequency of carbon–oxygen stretching vibrations. What are the ranges of wavelength ($\mu$m), frequency ($cm^{-1}$), and energy (J) associated with this absorption?

11. The stability of compounds in the stratosphere depends on the magnitude of the bond energies of the reactive part of the molecules. Using Appendix B.2 calculate bond energies of HF (g), HCl (g), and HBr (g), in order to determine the relative ability of these molecules to act as reservoirs for the respective halogen atoms.

12. Use the tabulated bond energy data in Appendix B.3 to estimate the enthalpy change of the gas-phase reaction between hydroxyl radical and methane.

13. Which of the following atmospheric species are free radicals?

$$OH, \ O_3, \ Cl, \ ClO, \ CO, \ NO, \ N_2O, \ NO_3^-, \ N_2O_5$$

14. In an indoor atmosphere, for $NO_2$ the value of the first-order rate constant has been estimated to be 1.28 $h^{-1}$. Calculate its residence time.

15. If the rate laws are expressed using mol $L^{-1}$ for concentrations, and Pa for pressure, what are the units of the second- and third-order rate constants, $k_2$ and $k_3$? Calculate the conversion factor for converting $k_2$ values obtained in the units above to ones using molecules per $cm^3$ for concentration and atm for pressure.

16. For the reaction

$$NO + O_3 \rightarrow NO_2 + O_2$$

the second-order rate constant has a value of $1.8 \times 10^{-14}$ molecule$^{-1}$ cm$^3$ s$^{-1}$ at 25°C. The concentration of NO in a relatively clean atmosphere is 0.10 ppbv and that of $O_3$ is 15 ppbv. Calculate these two concentrations in units of molecule per $cm^3$. Calculate the rate of the NO oxidation using concentration units of molecule cm$^{-3}$. Show how the rate law may be expressed in pseudo-first-order terms and calculate the corresponding pseudo-first-order rate constant.

17. At a particular temperature, the Arrhenius parameters for the reaction

$$\bullet OH + H_2 \rightarrow H_2O + \bullet H$$

are $A = 8 \times 10^{10}$ s$^{-1}$, and $E_a = 42$ kJ mol$^{-1}$. Given that the concentration of hydroxyl radical in the atmosphere is $7 \times 10^5$ molecules cm$^{-3}$ and that of $H_2$ is 530 ppbv, calculate the rate of reaction (units of molecule cm$^{-3}$ s$^{-1}$) for this process.

18. The equilibrium constant for the reaction

$$N_2O_4 \rightarrow 2NO_2$$

has a value of $K_c = 4.65 \times 10^{-3}$ L mol$^{-1}$ at 25°C and the corresponding standard enthalpy change is $\Delta H° = +57$ kJ.

Calculate the equilibrium concentration of $N_2O_4$ in an atmosphere where the concentration of $NO_2$ is 200 $\mu$g m$^{-3}$. If the temperature were to increase would you expect the relative concentration of $N_2O_4$ to increase or decrease?

19. It is postulated[5] that the hydroxyl radical concentration in the tropical atmosphere is directly proportional to the rate constant for photolysis of $O_3$ to $O^*$, and to the concentration of $O_3$ and $H_2O$. It is inversely proportional to $(C_{CO} + 0.03 \times C_{CH_4})$; concentrations are expressed as mixing ratios. Use this hypothesis to predict the relative hydroxyl concentrations in two tropical atmospheres with the following conditions.

| Air temperature / °C | 25−30 | 20−25 |
|---|---|---|
| Cloudiness / % | 50−75 | 25−50 |
| $O_3$ / ppbv | 10 | 19 |
| $H_2O$ / kPa | >25 | 20 |
| CO / ppbv | 38 | 50 |
| $CH_4$ / ppmv | 1560 | 1580 |
| $J(O_3)$ / s$^{-1}$ | 1.3 | 0.68 |

[5] Data taken from H. Rodhe and R. Herrera, *Acidification in tropical countries* (Scope 36), John Wiley and Sons, Chichester; 1988.

# Chapter 3
# Stratospheric chemistry—ozone

The focus of this chapter is the chemistry of ozone in the stratosphere. Here, we will discuss:

- how ozone is distributed within the atmosphere and the concerns about decreasing amounts of stratospheric ozone
- the oxygen-only synthesis and decomposition chemistry of ozone, and the catalytic cycles of destruction involving both natural and anthropogenic origins
- the properties, nomenclature, impacts, and replacements for chlorofluorocarbons (CFCs)
- the rate of ozone destruction reactions through kinetic calculations
- the process of ozone thinning, sometimes referred to as 'ozone hole' formation, at the poles.

In studying the chemistry of our atmospheric surroundings, for the most part we will limit ourselves to the troposphere—that part of the atmosphere where humans live and where most other life processes also occur. We make an exception in this chapter, however, as we examine important aspects of stratospheric chemistry. This is because gas molecules in the stratosphere absorb some of the solar radiation and therefore moderate what is transmitted to the Earth's surface. The qualitative and quantitative effect of this is an important determining factor with respect to life processes. Furthermore, unlike the Sun itself that continues to emit its characteristic radiation unaffected by human activities, the extent to which stratospheric gases interact with radiation can be significantly altered by activities that humans undertake. It is for these reasons that Earth-bound environmental chemists and scientists should have some knowledge of the chemistry of the stratosphere.

## 3.1 Concerns about stratospheric ozone

The stratosphere is the region of the atmosphere extending between approximately 15 and 50 km altitude above the Earth's surface (Fig. 2.1). It is distinguished from the troposphere below and the mesosphere above by having a temperature inversion, meaning that the temperature increases with altitude. With cooler heavier air below, convection is minimized creating a stable atmosphere with minimal mixing. Similar to the troposphere, the principal constituent gases are dinitrogen and dioxygen, but the higher-energy radiation striking these gases, oxygen in particular, leads to chemical reactions different from those found nearer the surface of the Earth.

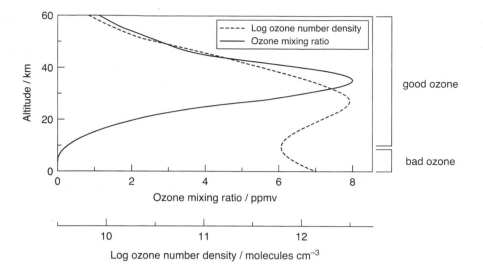

**Fig. 3.1** Concentration profile of ozone in the lower atmosphere, shown both as the mixing ratio (solid line) and as the log ozone number density (broken line). The terms 'good' and 'bad' ozone describe the human benefits / disadvantages derived from increased concentrations. (Data obtained from Wayne, R.P., *Chemistry of atmospheres*, Clarendon Press, Oxford; 1991. Reprinted with permission.)

The most important of these reactions relate to the synthesis and decomposition of ozone. Ozone is a naturally occurring gas found throughout the atmosphere, with a maximum mixing ratio at altitudes ranging from 15 to 30 km above the Earth. We can think of stratospheric ozone as being 'good ozone' for reasons that will become clear shortly. This region is frequently called the 'ozone layer'. Figure 3.1 plots the ozone distribution in two ways and it is important to note the distinction between graphs expressed as a mixing ratio and those as the logarithm of number density. In the latter case, you can see that the density of ozone molecules near the Earth's surface actually increases to within an order of magnitude of that found in the ozone layer. Near the Earth, elevated levels can be toxic to both plants and animals, and it is also a greenhouse gas and more will be said about this in Chapter 8. In contrast to stratospheric ozone, we can think of this as 'bad ozone' and we will discuss aspects of its tropospheric chemistry in Chapter 4.

---

**Ozone—environmental properties**

- Good ozone—ozone in the stratosphere protects plant and animal life on Earth by absorbing high-energy solar ultraviolet radiation.
- Bad ozone—ozone in the troposphere causes irritation and damage to plants and animals through its intense oxidizing ability, and it is also a greenhouse gas.

---

In the stratospheric ozone layer, the mixing ratio of the gas is in the part per million range, so even in that region it makes up only a small fraction of all molecules. However, this relatively small concentration is of great importance in screening out harmful ultraviolet (UV) radiation that would otherwise reach the surface of the Earth. In the stratosphere, ozone is an effective filter capable of absorbing ultraviolet radiation with wavelengths between 200 and 315 nm.

## Solar radiation and plant and animal life

The Sun emits a range of radiation over a broad region of the electromagnetic spectrum from wavelengths greater than 1000 μm to those less than 100 nm. The intensity maximum occurs at 550 nm, which is in the yellow region of the visible spectrum. As we shall see in Chapter 8, this radiation is near the optimum for photosynthesis and production of biomass—clearly, plant life has evolved to accommodate this radiation.

The relatively longer wavelength (visible and infra-red) photons therefore provide heat and light, defining the global climate and all manner of biological activity. The shorter-wavelength (UV) radiation comprises only a small portion of the total flux, yet is of great significance for a number of reasons—one being that it can be harmful, even lethal, to living organisms. In a biological classification,[1] this potentially dangerous UV radiation is subdivided into three categories:

- UV-A, 315 to 400 nm (the near-ultraviolet ranging into the visible, 7% of the total solar flux), which is, in the short term, not particularly harmful to living species;
- UV-B, 280 to 315 nm (1.5% of the total solar flux), which can be harmful to both plant and animal species, especially after prolonged exposure;
- UV-C < 280 nm (0.5% of the total solar flux), which rapidly damages living matter of all types.

The UV portion of the radiation that is emitted by the Sun is shown as the solid line in Fig. 3.2. The small flux of UV-C radiation below 280 nm penetrates into the upper atmosphere but is efficiently and completely absorbed by ozone (dotted line) and other atmospheric species (e.g. $O_2$, dashed line) before it reaches the Earth's surface. On the other hand, radiation in the

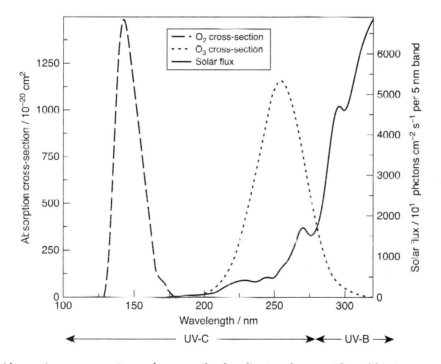

**Fig. 3.2** Absorption cross-sections of oxygen (broken line) and ozone (dotted line) compared to the solar flux density (solid line) over the region of biologically harmful ultraviolet radiation. The absorption cross-section is a measure of the ability of a molecule to absorb radiation, in the process limiting its penetration through the atmosphere. (Data from Chamberlain, J.W. and D.M. Hunten, *Theory of planetary atmospheres*, Academic Press, London; 1987. Reprinted with permission.)

---

[1] Miller, D.H., *Energy at the surface of the Earth*, Academic Press Inc., New York; 1981.

UV-A region and the 1.5% that is in the UV-B region are only partially absorbed. Ozone is the species responsible for intercepting these ultraviolet photons but, as shown in Fig. 3.2, it has a relatively small absorption cross-section in the UV-B wavelength range and almost no ability to absorb UV-A radiation. Where concentrations of ozone in the stratosphere are substantially reduced, dangerous levels of UV-B radiation can penetrate into the troposphere. This radiation can adversely affect a variety of biological processes and is largely responsible for the development of two forms of skin cancer, basal cell carcinoma and the more dangerous malignant melanoma.

## UV-B radiation dangers to humans

The UV-B radiation is of greatest concern to humans (most particularly those with fair skin) as it can cause cell damage within the midlayers of active skin. A positive response of the body to exposure to UV-B radiation is to produce melanin, the evidence of which is development of a tan, and this protects to some degree against further damage. UV-A radiation has usually been considered to pose less danger, but is known to penetrate skin efficiently and, through prolonged exposure to the Sun, is implicated in cell damage associated with the ageing process.

Sunscreens and sunblocks are pharmaceutical preparations that can be applied to the skin to reduce penetration of the dangerous forms of radiation. Sunscreens contain chemicals such as benzyl salicylate or cinnamate that absorb strongly in the 250 to 325 nm range, thus preventing the radiation from striking and passing into the body; they may also contain free-radical scavengers as additional protection. The effectiveness of sunscreen preparations is determined by their ability to absorb in the appropriate region of the electromagnetic radiation spectrum. As described by the standard Beer's law relation, this depends on the intrinsic nature of the compound, its concentration (defined by the percentage of active ingredient in the lotion), and pathlength (defined by the thickness of the applied layer). To measure effectiveness, a sun protection factor (SPF) is used. For example, an SPF value of 15 corresponds to an ability to absorb 93% of the UV-B radiation, using a standard application. The numerical value indicates that a person should be able to stay in the Sun's radiation $100 / (100 - 93) = 15$ times longer than without protection. Most of the absorbing materials used in sunscreens are ineffective in screening out UV-A radiation.

Sunblocks are related preparations, but contain solid particles, typically zinc oxide or titanium dioxide, that physically block radiation by reflecting it away from the skin surface. Because this is a physical process, such materials are effective in reducing the penetration of all types of radiation.

## Measuring ozone in the atmosphere

Measuring the amount of ozone in the atmosphere is by no means a simple task. It is not like taking a water or soil sample, where these can be easily brought back to the lab for analysis. In some cases of atmospheric analysis, a 'grab sample' of gas could be taken, but ozone is readily destroyed by reaction with the container materials. Even if a sample were obtained from high in the atmosphere it would be difficult to measure the amount of ozone at the low concentrations that are present. As an alternative to collecting gas samples, instrumentation is taken to the field and measurements are conducted in situ. This is done by instruments that look vertically through a column of the atmosphere and measure the total ozone in the column. Before continuing with a discussion of how measurements are conducted, it is instructive to first look at the units that are employed for ozone.

The Dobson unit (DU) is defined in terms of an equivalent thickness of pure ozone. An air column with a Dobson value of 300 DU (close to the global average) contains as much ozone as in a 3-mm thick layer of pure ozone at 0°C and $P°$. Using the ideal gas law, we can determine that a column of pure gaseous ozone, with a cross-sectional area of 1 m² that is 3 mm thick ($V = 0.003$ m³) would contain

$$\boxed{P} \qquad \boxed{V} \qquad \boxed{R} \qquad \boxed{T}$$
$$101\,325\text{ Pa} \times 0.00300\text{ m}^3 = n \times 8.315\text{ J mol}^{-1}\text{ K}^{-1} \times 273\text{ K}$$
$$n = 0.134\text{ mol}$$

and
$$0.134\text{ mol} \times 6.024 \times 10^{23}\text{ molecules mol}^{-1}$$
$$= 8.07 \times 10^{22}\text{ molecules of ozone}$$

Therefore, at a site where the ozone was measured to have a value of 300 DU, there would be a total of $8.07 \times 10^{22}$ molecules present in a column with a cross-sectional area of 1 m². These molecules are unevenly distributed through the atmospheric column, with their greatest number density being at an altitude of about 25 km, as depicted in Fig. 3.1.

Today, there are a number of 'field' stations that conduct ozone measurements on an on-going basis. Some of these are now permanent land-based monitoring sites, while others are mobile flight-based systems operating from balloons, aircraft, and satellites. The common feature in all of these systems is the instrumentation they use, which is based on measuring the absorption of light in the UV range. As we have seen, ozone absorbs radiation with wavelengths between 200 and 315 nm (Fig. 3.2) with maximum absorbance near 255 nm. It is this absorption capability that is put to use in determining ozone concentrations in the atmosphere. A few examples and simple descriptions of some of the systems currently used follow.

### Total Ozone Mapping Spectrometer (TOMS)

TOMS is a satellite-based ozone-monitoring system that has been in use since 1978 to observe ozone concentrations around the world (see Additional Resources 7). There have been four different satellites operated during this time including; Nimbus 7, Meteor-3, Earth Probe, and since 2004, Aura (formerly EOS / Chem-1).[2,3] The Aura houses the ozone-monitoring instrument (OMI) which measures ozone by observing the backscattered solar light from the atmosphere. The satellite information system is capable of providing global, as well as polar views of the Earth's ozone concentrations except during the dark winter months at the poles, when no sunlight is present.

### Ground-based ozone-measuring systems

Ground-based systems usually rely on laser light and a large telescope for observation, such as the Stratospheric Ozone Lidar Trailer Experiment (STROZ-LITE) system, in operation since 1988. It is a mobile Lidar (light detection and ranging) instrument housed in a large trailer. This system is capable of producing vertical profile measurements of temperature, and ozone and aerosol concentrations. When light with an appropriate wavelength (308 nm) is transmitted vertically through the atmosphere, it will be partially absorbed by the ozone, with the degree of attenuation being proportional to the total amount of ozone in the column of air. Particulate matter that is present in the column of air will also block the passage of the radiation, not by absorbing it, but rather by scattering it. In order to eliminate this problem, measurement at a second wavelength (351 nm) where ozone does not absorb is also used. The amount of ozone is determined by measuring the differential scattering. This method of determining ozone concentrations is known as DIAL—differential absorption Lidar.

Employing similar principles, the Dobson ozone spectrometer[4] has been used to study total ozone in an air column since its development in the 1930s. It has become the standard to which other systems are compared. The Dobson instrument measures the intensity of solar UV radiation

---

[2] European Space agency (ESA) Earthnet Online http://envisat.esa.int/object/index.cfm?fobjectid=5385, accessed October 2009.
[3] NASA, Goddard Space Flight Center, http://aura.gsfc.nasa.gov/index.html, accessed October 2009.
[4] Ozone Measurement at Fairbanks, Alaska, using the Dobson Spectrometer http://ozone.gi.alaska.edu/index.htm, accessed October 2009.

at two to six different wavelengths, some of which are absorbed by ozone radiation and some of which are not. Using the difference between the absorbed and non-absorbed values, the total amount of ozone in the column is calculated using 'Dobson units'.

## A global solar UV index

A global UV index (UVI) has been recommended by both the World Meteorological Organization (WMO) and the World Health Organization (WHO) (see Fig. 3.3). The index, given by equation 3.1 is based essentially on a sophisticated measurement of UV radiation (250 to 400 nm) reaching the Earth's surface, with the value weighted by an integrated erythema (sunburn) reference action factor. This factor takes into account the response of human skin to the UV radiation.

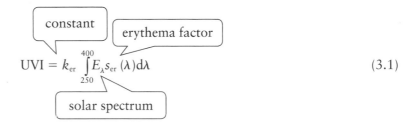

$$\text{UVI} = k_{er} \int_{250}^{400} E_\lambda s_{er}(\lambda)\,d\lambda \tag{3.1}$$

While ozone is a principal molecule controlling the amount of UV radiation reaching the Earth's surface (Fig 3.4), because other atmospheric species can also absorb or scatter radiation, the actual value of the UVI is determined by a complex series of factors.

## Stratospheric ozone concentrations

Stratospheric concentrations of ozone show variability both in position on the Earth and in time. The world average Dobson unit value is approximately 300 DU ranging from 250 in the tropics to 450 in parts of the more northern and southern regions of the planet. At any given location, daily fluctuations associated with short-term weather patterns are typically on the order of 20 to 30 DU. Seasonal changes are larger so, for example, in 2001–2 the daily average

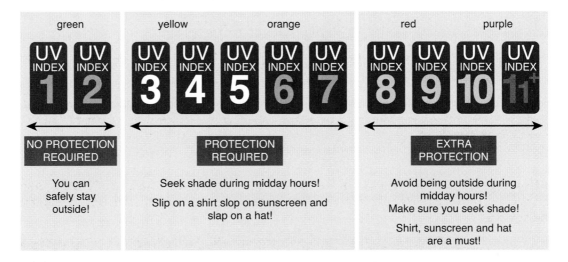

**Fig. 3.3** The UV index is scaled as shown above, along with recommendations for protection, especially important for light skinned persons. Reference: www.who.int/uv/en, accessed November 2009.

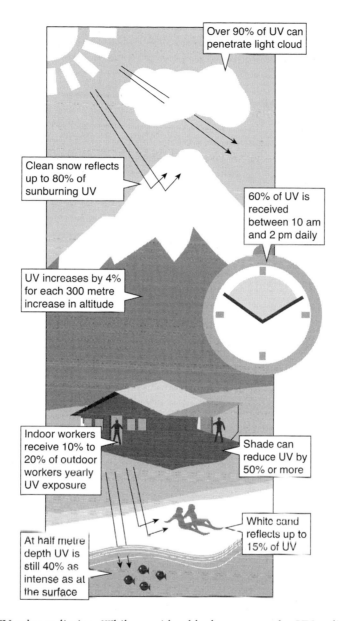

**Fig. 3.4** Fate of UV solar radiation. While considerable dangerous solar UV radiation is absorbed by ozone in the stratosphere, some reaches the Earth's surface. Reference: www.who.int/uv/en, accessed November 2009.

in Washington DC ranged from a low of 290 DU in November to a high of 370 DU in May. It is usual that ozone levels are higher in winter and spring, but then fall off during the summer and autumn.

**Fermi question**

For a person living in Europe, how much longer could (s)he stay unprotected in the afternoon sunlight before being sun-burned when the ozone layer is reported to register at 375 DU versus 325 DU?

The term 'ozone hole' is used to describe the thinning of a region of the ozone layer such as has been observed in recent years in the spring over the Antarctic (October) and, more recently, the Arctic (March). 'Ozone hole' is a partial misnomer as ozone is still present in the polar regions of the stratosphere, though at much reduced concentrations. For example, some measurements through the Antarctic 'ozone hole' have been found to be lower than 150 DU.

Two issues are therefore important with respect to ozone concentrations in the stratosphere. The first relates to the global average concentrations including their monthly and regional variations, and the factors that affect these values. The second issue concerns the apparently anomalous concentrations in specific regions of the globe, especially the Antarctic and Arctic areas. It is in these latter two locations in particular that human activities have had the greatest effect on the natural variations. Both of these issues will be discussed in further detail later in this chapter, but we will first investigate some of the details of 'natural' ozone chemistry.

> **Main point 3.1**   The stratosphere, the region of the atmosphere between 15 and 50 km altitude, contains elevated concentrations of ozone. This ozone is important in regulating the amount of solar ultraviolet radiation that reaches the Earth's surface and thereby offers a protecting effect to biological life. Significant losses of stratospheric ozone have the potential of causing serious harm to plant and animal life on the Earth.

## 3.2 Oxygen-only chemistry—formation and turnover of ozone

### The Chapman reaction sequence

Both synthesis and decomposition of ozone in the stratosphere may be described in terms of chemistry involving only oxygen-containing species. This was first done by Chapman[5] and involves four fundamental reactions. The enthalpies given along with the reactions are a combination of energy involved in making or breaking bonds between atoms and, in two of the reactions, the energy is derived from the Sun. In keeping with thermodynamic conventions, the solar energy is always expressed as having a negative value.

Synthesis $\qquad\qquad\qquad\qquad\qquad\qquad\qquad\qquad\qquad\qquad \Delta H° / \text{kJ mol}^{-1}$

$$O_2 + hv\ (\lambda < 240\ \text{nm}) \rightarrow O + O \quad \text{Slow} \quad -E\,(hv) + 498.4 \qquad (3.2)$$

$$O + O_2 + M \rightarrow O_3 + M \quad \text{Fast} \quad -106.5 \qquad (3.3)$$

Decomposition

$$O_3 + hv\ (\lambda \sim 200 - 315\ \text{nm}) \rightarrow O_2^* + O^* \quad \text{Fast} \quad -E\,(hv) + 386.5 \qquad (3.4)$$

$$O + O_3 \rightarrow O_2 + O_2 \quad \text{Slow} \quad -391.9 \qquad (3.5)$$

In reaction 3.3, M is a neutral third body, usually $N_2$ or $O_2$, the predominant ($>99\%$) species in the stratosphere. $E\,(hv)$ is the energy of the photon in excess of that required to bring about the reaction.

Looking at the oxygen-only reactions, we begin with the synthesis process. The photochemical reaction 3.2, is slow and results in the production of two 'odd oxygen' species—that is, species containing an odd number (in this case one) of oxygen atoms, which are both in the triplet

---

[5] Chapman, S.A., A theory of upper-atmosphere ozone, *Mem. Roy. Meteorol. Soc.* **3** (1930), 103.

ground state $O(^3P)$. The amount of energy and therefore the wavelength of radiation required for this process to occur is calculated as follows. We assume that the enthalpy change is independent of temperature.

---

**Example 3.1 Enthalpy change associated with dissociation of $O_2$**

$$\Delta H° \text{ (reaction 3.2)} = 2\Delta H_f° \text{ (O (g))} - \Delta H_f°(O_2 \text{ (g))}$$
$$= 2 \times 249.2 - 0$$
$$= 498.4 \text{ kJ mol}^{-1}$$

The result indicates how much energy is required for the reaction. Enthalpy values are obtained from Appendix B.2.

---

The formation of 2 mol of O (g) from 1 mol of $O_2$ (g) therefore requires 498.4 kJ of energy. The wavelength of light associated with this amount of energy is calculated as shown in the next example.

---

**Example 3.2 Relation between energy and the wavelength of electromagnetic ultraviolet radiation**

$$\lambda = \frac{hcN_A}{E}$$

$$\lambda = \frac{6.626 \times 10^{-34} \text{ J s} \times 2.998 \times 10^8 \text{ m s}^{-1} \times 10^9 \text{ nm m}^{-1} \times 6.022 \times 10^{23} \text{ mol}^{-1}}{498\ 400 \text{ J mol}^{-1}}$$

$$\lambda = 240.0 \text{ nm}$$

An energy value of 498.4 kJ is equivalent to solar radiation with a wavelength of 240 nm.

---

Because there is an inverse relation between wavelength and energy, radiation having wavelengths longer than 240 nm would not possess sufficient energy to cause the dissociation of the dioxygen molecule. This longer, non-absorbed radiation then penetrates through the stratosphere into the troposphere, potentially reaching the Earth's surface.

Reactions 3.3 (synthesis) and 3.4 (dissociation) are both very fast, and rapidly interconvert single oxygen atoms with ozone, another odd-oxygen species. Although the lifetimes of the individual species are very short, the overall lifetime of odd oxygen is much longer, on the order of months or years. The $\Delta H°$ (reaction 3.3) for formation of ozone from ground-state oxygen atoms and dioxygen molecules is calculated as shown.

---

**Example 3.3 Enthalpy change for the ozone-formation reaction**

$$\Delta H° \text{ (reaction 3.3)} = \Delta H_f° \text{ (O}_3 \text{ (g))} - \Delta H_f°(O \text{ (g))} - \Delta H_f°(O_2 \text{ (g))}$$
$$= 142.7 - 249.2 - 0$$
$$= -106.5 \text{ kJ mol}^{-1}$$

The negative result here indicates that energy will be released.

---

Reaction 3.4 appears to be the reverse of reaction 3.3, but a careful examination illustrates an important concept that must be taken into account. When ozone is photodissociated, according to spin-conservation theory the products dioxygen and atomic oxygen must both be in ground (triplet) or excited (singlet) states. The ground-state formation of the two species requires energy of only 106.5 kJ mol$^{-1}$ (the calculation is analogous to that in Example 3.3 for reaction 3.3) corresponding to a wavelength of 1123 nm, in the infra-red region. However, as shown in eqn 2.17 and reproduced here as eqn. 3.6, the rate constant of a photochemical reaction is dependent on the absorption cross-section ($\sigma_\lambda$) of the molecule, and the ability to absorb depends on the wavelength of the radiation.

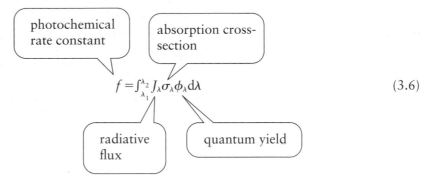

$$f = \int_{\lambda_1}^{\lambda_2} J_\lambda \sigma_\lambda \phi_\lambda \, \mathrm{d}\lambda \qquad (3.6)$$

If the curve for ozone in Fig. 3.2 were extrapolated to 1123 nm, it would be seen that the absorption cross-section is very small at that wavelength. Therefore, because ozone molecules do not significantly absorb 1123 nm photons, the value of the rate constant, $f$, for dissociation to ground-state products approaches zero.

On the other hand, calculation of the energy to produce the two excited species, $O_2$ ($^1\Delta_g$) and $O(^1D)$ takes into account the excitation energies ($E_e$) of $O_2$ (g) and O (g), which are approximately 90 and 190 kJ mol$^{-1}$, respectively.

---

**Example 3.4  Enthalpy change for the photodissociation of ozone into excited species**

$$\Delta H^\circ \text{ (reaction 3.4)} = \Delta H_f^\circ (\text{O (g)}) + E_e \text{ (O)} + \Delta H_f^\circ(O_2 \text{ (g)}) + E_e \text{ (O}_2) - \Delta H_f^\circ(O_3 \text{ (g)})$$

$$= 249.2 + 190 + 0 + 90 - 142.7$$

$$= 387 \text{ kJ mol}^{-1}$$

The solar energy required to produce excited species is greater than that to produce the equivalent ground-state species.

---

The calculated energy in Example 3.4 corresponds to a wavelength of 309 nm and radiation of this wavelength or less is capable of effecting the dissociation. The region between 315 and 200 nm is the part of the spectrum where ozone absorbs photons strongly. Absorption with consequent decomposition is the process that protects the Earth from harmful ultraviolet radiation.

Reaction 3.5 describes a mechanism for the destruction of odd oxygen species. This is an exothermic process ($\sim -400$ kJ mol$^{-1}$), but the relatively large activation energy (18 kJ mol$^{-1}$) and the very small concentrations of the two odd-oxygen species determine that it is a slow process in the stratosphere.

## The ozone layer

The existence of a layer of ozone in the stratosphere is explained in terms of the reactions above. In the upper stratosphere the intensity of high-energy solar radiation is strong, so that dioxygen has a short lifetime in terms of dissociation into oxygen atoms (at 50 km the lifetime of dioxygen is approximately 1 h). Therefore the ratio, O to $O_2$, is relatively large but, because the absolute concentration of oxygen is low, ozone production is limited by reaction 3.3. At low altitudes, there is a plentiful supply of molecular oxygen, but little radiation sufficiently energetic to cause it to dissociate and so the limitation is reaction 3.2. At 20 km the lifetime of dioxygen is about 5 years. In the intermediate range there are both dioxygen and intense high-energy radiation sufficient to maximize the rate of oxygen-atom (and therefore ozone) production. The maximum ozone number density is therefore predicted and found to be at an altitude of approximately 23 km (Fig. 3.1). The actual height of the maximum varies with time of year and latitude on the Earth.

Although oxygen-only chemistry correctly describes the shape and altitude of the ozone profile, it predicts an absolute concentration of ozone that is high by a factor of about two. This indicates that other processes must also contribute to ozone destruction. It has been suggested that reaction 3.5 accounts for only 20% of the odd oxygen removal from the stratosphere. The reasons for the 80% excess destruction have been of great interest to atmospheric chemists, especially since they are in part connected with human activities.

> **Main point 3.2** Ozone, an allotrope of oxygen occurs naturally, with its synthesis and destruction being controlled by oxygen concentrations and photolysis reactions. These factors are such that a maximum concentration of ozone occurs at approximately 23 km above the surface of the Earth—the so-called ozone layer.

## 3.3 Processes for catalytic decomposition of ozone

### Ozone-destruction catalysts

In the previous section, we have seen that there are natural photochemical and chemical processes that lead to the continuous formation and destruction of ozone in the stratosphere. The balance between these processes would result in a steady-state situation with a well-defined and stable ozone layer. However, oxygen-only chemistry does not adequately explain the present situation. Other chemical species also play a role in determining actual concentrations and, in particular, can lead to enhanced ozone destruction. Some of these species occur naturally, while others are of recent industrial and agricultural origin. Small amounts of these chemicals, after they are released in the troposphere, move up into the stratosphere and take part in catalytic ozone-consuming processes. We will look at several catalytic routes that have been shown to take part in additional removal of ozone in the stratosphere. Many of the processes share a general mechanism as follows:

$$X + O_3 \rightarrow XO + O_2 \tag{3.7}$$

$$\underline{XO + O \rightarrow X + O_2} \tag{3.8}$$

net reaction $$O + O_3 \rightarrow 2O_2 \tag{3.9}$$

where reaction 3.9 is the sum of reactions 3.7 and 3.8.

It is important to recognize that the reaction between O and $O_3$ does occur, but only at a very slow rate compared to the catalysed route shown. The $X$ species reacts rapidly with $O_3$ (reaction 3.7) and is regenerated in reaction 3.8 with the next result being the destruction of ozone, while the catalyst can go on to repeat its reaction with ozone again.

The most important of the catalytic species have been identified to be free radicals and are symbolized in three categories:

$HO_x$, which includes •H, •OH, and HOO•;

$NO_x$, which includes •NO and •$NO_2$;

$ClO_x$, which includes •Cl and ClO•.

Depending on altitude and the mixing ratio, each of these species has varying ability to destroy ozone. For example, at an altitude of 50 km near the stratopause, the $HO_x$ radicals may account for as much as 70% of the total mechanism of ozone destruction. Lower in the stratosphere around 30 km, the $NO_x$ catalytic decomposition cycle dominates the removal mechanism of ozone. The $NO_x$ cycle in the region just above the tropopause is usually considered to account for 70% of the destruction of ozone, although some recent work indicates that this figure may be high.[6] There are also questions about the relative involvement of the other catalytic cycles in the region near 30 km; they are generally thought to account almost equally for the remaining destruction.

## Catalysis by hydrogen-containing (HO$_x$) species

This set of catalytic reactions depends on the availability of a source of hydrogen to combine with the plentiful supply of oxygen throughout the stratosphere. The most important of these sources are water or methane. We have seen that the water content of the troposphere is highly variable, but it can be quite large, having mixing ratios up to several per cent. However, very little of this water gets into the stratosphere. The low temperature at the tropopause, typically around $-50°C$, ensures that any humidity at that altitude is present in the form of ice crystals, and these do not readily cross the boundary into the stratosphere. The stratosphere therefore is a relatively dry portion of the atmosphere.

Methane is much less abundant in the troposphere than is water, but it does not freeze out and, as a result, some of it is drawn into the stratosphere above. In the upper part of the stratosphere, through a series of photochemical reactions involving oxygen species, the hydrogen is extracted from methane finally reacting to form water molecules. It is this methane-derived material that is responsible for most of the water in the stratosphere.

The water can then take part in either of two processes to produce hydroxyl radical species.

$$O(^1D) + H_2O \rightarrow 2 \cdot OH \tag{3.10}$$

$$H_2O + h\upsilon \rightarrow \cdot H + \cdot OH \tag{3.11}$$

Methane can also react directly with the ground-state single oxygen atom to form a hydroxyl radical, but no matter the route by which the hydroxyl radical is formed, it can then catalytically decompose ozone as described by the general cycle in reactions 3.7 to 3.9, that are reproduced here in reactions 3.12 to 3.14.

$$\cdot OH + O_3 \rightarrow HOO \cdot + O_2 \tag{3.12}$$

$$HOO \cdot + O \rightarrow \cdot OH + O_2 \tag{3.13}$$

net reaction $\qquad\qquad O + O_3 \rightarrow 2O_2 \tag{3.14}$

The production of hydroxyl radicals is largely due to natural processes. Nevertheless, because it is also influenced by the availability of methane from tropospheric sources it is, in part, affected by human activities.

The hydrogen radical (formed in reaction 3.11) also can participate in ozone-removal cycles by reacting with ozone to form a hydroxyl radical and molecular oxygen. In this way the end result is the same as in reaction 3.10, where one molecule of water contributes two hydroxyl radicals that then participate in the cycle of ozone destruction. Similarly to the other radicals, the hydrogen radical can also be shown in an ozone-destruction catalytic cycle.

$$\cdot H + O_3 \rightarrow \cdot OH + O_2 \tag{3.15}$$

$$\cdot OH + O \rightarrow \cdot H + O_2 \tag{3.16}$$

net reaction $\qquad\qquad O + O_3 \rightarrow 2O_2 \tag{3.17}$

[6] Rowland, F. Sherwood, Stratospheric ozone depletion, *Philos. Trans. R. Soc.* B **361**, (2006) 769–790.

## Catalysis by nitrogen-containing ($NO_x$) species

As we noted in the previous chapter, the two nitrogen oxides, NO and $NO_2$ (together called $NO_x$), are present in the troposphere as a consequence of being produced in various combustion processes—in heating and power plants, in vehicles, and so on. However, these two radical species have very short atmospheric lifetimes, about 4 days, being ultimately converted to nitric acid and removed in rainfall (see Chapter 5). As a consequence, only a small fraction of the tropospheric $NO_x$ drifts into the stratosphere.

The advent in the 1960s of supersonic aircraft that fly in the stratosphere at altitudes of about 17–20 km raised fears that they would contribute to considerable ozone destruction. Such aircraft typically consume about $20\,000$ kg $h^{-1}$ of fuel and in the process generate 160 kg of $NO_x$. Twenty-five thousand kilograms of water are simultaneously released and it is possible that this could contribute to increased production of hydroxyl radicals (reactions 3.10 and 3.11) that would then also enhance ozone decomposition. The fact that these species are injected directly into the stable stratosphere is of particular concern. However, commercial flights using such aircraft have been disbanded and, up to the present time, it is thought that their contribution to enhanced ozone destruction has been small.

Strap-on rocket engines used in various national space programs emit large quantities of HCl and other reactive chlorine-containing products over a large vertical distance in the stratosphere. While at present launch levels are low enough so that these exhaust gases may not contribute significantly to ozone loss, future increases in space activity may dictate a need for more benign rocket-propulsion systems.

### Nitrous oxide

It is now known that there are sources of stratospheric $NO_x$ other than the small amount that originates in the troposphere or is produced directly in the stratosphere. The principal source is another nitrogen-oxide compound, nitrous oxide ($N_2O$).

The nitrous oxide originates from soil and water environments as we will consider in Chapter 15, when we look at the nitrogen cycle, which is one of the most important natural cycles with respect to growth of plants and other organisms. It is produced by a natural process called denitrification, a process that requires reducing conditions and the nitrate ion as one of the reactants. Although denitrification occurs naturally, the extent of the reaction depends on availability of nitrate, and so is augmented in soils supplied with large quantities of nitrogen fertilizers. The agricultural sources are thought to be the principal causes of the slow but steady increase in tropospheric mixing ratios of nitrous oxide. The pre-industrial era level of nitrous oxide was about 275 ppbv; the 2007 mixing ratio of 319 ppbv is increasing at a rate of 0.6 ppbv per year.

Unlike $NO_x$, nitrous oxide is not a radical, does not absorb visible light and photolyse, and is not very water soluble. In the troposphere, it is therefore a stable species with a residence time estimated to be about 120 years, and much of it migrates to the stratosphere where much of it is destroyed by the photochemical reaction 3.18.

$$N_2O + h\nu \rightarrow N_2 + O \qquad (3.18)$$

A smaller but significant amount, however, reacts with excited-state oxygen to produce nitric oxide (reaction 3.19).

$$N_2O + O(^1D) \rightarrow 2NO \qquad (3.19)$$

Yet another means by which nitric oxide is produced in the stratosphere becomes important at altitudes much greater than 30 km. At those altitudes, photochemical decomposition of dinitrogen is possible, producing both an electronically excited $N(^2D)$ and a ground state (but with high translational energy) $N(^4S)$ atom. The ground-state nitrogen subsequently reacts with dioxygen resulting in another natural source of nitric oxide.

$$N_2 + h\upsilon(\lambda < 126 \text{ nm}) \rightarrow N(^4S) + N(^2D) \tag{3.20}$$

$$N(^4S) + O_2 \rightarrow NO + O \tag{3.21}$$

Once nitric oxide has been produced, it can take part in the catalytic processes according to the usual pattern

$$NO + O_3 \rightarrow NO_2 + O_2 \tag{3.22}$$

$$\underline{NO_2 + O \rightarrow NO + O_2} \tag{3.23}$$

net reaction $\qquad\qquad\qquad O + O_3 \rightarrow 2O_2 \tag{3.24}$

The involvement of nitrogen oxides in ozone chemistry goes beyond that described by the catalytic cycle. Nitric oxide also reacts with hydroxyl radicals to produce nitrous acid.

$$\bullet NO + \bullet OH + M \rightarrow HNO_2 + M \tag{3.25}$$

Since both reactants (but not the product) are potential catalysts for ozone destruction, the consequence of their reaction together is reduced removal of ozone. This is a good example of the fact that catalytic effects of individual species are not necessarily additive.

## Catalysis by chlorine-containing (ClO$_x$) species

Chlorine and chlorine-containing radicals ($\bullet$Cl and ClO$\bullet$, and their bromine analogues) are the most reactive of all the stratospheric species that catalyse ozone destruction. These radicals derive from both natural and anthropogenically produced compounds. The most important natural precursor of atomic chlorine is methyl chloride ($CH_3Cl$), which is produced and released by biological reactions throughout the oceans and in much smaller amounts from the burning of vegetation and from volcanic emissions. Much of the ocean-derived methyl chloride is removed at lower altitudes and never reaches the stratosphere. The first step of its destruction is a reaction with hydroxyl (reaction 3.26). Its tropospheric lifetime is about 1 year.

$$CH_3Cl + \bullet OH \rightarrow \bullet CH_2Cl + H_2O \tag{3.26}$$

The small amount of this compound that moves into the stratosphere photolyses to release the reactive atomic chlorine radical.

$$CH_3Cl + h\upsilon \rightarrow \bullet CH_3 + \bullet Cl \tag{3.27}$$

Other natural chlorine sources, such as hydrochloric acid released by volcanoes or chloride ion arising from sea-salt spray, are short-lived in the troposphere. In these sources, the element is in the chloride form, and most is washed out through rainfall before it reaches the stratosphere.

Atomic chlorine from sources like methyl chloride can then take part in ozone destruction. Similar to the OH and NO cases, an ozone-destruction mechanism involving atomic chlorine can be described in the following sequence:

$$\bullet Cl + O_3 \rightarrow ClO\bullet + O_2 \tag{3.28}$$

$$\underline{ClO\bullet + O \rightarrow \bullet Cl + O_2} \tag{3.29}$$

giving the usual net reaction $\qquad\qquad O + O_3 \rightarrow 2O_2 \tag{3.30}$

There are other catalytic cycles involving chlorine and other halogen radical species:

$$HOO\bullet + ClO\bullet \rightarrow HOCl + O_2 \tag{3.31}$$

$$HOCl + h\upsilon \rightarrow \bullet OH + \bullet Cl \tag{3.32}$$

$$•Cl + O_3 \rightarrow ClO• + O_2 \tag{3.33}$$

$$•OH + O_3 \rightarrow HOO• + O_2 \tag{3.34}$$

net reaction
$$2O_3 \rightarrow 3O_2 \tag{3.35}$$

and

$$BrO• + ClO• \rightarrow •Br + •Cl + O_2 \tag{3.36}$$

$$•Br + O_3 \rightarrow BrO• + O_2 \tag{3.37}$$

$$•Cl + O_3 \rightarrow ClO• + O_2 \tag{3.38}$$

net reaction
$$2O_3 \rightarrow 3O_2 \tag{3.39}$$

In reactions 3.31–3.33 bromine can replace chlorine in the cycle and together these reactions account for nearly 60% of the routes by which ozone is destroyed by halogens.

Although the natural sources produce small amounts of atomic chlorine in the stratosphere, of greater significance to the destruction of the ozone is chlorine derived from the anthropogenically produced chlorofluorocarbons (CFCs).

---

**Main point 3.3** In addition to the Chapman 'oxygen only' chemistry, other radical species are involved in ozone destruction. The presence in the atmosphere of two of these—the chlorine radical and nitric oxide—is to a large extent due to human activities, and their elevated levels in the stratosphere have caused a general global reduction in ozone concentrations, estimated at about 5 to 7%. Even greater reductions are observed in the northern midlatitudes over heavily populated areas of Asia, Europe, and North America.

---

## 3.4 Chlorofluorocarbons (CFCs) and related compounds

### Properties of chlorofluorocarbon compounds

Chlorofluorocarbons (CFCs) are a class of compounds initially developed in the 1930s, and since the early 1990s their production has been discontinued, but their legacy will remain with us for years to come. The CFCs have properties that make them particularly useful in a number of applications. The desirable properties include low viscosity, low surface tension, low boiling point, and chemical and biological inertness, which accounts for them being non-toxic and non-flammable. When compressed at room temperature, the gases become liquids; when the pressure is released, they revert to gaseous form taking up large amounts of heat in the process. Applications that take advantage of these properties include use as refrigerants, blowing agents for polymer foams, and solvents for cleaning electronic and other components.

To begin to understand the environmental importance of CFCs, we begin by looking at how they are named. The nomenclature for CFCs is as follows:

$$CFC - xyz \begin{cases} \text{number of C atoms} - 1 \text{ (omitted if } x = 0) \\ \text{number of F atoms} \\ \text{number of H atoms} + 1 \end{cases}$$

The balance of atoms required for a saturated carbon is then made up of chlorine.

A simple way of determining the chemical formula from the numerical symbol for a specific CFC is to add 90 to the number. The modified three-digit number then gives the number of carbon, hydrogen, and fluorine atoms in sequence.

---

**Example 3.5 Relating a common CFC name to its chemical formula**

The chemical formula for CFC-115 would be determined by the '90 rule' in the following manner:

$$115 + 90 = 205$$

Following the scheme described above, the number 205 then indicates that there are 2 carbons, 0 hydrogens, and 5 fluorines. With a two-carbon chain, 6 atoms are required for saturation, so the one remaining, unaccounted for atom is chlorine. The chemical formula for CFC-115 is therefore $CF_3CF_2Cl$.

---

Postscript letters are used to indicate different structural isomers. CFC-114a is $CFCl_2CF_3$ to distinguish it from $CF_2ClCF_2Cl$, which is named CFC-114b.

Despite the fact that CFC molecules are much heavier than most other molecules in air, they tend to be well mixed and homogeneously distributed throughout the troposphere due to their long lifetime, during which they are subjected to strong convective mixing. As noted above, an important property of CFCs is that they are almost completely inert both biologically and chemically in the Earth's environment including in the troposphere. Because they do not react, they circulate through the troposphere until they escape into the stratosphere. While unreactive in the troposphere, at higher altitudes they are able to undergo photolytic decomposition as a consequence of being exposed to the intense flux of more energetic ultraviolet radiation. For CFC-11, which contains three chlorine atoms, the first photolytic step is:

$$CFCl_3 + h\upsilon(\lambda < 290) \rightarrow \cdot CFCl_2 + \cdot Cl \tag{3.40}$$

The released chlorine radical is now able to take part in the catalytic cycles described by reactions 3.28–3.30. A second and possibly third chlorine radical could also be produced by further decomposition of the remnant $\cdot CFCl_2$ radical, generating additional potential for ozone depletion.

Table 3.1 lists some properties and environmental characteristics of CFCs that were widely used up to the 1990s. For our purposes, one of the important properties is the ozone-depletion potential (ODP), which is defined as the ratio of the long-term impact on ozone from a specific chemical to the impact from an equivalent mass of CFC-11, the standard against which others are calculated. The ODP values generated are then useful as a guide to predict the overall impact on the ozone layer from certain chemical species. The ODP takes into account the reactivity of the species, its atmospheric lifetime, and its molar mass. The amount of chlorine in the chemical species is also important. CFC-114b ($CF_2ClCF_2Cl$) contains two chlorine atoms and has an ODP of 0.94, while the similar compound CFC-115 ($CF_2ClCF_3$) with a single chlorine has a substantially lower ODP of 0.44. Similarly, carbon tetrachloride ($CCl_4$) containing four atoms of chlorine per molecule has a high (1.1–1.2) ODP. Most CFCs have ODP values between 0.1 and 1.0, while some of the related HCFCs (see below) have ODP values that are about ten times lower (0.01–0.1). Hydrofluorocarbons, which contain no chlorine, have a zero ODP value.

A large portion of the CFCs produced over the past 70 years have been released and have migrated into the stratosphere. These CFCs are the major source of increased stratospheric chlorine concentrations. It is now believed that the observed decreases in concentration of ozone throughout the stratosphere are in large part due to catalytic destruction by the chlorine cycle. It is for this reason that major international efforts have been directed toward limiting the use of CFCs. The efforts have resulted in a number of agreements of which the Montreal Protocol of 1987 is the central feature.

**Table 3.1** Properties of common CFCs[a]. Mixing ratios are in parts per trillion by volume (pptv).

| CFC | Formula | Tropospheric lifetime / y | ODP[b] | Release rate / kt y$^{-1}$ | | Tropospheric mixing ratios / pptv | | | | Contribution to O$_3$ loss / %[c] |
|---|---|---|---|---|---|---|---|---|---|---|
| | | | | 1988 | 2005 | 1980 | 1990 | 2000 | 2005 | |
| CFC-11 | CFCl$_3$ | 45 | 1.0 | 390 | 85 | 164 | 257 | 263 | 253 | 31 |
| CFC-12 | CF$_2$Cl$_2$ | 100 | 1.0 | 520 | 100 | 295 | 474 | 538 | 539 | 36 |
| CFC-113 | CF$_2$ClCFCl$_2$ | 85 | 0.8 | 250 | 5 | 20 | 67 | 82 | 79 | 14 |
| CFC-114b | CF$_2$ClCF$_2$Cl | 300 | 0.94 | 18 | 0.5 | 10 | 15 | 17 | 17 | 17 |
| CFC-115 | CF$_2$ClCF$_3$ | 1700 | 0.44 | — | — | 1.5 | 3.2 | 8.7 | 9.1 | — |

[a] Ozone-depletion potential, tropospheric lifetime and mixing-ratio values were obtained from the World Meteorological Organization's Scientific Assessment of Ozone Depletion, 2006, Chapter 8. Release rate values were estimates from figures in the same document. 1988 was the year of maximum release of most of these compounds.
[b] ODP = Ozone-depletion potential
[c] The percentage contribution to ozone depletion is based on the major halogen-containing species only.

## The Montreal Protocol—a partial success story in international negotiations

Taking into account all the well-known environmental issues, the Montreal Protocol on Substances that Deplete the Ozone Layer stands out as a landmark international agreement (see Additional Resources 1 and 8). The treaty is the central feature of on-going studies and discussions at scientific and political levels, activities that began many years before the signing in 1987 and continue up to the present.

A small number of scientists had suggested as early as the 1960s that human activities were damaging the ozone layer, but 1974 was the year of the first substantial hypothesis that CFCs and related compounds could destroy substantial amounts of stratospheric ozone. These findings launched international discussions that led to the Vienna Convention for the Protection of the Ozone Layer (1985), which outlined states' responsibilities for protecting human health and the environment against the adverse effects of ozone depletion. The Vienna Convention became the framework under which the Montreal Protocol was negotiated. Twenty-seven countries signed the Protocol in Montreal in September 1987. In its original version, it committed every signatory state by 1999 to reduce their use of certain CFCs to 50% of their level of use in 1986. Reductions were to be measured using CFC-11 as the standard value (ODP defined as 1.0) and normalizing amounts of other compounds to this standard by multiplying the total amount of each by its ODP value. Problem 13 makes use of this method of assessment.

Since 1987, the Protocol has undergone several updates, most recently in Beijing in 1999. In the intervening years, the World Meteorological Organization (WMO) had recognized that ozone-layer 'thinning' was occurring at an unexpectedly rapid rate. Their strong statements led to several rounds of intense negotiating, finally leading to new and strengthened provisions. The amendments have, however, allowed for continued use of related but less potent compounds, principally the HCFCs, in order to provide time to develop suitable replacements. The HCFCs are also scheduled to be phased out beginning in the second decade of this century. Some provision has been made permitting low-income countries a longer period of time to eliminate use of these compounds. The original Protocol has now been ratified by 196 countries, although with a smaller number (158) signed on to the most recent Beijing amendment.[7]

It is encouraging that some real action has followed the signing of the treaty and its amendments. Release rates for the traditional CFCs such as CFC-11, 12 and $CCl_4$ reached a peak around 1988, and have declined significantly since then. For example the maximum release rate for CFC-12 was 390 million tonnes in 1988,[8] but fell steadily to 85 million tonnes in 2005. This pattern has been followed by the other most harmful compounds and is expected to continue. Production of CFC-11 and 12 is scheduled to cease in 2010. At least in the short term, there are expected to be increases in production and release of some of the replacement compounds.

Stratospheric chlorine concentrations (essentially the sum of contributions by CFCs, HCFCs, and methyl bromide) had been rising steadily over the years from a background level of 2 ppbv to just below 4 ppbv in 1998. Since that time, in agreement with predictions, there has been a slight decrease and this is expected to continue until around 2050, before levelling off near the original background value.

## Replacements for CFCs

The Montreal Protocol has achieved a large measure of success, but there has not been unanimous acceptance of all its provisions. Chlorinated fluorocarbons are very useful materials and

---

[7] United Nations Environment Programme, Ozone Secretariat, Status of Ratification http://ozone.unep.org/Ratification_status/ accessed October 2009.

[8] United Nations Framework Convention on Climate Change, Production and Atmospheric Release Table for CFC-12, http://unfccc.int/files/methods_and_science/other_methodological_issues/interactions_with_ozone_layer/application/pdf/cfc1200.pdf, accessed October 2009.

are less expensive than the substitutes that have been developed. Low-income nations, in particular, have argued that they need to take advantage of this well-established low-cost technology. Current research is aimed at developing replacement compounds that retain the desirable properties of the original CFCs, but will not be involved in stratospheric ozone decomposition. In a way, there is a contradiction here, as the essential property of chemical inertness is also the property that leads to tropospheric stability and slow leakage into the stratosphere. Furthermore, the fact that both new and old compounds are greenhouse gases must also be taken into consideration.

Much of the research is centred on modifying the relative amounts of fluorine, chlorine, and hydrogen in new compounds. A common approach has been to incorporate hydrogen into the structure—then called hydrochlorofluorocarbons (HCFCs)—or to replace chlorine altogether and produce what are known as hydrofluorocarbons (HFCs). Increasing the hydrogen content reduces inertness, in this way making the tropospheric lifetime shorter. Recall that the $\cdot$OH radical is able to abstract hydrogen atoms from gaseous organic compounds, leaving behind a reactive radical fragment that can undergo further reactions. This kind of reaction was not possible with the original CFCs. Unfortunately, the higher reactivity also means that HCFCs are less stable and more flammable and cannot be used in some applications. Nevertheless, the shorter tropospheric lifetime and the smaller number of chlorine atoms both contribute to the much reduced ODP of the HCFCs. Another approach is to increase the proportion of fluorine at the expense of chlorine in order to produce a highly stable compound. The C–F bond has a large bond enthalpy (484 kJ mol$^{-1}$ compared with 338 kJ mol$^{-1}$ for the C–Cl bond) and stratospheric photolysis does not occur to any significant extent. When all chlorines are replaced by fluorine, highly stable hydrofluorocarbons are formed and, without chlorine, these compounds have ODP values of zero. Unfortunately again, there is a downside to this excellent stability. The HFCs frequently have long lifetimes and accumulate in the troposphere. While they do not deliver chlorine to the stratosphere, they are excellent absorbers of infra-red radiation and can contribute to global warming (as can CFCs and HCFCs).

Table 3.2 lists some CFC alternatives and indicates the outlook as far as regulation in the USA is concerned. GWP means global-warming potential, a term that is defined in Chapter 8.

From a combined industrial–environmental perspective, one of the more promising new compounds is HFC-134a ($CF_3CH_2F$). This HFC has intermediate stability; it is oxidized by the hydroxyl radical in the troposphere and has a relatively short residence time of 14 years. It is not very flammable and, having no chlorine, will have no ozone-depletion potential. Unfortunately, it is difficult and expensive to manufacture. Another CFC replacement is HCFC-123 ($CF_3CHCl_2$), which contains chlorine but has only about 2% of the ODP of CFC-11, largely because of its very short tropospheric lifetime. Despite these beneficial elements of these compounds they also exhibit global-warming potentials of great concern (see Chapter 8).

## Bromine-containing compounds

In terms of destroying ozone, chemicals containing bromine are even more reactive, estimated to be a factor of about 60 times higher, than their chlorine analogues. As a result, some of the ODP values for bromine compounds are very high.

Halons are a class of chemical compounds derived from $C_1$ and $C_2$ hydrocarbons by exchanging one or more hydrogen atoms with bromine, as well as exchanging other hydrogen atoms with fluorine, chlorine or iodine atoms. In a simple comparison, the bromine-containing halons could be considered the bromine analogues of CFCs. Table 3.3 shows a list of some commonly used bromine-containing halons. These compounds are widely used for extinguishing fires. Other bromine-containing compounds are used as fire retardants in clothing and other fabrics (see PBDE compounds in Section 7.3).

**Table 3.2** CFC alternatives, applications, and regulations[a].

| Substance | Formula | Atmospheric lifetime / y | ODP[b] | GWP[b] | Major uses | Regulatory outlook[c] |
|---|---|---|---|---|---|---|
| HCFC-22 | $CHClF_2$ | 12 | 0.05 | 1810 | Air-conditioning, refrigeration, foams, aerosols | US Clean Air Act bans aerosol use in new equipment after 2005 |
| HCFC-142b | $CH_3CClF_2$ | 17.9 | 0.07 | 2310 | Foams, refrigerants | EPA likely to ban use in new equipment after 2005 |
| HCFC-141b | $CH_3CCl_2F$ | 9.3 | 0.12 | 725 | Foams, solvents | EPA likely to approve for foam use only and ban use in new equipment after 2005 |
| HCFC-123 | $CHCl_2CF_3$ | 1.3 | 0.02 | 77 | Air-conditioning, foams, fire fighting | EPA likely to approve only air-conditioning use; US Clean Air Act bans use in new equipment after 2015 |
| HFC-134a | $CH_2FCF_3$ | 14.0 | 0.0 | 1430 | Refrigeration, air-conditioning | No restrictions anticipated |
| HCFC-124 | $CHClFCF_3$ | 5.8 | 0.022 | 609 | Refrigeration, sterilant | US Clean Air Act bans use in new equipment after 2005 |
| HFC-125 | $CHF_2CF_3$ | 29 | 0.0 | 3500 | Refrigeration | No restrictions anticipated |
| HFC-32 | $CH_2F_2$ | 4.9 | 0.0 | 675 | Refrigeration, air-conditioning | No restrictions anticipated |

[a] Unreferenced data in this table is reproduced with permission from Zurer, P.S., Industry consumers prepare for compliance with pending CRC ban, *Chem. Eng. News*, **70** (1992), 7–13.
[b] Ozone-depletion potential (ODP) and global-warming potential (GWP) values were obtained from the World Meteorological Organization's Scientific Assessment of Ozone Depletion, 2006. GWP is estimated over a 100-year time horizon. Note: GWP is discussed further in Chapter 8.
[c] EPA, US Environment Protection Agency.

The ability of halons to extinguish fires derives from two processes. Because of their high density they tend to settle on and smother fire at ground level; the bromine atom is also capable of terminating the radical chain reactions that propagate the combustion process. In a general sense, organohalides release the halogen in the HX form, species that can then quench the flame by reacting with high-energy •H and •OH radicals to produce lower-energy products.

**Table 3.3** Examples of bromine-containing halons.

| Bromomethane | $CH_3Br$ | Halon 1001 |
|---|---|---|
| Bromochloromethane | $CH_2BrCl$ | Halon 1011 |
| Tribromofluoromethane | $CFBr_3$ | Halon 1103 |
| Bromodifluoromethane | $CHBrF_2$ | Halon 1201 |
| Dibromodifluoromethane | $CBr_2F_2$ | Halon 1202 |
| Bromochlorodifluoromethane | $CBrClF_2$ | Halon 1211 |
| Bromotrifluoromethane | $CBrF_3$ | Halon 1301 |
| Dibromotetrafluoroethane | $CBrF_2CBrF_2$ | Halon 2402 |

$$HX + \cdot H \rightarrow H_2 + \cdot X \tag{3.41}$$

$$HX + \cdot OX \rightarrow H_2O + \cdot X \tag{3.42}$$

For the different halogen elements, effectiveness increases in the sequence

$$F < Cl < Br < I$$

Among the various organohalogens, the carbon–fluorine bond is so strong that fluorine is not readily released during combustion, while the carbon–iodine bond is relatively weak and iodine is released before the high temperatures of the flame are reached. Organobromine compounds release bromine within an appropriate temperature range, and the HBr molecule is highly efficient at capturing combustion-supporting radicals.

The halons, bromochlorodifluoromethane, $CBrClF_2$ (1211), bromotrifluoromethane, $CBrF_3$ (1301), and dibromotetrafluoroethane $CBrF_2CBrF_2$ (2402), have ODP values of 7.1, 16, and 11.5, and GWP values of 16, 65 and 20, respectively. The nomenclature for the halons is straightforward. The first digit indicates the number of carbon atoms; the second the number of fluorine atoms; the third the number of chlorine atoms; and the fourth the number of bromine atoms. Additional atoms needed for a saturated carbon are assigned to hydrogen. As fire extinguishers, it is, of course, required that halons be released directly into the environment. This, combined with their very large ODP values, means that they are very harmful in terms of their ability to enhance stratospheric ozone destruction.

Like methyl chloride, small amounts of methyl bromide ($CH_3Br$) are also emitted by natural processes occurring in the oceans. More important is synthetic methyl bromide produced for use as a crop fumigant. Most commonly, this compound finds application on crops like strawberries that grow close to the ground and are subject to a number of pests. To use it, the field is covered with a large plastic sheet and methyl bromide is injected underneath. This eliminates all the pests including pathogens, insects, and weeds, and the crop is then planted. Obviously, some of the methyl bromide escapes into the surrounding atmosphere. The Montreal Protocol included provisions for a phase-out of methyl bromide by 2005. This component of the protocol has, however, met with stiff opposition because manufacturers do not want the expense and risk of registering new substitute products, and agriculturalists have been largely unwilling to adopt a more complex integrated pest-management approach.

**Main point 3.4** The chlorofluorocarbons are highly stable compounds in the troposphere, but in the stratosphere they photochemically break down to form ozone-destroying chlorine radical species. Bromine-containing fluorocarbons are even more potent ozone destroyers.

## 3.5 Kinetic calculations describing catalytic ozone destruction

We have presented the main cycles describing stratospheric ozone decomposition, and made general comments on their relative importance. The percentage involvement for each catalyst given earlier is based on estimates from kinetic data.

As an example, the overall rate of destruction of ozone by the $NO_x$ cycle (reaction 3.22–3.24) at an altitude of 20 km ($T \cong 220$ K) is shown. The rate expressions for reactions 3.22 and 3.23 are

$$\text{rate} = k_{22}[NO][O_3]$$

$$\text{rate} = k_{23}[NO_2][O]$$

where $k_{22}$ and $k_{23}$ represent the respective second-order rate constants.

The overall rate of reaction is determined not by the sum of the rates of reaction but by the rate-determining step. In order to determine which of the two is rate limiting, we must take into account the concentrations of the species involved and calculate the rate constants under conditions obtaining in a particular region of the atmosphere.

The rate constants can be calculated using the Arrhenius expression

$$k = Ae^{-E_a / RT}$$

and values at 220 K are given in the table below.

|  | $E_a$ / kJ mol$^{-1}$ | $A$ / cm$^3$ molecule$^{-1}$ s$^{-1}$ | $k$ / cm$^3$ molecule$^{-1}$ s$^{-1}$ |
|---|---|---|---|
| Reaction 3.22 | 10.4 | $1.8 \times 10^{-12}$ | $3.5 \times 10^{-15}$ |
| Reaction 3.23 | 0 | $9.3 \times 10^{-12}$ | $9.3 \times 10^{-12}$ |

At an altitude of 20 km the number concentrations of the four species involved in the two reactions are as follows[9]

$$C_O = 2.0 \times 10^7 \text{ molecules cm}^{-3} \qquad C_{O_3} = 3.0 \times 10^{12} \text{ molecules cm}^{-3}$$

$$C_{NO} = 2.0 \times 10^9 \text{ molecules cm}^{-3} \qquad C_{NO_2} = 8.0 \times 10^9 \text{ molecules cm}^{-3}$$

The rates of the individual reactions are then determined as shown in Example 3.6.

---

### Example 3.6 Determining the rate of ozone-destruction reactions

For reaction 3.22

$$\begin{aligned}
\text{rate}_{3.22} &= k_{22}[NO][O_3] \\
&= 3.5 \times 10^{-15} \times 2.0 \times 10^9 \times 3.0 \times 10^{12} \\
&= 2.1 \times 10^7 \text{ molecules cm}^{-3} \text{ s}^{-1}
\end{aligned}$$

For reaction 3.23

$$\begin{aligned}
\text{rate}_{3.23} &= k_{23}[NO_2][O] \\
&= 9.3 \times 10^{-12} \times 8.0 \times 10^9 \times 2.0 \times 10^7 \\
&= 1.5 \times 10^6 \text{ molecules cm}^{-3} \text{ s}^{-1}
\end{aligned}$$

---

[9] Values were taken from plots in Chamberlain, J.W. and D.M. Hunten, *Theory of planetary atmospheres*, Academic Press, Inc., London; 1987.

The overall rate of reaction for the $NO_x$ catalytic cycle is limited by this latter value at $1.5 \times 10^6$ molecules cm$^{-3}$ s$^{-1}$. It is interesting that reaction 3.23, in spite of having a lower activation energy than reaction 3.22, is rate limiting.

The same type of calculation repeated at an altitude of 40 km ($T \cong 250$ K) indicates that reaction 3.22 becomes rate-limiting with the rate $= 1.7 \times 10^6$ molecules cm$^{-3}$ s$^{-1}$.

Rate parameters for other catalytic cycles are shown in Table 3.4 below.

## 3.6 Other reactions involving stratospheric ozone

### Null cycles—'do nothing' chemistry

Calculations of rates of individual reactions are readily carried out but their accuracy depends on the quality of the analytical data and rate constants. Furthermore, as was indicated by reaction 3.25, there are other reactions that occur in the stratosphere and are in competition with the catalytic cycles. Their relative importance further complicates our ability to make predictions about the extent of ozone destruction that may occur under various conditions. Null and holding cycles are two types of reaction sequence that, at least temporarily, prevent species from taking part in catalytic processes.

**Table 3.4** Kinetic data[a] for various reactants involved in the catalytic decomposition of ozone calculated at 235 K.

| X + O$_3$ | | | | |
|---|---|---|---|---|
| X | Concentration / molecules cm$^{-3}$ | $A$ / cm$^3$ molecule$^{-1}$ s$^{-1}$ | $E_a$ / kJ mol$^{-1}$ | $k^{235}$ / cm$^3$ molecule$^{-1}$ s$^{-1}$ |
| O | $1.0 \times 10^9$ | | | |
| H | $2.0 \times 10^5$ | $1.4 \times 10^{-10}$ | 3.9 | $1.9 \times 10^{-11}$ |
| OH | $1.0 \times 10^6$ | $1.6 \times 10^{-12}$ | 7.8 | $3.0 \times 10^{-14}$ |
| NO | $5.0 \times 10^8$ | $1.8 \times 10^{-12}$ | 11.4 | $5.3 \times 10^{-15}$ |
| Cl | Very small | $2.8 \times 10^{-11}$ | 21 | $9.6 \times 10^{-12}$ |
| XO + O | | | | |
| XO | Concentration / molecules cm$^{-3}$ | $A$ / cm$^3$ molecule$^{-1}$ s$^{-1}$ | $E_a$ / kJ mol$^{-1}$ | $k^{235}$ / cm$^3$ molecule$^{-1}$ s$^{-1}$ |
| O$_2$ | $5.0 \times 10^{16}$ | $8.0 \times 10^{-12}$ | 17.1 | $1.3 \times 10^{-15}$ |
| HO | $1.0 \times 10^6$ | $2.3 \times 10^{-11}$ | 0 | $23 \times 10^{-11}$ |
| HO$_2$ | $2.5 \times 10^7$ | $2.2 \times 10^{-11}$ | $-0.1$ | $2.3 \times 10^{-11}$ |
| NO$_2$ | $5.0 \times 10^9$ | $9.3 \times 10^{-12}$ | 0 | $9.3 \times 10^{-12}$ |
| ClO | $2.0 \times 10^7$ | $4.7 \times 10^{-11}$ | 0.4 | $3.8 \times 10^{-11}$ |

[a] The concentrations of species shown are for an altitude of 30 km with the exception of that for ClO which is for 35 km. The ozone concentration at 30 km is $2.0 \times 10^{-12}$ molecules cm$^{-3}$. No value is given for chlorine. The mixing ratio of chlorine-containing compounds throughout the stratosphere is about 3 ppbv at all altitudes. This corresponds to a concentration of $5.1 \times 10^{-13}$ molecules cm$^{-3}$ at 30 km. Only a small fraction of this is free chlorine. Concentrations are derived from plots in, Chamberlain, J.W. and D.M. Hunten, *Theory of planetary atmospheres*, Academic Press, London; 1987. The Arrhenius parameters are from, Wayne, R.P., *Chemistry of atmospheres*, Clarendon Press, Oxford; 1991.

Null (do nothing) cycles interconvert the species X and XO while effecting no net odd oxygen removal. Null cycles involving nitrogen oxides are shown below.

$$NO + O_3 \rightarrow NO_2 + O_2 \tag{3.43}$$

$$\underline{NO_2 + h\nu \rightarrow NO + O} \tag{3.44}$$

net reaction $\qquad\qquad O_3 + h\nu \rightarrow O_2 + O \tag{3.45}$

This sequence competes with the catalytic $NO_x$ cycle and is important only during daytime as it requires near-ultraviolet radiation for the photolytic step. While the net effect is ozone photolysis, it does not change the concentration of odd-oxygen species and ozone is rapidly and stoichiometrically resynthesized by reaction 3.3. For that reason it is termed a do-nothing or null cycle.

Another reaction involving $NO_2$ results in production of $NO_3$ and the establishment of another null cycle.

$$NO_2 + O_3 \rightarrow NO_3 + O_2 \tag{3.46}$$

$$\underline{NO_3 + h\nu \rightarrow NO_2 + O} \tag{3.47}$$

net reaction $\qquad\qquad O_3 + h\nu \rightarrow O_2 + O \tag{3.48}$

## Holding cycles and production of reservoir species

In addition to taking part in the null cycle, some of the $NO_3$ reacts in a three-body process to produce $N_2O_5$.

$$NO_3 + NO_2 + M \rightleftharpoons N_2O_5 + M \tag{3.49}$$

---

**Types of reactions involving radicals that interact with ozone in the stratosphere**

- Catalytic reactions enhance the rate of ozone destruction.
- Null cycles interconvert reactive species from one form to another, but do not react with ozone.
- Holding cycles—species combine to create non-reactive forms that can later be released and take part in ozone destruction.

---

$N_2O_5$ is a relatively stable species and is itself not a catalyst for ozone destruction. It therefore behaves as an unreactive reservoir of $NO_x$, at any one time containing 5 to 10% of the total $NO_x$ budget. Formation of $N_2O_5$, however, does not constitute a permanent loss of odd-nitrogen species as the reaction is reversible and it ultimately decomposes back to $NO_2$ and $NO_3$. Reaction 3.49 therefore acts as a *holding cycle*, temporarily limiting the availability of $NO_x$ for catalysing ozone decomposition in the stratosphere.

Two other very important reservoir compounds are formed in the stratosphere as follows

$$\cdot NO_2 + \cdot OH + M \rightarrow HNO_3 + M \tag{3.50}$$

$$\cdot Cl + CH_4 \rightarrow HCl + \cdot CH_3 \tag{3.51}$$

Almost 50% of $NO_x$ is stored in the nitric acid reservoir, while 70% of the stratospheric chlorine is present as hydrochloric acid. The nitric acid photolyses in daylight, producing nitrogen dioxide in the reverse of reaction 3.50, and the hydrochloric acid releases chlorine and water after reaction with hydroxyl radical.

In addition, several lesser-known species have recently been identified as reservoirs of chlorine and $NO_x$ in the stratosphere. Reactions producing these species include

$$ClO\cdot + HOO\cdot \rightarrow \underset{\text{hypochlorous acid}}{HOCl} + O_2 \tag{3.52}$$

$$HOO\bullet + \bullet NO_2 + M \rightarrow HO_2NO_2 + M \qquad (3.53)$$
pernitric acid

$$ClO\bullet + \bullet NO_2 + M \rightarrow ClONO_2 + M \qquad (3.54)$$
'chlorine nitrate'

These compounds serve to store catalytic species until they leak back into the troposphere or are released as active catalysts. A consequence of their release is the development in recent years of the Antarctic 'ozone hole'.

---

**Main point 3.5** The nitrogen and chlorine radical species that are able to catalyse ozone destruction can also take part in other cycles that, at least temporarily, do not contribute further to the loss of ozone.

---

## 3.7 Antarctic and Arctic 'ozone hole' formation

Because ozone has long been recognized as an important chemical species with respect to the absorption of biologically damaging solar radiation, concentrations in the stratosphere have been monitored worldwide since the mid-1950s. One of the monitoring sites associated with the British Antarctic Survey is at the Halley Bay Station, Antarctica and, in the early 1980s, a noticeable decline in ozone levels in the early spring at that location was observed. By 1984 it became evident that declining ozone was an annual event recurring each year in the early spring, reaching a maximum loss in September and October. During these months the average ozone column thickness in the years between 1956 and 1966 had been 314 DU, while in the first 3 years of the new millennium, the same average dropped to less than 150 DU. The 'hole' extends over a maximum area of the Earth's surface of around 28 million square kilometres, an area that touches on small cities in Chile, Argentina, and the Falkland Islands. The observations of steady decline over the years have led to a major research effort for the purpose of understanding this unexpected phenomenon.

In the Antarctic (and Arctic), conditions exist where seasonal (springtime) ozone depletion events can occur, resulting in a major loss of ozone over a relatively short time. The conditions include a complex combination of climatic factors and the accumulation of reservoir species that, at the onset of the first daylight of the polar spring, react vigorously to bring about a major loss of ozone.

During the long dark winter in the Antarctic, due to the intense cold and the Earth's rotation, a stream of air is drawn toward the South Pole, creating a giant vortex (Fig. 3.5). The area within the vortex acts like a self-contained chemical reactor in which important and unique chemical processes occur. For one thing, inside the reactor, stratospheric clouds form as a result of exceptionally low temperatures, typically around $-80°C$, that exist under conditions of no sunlight. The clouds are categorized into two types. Type 1 (the most common) polar stratospheric clouds (PSC) form at about 193 K and consist of approximately 1-$\mu$m diameter particles of nitric acid and water in a ratio of 1 to 3. Type 2 PSC occur when the temperature has dropped to 187 K, are up to 10 $\mu$m in size, and consist of relatively pure water-ice. Also present within the vortex are the accumulated gases-reservoir chlorine- and nitrogen-containing species, principally hydrochloric acid (reaction 3.51) and chlorine nitrate (reaction 3.54). During the winter, on the surface of the PSC, these species take part in heterogeneous reactions that release chlorine molecules (reaction 3.55) and hydrogen hypochlorite (reaction 3.56).

$$HCl + ClONO_2 \xrightarrow{\text{on PSCs}} Cl_2 + HNO_3 \qquad (3.55)$$

$$H_2O + ClONO_2 \xrightarrow{\text{on PSCs}} HOCl + HNO_3 \qquad (3.56)$$

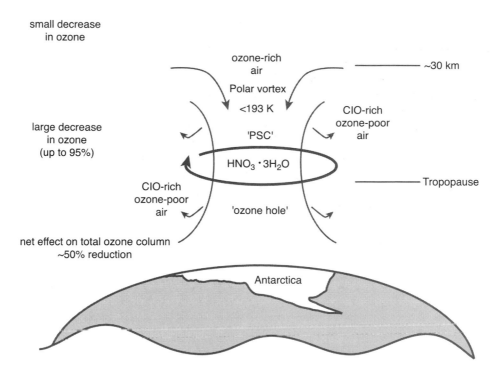

**Fig. 3.5** The Antarctic 'ozone hole' illustrating the polar vortex and the location and relative amounts of ozone loss during the polar sunrise. (Redrawn with permission from Wayne, R.P., *Chemistry of atmospheres*, Clarendon Press, Oxford; 2000.)

This relatively stable situation persists until after the onset of sunlight in late October. At that time, the solar radiation provides energy for photolysis of chlorine and hydrogen hypochlorite to produce the chlorine radical species.

$$Cl_2 + h\nu \rightarrow 2 \cdot Cl \tag{3.57}$$

$$HOCl + h\nu \rightarrow \cdot Cl + \cdot OH \tag{3.58}$$

The chlorine radical is then available to deplete ozone by the standard catalytic cycle or according to the reactions shown in 3.59–3.63. Destruction of ozone occurs rapidly so that, in a matter of days, ozone levels fall dramatically to half or less of their winter value.

Unlike the cycle 3.28 to 3.30 involving chlorine, atomic oxygen is unnecessary for depletion to occur.

$$2(\cdot Cl + O_3 \rightarrow ClO\cdot + O_2) \tag{3.59}$$

$$ClO\cdot + ClO\cdot + M \rightarrow ClOOCl + M \tag{3.60}$$

$$ClOOCl + h\nu \rightarrow ClOO\cdot + \cdot Cl \tag{3.61}$$

$$ClOO\cdot + M \rightarrow Cl\cdot + O_2 + M \tag{3.62}$$

net reaction

$$2O_3 + h\nu \rightarrow 3O_2 \tag{3.63}$$

This situation persists until the air temperature rises causing the vortex to break up and the polar stratospheric clouds to dissipate. When that occurs in mid to late spring, the chlorine radicals again become tied up by the formation of hydrochloric acid and chlorine nitrate, and the ozone level begins to recover to 'pre-hole' levels. There is concern that the air mass, while low in

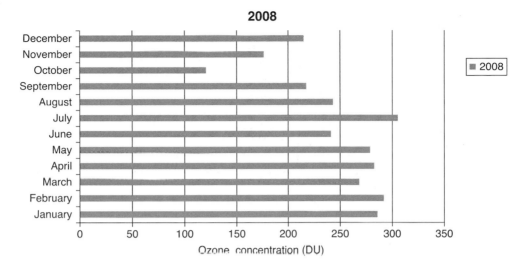

**Fig. 3.6** Typical monthly ozone data (in Dobson units) over Antarctica. Data obtained from http://jwocky.gsfc.nasa.gov/, accessed October 2009.

ozone concentration, will continue to extend out over the most southern land masses, exposing additional people in the southern parts of South America or Australia to unusually high levels of UV-B radiation.

Ozone data and global ozone images of the current (near-real-time) and past ozone concentration have been collected from the Total Ozone Mapping Spectrometer (TOMS) and are available from the National Aeronautics and Space Administration website (Additional Resources 7). Global images centred on the equator as well as both poles are available. The entire data set includes DU values for each 1 degree change in longitude and latitude and you are then able to search any location on the Earth's surface. It should be pointed out that there are no data available during the dark periods at night and at both poles when the Sun has set for half of the year. Figure 3.6 shows the typical yearly trend of ozone concentrations centred over the Antarctica (see Problems 17 and 18).

Events such as those described and shown in Fig. 3.6 have been observed through measurement since 1984. Analogous observations of 'ozone-layer thinning' have been made in the Arctic during the northern spring in March, but not nearly to the same extent. Figure 3.7 shows ozone number density measurements obtained from balloon-borne sensors above Spitsbergen, Norway (79 North) in: (a) March of 1992; and (b) 1995. The 1995 data at 18 km altitude are only about half of their 'normal' concentration. This was a particularly large loss of ozone, and March levels in the first years of the new millennium are more commonly about 30% lower than normal. At other Arctic locations, there is similar evidence of significant springtime ozone loss. Compared with normal values of about 450 DU, occasional values lower than 250 DU have been observed at stations like Resolute (Canada) and Lerwick (Scotland). In all cases, the decreases are not as great as those near the South Pole. In the south, the polar vortex is stronger and more persistent, and the temperatures within it are as much as 10 K lower than in the Arctic.

As a result of the eruption of Mt Pinatubo in the Philippines in 1991, sulphate in the Arctic atmosphere was found to be present at higher concentrations than usual. Stratospheric sulphate is usually in the form of an aerosol (see Chapter 6) and acts as a catalyst in removing $N_2O_5$ gas by forming nitric acid.

$$N_2O_5 + H_2O \xrightarrow{\text{sulphate aerosol}} 2HNO_3 \qquad (3.64)$$

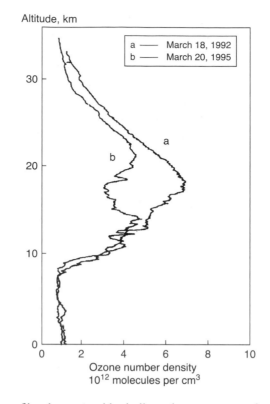

**Fig. 3.7** Ozone vertical profiles determined by balloon-borne sensors above Spitsbergen, Norway (79 N). (Reprinted with permission from vonder Gathen, P., Complexities of ozone loss continue to challenge scientists, *Chem. Eng. News*, **73** (1995), 24.

The ultimate consequence of reaction 3.64 is that it becomes a means by which $NO_x$ species are removed from the polar stratosphere, thus eliminating one of the species that is associated with holding cycles that tie up chlorine-monoxide radicals (reaction 3.54). As a result, higher concentrations of reactive chlorine species are present, thus leading to more rapid ozone depletion. In addition, there is concern that this reaction could occur to a large extent throughout the stratosphere and not just at the poles, resulting in a lowering of the worldwide ozone concentration.

---

**Main point 3.6** The ozone-layer thinning and ozone hole phenomena are indicative of human activities such as the release of stable, volatile chlorinated hydrocarbons and the excessive use of nitrogen-containing fertilizers having affected the global environment. Significantly, whether these activities are local or widespread, our Earth's surroundings are altered over broad regions of space and time.

---

## ADDITIONAL RESOURCES

1. World Meteorological Organization, Scientific Assessment of Ozone Depletion: Global Ozone Research and Monitoring Project – Report no. 50, Geneva 2007
2. Rowland F.S., Stratospheric ozone depletion, *Philos. Trans. R. Soc. B*, **361** (2006), 769–790.

3. Bramstedt, K., J. Gleason, *et al.*, Comparison of total ozone from the satellite instruments GOME and TOMS with measurements from the Dobson network 1996–2000, *Atmos. Chem. Phys.*, **3** (2003), 1409.

4. de Gruijl, F.R., J. Longstreth, *et al.*, Health effects from stratospheric ozone depletion and interactions with climate change, *Photochem. Photobiol. Sci.*, 2 (2003), 354.

5. Harris, J.M., S.T. Oltmans, *et al.*, Long-term variations in total ozone derived from Dobson and satellite data, *Atmos. Environ.*, **37** (2003), 3167.

6. Brasseur G. and C. Granier, Mt. Pinatubo aerosols, chlorofluorocarbons, and ozone depletion, *Science*, **257** (1992), 1239.

7. NASA Total Ozone Mapping Spectrometer (TOMS), Ozone Index, http://jwocky.gsfc.nasa.gov/ accessed October 2009, see note at the end of problem 16.

8. United Nations Environment Programme, Ozone Secretariat, http://ozone.unep.org/, accessed October 2009.

## PROBLEMS

1. Draw Lewis structures for ozone and for dioxygen. Using the data given below, qualitatively compare the bond enthalpies, bond orders, and bond lengths of these two compounds.

$$O_2 \text{ (g)} \rightarrow 2O \text{ (g)} \quad \Delta H° = +498 \text{ kJ}$$
$$O \text{ (g)} + O_2 \text{ (g)} \rightarrow O_3 \text{ (g)} \quad \Delta H° = -105 \text{ kJ}$$

2. It has been suggested that the loss of ozone in the stratosphere could lead to a negative feedback that might allow more ozone to be produced. Explain why such a feedback is possible. (This 'self-healing' does in fact, occur, but only to a very small extent.)

3. Average ozone concentrations in Jakarta, Indonesia have been reported to be 0.015 mg m$^{-3}$ and in Tokyo, Japan are 20 ppbv. What is the approximate ratio of these two values when expressed in the same units?

4. Using data from Table 3.4, identify the rate-determining step in the catalytic cycle involving hydrogen and hydroxyl radicals, and determine the overall rate of ozone destruction as a consequence of this cycle. (Note that the calculation applies to reactions occurring at 30 km only.)

5. A catalyst may be defined as a substance that enhances the rate of a chemical reaction without being consumed in the process. By this definition, a catalyst would have an infinite lifetime. The ozone decomposition catalysts, however, have finite lifetimes. What are possible sinks for removal of the stratospheric catalysts, NO and •Cl?

6. Calculate the enthalpy change (Appendix B.3) in the following reaction:

$$N_2O \text{ (g)} + O \text{ (g)} \rightarrow 2NO \text{ (g)}$$

   (a) when the oxygen is derived from photolysis of nitrogen dioxide;

   (b) when it is derived from photolysed ozone.

   Comment on the results in terms of the tropospheric and stratospheric lifetimes of nitrous oxide.

7. Using bond energies (Appendix B.3), explain the reaction sequence of ozone-destroying capability in the stratosphere of hydrocarbons containing the halogens: Br > Cl > F.

8. HCFC-123 has been proposed as a substitute for CFC-11. How would you expect the following environmental properties to compare:

   (a) tropospheric lifetime;

   (b) combustibility;

   (c) ozone-depletion potential (ODP);

   (d) greenhouse gas properties (this can be answered after reading Chapter 8)?

9. CFC-114b has a tropospheric lifetime of 236 years. Would you expect CFC-115 to have a longer or shorter lifetime? Why?

10. HCFC-22 is one of the compounds recommended as a replacement for CFCs. Use the following information to determine its reaction rate and tropospheric residence time. Compare the residence time with that of CFC–12 which is about 100 y.

    $$[\bullet OH] = 6.6 \times 10^5 \text{ molec cm}^{-3} \text{ (remains constant)}$$
    $$[HCFC\text{-}22] = 0.10 \text{ ppbv}$$
    $$k = 4.0 \times 10^{-15} \text{ cm}^3 \text{ molec}^{-1} \text{ s}^{-1}$$

11. The following proposal to repair the ozone layer has been made.[10] The suggestion is to inject 'negative charges' into the lower stratosphere, and these would react with CFCs to produce harmless products. From your knowledge of basic chemistry, indicate if this process would be theoretically possible, and discuss the practical requirements of it.

12. Consider the following data for total atmospheric column ozone measurements (as DU) at three locations around the Earth, obtained using the TOMS system in 2001.

| | January 15 | April 15 | July 15 | October 15 |
|---|---|---|---|---|
| Tierra del Fuego (Chile, Argentina) | 323 | 261 | 339 | 206 |
| Nairobi (Kenya) | 234 | 273 | 266 (Aug. 15) | 266 |
| Kiev (Ukraine) | 321 | 420 | 314 | 273 |

    Assume that these are typical of values that might be obtained in any other year, and discuss the trends as you move down the columns and along the rows, in terms of your knowledge of stratospheric ozone behaviour.

13. According to the original Montreal Protocol, Australia's ozone cap on CFCs and related compounds was set at 548 tonnes (t) after 1996 (this value was later modified as described in the text). In 1998, use of CFCs had been completely phased out and amounts of HCFCs were reported to be:

| | |
|---|---|
| HCFC-22 | 2820 t |
| HCFC-134a | 1700 t |
| HCFC-141b | 442 t |
| HCFC-123 | 35 t |

    Do these quantities meet the original terms of the agreement?

14. The synthesis and decomposition of ozone can be described using 'oxygen-only chemistry' as shown in the sequence of reactions 3.1–3.4. Explain the nature and significance of the change from $O_2$ and O species in reaction 3.2 to $O_2^*$ and $O^*$ in reaction 3.3.

15. What are the main natural and anthropogenic sources for the production of $HO_x$, $NO_x$, and $ClO_x$ radical species? Describe any connections between the natural and anthropogenic sources of the various species.

16. Although the name of a null or 'do nothing' cycle, related to ozone depletion, sounds in effect harmless, describe the potentially harmful effects that are associated with these types of cycles.

---

[10] *Chem. Eng. News*, May 23 (1994), 36 and Wong, A.Y., D.K. Sensharma, A.W. Tang, R.G. Suchannek, and D. Ho, Observation of charge-induced recovery of ozone concentration after catalytic destruction by chlorofluorocarbons, *Phys. Rev. Lett.*, **72** (1994), 3124.

17. Using the webpage given in Additional Resources 7 and the note below, acquire South Pole ozone data for the past 24 months and plot a similar but updated version of the graph shown below.

The data for the images shown below were obtained from the NOAA Climate Monitoring and Diagnostics Laboratory, Boulder, Colorado from their website, http://www.cmdl.noaa.gov/ in 2004, now identified as US Department of Commerce, National Oceanic and Atmospheric Administration, Earth Research Systems Laboratory, Global Monitoring Division, at the web address http://www.ersl.noaa.gov/gmd/, accessed October 2009.

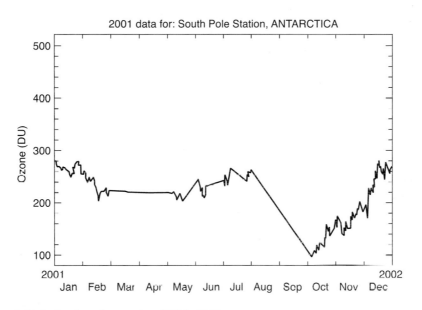

**Graph 1** South Pole Station, Antarctica, 2001–2002.

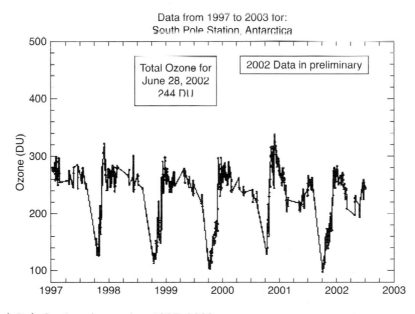

**Graph 2** South Pole Station, Antarctica, 1997–2003.

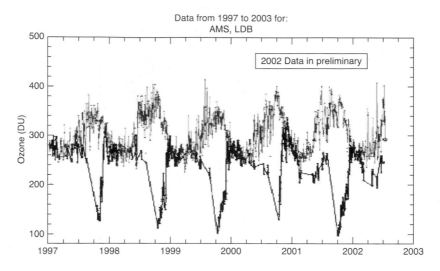

**Graph 3** South Pole Station, Antarctica (bottom trace) and Lauder, New Zealand (top trace), 1997–2003.

Note: *Instructions for the TOMS website* From the main page select the 'Ozone' link under 'Product' (on the left-hand side of the page), then on the new page choose the 'output' format as 'South Pole image—PNG', the OMI satellite, and the date of interest, then select request. You will notice that there are plenty of other options. To get only data, choose the 'output' format as Data file (default), as of October 2009.

18. Using the webpage given in Additional Resources 7 and the note above, investigate the state of ozone above where you live. Create a graph showing the variation of ozone concentration during the past year. How does this compare to the variation seen each year at the South Pole?

# Chapter 4
# Tropospheric chemistry—smog

The focus of this chapter is on the events and chemical processes that lead to the formation of smog. Our examination of this increasingly serious global issue will be done by:

- defining what smog is, identifying the types of smog that form, and the chemical compounds and climatic conditions that ultimately lead to their formation
- describing the chemistry of photochemical smog production, introducing the central species to this process, the hydroxyl radical, and outlining the key reaction sequence for oxidation of volatile organic compounds (VOCs)
- exploring the emission concerns connected with different types of internal combustion engines, fuel types and additives, catalytic converters, and control of specific pollutants such as carbon monoxide, VOCs, nitrogen oxides and ozone.

In the next several chapters, we will be looking at aspects of tropospheric chemistry. Because the troposphere is where we live, chemical reactions involving gases in this region of the atmosphere directly and immediately affect human life and the total environment around us. (See Additional Resources 5 for an example website that introduces many issues regarding air pollution.)

Urban smog is a highly visible problem encountered in major metropolises on every continent of the Earth. And poor air quality is not just a modern problem. In 1661, John Evelyn described the atmosphere in London, England in the book *Fumifugium: or, The Inconveniencie of the Aer and Smoak of London Dissipated* as '. . . a cloud of sea-coal, as if there be a resemblance of hell upon Earth, it is in this volcano in a foggy day: this pestilent smoak, which corrodes the yron, and spoils all the moveables leaving soot on all things that it lights and so fatally siezing on the lungs of the inhabitants, that cough and consumption spare no man . . .' But life in the twenty-first century has, in some ways changed the nature and increased the complexity of urban-smog situations. The compositional complexity, and particularly the chemical reactions involved in creating modern types of smog and other air-quality problems, are the subjects of this chapter.

We will begin by examining the nature of smog, particularly photochemical smog; the chemical processes involved in creating this type of smog event are well known. Moreover, the same reactions that produce smog operate to a more limited degree throughout the globe, even in regions we would identify as being free of pollution. Consequently, in a general sense, the topic of this chapter could be broadly described as gas-phase tropospheric chemistry of organic compounds.

## 4.1 **What is smog?**

Smog is a general term referring to forms of air pollution in which atmospheric visibility is partially obscured by a haze consisting of solid particulates and / or liquid aerosols. The name in its original definition is derived by combining the words smoke and fog. Within the broad range of smog characteristics are two well-defined types—the so-called *classical* or *London smog* and the *photochemical* or *Los Angeles smog*.

### Classical smog

Classical smog is the type described by John Evelyn and is so named because it is associated with the use of the traditional fuel, coal. It is characterized by a high concentration of unburned carbon soot as well as elevated levels of atmospheric sulphur dioxide.[1] Due to the presence of sulphur dioxide, a mild reducing agent and the precursor of a weak acid, the overall chemical properties are reducing and acidic. Where the atmosphere is humid, the carbon particles may serve as nuclei for condensation of water droplets forming an irritating fog so graphically described above.

The classical smog situation was encountered in many nineteenth-century heavily populated industrial centres including London and many other urban areas of Europe. In these cities high-sulphur coal was used both for domestic heating and as an energy source for industry. There were no pollution controls and, in many cases, the emissions were released near ground level. Depending on climatic conditions, smog episodes were frequent and continued to occur well into the twentieth century. As late as 1952 in London, a severe smog lasting several weeks led to the death of more than 4000 persons, mostly by the acute aggravation of pre-existing respiratory problems. Over the years, technical improvements and strict legislation have combined to almost eliminate this form of smog in London and in some other locations.

Unfortunately, there are still recent and present-day examples where classical smog situations persist. Up to 1990, much of Eastern Europe was the site of coal-burning industries that operated without emissions control. Added to the dust and soot from these sources were domestic heating emissions and those from automobiles and other vehicles, many of which use two-stroke engines that emit several times higher concentrations of volatile hydrocarbons as well as substantially more carbon monoxide than conventional four-stroke engines. One area where the problem was especially acute is in Poland's Upper Silesia region adjacent to the city of Krakow. This industrial region is home to some four million people and for many years, airborne dust, carbon monoxide, sulphur dioxide, nitrogen oxides, and lead all exceeded the safe limit, sometimes by factors of ten times or more. Since 1990, closing down or upgrading offending industries, and other major pollution-control efforts, have substantially reduced the problem. Recent records show that on most, but not all days, air-quality standards meet most guideline values.

In the new millennium, classical smog problems remain in some other rapidly growing economies—China and India being prime examples. China produces about 65% of its electricity using coal-fired units and their number grows monthly. Most of the plants have minimal emission controls allowing much of the black soot, sulphur dioxide and other compounds to be released unchecked. The heavy smog that blanketed Beijing in the weeks prior to the 2008 Olympics (in part photochemical smog, see below) was visible evidence of the magnitude, if not the complexity, of this problem. Fortunately, there is growing recognition of the problems associated with coal combustion and preliminary plans and steps have been taken to reduce some of the most harmful emissions.

### Photochemical smog

In contrast to classical smog, photochemical smog is based on emissions from petroleum combustion, principally from motor vehicles, followed by a sequence of chemical and photochemical reactions occurring under specific conditions. The smog contains elevated levels of oxidants

---

[1] The oxidation chemistry of the reduced forms of sulphur are described in detail in Section 5.3.

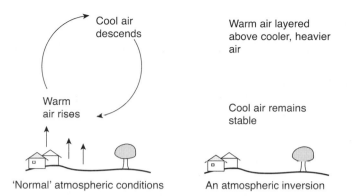

Cool air
descends

Warm air layered
above cooler, heavier
air

Warm
air rises

Cool air remains
stable

'Normal' atmospheric conditions      An atmospheric inversion

**Fig. 4.1** A well-mixed atmosphere occurs when warm air at the surface of the Earth rises and is replaced by cooler air descending from above. In a less-common situation, the atmosphere is stable during an inversion event, when the surface air is cooler (and more dense) than that above.

and carbon-containing reaction products. Photochemical smog is a twentieth- and twenty-first-century phenomenon as it requires the presence of unburned gaseous hydrocarbons and nitrogen oxides, both of which are emitted by internal combustion engines. Specific climatic conditions are also required. Some of the smog-producing reactions are thermal reactions and are therefore favoured by warm temperatures; others are photochemical, so sunlight is a requirement. A stable atmosphere ensures that the released gases remain in the same location where they are able to react. Stability is achieved when there is a temperature inversion in the troposphere. During an inversion, cooler air remains close to the surface of the Earth. Having greater density than the air above, it does not rise and there is little convective mixing. A variety of topographical and climatic features can produce an inversion situation (Fig. 4.1).

The requisite conditions are found frequently in the automobile-rich city of Los Angeles, but the phenomenon is now very widespread, appearing during hot, stable weather in many North American and western European conurbations. In other major urban centres—large cities such as Mexico City, Cairo, Lagos, Jakarta, and Johannesburg—air pollution results from a complex suite of emissions. For example, Indian cities have increasing numbers of motorized vehicles, with annual domestic sales doubling in the period since 2001 to almost 1.4 million cars in 2007. This clearly affects air quality in cities like Delhi. While that municipality has enacted regulations that require the use of natural gas as the fuel for autorickshaws and buses, the sheer number of vehicles ensures that there are significant emissions of smog precursors. In addition, considerable domestic energy supply comes from kerosene, charcoal, and a range of biomass sources. Usually these fuels are burned in low efficiency units at ground level. In addition, during the nine dry months from October through June, the atmosphere is laden with dust from the clay-rich alluvium that makes up the soil in the Indo-Gangetic plains. The combination of gaseous and particulate pollutants together forms a uniquely characteristic haze that is especially evident during hot, dry summer evenings. In this dusk period from 6 to 8 p.m., the heavy smoky haze hangs over cities and the surrounding countryside.

For persons around the world, the Beijing Olympics in August 2008 brought smog issues to the fore. The air pollutants of most concern for athletes and spectators alike were high levels

> **Conditions required for photochemical smog formation**
>
> - source of precursors, hydrocarbons and $NO_x$ compounds—principally associated with internal combustion engines;
> - a stable atmosphere to hold the reactants in place;
> - high temperature to increase thermal reaction rates;
> - intense sunlight to facilitate photochemical reactions.

of ozone and particulate matter, both of which could affect performance. In Beijing, high summer temperatures, sunlight and nitrogen-oxide emissions from cars promote photochemical formation of ozone during the late summer season. Particulate matter also derives from vehicle emissions and from power plants as well as dust from the surrounding dry areas. Beijing's average concentrations of both particulate matter and ozone over the past five years have exceeded China's standards and in the days immediately preceding the Olympics, heavy smog was clearly visible in and around the city. Fortunately, during the two week period of the games, to a large extent the smog cleared. This followed enforcement of stringent restrictions on automobile use and scaling back production in nearby fuel-consuming factories. It was also fortunate that, during several days of the competitions, rain and steady winds disturbed the stable atmosphere, washing out particulates and carrying away reactants and products of the photochemical smog-producing process.

---

**Main point 4.1** Smog is an urban air pollution phenomenon and there are a variety of types depending on the local situation. Two general classes have been identified. Classical smog consists of carbon-based soot and other solid particulates along with sulphur dioxide giving it reducing and acidic properties. Its most important source is combustion of coal. Photochemical smog is produced by atmospheric chemical reactions involving nitrogen oxides and hydrocarbons emitted in large part from vehicular exhaust.

---

## 4.2 The chemistry of photochemical smog

The chemistry involved in the formation of photochemical smog has been studied in detail and the processes can be summarized graphically as in Fig. 4.2. Figure 4.2(a) depicts an *idealized* photochemical smog event, while Fig. 4.2(b) plots *actual data* for one such occurrence in Toronto, Canada on May 21, 1992.

The measured chemical features are the following. Beginning at about 6:00 a.m. on what is to be a sunny, hot day as the morning traffic takes to the streets, a simultaneous increase in the atmospheric concentrations of volatile hydrocarbons and nitric oxide is observed. The nitric oxide concentration rapidly reaches a maximum and then decreases while, at the same time, nitrogen dioxide levels begin to rise. Later in the morning, both hydrocarbon and nitrogen-dioxide concentrations fall off and elevated levels of oxidizing agents and aldehydes are detected. Some aspects of the pattern are repeated on a reduced scale during the period of heavy evening traffic, but there is a general lowering of concentrations of all the aforementioned species to background levels that remain constant during the night.

The smog, consisting of a mixture of partially oxidized hydrocarbons, ozone (this is the 'bad ozone' to which we referred earlier), and other oxidants, is observable from midday to late afternoon. In addition to producing a visible haze, it also causes eye and other membrane irritation, can adversely affect plant growth, and is implicated in other serious ecotoxicological problems.

### Hydroxyl radical production

The chemical reactions that lead to smog formation centre around the hydroxyl radical. As we showed in Chapter 2, hydroxyl-radical formation takes place via a reaction sequence that begins with the production of nitric oxide.

$$N_2 + O_2 \rightleftharpoons 2NO \qquad (4.1)$$

Because the left-to-right reaction is endothermic and endergonic, nitric oxide is produced to a significant extent only under high-energy conditions, including those provided by internal combustion engines. A calculation illustrating this was done in Section 2.3.

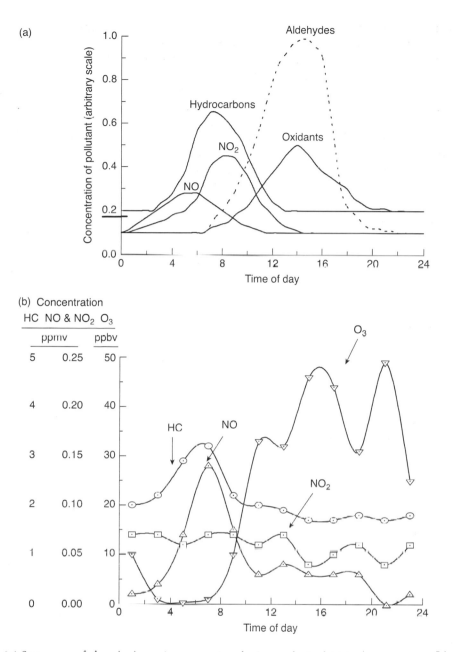

**Fig. 4.2** (a) Sequence of chemical species appearing during a photochemical smog event. Idealized plot based on results from laboratory studies in a smog chamber. (b) Sequence of chemical species during a photochemical smog event. Actual data obtained from measurements on the corner of Bay and Grosvenor Streets in Toronto, Canada on May 21, 1992. This was a cloudless, still day with maximum temperature of 26°C. Data obtained from the Ministry of Environment and Energy, Government of Ontario.

When it is exhausted to the open atmosphere, the nitric oxide is oxidized to nitrogen dioxide by oxygen or, most importantly, by other oxidants.

$$2NO + O_2 \rightarrow 2NO_2 \tag{4.2}$$

The three-body reaction with oxygen proceeds very slowly and accounts for only a small portion of the $NO_2$ generated. A second and more important mechanism for oxidation of $NO_2$ is

$$NO + O_3 \rightarrow NO_2 + O_2 \tag{4.3}$$

The reaction with ozone is rapid but ozone itself is a by-product of the formation of $NO_2$ (see reaction 4.6). It therefore does not appear in the atmosphere in significant quantities until after a substantial concentration of $NO_2$ has already been produced.

A third route for oxidation of nitric oxide involves reactions with peroxyl radicals

$$ROO\bullet + NO \rightarrow RO\bullet + NO_2 \tag{4.4}$$

Later, it will be shown that, like ozone, these peroxyl species are also generated as part of the overall sequence of steps in the oxidation of hydrocarbons. Along with nitric oxide, the hydrocarbons are a product of vehicular emissions—either due to evaporation from fuel tanks or as unburned species in the exhaust.

The nitrogen dioxide produced in reaction 4.4 absorbs visible and ultraviolet radiation from the sunlight ($\lambda < 400$ nm) and this leads to photolysis with production of atomic oxygen in the ground state ($O(^3P)$). Note that the reaction sequence that follows here is the same as that described earlier in reactions 2.20 to 2.23)

$$NO_2 \xrightarrow{h\nu,\, \lambda < 400\text{ nm}} NO + O \tag{4.5}$$

Analogous to the reaction in the stratosphere, atomic oxygen reacts rapidly with molecular oxygen in the presence of a third body—usually another $O_2$ or $N_2$ molecule—to produce ozone.

$$O + O_2 + M \rightarrow O_3 + M \tag{4.6}$$

Ozone then photolyses due to ultraviolet radiation from sunlight.

$$O_3 \xrightarrow{h\nu,\, \lambda < 315\text{ nm}} O_2^* + O^* \tag{4.7}$$

The photolytic products include one oxygen molecule $O_2(^1\Delta_g)$ and one oxygen atom ($O(^1D)$), both in an excited state. A large fraction of the excited-state oxygen atoms is deactivated by collisions with ground-state dioxygen or dinitrogen, but that which retains its additional energy can react with water vapour to produce hydroxyl radicals.

$$O^* + H_2O \rightarrow 2\bullet OH \tag{4.8}$$

Accordingly, a single nitrogen-dioxide molecule is able to create two hydroxyl radicals. The sum of reactions 4.5–4.8 is

$$NO_2 + H_2O \rightarrow NO + 2\bullet OH \tag{4.9}$$

Quantitatively, this is the most important means by which the hydroxyl radical is produced.

A second mechanism of generating hydroxyl arises from reactions involving nitric oxide and nitrogen dioxide as follows

$$NO + NO_2 + H_2O \rightarrow 2HONO \tag{4.10}$$

$$2HONO \xrightarrow{h\nu,\, \lambda < 400\text{ nm}} 2NO + 2\bullet OH \tag{4.11}$$

Reactions 4.10 and 4.11 are especially important in heavily polluted atmospheres and it is evident that in the process two hydroxyl radicals are again produced by one nitrogen-dioxide molecule.

Therefore, once again, the sum of reactions 4.10 and 4.11 is equal to the following overall reaction.

$$NO_2 + H_2O \rightarrow NO + 2 \cdot OH \qquad (4.12)$$

It is again worth emphasizing that the accelerated synthesis of hydroxyl radicals in the urban atmosphere is substantially a consequence of production of nitrogen oxides in an internal combustion engine. The actual atmospheric concentration of hydroxyl is very small and difficult to measure, but has been estimated to be on the order of $2.5 \times 10^6$ molecules $cm^{-3}$ in a polluted urban environment in contrast to a concentration of about $1.0 \times 10^5$ molecules $cm^{-3}$ in a relatively clean rural area in a temperate zone. Other factors that contribute to enhanced hydroxyl radical concentration are high temperature and intense sunlight, so values tend to be higher in tropical compared to temperate regions.

## Oxidation of hydrocarbons—alkanes

Besides being a source of nitrogen oxides, internal combustion engines simultaneously emit unburned volatile hydrocarbons and these are oxidized through reactions initiated by the highly reactive hydroxyl radical. Considering a generic aliphatic hydrocarbon, one version of the hydroxyl radical-initiated oxidation sequence is summarized in the following way.

hydrogen abstraction $\qquad \cdot OH + RCH_3 \rightarrow R\overset{\bullet}{C}H_2 + H_2O \qquad (4.13)$
                         alkyl

dioxygen addition $\qquad R\overset{\bullet}{C}H_2 + O_2 + M \rightarrow RCH_2OO \cdot + M \qquad (4.14)$
                         per oxyalkyl

oxygen abstraction by NO $\qquad RCH_2OO \cdot + NO \rightarrow RCH_2O \cdot + NO_2 \qquad (4.15)$
                         alkoxyl

formation of aldehyde $\qquad RCH_2O \cdot + O_2 \rightarrow RCHO + HOO \cdot \qquad (4.16)$
                         aldehyde  hydro peroxyl

$NO_2$ regeneration $\qquad HOO \cdot + NO \rightarrow NO_2 + \cdot OH \qquad (4.17)$

Note that each step in the sequence produces a new radical. The sum of the reactions is

$$RCH_3 + 2O_2 + 2NO \rightarrow RCHO + 2NO_2 + H_2O \qquad (4.18)$$

If we incorporate the fact that nitrogen dioxide, besides being a precursor of ozone, is the principal source of hydroxyl radical in the atmosphere through the overall reaction 4.9, and include this with reaction 4.18, the oxidation of the hydrocarbon is now written as

$$RCH_3 + 2O_2 + H_2O \rightarrow RCHO + 4 \cdot OH \qquad (4.19)$$

In this overall reaction, the hydroxyl radical as well as other radicals play a catalytic role (there is actually the potential for net production of radical species) so that a very small amount of these radicals produces a large amount of product. As was noted in Chapter 2, the central role of hydroxyl radical in tropospheric chemistry cannot be overemphasized.

Were it not for various means of removing hydroxyl radicals from the atmosphere, the concentration of hydroxyl would continue to increase and the rate of oxidation of hydrocarbons would accelerate. Several termination reactions serve to remove hydroxyl, its precursor nitrogen dioxide, and the hydroperoxyl radical.

$$\cdot OH + \cdot NO_2 + M \rightarrow HNO_3 + M \qquad (4.20)$$

$$2HOO \cdot \rightarrow H_2O_2 + O_2 \qquad (4.21)$$

$$\cdot OH + HOO \cdot \rightarrow H_2O + O_2 \qquad (4.22)$$

The products of each of these reactions are relatively stable. The nitric acid and hydrogen peroxide are water soluble and are removed from the atmosphere by precipitation.

Keep in mind also that hydroxyl production involves two photochemical steps. Consequently, night-time also brings the smog-forming reactions to a close.

## Secondary reactions

There are secondary reactions that occur simultaneously with the oxidation of hydrocarbons. The first two have been noted previously.

$$NO_2 \xrightarrow{h\upsilon,\ \lambda\ <\ 400\ nm} NO + O \qquad (4.23)$$

$$O + O_2 + M \rightarrow O_3 + M \qquad (4.24)$$

Other very important reactions involve the aldehyde product of oxidation of hydrocarbons. Using acetaldehyde as an example, the reactions[2] are:

$$CH_3CHO + \bullet OH \rightarrow CH_3\overset{\bullet}{C}O + H_2O \qquad (4.25)$$

$$CH_3\overset{\bullet}{C}O + O_2 + M \rightarrow \underset{\text{acetylperoxy}}{CH_3\ C(O)OO\bullet} \qquad (4.26)$$

$$CH_3C(O)OO\bullet + \bullet NO_2 \rightleftharpoons \underset{\text{peroxyacetic nitric anhydride (PAN)}}{CH_3C(O)OONO_2} \qquad (4.27)$$

Peroxyacetic nitric anhydride and related compounds, PANs,[3] are the major eye irritants in a photochemical smog. The importance of PANs as reaction products is that they act as reservoirs for nitrogen-oxide species. PAN is a relatively stable molecule especially at low temperatures and therefore may be transferred over long distances by air currents. In warmer locations often remote from the source, PAN breaks down via the reverse process of reaction 4.27 so that nitrogen dioxide is released with the potential of producing additional ozone and hydroxyl radicals. This allows for carry-over of smog conditions through both space and time.

## Nature of photochemical smog

From this complex series of thermal and photochemical reactions it is seen that a number of chemicals would be expected to be present at elevated atmospheric concentrations during a photochemical smog event. These include the nitrogen oxide and hydrocarbon precursors as well as products such as aldehydes and oxidants, ozone, and PANs. Some of the chemicals are gases but others, particularly aldehydes, are present as small liquid droplets in the form of an aerosol. This is the cause of the hazy appearance present during an intense smog. The yellowish colour is due to nitrogen dioxide. The temporal order of reactions described throughout this section is also consistent with the sequence as depicted in Fig. 4.2.

Typical values for some smog chemicals found under polluted and non-polluted conditions confirm expectations regarding higher concentrations of particular chemicals (Table 4.1). A summary of the reactions that we have been discussing is given in Fig. 4.3.

---

[2] The brackets in the formulae of some of the species indicate that both the bracketed atom and the next one are bonded to the adjacent carbon. In these particular formulae, the single oxygen in brackets is joined to carbon by a double bond. This allows us to write the formulae in a linear fashion.

[3] Peroxyacetic nitric anhydride is the name we use here. Peroxyacetyl nitrate is the normal name used, but does not conform to IUPAC conventions. Using the IUPAC rules it should be called ethane peroxoic nitric anhydride, and a compromise name is peroxyacetic nitric anhydride (PAN). Other carbonyl precursors result in the production of related compounds—for example, from propionaldehyde, peroxypropionic nitric anhydride (PPN) is produced. The entire class of compounds is usually referred to as PANs.

**Table 4.1** Typical atmospheric concentrations of selected species characteristic of a photochemical smog[a].

| Species | Concentration / ppbv | |
| --- | --- | --- |
| | Polluted area | Unpolluted area |
| Carbon monoxide | 10 000–30 000 | <200 |
| Nitric oxide plus nitrogen dioxide ($NO_x$) | 100–400 | <20 |
| Hydrocarbons (excluding methane) | 600–3000 | <300 |
| Ozone | 50–150 | <5 |
| PANs | 50–250 | <5 |

[a] Most values are estimates based on data in *Air quality in Ontario 1991*, Environment Ontario, Queen's Printer for Ontario; 1992.

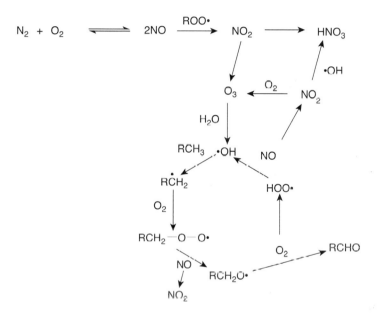

**Fig. 4.3** Reaction series during a photochemical smog event. It should be emphasized that the initial equilibrium (top left) in an ambient atmosphere would lie extremely far to the left, and it is only during specific combustion conditions or lightning strikes where this equilibrium shifts to the product side.

---

**Main point 4.2** Photochemical smog oxidation reactions are initiated by the hydroxyl radical (produced in large part due to the presence of nitric oxide from combustion emissions). Hydrocarbons and other volatile organic compounds are the oxidizable substrates. The products of the smog-producing reactions include partially oxidized hydrocarbons such as carbon monoxide, aldehydes and ketones, residual nitrogen oxides, and ozone.

---

## Oxidation of other volatile organic compounds

In discussing a reaction sequence for the oxidation of organic compounds in the gas phase and the generation of photochemical smog, our example of a hydrocarbon substrate was a general aliphatic saturated hydrocarbon. Of course, there are many other types of volatile organic

**Table 4.2** Mean concentration of volatile organic compounds in the atmosphere of Taipei City[a].

| Compound | Atmospheric concentration / $\mu g\ m^{-3}$ |
|---|---|
| Toluene | 980 |
| *m,p*-xylene | 910 |
| *o*-xylene | 510 |
| Benzene | 370 |
| Ethylbenzene | 310 |
| 1,3,5-trimethylbenzene | 230 |
| 1-ethyl,4-methylbenzene | 200 |
| Hexane | 150 |
| Heptane | 130 |
| 1-ethyl,2-methylbenzene | 120 |

[a] Measurements were made in the breathing zone of cyclists and pedestrians in three parts of the city frequented by commuters (from Chan, C-C., S-H. Lin, and G-R. Her, Student's exposure to volatile organic compounds while commuting by motorcycle and bus in Taipei City. *Air Waste*, **43** (1993), 1231–8).

compounds (VOC) that are released into the atmosphere from both anthropogenic and natural sources. For vehicular emissions, the list of compounds is long and variable depending on the fuel, type of engine, and operating conditions. Hydrocarbons such as ethene, ethyne, higher aliphatic hydrocarbons, benzene, toluene, and xylenes are important emissions in almost all cases. Each of these compounds can be released unreacted or can undergo oxidation reactions and, in fact, reactions of alkenes are generally much more rapid than those of saturated hydrocarbons.

In a large city, the atmospheric levels of fuel-derived gases can be very high. For example, measurements of personal exposure to VOCs while commuting by motorcycle in Taipei City (Taiwan) showed large concentrations of several hydrocarbons, especially aromatic compounds. The 10 hydrocarbons found in highest concentration are listed in Table 4.2.

In their unreacted state, the compounds may have undesirable ecotoxicological properties. Benzene, besides causing annoying physiological reactions such as dizziness and membrane irritation, is known to be a human carcinogen. Furthermore, all of the compounds can undergo oxidation producing a range of products directly and by related secondary reactions.

Considering the various hydrocarbon reactants, there are two principal mechanisms by which hydroxyl radicals initiate oxidation. The first mechanism is the type that we have already seen in the case of saturated aliphatic hydrocarbons. It begins when a hydroxyl abstracts a hydrogen to form water and an organic radical. The abstraction reaction can occur whenever there is a hydrogen atom available to be removed but the rate depends on the strength of the carbon–hydrogen bond. The general order of strength of carbon–hydrogen bonds is primary > secondary > tertiary carbons and so the rate of hydrogen abstraction is in the reverse order.

## Unsaturated compounds

### Alkenes

Earlier, we saw that a first step in a hydroxyl-radical-initiated reaction was the *abstraction of a hydrogen atom* from a hydrocarbon species. A second type of initiation process is expressed with olefins where the electrophilic hydroxyl *adds to a double or other multiple bond*, a region of high electron density (reaction 4.28). It is this reaction that leads to the high reactivity of this class of compounds.

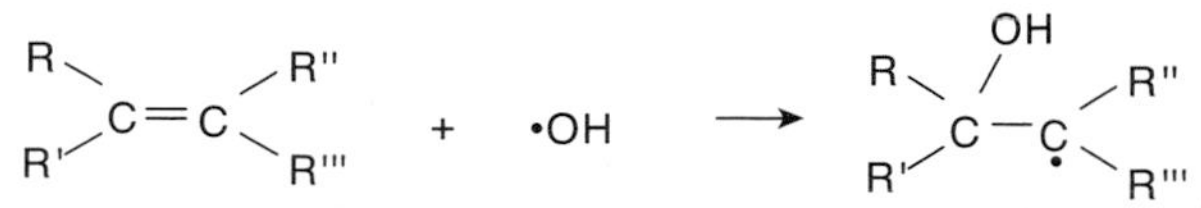

$$\text{R--C=C--R''} \quad + \quad \text{•OH} \quad \longrightarrow \quad \text{R--C--C--R'''} \qquad (4.28)$$

Following this step, one mechanism continues with the addition of dioxygen to the hydroxylated species. The β-hydroxy peroxyl compound then transfers an oxygen atom to nitric oxide and the resultant β-hydroxyalkoxyl radical usually undergoes decomposition. Finally, a dioxygen molecule abstracts a hydrogen from the alkoxyl radical (one of the decomposition products) forming a hydroperoxyl radical and a ketone, (reaction 4.32) and, in the overall reaction, two ketones are produced. For the common situation of a straight-chain alkene (R′ and R″ = H), instead of ketones the products are aldehydes. The entire process is shown below in its most general form.

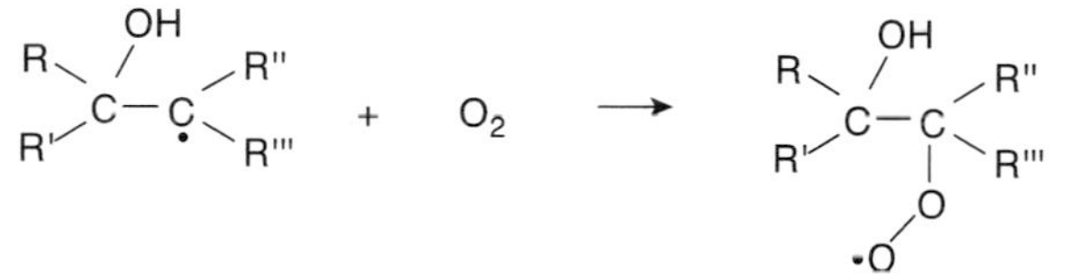

$$\text{R--C--C--R''} \quad + \quad O_2 \quad \longrightarrow \quad \text{R--C--C--R'''} \qquad (4.29)$$

$$\text{R--C--C--R''} \quad + \quad NO \quad \longrightarrow \quad \text{R--C--C--R'''} \quad + \quad NO_2 \qquad (4.30)$$

$$\text{R--C--C--R''} \quad \longrightarrow \quad \text{R--C--OH} \quad + \quad \text{O--C--R'''} \qquad (4.31)$$

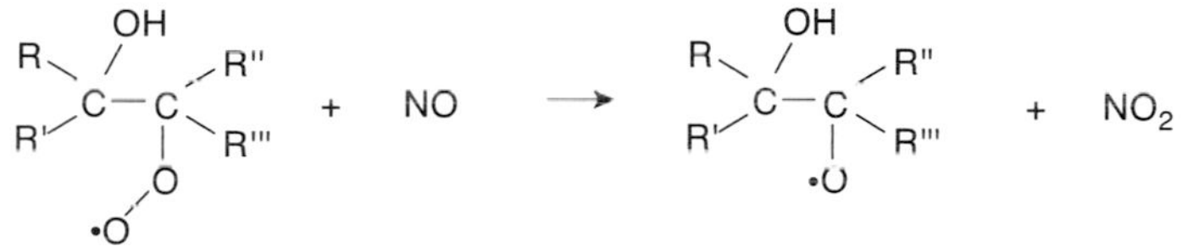

$$\text{R--C--OH} \quad + \quad O_2 \quad \longrightarrow \quad \text{R--C=O} \quad + \quad HOO• \qquad (4.32)$$

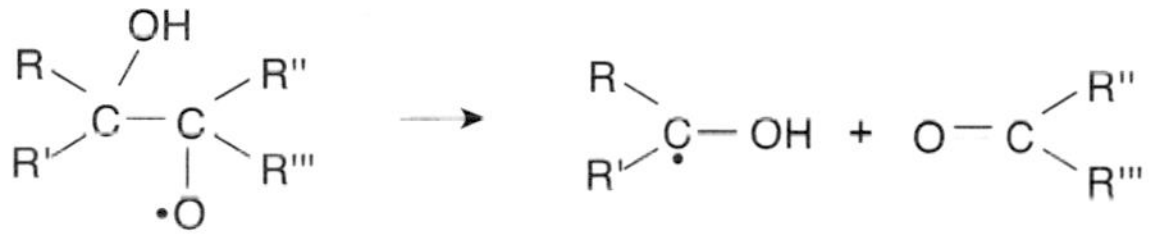

### Alkynes

For alkynes, the initial step is also addition of hydroxyl, in this case to the electron-rich triple bond. For acetylene, after further reaction, the principal products are glyoxal $(CHO)_2$ and formic acid (HCOOH).

### Aromatics

Like other hydrocarbons, aromatic compounds are oxidized in reactions initiated by the hydroxyl radical. Where the aromatic is alkyl-substituted, a hydrogen atom is first abstracted and the sequence continues in the same way as for other alkyl compounds. The final products are aldehydes.

An alternative route that can also apply where there is no alkyl substituent involves addition of hydroxyl radical to the ring. The reaction is shown below.

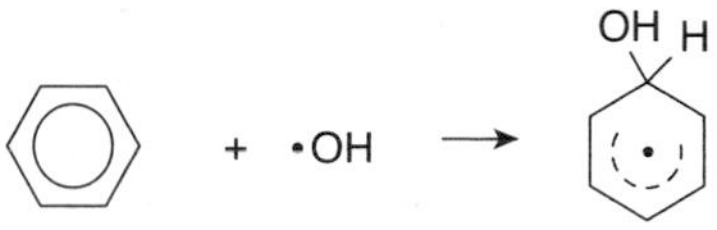

$$\text{benzene} \quad + \quad \text{•OH} \quad \longrightarrow \quad \text{hydroxycyclohexadienyl radical} \qquad (4.33)$$

By reaction with molecular oxygen, the hydroxybenzene radical forms phenol. When the initial substrate is a substituted aromatic compound, corresponding phenols and other types of products are also formed.

(4.34)

In addition to this sequence, it is observed that ring cleavage of the peroxidized aromatic also occurs to a significant extent, producing a variety of poorly characterized products.

An environmentally significant class of compounds is the fused-ring polynuclear aromatic hydrocarbon (PAH) class. Except for the simplest PAH, naphthalene and its derivatives, these species are stable to oxidation and therefore have long atmospheric lifetimes. The PAHs form an important constituent of some atmospheric aerosols and more will be said about them in Chapter 6.

## Nearly final products of photochemical oxidation

### Aldehydes and ketones

We have seen that aldehydes (and ketones) are important products of several of the oxidation schemes. These are somewhat stable species, but they do undergo reactions of their own. As shown in reactions 4.25–4.27, one possibility is yet another reaction that is initiated by hydroxyl radicals, followed by the addition of dioxygen and then reaction with nitrogen dioxide to form a member of the PAN family of compounds. A second possible sequence also starts with reactions 4.25 and 4.26. The acetylperoxy radical then gives up an oxygen to nitric oxide followed by cleavage of the C–C bond to form a methyl radical and carbon dioxide (reactions 4.35 and 4.36).

$$CH_3C(O)OO\bullet + NO \rightarrow CH_3C(O)O\bullet + NO_2 \qquad (4.35)$$

$$CH_3C(O)O\bullet \rightarrow \bullet CH_3 + CO_2 \qquad (4.36)$$

A third important route for decomposition of aldehydes is by photolysis. Aldehydes are capable of absorbing UV radiation that has wavelengths longer than 290 nm and this leads to their photochemical degradation. For acetaldehyde, observed photolytic reactions are

$$CH_3CHO \xrightarrow{hv,\ \lambda\ \sim\ 290\ nm} \bullet CH_3 + H\overset{\bullet}{C}O \qquad (4.37)$$

$$CH_3CHO \xrightarrow{hv,\ \lambda\ \sim\ 290\ nm} CH_4 + CO \qquad (4.38)$$

The lifetime of aldehydes in the atmosphere is in the range of 24 h.

In summary, there are at least three possible reactions consuming aldehydes that have been formed in a photochemical smog:

- reaction to form a PAN chemical;
- reaction to form an alkyl radical and carbon dioxide;
- photolysis, producing an alkane and carbon monoxide.

## The most abundant atmospheric hydrocarbon

### Methane

We have left a discussion about the tropospheric oxidation of methane to the end. Quantitatively speaking, methane, derived from several sources and with a mixing ratio of 1.8 ppmv, is present in the global atmosphere at the largest concentration of any hydrocarbon. At most locations, the usual situation is that higher molar mass hydrocarbons and partially oxidized species are present in much smaller concentrations. There are, however, situations in urban areas where non-methane hydrocarbon (NMHC) mixing ratios may reach values as high as 5–10 ppmv C as a result of vehicular and other emissions.[4] As has been noted, in cities these compounds derive mostly from unburned petroleum-based fuel.

---

### Example 4.1 Concentration units for hydrocarbon gases

The concentration of benzene in an unventilated laboratory is found to be 220 µg m$^{-3}$. Calculate the mixing ratio in ppbv (benzene) and ppbv C. What is the significance of expressing results in terms of ppbv C?

220 µg m$^{-3}$ is equivalent to 220 / 78.1 = 2.82 µmol m$^{-3}$

Using the ideal gas law, $PV = nRT$, 1 m$^3$ of air at $P°$ and 25°C contains 40.9 mol of gas, and the mixing ratio expressed in terms of benzene is

$$2.82 \times 10^{-6} / 40.9 = 6.89 \times 10^{-8} \times 10^{9} = 69 \text{ ppbv (benzene)}$$

Because benzene contains six carbon atoms, the mixing ratio in terms of carbon itself = 6 × 69 = 414 ppbv C.

---

The significance of using mixing ratios expressed in terms of carbon is that a hydrocarbon compound that contains $n$ carbon atoms can go through a reaction sequence such as that described in reactions 4.13–4.17 a total of $n$ times.

In rural areas, the hydrocarbon background is provided by emissions of terpenes and other VOCs wherever there are trees, especially in heavily forested regions. In total, this background level of hydrocarbons other than methane is typically 10 to 20 ppbv C, two orders of magnitude lower than that of methane.

There are several natural and anthropogenic sources of methane. It is released during extraction, production, and transport of natural gas, which is itself mostly methane. It is produced by biological reactions in submerged soils like swamps and rice fields, and in landfills. The means by which this happens will be described later when we consider conditions under which microorganisms can be agents of degradation of organic matter. Methane is a greenhouse gas and its role in absorbing infra-red radiation is described in Chapter 8.

All of the hydrocarbons are susceptible to oxidation via sequences initiated by the hydroxyl radical, but the rate of oxidation depends on the particular substrate. The oxidation reaction rate of methane is considerably slower than that of most other hydrocarbons, giving it an atmospheric lifetime of about 12 years. Therefore, the presence of methane provides a substrate that undergoes a continuous slow sequence of oxidation reactions but local emissions of other hydrocarbons can have a larger effect on the daily cycle of hydrocarbon oxidation. On a global scale, recall that methane and carbon monoxide are the two principal consumers of hydroxyl radicals.

The reactions involving methane are as follows.

$$CH_4 + \cdot OH \rightarrow \cdot CH_3 + H_2O \tag{4.39}$$

$$\cdot CH_3 + O_2 + M \rightarrow CH_3OO\cdot + M \tag{4.40}$$

---

[4] When considering non-methane hydrocarbons as a group, it is useful to report the mixing ratio in ppbv C. This is done by multiplying the normal mixing ratio (ppbv) of an individual compound by the number of carbon atoms. Therefore, one ppbv ethane is equivalent in carbon content to two ppbv methane.

For conversion of NMHC concentrations in units such as µg m$^{-3}$ to mixing ratios in ppbv C, the number of µmoles of carbon is calculated by dividing the mass (µg) by 14, equivalent to the mass of a methylene group.

$$CH_3OO\bullet + NO \rightarrow CH_3O\bullet + NO_2 \tag{4.41}$$

$$CH_3O\bullet + O_2 \rightarrow CH_2O + HOO\bullet \tag{4.42}$$

Up to this point, the oxidation of methane has produced formaldehyde and the sum of the reactions is

$$CH_4 + \bullet OH + 2O_2 + NO \rightarrow CH_2O + H_2O + HOO\bullet + NO_2 \tag{4.43}$$

The formaldehyde is involved in additional reactions, with photolysis being most important.

$$CH_2O \xrightarrow{h\upsilon,\,\lambda\,<\,330\,nm} HC\overset{\bullet}{O} + \bullet H \tag{4.44}$$

$$HC\overset{\bullet}{O} + O_2 \rightarrow HOO\bullet + CO \tag{4.45}$$

$$CO + \bullet OH \rightarrow CO_2 + \bullet H \tag{4.46}$$

The hydrogen and the hydroperoxyl radical then undergo subsequent reactions.

$$2(\bullet H + O_2 \rightarrow HOO\bullet) \tag{4.47}$$

$$2(HOO\bullet + NO \rightarrow \bullet OH + NO_2) \tag{4.48}$$

In sum, the net reaction is

$$CH_4 + 5O_2 + 2H_2O \rightarrow 2HOO\bullet + 6\bullet OH + CO_2 \tag{4.49}$$

Therefore, the oxidation of methane ultimately produces carbon dioxide as the final stable carbon compound. The reaction is initiated by hydroxyl radicals but in the process, additional radicals are generated. Clearly, these reactions are self-sustaining but the variety of termination processes serves to keep them in balance.

---

### Example 4.2 **Reaction rates for hydrocarbon oxidation**

Consider an urban situation with the following atmospheric concentrations. The mixing ratio of methane has the usual background value of 1.8 ppmv. This is equivalent to a concentration of $4.9 \times 10^{13}$ molecules $cm^{-3}$. The concentration of one of the higher hydrocarbons, hexane, is 100 µg $m^{-3}$ which equals $7.0 \times 10^{11}$ molecules $cm^{-3}$. The concentration of hydroxyl radical is $2.0 \times 10^6$ molecules $cm^{-3}$. The rate constants for the second-order reactions between hydroxyl and these hydrocarbons are $8.36 \times 10^{-15}$ and $5.61 \times 10^{-12}$ $cm^3$ molecule$^{-1}$ s$^{-1}$ for methane and hexane, respectively (see Additional Resources 3 at the end of this chapter).

Using the atmospheric data and rate constants given and an equation for the instantaneous rate of disappearance of the two hydrocarbons,

$$Rate_{methane} = k_2[\bullet OH][methane]$$
$$Rate_{methane} = 8.36 \times 10^{-15}\,cm^3\,molecule^{-1}s^{-1} \times 2.0 \times 10^6\,molecule\,cm^{-3}$$
$$\times\,4.9 \times 10^{13}\,molecule\,cm^{-3}$$
$$= 8.2 \times 10^5\,molecule\,cm^{-3}\,s^{-1}$$

Likewise

$$Rate_{hexane} = 7.8 \times 10^6\,molecule\,cm^{-3}\,s^{-1}$$

Note that these are instantaneous values and assume a steady-state concentration of hydroxyl radical (implying its very rapid synthesis). Under these circumstances, hexane is a more important reactant for hydroxyl than methane, even though its concentration is two orders of magnitude lower. This is a consequence of the rate of reaction being much greater for hexane. In a non-polluted atmosphere, however, where the hexane concentration would be several orders of magnitude lower than the urban value, it would make a negligible contribution to hydroxyl radical destruction compared with methane.

---

## General principles describing VOC oxidation

In this survey of hydroxyl-initiated reactions of VOCs, we have observed a number of important general steps.

1.  The initiation begins with either hydrogen abstraction or hydroxyl addition.
2.  The radical produced by step 1 adds an oxygen molecule forming a peroxyl species or, in the case of aromatics, the dioxygen abstracts a hydrogen.
3.  The peroxyl species transfers an oxygen atom to a molecule of nitric oxide.
4.  The product molecule now loses a hydrogen atom to another oxygen molecule, or it splits into two smaller species. In either case, aldehydes (or, less commonly, ketones) are formed. The hydroperoxyl radical is the other product.
5.  The aldehydes react with nitrogen dioxide to form PANs, undergo further hydroxyl-initiated oxidation, or photochemically decompose.
6.  The decomposition products are again subject to a repeat sequence of oxidation steps and the ultimate stable products are carbon dioxide and water.
7.  The reactions indicated in this section are a summary of some of the more important reactions occurring in the troposphere. The situation is highly complex and many alternative processes occur. For those who model atmospheric behaviour in a polluted environment, hundreds of simultaneous reactions must be taken into account, rate constants measured, and concentrations estimated. Assumptions are involved at every step and the natural climatic variations make quantitative calculations even more difficult.

> **Main point 4.3** There are diverse sources of the many hydrocarbons emitted into the atmosphere. The hydroxyl radical plays a major role in oxidizing these compounds. The ultimate oxidation product is carbon dioxide, but many intermediate species have substantial lifetimes and these secondary products are found in the atmosphere at locations around the globe.

## 4.3 Exhaust gases from the internal combustion engine[5]

The point has been made that a major anthropogenic source of organic chemicals and other atmospheric pollutants is vehicles of various kinds. In many countries, especially Australia, Japan, and countries in Europe and North America, the gasoline (petrol)-powered four-stroke internal combustion engine is the predominant type of vehicle engine. Diesel engines are used more commonly for larger vehicles such as buses, trucks, locomotives, and ships and in some power generators, for example, those used in mines. Two-stroke gasoline-powered engines are less commonly employed in larger vehicles, but are important sources of power for chainsaws, lawn mowers, mopeds, motorcycles, and in outboard motors for boats.

In low-income countries, there is a smaller proportion of gasoline-powered automobiles, while highly developed public transportation systems involving diesel buses and trains are very important. Scooters, motorcycles, and autorickshaws are also common and frequently are powered by small two-stroke engines. In Shanghai, China, for example, in 2005 there were estimated to be around 1.2 million scooters of various designs, many operating without significant pollution controls.

The nature and quantity of emissions from each of these engine / fuel combinations are different. However, in a quantitative sense, it is very difficult to compare emissions from the various engine types. The size of vehicle on which each type is used is usually different. Some engines

---

[5] Much of the discussion in this section is based on material in reference 4 from the Additional Resources list.

have catalysts that effectively remove a part of the harmful gases before they are sent to exhaust. Furthermore, in practice many vehicles operate under far less than ideal conditions. Laboratory data obtained using controlled loads and a well-tuned engine are therefore not reliable indicators of the real situation on the road. We will look briefly at the nature of emissions emanating from engines of the three types.

### Gasoline-powered four-stroke engines—the power source for many automobiles

#### The Otto cycle

The Otto cycle (Fig. 4.4) describes the sequence followed in the cylinders of a gasoline-powered four-stroke engine. It is similar but not identical to the well-known Carnot cycle.

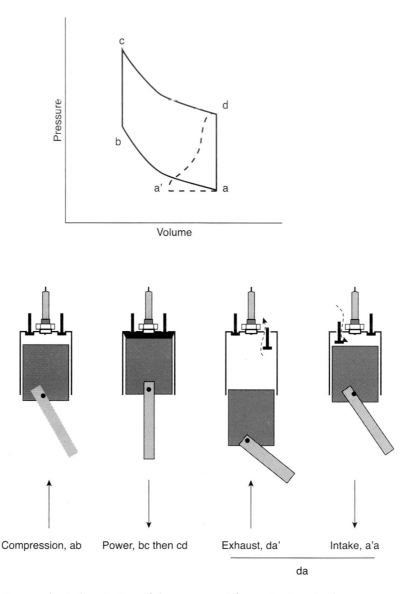

**Fig. 4.4** The Otto cycle. A description of the sequence of steps is given in the text.

When the piston is at the bottom of its intake stroke, it is at position 'a' on the pressure–volume diagram. As the piston rises rapidly, the fuel / oxidant mixture is compressed and there is little time for heat exchange, so this may be considered an adiabatic process, one in which no heat enters or leaves the system. This is represented by the line 'ab'. At 'b', the spark ignites the fuel and there is almost instantaneous combustion of the contents of the cylinder. In this brief instant before the piston has moved, there is a constant volume addition of heat corresponding to line 'bc'. However, as a result of the explosive combustion, an instant later the piston moves downwards in the power stroke. Again, there is negligible heat transfer and the process is represented by the adiabatic curve 'cd'.

At this point, the exhaust valve opens and the hot gases are expelled (path 'da''') followed by intake of fresh, cool gases (path 'a'a'). This complicated sequence of processes is represented in simplified form on the figure by the line 'da'. The cycle then repeats itself.

The efficiency of the Otto engine is given by the relation

$$\frac{W_{cycle}}{Q_{bc}} = 1 - \frac{T_d}{T_c} = 1 - \left(\frac{V_c}{V_d}\right)^{R/C_v} \tag{4.50}$$

In this relation, $W_{cycle}$ is the work produced by the engine and $Q_{bc}$ is the heat generated in the cylinder during combustion. The ratio given in the left-hand side of the equation gives the thermodynamic efficiency of the engine. $R$ is the gas constant and $C_v$ is the heat capacity of the gas at constant volume. $T_c$ and $T_d$ are the temperatures at the beginning and end of the power stroke and $V_c$ and $V_d$ are the corresponding volumes. The ratio of cylinder volume at the bottom to the top of the stroke equals

$$\frac{V_d}{V_c} = \text{the comparison ratio, } r_c$$

$$\therefore \frac{W_{cycle}}{Q_{bc}} = 1 - \left(\frac{1}{r_c}\right)^{R/C_v} \tag{4.51}$$

There is environmental significance to this thermodynamic calculation.

From eqn 4.51 we can see that good fuel efficiency, that is, a large value of $W_{cycle} / Q_{bc}$, in the internal combustion engine requires an engine designed to have a high compression ratio, $r_c$. High compression ratios, however, lead to engine knocking,[6] which needs to be overcome in one of two ways.

### The octane number and its enhancement

One way is to increase the octane number of gasoline. Straight-chain hydrocarbons tend to produce knocking at relatively low compression ratios, while aromatic and branched-chain hydrocarbons combust quietly even at somewhat higher ratios. An empirical scale has therefore been devised based on binary mixtures of n-heptane and 2,2,4-trimethylpentane ('isooctane') in which the compression ratio where a certain level of knocking occurs is plotted against the percentage of isooctane in the mixture. An octane number of 87 corresponds to an 87:13 ratio of isooctane to n-heptane. An actual gasoline sample is a mixture of a wide range of hydrocarbons and, if its octane number is defined as 87, this means that it behaves as does a binary mixture of the above composition. Higher octane numbers correspond to mixtures with more branched chain, aromatic, and oxygenated molecules. Incorporating oxygen-containing species such as ethanol or methyl tertiary butyl ether (MTBE) in the gasoline mixture, for example, increases the octane number. Such compounds tend to be relatively expensive to produce.

A second way of effectively increasing the octane number of gasoline is by including a small amount of an appropriate additive in the fuel formulation. One such additive is tetraethyl lead

---

[6] Knocking is a phenomenon in which pre-ignition occurs in the cylinder leading to noisy and potentially damaging engine operation. Besides engine design, it also depends on the nature of the fuel.

$((C_2H_5)_4Pb)$. The addition of this substance at a concentration of about 1 g L$^{-1}$ of gasoline substantially improves anti-knock properties by augmenting the octane number of the fuel. Until recently, this additive was widely used to enhance engine performance. The added lead reacts with halogenated compounds (also added to the fuel) to produce a variety of lead halides that are sufficiently volatile to be emitted in the gaseous combustion products, but condense in the ambient atmosphere and form an aerosol that is deposited on the surrounding vegetation, soil, and water. Legislation in many countries now requires that tetraethyl lead not be used for this purpose. It is still, however, marketed in some parts of South America, Africa and Asia and vehicular lead emissions continue. In Nigeria, a country with major high-quality oil resources, the specification for tetraethyl lead in gasoline is a maximum concentration of 0.7 g L$^{-1}$ although it is added at an actual average amount of 0.25 g (Pb) L$^{-1}$. In 2002 about 9 billion L of gasoline were consumed in Nigeria, so that approximately 2250 t of lead was released to the atmosphere throughout the country, mostly in areas close to traffic corridors.[7] Deposition rates are especially high in a city like Lagos, which has over one million vehicles and where congestion results in peak hour average operating velocities of less than 10 km per hour. Excessive concentrations of carbon monoxide and major smog episodes are also frequent occurrences in that city.

Other additive formulations have been developed more recently. Methylcyclopentadienyl manganese tricarbonyl, $CH_3(C_5H_4)Mn(CO)_3$, commonly called MMT, has become available and is added at a level of 0.1 g L$^{-1}$.

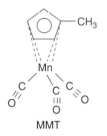

MMT

MMT is itself toxic, but it combusts completely in the engine, and the only significant emission products are $Mn_3O_4$ and other manganese oxides. Manganese is an essential nutrient for plants and animals, but in higher doses it can be toxic, causing neurological dysfunction. It is claimed that the solid emission products from fuels do not add significantly to the amount of Mn ingested in normal situations.

It is clear that the fuel efficiency–compression ratio–octane rating issue has well-established technological solutions, but each one of the solutions has environmental and economic implications that must be considered.

### Fuel combustion and reformulated gasoline—oxygenating additives

In addition to the search for improved efficiency, there is an overlapping companion problem—the need to reduce emissions of carbon monoxide, unburned hydrocarbons and nitric oxide. Two approaches are applied to bring about better removal of carbon monoxide and smog-producing chemicals. The first focuses on fuel and the other on engine design.

The complete combustion of octane is described in simple overall form as

$$C_8H_{18} + 12.5O_2 \rightarrow 8CO_2 + 9H_2O \qquad (4.52)$$

For this or any other saturated hydrocarbon, the stoichiometric oxygen to fuel ratio (on a mass basis) is about 3.5:1 corresponding to an air to fuel ratio of 17:1. With insufficient oxygen, incomplete combustion occurs and the reaction products include carbon monoxide and a range

[7] C.C. Chikwendu, *Gasoline as a key source of lead in the atmosphere*. Friends of the Environment, Nigeria, http://www.cleanairnet.org/ssa/1414/articles-36406_pdf.pdf, accessed October 2009.

of partially reacted or unreacted hydrocarbons. Carbon monoxide is well known for its ability to inhibit the oxygen-transport properties of haemoglobin, and some jurisdictions have established stringent maximum allowable levels, typically 10 to 30 ppmv in urban atmospheres. The partially or unreacted hydrocarbons, as we have seen, are precursors in the photochemical smog production sequence of reactions. It should be noted that emission of hydrocarbons also occurs to a significant extent by evaporation from the fuel tank or from the carburetor. These losses to the atmosphere are especially severe when the ambient temperature is very high.

Because of the contribution of carbon monoxide and gaseous hydrocarbons to an unhealthy atmosphere, it is important to minimize their release. Oxygenating compounds added to the hydrocarbon fuel are one means by which more complete combustion is achieved. This brings up the subject of reformulated fuels. Oxygenating compounds were mentioned above as being able to increase the octane number of a gasoline mixture. Furthermore, because these compounds are oxygen suppliers they also have the ability to promote more complete combustion to the final products, carbon dioxide and water. For this reason, legislation in the USA (the Clean Air Act, 1990) and elsewhere has been enacted. In the American case, the legislation requires year-round use of reformulated oxygen-enriched gasoline in nine major American conurbations that are locations of severe air-pollution problems. These areas include greater Connecticut, New York City, Philadelphia, Baltimore, Chicago, Milwaukee, Houston, Los Angeles, and San Diego. Other regions are subject to less stringent restrictions. One formulation recommended for compliance with the Act includes 10% oxygenating compound in the fuel.

## Methyl tertiary butyl ether (MTBE)

One widely used oxygenating compound is methyl tertiary butyl ether—an industrial product based on a petrochemical feedstock. MTBE, typically added as 15% of the fuel volume, is a cost-effective oxygenating agent that not only increases completeness of combustion but also augments the octane number of the gasoline. The major problems with MTBE occur when it gets into water supplies. Leakage of gasoline from underground storage tanks is a common problem worldwide. The underground plume of fuel consists largely of insoluble and slowly degradable gasoline but, if MTBE is present, it readily dissolves in the associated groundwater. In California and elsewhere in the United States, instances of serious contamination of drinking water sources have occurred. MTBE gives water an offensive taste and, more importantly, is known to be a carcinogen. It has now been banned in at least 25 states, creating a demand for a safer alternative.

## Ethanol

Ethanol has become a favoured alternative to MTBE as an oxygenating agent but its use too has serious environmental drawbacks. Ethanol is more costly to produce. It cannot be shipped by pipeline, requiring dedicated sealed trucks to transport it separately from gasoline, and it needs to be blended with gasoline near the retail outlets. While ethanol contributes to improvements in terms of carbon-monoxide emissions, it also increases the volatility of the fuel by about 7 kPa measured as Reid vapour pressure (RVP). The RVP is the vapour pressure measured at a temperature of 34.4°C which is 100°F. The increased volatility operates against other sections of the 1990 Clean Air Act that require reductions in volatility from levels of 62 kPa for conventional gasoline to 56 kPa in northern cities and 50 kPa in the south. These regulations were introduced because volatility of fuel is a major contributor to release of VOCs, which are, along with carbon monoxide, significant factors in urban air pollution. Appropriate legislation then requires weighting of the carbon monoxide and VOC emission factors.

Even with these limitations, ethanol is being increasingly employed as an oxygenating fuel additive, often used at a rate of 5 or 10% in a gasoline formulation. In some countries, most notably Brazil, it is used in much higher concentrations with gasoline or in 100% (aqueous ethanol) form as the main component of the fuel. While MTBE is a manufactured petrochemical product, ethanol is a manufactured agricultural product and is therefore called a biofuel. Various plants that produce large quantities of starch or sugar as part of their biomass, can serve as feedstocks for the manufacture of ethanol. Sugarcane and maize (corn) are two of the crops commonly grown for this purpose. The argument is made that plant-based fuels are renewable in the sense that the starting material for their manufacture is derived from a crop that is grown year after year. One must, however, also consider the effects of devoting vast areas of prime agricultural land to continuous production of a crop like maize through high-intensity agriculture. Maize is the major feedstock in many ethanol-production facilities and growing it with high yields depends on heavy application of synthetic fertilizer whose manufacture in turn depends on fossil fuels as well as other raw materials. The environmental consequences of producing and using such chemicals is yet another factor that must be included in any calculation of problems and benefits of gasoline reformulation strategies.

## Case study—an energy budget when using maize as a source of fuel

Maize is a particularly good substrate for alcohol production. The process involves two main steps:

- fermentation under conditions maintained to enhance the microbial reaction;
- distillation to separate the alcohol from the mixed slurry.

The fermentation reaction, in its simplest form can be represented as

$$3\{CH_2O\} \rightarrow C_2H_5OH + CO_2$$

where $\{CH_2O\}$ is used as a simple formula giving approximate atomic ratios of the major elements in biomass (see Chapter 12).

The by-products of fermentation include milling wastes that can be used as animal feeds. From the distillation, the residue is called stillage and this can serve as a feedstock for a variety of useful chemicals and is sometimes also used as an animal feed or soil amendment.

Each of these products can be assigned an energy value, although only the ethanol will be used as a fuel. The following set of data illustrates potential energy benefit from growing corn for ethanol production.

Energy output per litre of fuel produced:

| | |
|---|---|
| ethanol fuel value | $19.0 \times 10^6$ J |
| coproduction products | $8.2 \times 10^6$ J |
| *total output* | $27.2 \times 10^6$ J |

Likewise, the operations required to produce the products all require energy, including energy associated with the chemical inputs employed in growing the corn, for machinery, transport, etc. Then there are the energy inputs associated with the industrial fermentation and distillation processes.

Energy input per litre of fuel produced:

| | |
|---|---|
| corn production | $5.4 \times 10^6$ J |
| processing | $9.1 \times 10^6$ J |
| *total input* | $14.5 \times 10^6$ J |

Using MJ (J $\times 10^6$)

$$\text{Total energy gain} = (27.2 - 14.5) / 14.5 \times 100 = 88\%$$

$$\text{Fuel energy gain} = (19.0 - 14.5) / 14.5 \times 100 = 31\%$$

Clearly, in terms of production of fuel, a relatively large amount of energy, most of which is ultimately in the form of fuel, is required to produce a 31% gain in fuel energy. The efficiency appears to be much more substantial if one includes the energy value of the by-products produced.

With fuel-based energy consumption comes carbon dioxide release. A comparison of emissions from regular and reformulated gasolines is given in *Environ. Sci. Technol.* 26 (6), (1992), 206.

> **Fermi question**
> What area of land would be required to provide fuel for an automobile engine designed to run on pure ethanol for 1 year?

## Engine design and catalytic converters for emission control

Another approach taken to maximize complete combustion of fuel is to ensure that there is a plentiful supply of oxygen in the fuel / air mixture in the cylinder. When the air to fuel ratio is large, this is referred to as a lean mixture, meaning lean in terms of fuel. Under these conditions, the combustion temperature is elevated and therefore this also increases the compression ratio, thereby maximizing fuel efficiency. Unfortunately, there is a negative environmental side-effect that results from this practice.

The increased supply of air not only elevates the combustion temperature but also augments the concentration of nitrogen and it supplies excess oxygen over that required to react with the fuel. All of these factors, while reducing emissions of hydrocarbons, increase the formation and release of nitric oxide (Fig. 4.5).

We have seen that nitric oxide product is an essential component in photochemical smog formation and is also a precursor for atmospheric generation of nitric acid, one of the constituents of acid rain. Initial attempts at minimizing $NO_x$ production involved using lower-temperature combustion and lower compression ratios along with recycling partially burned exhaust gases as a diluent in the fuel–air mixture. While this practice lowered emission levels of nitrogen oxides, it also reduced fuel efficiency. Therefore, a number of more complex strategies of exhaust gas clean-up have been developed. Current technology centres around the use of a *three-way catalytic system* capable of reducing levels of three types of emission products, namely, non-methane hydrocarbons (NMHCs), carbon monoxide, and nitric oxide. This system involves combustion

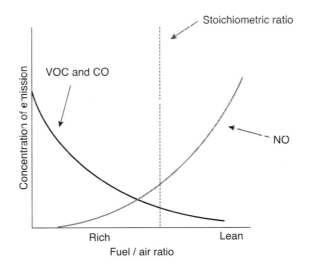

**Fig. 4.5** Relative amounts of emission products as a function of fuel to air ratio.

carried out under near-stoichiometric conditions and then allowing the exhaust gases to interact with two catalysts located in series. The catalysts are mounted on a high surface area ceramic material through which the exhaust gases pass. The first catalyst is typically rhodium, which facilitates the reduction of the nitrogen oxides to form dinitrogen; in this case, the reducing agent is one of the incompletely oxidized components of the exhaust gas stream. Unreacted hydrocarbons, hydrogen gas, and carbon monoxide can all react in this way.

$$2NO + 2CO \rightarrow N_2 + 2CO_2 \tag{4.53}$$

As the process proceeds, some ammonia is co-generated due to reaction of nitric oxide with hydrogen that is present as a result of decomposition of some components of the hydrocarbon fuel.

$$2NO + 5H_2 \rightarrow 2NH_3 + 2H_2O \tag{4.54}$$

To oxidize the residual hydrocarbons, carbon monoxide, and ammonia, air is supplied to the system and is passed over a second catalyst, either palladium or platinum or non-stoichiometric oxides such as $Fe_2O_3$ or $CoO \cdot Cr_2O_3$, to favour oxidation of the reduced components.

$$RH + O_2 \rightarrow CO_2 + H_2O \tag{4.55}$$

$$CO + \tfrac{1}{2}O_2 \rightarrow CO_2 \tag{4.56}$$

$$2NH_3 + \tfrac{3}{2}O_2 \rightarrow N_2 + 3H_2O \tag{4.57}$$

The use of efficient three-way catalyst systems serves to reduce emissions of the three problem gases by 80% or more. There are, however, outstanding technical problems that need to be solved in order to achieve even further reductions. Principal among these is the need to reduce emissions during the 'warm-up' period of vehicle operation, when the temperature in the catalytic chamber is not high enough to bring about efficient reaction of the exhaust gases.

### Gasoline-powered two-stroke engines—power sources for small vehicles and other equipment

The two-stroke engine is used to power small vehicles and other labour-saving devices such as chain saws and power mowers. In simplified form, the cycle is shown in Fig. 4.6.

In part a of the figure, the piston is near the bottom of its stroke and it is about to move upward compressing the fuel / air mixture. As it moves up to position b, a fresh fuel / air mixture

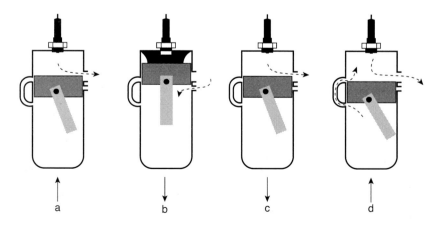

**Fig. 4.6** The two-stroke engine cycle. A description of the steps in the cycle is given in the text. The key feature is that exhaust of the combustion products and intake of a new air:fuel mixture occur simultaneously as shown in d.

is drawn in below the piston while the mixture above is compressed. At sufficiently high compression, the spark plug fires, igniting the fuel and, in the power stroke, the piston is forced down (part c). As it nears the bottom, exhaust gases begin to escape through the exhaust vent. This continues to occur while fresh fuel / air mixture is forced into the cylinder simultaneously. The cycle then repeats itself.

The engine described here is a simple two-stroke design, and many design modifications have been made to improve efficiency and other characteristics. Nevertheless, it should be clear that one feature of the design—namely, the simultaneous introduction of fuel and release of exhaust gases—can lead to problems of loss of unburned hydrocarbons. The problem is enhanced because two-stroke engine fuel normally consists of a mixture of gasoline and a petroleum lubricant; the latter material has higher average molar mass and is less efficiently oxidized during the combustion process than the lighter gasoline.

Another aspect of the engine design is that combustion takes place at a lower temperature than in the conventional four-stroke engine. Taken together, these features lead to an exhaust-gas composition that is relatively low in nitric oxide but can be very high in carbon monoxide and especially hydrocarbons, even when the engine operates under stoichiometric conditions.

The nature, but not the amounts, of the hydrocarbons emitted from a marine outboard engine burning gasoline is similar to that released from road vehicles burning the same fuel.[8] The components include aromatic hydrocarbons, especially alkyl aromatics. From an outboard motor, the exhaust is released directly into water. Although the solubility of these hydrophobic molecules is small, their toxicological effects on aqueous organisms could be significant.

Two- and four-stroke outboard engines with the same power output are now manufactured and it is therefore possible to make comparisons of emissions under similar operating conditions. Carrying out this kind of test using 7.3 kW engines burning gasoline,[8] the exhaust was found to have the composition shown in Table 4.3.

The two previously noted effects are confirmed in these data—namely, that two-stroke engines emit relatively large amounts of hydrocarbons, but relatively small amounts of nitrogen oxides. Emissions catalysts can be used to reduce the hydrocarbon emissions and, clearly, the catalyst must be of the oxidative variety. Because most two-stroke engines are physically small, gases are emitted close to the combustion chamber and tend to be very hot under running conditions. It may therefore be necessary to cool them to a temperature at which the catalyst can be effective and one means of doing this is to introduce additional cooling air just prior to passage through the catalyst system.

Catalysts are able to successfully reduce emissions to a much lower level. Table 4.4 gives data for a 125-cc high-performance motorcycle engine operating in three formats—as a standard production engine; with optimization of the carburettor and output-gas scavenging conditions;

**Table 4.3** Exhaust gas output of CO, $NO_x$ and hydrocarbons, from 7.3-kW two- and four-stroke engines operating under the same conditions[a].

|  | Exhaust gas output / $10^{-8}$ g J$^{-1}$ | | |
|---|---|---|---|
|  | CO | $NO_x$ | Hydrocarbons |
| Two-stroke engine | 165 | 0.3 | 89 |
| Four-stroke engine | 127 | 0.7 | 7 |

[a] Juttner, F., D. Backhaus, U. Matthias, U. Essers, R. Greiner, and B. Mahr, Emissions of two- and four-stroke outboard engines. I. Quantification of gases and VOC. *Water Res.*, **29** (1995), 1976–82.

[8] Juttner, F., D. Backhaus, U. Matthias, U. Essers, R. Greiner, and B. Mahr, Emissions of two- and four-stroke outboard engines. I. Quantification of gases and VOC, *Water Res.*, **29** (1995), 1976–82.

**Table 4.4** The effects of engine optimization and catalysts on release of CO, NO$_x$, and hydrocarbons from a 125-cc two-stroke motorcycle engine.

|  | Output / g km$^{-1}$ | | |
| --- | --- | --- | --- |
|  | CO | NO$_x$ | Hydrocarbons |
| Production engine | 21.7 | 0.01 | 16.9 |
| Optimized engine | 1.7 | 0.03 | 10.4 |
| Engine with catalyst | 0.8 | 0.02 | 1.9 |
| Swiss standards | 8 | 0.1 | 3 |

and, finally, with the addition of a catalyst to the exhaust system. In the final configuration, the emissions met the strict standards set by the Swiss government.[9]

## Diesel-powered engines—power source for heavy-duty use

Diesel engines are very commonly used in large vehicles—locomotives, buses, trucks, ships and in stationary engines such as those that are used for electricity generation. Increasingly, diesel engines are also employed in automobiles. Diesels are efficient four-stroke engines that operate with higher compression ratios than the standard Otto engine. The large pressure in the cylinder results in a high temperature so that the fuel mixture spontaneously ignites and burns in the absence of a spark ignition source. The cycle takes the following form (Fig. 4.7).

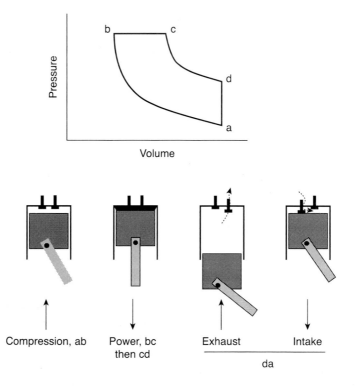

**Fig. 4.7** The diesel cycle. The steps are similar to those in the Otto cycle and are described in the text.

[9] Laimbock F. quoted in Blair, G.P., *The basic design of two-stroke engines*, p. 328. Society of Automotive Engineers, Inc., Warrendale, Pennsylvania; 1990.

In the cylinder filled with air, adiabatic compression takes place (ab) with a compression ratio of about 15:1. At the end of step ab, fuel is introduced. Compression to a very small volume results in a temperature increase to about 500°C in the cylinder, and the fuel ignites spontaneously and instantaneously, burning continuously as the piston is forced downwards in the power stroke (bc and cd). The time taken for fuel injection and combustion is sufficiently long that cylinder pressure remains nearly constant in the initial stage (bc) of expansion. Finally, exhaust of the combustion products occurs, followed by intake of fresh air (da).

The diesel engine uses a lower-grade higher-boiling fuel than the gasoline that is employed in the previously described engines. The high compression ratio leads to a more efficient power source (eqn 4.51) but also requires that the engine be heavily constructed to withstand the pressures generated within the cylinders. Because fuel is injected at the completion of the compression stroke, to some extent mixing of fuel and air is incomplete. Consequently, one of the principal problems associated with diesel engines is release of particulates that are made up of unburned carbon particles (soot) and a soluble organic fraction (SOF). The SOF has at least two components—unburned or partially burned fuel that forms a liquid aerosol, and other components that are sorbed on to the soot particles. Gaseous hydrocarbon emissions are less than from gasoline-powered engines because of the higher boiling nature of diesel fuel. Inorganic sulphates are also present in the aerosol as the sulphur content of diesel fuel is somewhat higher (typically 0.1 to 0.3%) than that found in most gasolines.

The other important constituent of exhaust gas is nitric oxide, which is present as a consequence of the high combustion temperature and the lean conditions that exist for at least a part of the combustion process.

Soot traps, often employing polytetrafluoroethene (PTFE) or quartz fibre filters, are one means of controlling emission of the aerosol particulates. However, as the soot accumulates, resistance to flow increases and it is necessary to regenerate the filter by burning off the accumulated material. The technology for accomplishing regeneration is complex, usually requiring an energy source, a second trap to be used during regeneration, and a control package. An alternative approach is to use an oxidation catalyst that can accommodate liquid and solid emissions as well as unburned gaseous hydrocarbons. One type of catalyst consists of platinum or palladium deposited on a high surface area silica substrate. Alumina is less desirable as a support because of its strong specific interaction with sulphate, which is also a component of the exhaust gas. The usual reduction catalysts are suitable for removing nitric oxide from the gas stream.

While removal of pollutant gases after their production in the engine remains a requirement, an even better approach is to prevent the generation of the gases in the first place. This is difficult, but some success has been achieved by engine design, wherein recirculation of exhaust gases, altering injection timing, and ensuring good fuel–air mixing are used to bring about complete combustion while minimizing nitric oxide production.

**Literature link** Diesel versus natural gas as a fuel

In an attempt to reduce the incidence of smog in some large cities such as Los Angeles in the USA and Delhi in India regulations have been enacted that promote the conversion of engines used for transit and other public utilities from diesel fuel to compressed natural gas (CNG). Clearly, there is visual evidence that exhaust from gas-powered vehicles is much cleaner than that from a diesel engine. But what about other emission products that are not visible? A study of transit and school buses and other vehicles used in California (Hesterberg, T.W., Lapin, C.A. and Bunn, W.B., *A Comparison of Emissions from Vehicles Fuelled with Diesel or Compressed Natural Gas.* Environ. Sci. Technol. **42**, (2008), 6437–6445), provides some interesting results.

For transit buses, diesel fuel not surprisingly gave the highest emissions of particulate matter and polycyclic aromatic hydrocarbons (PAHs). On the other hand, CNG buses produced the highest emissions of carbon monoxide, hydrocarbons, non-methane hydrocarbons (NMHC), volatile organic compounds (VOCs including benzene, butadiene, ethylene), and carbonyl compounds (formaldehyde, acetaldehyde, acrolein). This was for engines without provision for after-treatment of emissions. Treatment of the exhaust with soot traps and oxidation catalysts in the case of diesel engines and with three-way catalysts for CNG engines reduced emissions considerably, resulting in statistically similar levels in all cases. There were no differences in nitrogen oxide ($NO_x$) levels for untreated diesel and CNG emissions, but the three-way catalyst reduced the concentration of $NO_x$ substantially for gas-powered vehicles.

Besides transit buses, results for other vehicles were also obtained. In many, but not all cases, different vehicles exhibited similar behaviour. In general, the data obtained in the paper are consistent with the chemistry described in this chapter. It is worthwhile to study the tables carefully and to relate the information to what you know about fuel, combustion conditions and after-treatment technologies. Several important conclusions arise from the study.

- While untreated diesel exhaust is obviously 'dirty', untreated CNG exhaust also carries important smog-producing chemicals.

- Efficient after-treatment of exhaust gives much improved emissions for both types of engines – resulting in similar emissions characteristics. After-treatment appears to be more important than fuel type.

- Especially important was that emissions of minor, but toxic gases like benzene, formaldehyde and PAHs were also reduced after fitting engines with oxidative catalysts.

## Biodiesel fuels

A recent fuel-design innovation is the production and use of diesel fuels that are based on chemical modification of fats and oils. These feedstocks can be derived from crops such as soybeans that are grown specifically for fuel production. Claims are also made that wasteland can be set aside to grow oilseed-producing trees. Jatropha is one such plant that grows widely in the Caribbean, Africa and Asia. It is actually a genus of some 175 trees and shrubs and is a source of seeds that are being promoted as an excellent feedstock for biodiesel production. Finally, it is also possible to collect waste fats and oils from restaurants and other sources and use these to produce biodiesel. A number of processes have been developed to bring about transesterification of the materials producing glycerol and the fuel itself. A base-catalysed reaction is usually employed to produce the esterified fatty acids.

$$
\begin{array}{ccccc}
CH_2O(O)CR & & CH_2OH & & \\
| & & | & & \\
CH_2O(O)CR & + 3C_2H_5OH \rightarrow & CH_2OH & + 3C_2H_5O(O)CR \\
| & & | & & \\
CH_2O(O)CR & & CH_2OH & & \\
\text{fat or oil} & \text{ethanol} & \text{glycerol} & \text{biodiesel fuel} &
\end{array}
$$

where $R = CH_3(CH_2)n-$; $n$ is typically 14 to 18 carbons.

The fuel has the debatable advantage of being derived from a renewable resource. There is added benefit if it can be manufactured from otherwise waste products. Furthermore, it has been shown that some emissions are substantially reduced compared to emissions from petroleum-based diesel. Most notably, the 10–12% oxygen content of the fuel leads to more complete combustion and, consequently, reduced release of unburned carbon-containing products. Plant-derived oils also contain little sulphur in their structure, so that sulphur dioxide emissions are almost zero.

**Table 4.5** Change in emission products when biodiesel fuel is used as a replacement for conventional petroleum-based diesel.

| Emission type | Change in emission products / % when using | |
|---|---|---|
| | B100 | B20 |
| CO | −48 | −12 |
| Particulate matter | −47 | −12 |
| $NO_x$ | +10 | +2 |
| Sulphates | −100 | −20 |
| Ozone-production potential | −50 | −10 |

Currently, most biodiesel is used in commercial vehicles such as buses. Two formulations are common: B100 is pure biodiesel, while B20 is a mixture of 20% biodiesel along with 80% conventional diesel. Table 4.5 compares emissions from biodiesel fuels with those from the petroleum-based formulation.

**Fermi question**

Use information from a fast-food restaurant in the city where you live to estimate how many litres of biodiesel could be produced each day from the waste cooking oil.

## Ozone production from engine emissions

In the mix of gases emitted as products of combustion in any engine we have shown that there are a number of species with undesirable, sometimes toxicological properties. We have also seen that secondary reactions bring about chemical changes that produce other gases and aerosol particulates. Among these, ozone is of considerable concern as it can have adverse effects on both plants and animals; hence we have referred to it earlier as 'bad ozone'. In photochemical smog situations, elevated levels of ozone are frequently observed. In Chapter 7, we will note some urban areas where excessive atmospheric ozone levels have been recorded.

Ozone is one of the by-products of the sequence of reactions by which a hydrocarbon is oxidized in reactions initiated by the hydroxyl radical. In order to reduce the extent of ozone production in an urban area then, it is necessary to reduce the atmospheric concentration of the reactants, nitric oxide (the precursor of hydroxyl) and the hydrocarbons themselves. Ideally, one would wish to reduce the concentration of both types of gases. We have seen, however, that this is a technological challenge. Fortunately, it is possible to reduce ozone synthesis by considering which of the two reactants is the *limiting reagent*. Figure 4.8 plots ozone isopleths (contours of constant concentration) as a function of both nitric oxide and hydrocarbon concentration.[10]

Consider a situation where the number of two-stroke engines is large and emissions are therefore highly enriched in hydrocarbon content. In this location, represented by point 'a', a region of large hydrocarbon:nitrogen oxides ratios on the plot, reducing the hydrocarbon emissions by 50% would change the conditions to point 'b'. In spite of the dramatic decrease in hydrocarbons, the ozone concentration at this point is seen to be little improved. On the other hand, a

[10] Finlayson-Pitts, B.J. and J.N. Pitts Jr., Atmospheric chemistry of tropospheric ozone formation: scientific and regulatory implications, *Air Waste*, 43 (1993), 1091–1100.

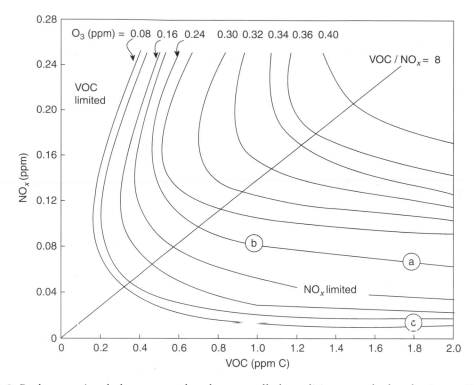

**Fig. 4.8** Peak ozone isopleths generated under controlled conditions as calculated using an EKMA (Empirical Kinetic Modelling Approach). (Redrawn from Finlayson-Pitts, B.J. and J.N. Pitts Jr., Atmospheric chemistry of tropospheric ozone formation: scientific and regulatory implications, *Air Waste*, **43** (1993), 1091–1100.)

reduction in the already low nitric oxide levels might alter the situation to point 'c' on the plot, with significant reduction in ozone production. This is due to the fact that the amount of ozone produced is limited by the availability of hydroxyl radical, which, in turn, depends on a plentiful supply of nitrogen oxides.

It is important to understand these factors but, at the same time, they apply only to the production of a single pollutant gas—ozone. It may also be important to reduce emissions of hydrocarbons because, in themselves, compounds like benzene have harmful properties.

---

**Main point 4.4** Engines of various kinds release the nitric oxide and carbon compounds essential to generating photochemical smog. The amounts of emission products generated depend on fuel composition and engine design. Emission-control strategies involve catalytic oxidation of unburned hydrocarbons and reduction of nitric oxide. Current technology can reduce the emissions substantially.

---

## ADDITIONAL RESOURCES

1. Atkinson, R., J. Arey, and S.M. Aschmann, Atmospheric chemistry of alkanes: Review and recent developments. *Atmos. Environ.* **42** (2008), 5859–5871.

2. Farrauto, R.J., R.M. Heck, and B.K. Speronello, Environmental catalysts. *Chem. Eng. News*, **70** (1992), 34–44.

3. Atkinson, R., Gas-phase tropospheric chemistry of organic compounds: a review, *Atmos. Environ.*, **24A** (1990), 1–41.

4. Fenn, J.B., *Engines, energy, and entropy*. W.H. Freeman and Company, New York; New York, 1982.

5. Environment Canada, Clean air on line, http://www.ec.gc.ca/cleanair-airpur/Home-WS8C3F7D55-0_En.htm, accessed October 2009.

## PROBLEMS

1. A general formula for gasoline is $C_7H_{13}$. Calculate the air to gasoline mass ratio required for stoichiometric combustion.

2. Write out a general sequence for the photochemical oxidation of butane.

3. Suppose propene ($CH_2=CH–CH_3$) is the hydrocarbon that reacts with the hydroxyl radical •OH. Beginning with addition of •OH, write the set of chemical reactions that ultimately produces an aldehyde. What is this final aldehyde?

4. Peroxyacetic nitric anhydride (PAN) can be thought of as related to hydrogen peroxide. Draw its structure and indicate why you suppose that it is a powerful oxidizing agent.

5. The atmospheric ratio of PAN:PAN + inorganic nitrate varies from less than 0.1 to 0.9. High values of the ratio are associated with 'photochemically aged' air masses, situations where precipitation has recently occurred, and situations where unusually high night-time nitrogen oxide concentrations are found.[11] What do these observations tell us about factors affecting the relative rates of formation and removal of nitrogen-oxide compounds?

6. During production, transport and use of natural gas, some is inevitably leaked into the atmosphere where it acts as a greenhouse gas (Chapter 8). Likewise, the commonly used fuel propane suffers the same fate. The principal removal mechanism is through a second-order reaction with hydroxyl radical whose mixing ratio is assumed to remain constant at $1.0 \times 10^6$ molec cm$^{-3}$. Use the following information:

    For methane, $k = 8.4 \times 10^{-15}$ cm$^3$ molec$^{-1}$ s$^{-1}$

    For propane, $k = 1.1 \times 10^{-12}$ cm$^3$ molec$^{-1}$ s$^{-1}$

    Calculate the half-life and residence time of methane and propane and use this information to suggest whether major regulatory initiatives should be directed toward methane or non-methane hydrocarbons.

7. Calculate the concentration of atmospheric carbon compounds in ppbv C using the data in Table 4.2.

8. A Canadian company is developing a process for converting agricultural wastes (straw, wood chips, etc.) to ethanol. This is an advance on the current technology that uses high-value feedstocks (corn, cane sugar, etc.) for the process. The company claims that there is the potential to produce 300 L of ethanol from 1 t of agricultural waste. Use the information on the fermentation of biomass to calculate the efficiency (in terms of mass) of this process. The density of ethanol is 0.79 g mL$^{-1}$.

9. For internal-combustion engines, ethanol is used either as a gasoline additive to increase the oxygenating ability of the fuel, or it can be used alone as a fuel. Vehicles in Brazil have engines that are modified so that they can burn ethanol as the fuel. Compare on a mass basis mass fuel to air ratio required for complete combustion of heptane and ethanol.

10. Methanol can be used as a fuel in specifically designed engines. Calculate the energy density (energy produced on combustion per mass of fuel) with that of octane. Comment on the relative merits of these two fuels in terms of environmental and practical issues.

---

[11] Roberts, J.M., The atmospheric chemistry of organic nitrates. *Atmos. Environ.*, **24A** (1990), 243–87.

11. The oxygen content of a fuel that is recommended in order to ensure complete combustion is about 2.7%.

    (a) Consider a case where methyl tertiary butyl ether (MBTE) is to be added to conventional gasoline. What percentage of MTBE would be required to achieve the 2.7 mass% oxygen content in the mixture?

    (b) Calculate the oxygen content of a typical biodiesel fuel.

# Chapter 5
# Tropospheric chemistry—precipitation

The focus of this chapter is on the activities and chemical processes that influence the chemistry of precipitation. In particular the issues surrounding nitrogen and sulphur atmospheric chemistry, including methods used to control their emissions, are of central importance. Here, we will discuss:

- the composition of natural precipitation and how the chemistry of various forms—rain, snow, and fog—is affected by both natural (e.g. $CO_2$, biogenic processes) and anthropogenic (e.g. fuel combustion and smelting) influences;

- the atmospheric production and removal of nitric and sulphuric acids, the two most prominent anthropogenic emissions influencing the pH of precipitation;

- the global situation, the sources and sinks of nitrogen and sulphur compounds, using aggregate fluxes, and their influence on the chemistry of precipitation around the globe;

- some of the technology and related chemistry for controlling emissions of nitrogen and sulphur oxides.

One of the early environmental issues to generate widespread publicity and public attention was the problem of acid rain. Emissions from energy and smelting industries were shown to be a source of sulphuric and nitric acid in rain and, at various locations, precipitation pH values of 4 or even lower were measured. Studies showed that this low pH rainfall, especially in western Europe and eastern North America, was causing acidification of lakes, affecting growth in forests, and causing damage to stone buildings and other structures. This was a much discussed subject of the 1960s and 1970s and the research, publicity, and political pressure led to positive action on the part of many industries. While the problem is now much less severe in those areas, it has not been totally solved and has also moved on to other parts of the world. Furthermore, in terms of the environment, there is more to precipitation than its acidity. It is for this reason that we emphasize that the subject of this chapter is *precipitation chemistry*, not only acid rain. Importantly, also, there are other causes of environmental acidification beyond precipitation. This is a subject that will be discussed in Chapter 18.

Rain and snow are the forms of precipitation that are most familiar to all of us. In an environmental context, however, the term precipitation has a broader definition and can be subdivided into two categories. Precipitation by *wet deposition* refers to deposition on the Earth's surface of water-based particles in liquid or solid form—rain, snow, and various types of ice such as sleet and hail. In its wider meaning, precipitation also includes surface deposition of *dry* species, including gases such as sulphur dioxide and solid particles such as various components of dust. Without specifically using the term *dry deposition*, in the next chapter we will discuss the chemical and physical behaviour of dry species in the atmosphere. In this chapter, the emphasis is on wet deposition.

When water condenses to form clouds, various chemical species are incorporated into the cloud droplets and sometimes further transformed there. During a precipitation event, the accumulated species are carried to the Earth's surface. When chemicals accumulate and are removed from the atmosphere in this way, it is referred to as *rainout*, while *washout* is the term used when chemicals are taken up by water droplets as they pass though air below the clouds.

## 5.1 **The composition of rain**

If rainwater is in equilibrium with gaseous species in the air, it contains soluble forms of the atmospheric gases with concentrations determined by Henry's law. Among the gases is carbon dioxide, whose atmospheric mixing ratio in the Northern Hemisphere averaged approximately 387 ppmv in 2009. Dissolved carbon dioxide is a weak acid and an aqueous solution in equilibrium with the unpolluted atmosphere has a pH of 5.7. The significance of this value will become clear when we carry out calculations of gas solubilities in water and the effect on pH in Chapter 11. Rainwater containing dissolved carbon dioxide is therefore slightly acidic, with a hydronium ion concentration approximately 20 times greater than that of pure water. Since much of the carbon dioxide is of 'natural' origin, it is frequently suggested that the pH of unpolluted rain is 5.7.

In fact, other natural acid-producing species, including organic acids and sulphur compounds that are produced by microbiological processes from living and dead biomass, can be a source of enhanced acidity in rain in areas remote from human influence. Still other 'natural' chemicals

**Table 5.1** The composition of rain from several locations[a].

| | Concentration / $\mu$mol $L^{-1}$ in rain samples from | | | | |
|---|---|---|---|---|---|
| | Urban Guiyang, Guizhou, PRC[b] | Birkenes, Southern Norway[c] | Katherine, Northern Territories, Australia[d] | Pune, Maharashtra State, India[e] | St Georges, Bermuda[d] |
| $H^+$ | 112 | 57 | 16.6 | 0.04 | 16.2 |
| (pH) | (3.95) | (4.2) | (4.8) | (7.4) | (4.8) |
| $Cl^-$ | | 58 | 11.8 | 155 | 175 |
| $NO_3^-$ | 10.3 | 38 | 4.3 | 185 | 5.5 |
| $SO_4^{2-}$ | 222 | 68 | 6.3 | 11 | 36.3 |
| $Ca^{2+}$ | 128 | 9 | 2.5 | 55 | 9.7 |
| $Mg^{2+}$ | | 13 | 2.0 | 35 | 34.5 |
| $Na^+$ | | 56 | 7.0 | 150 | 147 |
| $K^+$ | | 4 | 0.9 | 36 | 4.3 |
| $NH_4^+$ | 57 | 38 | 2.4 | 28 | 3.8 |

[a] The sites in China and Norway are considered to contain anthropogenic-source chemical species. The others are influenced to a smaller or negligible extent by human activity.
[b] Dianwu, Z. and X. Jiling, *Acidification in southwestern China*, In *Acidification in tropical countries* (ed. Rodhe, H. and R. Herrera), John Wiley and Sons, Chichester; 1988. No data are provided for Cl, Mg, Na, and K.
[c] Overrein, L.N., H.M. Seip, and A. Tollan, *Acid precipitation—effects on forest and fish*, Final report of the SNSF project 1972–1980, Oslo; 1980.
[d] Legge, A.H. and S.V. Krupa, *Acid deposition, sulfur and nitrogen oxides*. Lewis Publishers, Chelsea, Michigan; 1990.

such as dust containing calcium carbonate, when incorporated into rain, can result in a mildly alkaline solution. Depending on the setting therefore, a variety of anionic and cationic species, some with acidic or basic properties, accumulate in rain and the distinction between natural and anthropogenic origin is not always clear. As a consequence of the combined factors, the pH of unpolluted rain might range from below 5.5 to 8 or higher. In any case, rain and snow are not pure water and their chemistry is determined by a range of complex, interrelated factors. The concentration of major species in rain taken at a variety of locations is given in Table 5.1.

The same dissolved species as those listed in the table are also important components of precipitation occurring at other locations around the globe. While individual species cannot be assigned with certainty to a specific origin, sodium and chloride are to a large extent derived from the oceans (as, for example, in the St. Georges, Bermuda rainfall). Accompanying sodium and chloride are somewhat smaller concentrations of other ions found in seawater including potassium, calcium, magnesium, and sulphate. In less-common situations, elevated concentrations of sodium chloride and / or calcium and chloride in the atmospheric aerosol originate from salts used to melt ice and snow on highways during the winter in countries like Russia and Canada. From dust, especially in areas with limestone bedrock, calcium and magnesium are incorporated into water droplets and, consequently, high levels of these elements are usually associated with a terrestrial origin. An example is the Pune, India sample. Ammonium ion originates from biological processes involving nitrogen from plant and animal residues and from inorganic fertilizers. As well as the principal species, smaller concentrations of other elements are incorporated into rain, especially in heavily industrialized regions.

---

### Example 5.1  Charge balance in a rainwater sample

As a check on the analytical results, it can be useful to calculate the charge balance in a water sample such as those given for the rain samples in Table 5.1.

For example, determine the magnitude of positive and negative charge associated with the major elements in the rain sample obtained at Birkenes, Norway.

| Positive (cationic) charge | Concentration of ion / $\mu$mol L$^{-1}$ | Concentration of charge / $\mu$mol L$^{-1}$ |
|---|---|---|
| H$^+$ | 57 | 57 |
| Ca$^{2+}$ | 9 | 18 |
| Mg$^{2+}$ | 13 | 26 |
| Na$^+$ | 56 | 56 |
| K$^+$ | 4 | 4 |
| NH$_4^+$ | 38 | 38 |
| Total | | 199 |

| Negative (anionic) charge | Concentration of ion / $\mu$mol L$^{-1}$ | Concentration of charge / $\mu$mol L$^{-1}$ |
|---|---|---|
| Cl$^-$ | 58 | 58 |
| NO$_3^-$ | 38 | 38 |
| SO$_4^{2-}$ | 68 | 136 |
| Total | | 232 |

In this case there is a $232 - 199 = 33$ $\mu$mol L$^{-1}$ or approximately 15% difference in the total concentrations of positive and negative charge. Given that this calculation is based on nine separate analytical measurements all at quite low concentrations, the difference is probably not significant. Furthermore, the deficit is on the positive ion side; there are no environmentally common cations that would be likely to account for the difference. In other cases where this kind of calculation is done, however, a major discrepancy in charge balance can indicate either a significant analytical error or that some important species in the sample has not been measured.

---

Since the sodium, potassium, calcium, and magnesium cations and the chloride anion are all common, of natural origin, and generally benign species at the $\mu mol\ L^{-1}$ concentrations found in precipitation, their presence has a relatively minor environmental significance. It is the sulphur and nitrogen species that generate acidity and are therefore of particular concern. To a large extent these species derive from anthropogenic sources and together they create the phenomenon known as acid precipitation. Table 5.2 lists sources and estimates of mixing ratios and residence times for sulphur and nitrogen species present in the atmosphere.

As with other compounds, distinctions between natural and anthropogenic sources are not clear, but there is no doubt that a major contribution of both sulphur and nitrogen atmospheric species arises from combustion and a variety of industrial processes. It is these emissions

**Table 5.2** Nitrogen and sulphur species present in the atmosphere.

| | Approximate atmospheric mixing ratio / ppbv[a] | Approximate residence time / days | Source |
|---|---|---|---|
| **Nitrogen species** | | | |
| Nitrogen oxides, $NO_x$ | 1– >10 (urban) 0.1–1 (remote) | 0.2 (urban, summer) to 10 (remote, winter) | Fossil fuel, biomass, combustion; lightning; microbiological release |
| Ammonia, $NH_3$ | 0.1–1 | 2–70 | Animal excreta, fertilizers, microbiological release |
| **Sulphur species** | | | |
| Sulphur dioxide, $SO_2$ | 0.01–0.3 | 3–5 | Fossil fuel, biomass combustion; sulphide ore smelting |
| Hydrogen sulphide, $H_2S$ | 0.05–0.3 | 1–2 | Submerged soils, wetlands |
| Carbon disulphide, $CS_2$ | 0.02–0.5 | $\approx 50$ | Submerged soils, wetlands |
| Dimethyl sulphide, $(CH_3)_2S$ | 0.01–0.07 | $\approx 1$ | Oceans |
| Carbonyl sulphide, COS | 0.3–0.5 | 200–2500 | Oceans, soils |
| Methyl mercaptan, $CH_3SH$ | | | Oceans, soils |
| Dimethyl disulphide, $CH_3SSCH_3$ | | | Oceans, soils |

[a] Mixing ratios are for unpolluted areas unless otherwise noted. Data are from various sources.

that ultimately contribute to enhanced production of the strong acids found in wet and dry precipitation.

---

**Main point 5.1** The forms of liquid and solid water in the atmosphere include rain, snow, and fog. These and other types of precipitation always contain various dissolved species derived from both natural and anthropogenic processes. Of these substances, acids produced from the oxides of nitrogen and sulphur are of widespread concern. Their presence in excessive amounts can cause the rainfall pH to be depressed to values as low as 4 or even less.

---

## 5.2 Atmospheric production and removal of nitric acid

The principal reaction sequence contributing to production of nitric acid is relatively straightforward and has actually been discussed in Chapters 3 and 4. It begins with nitric oxide emissions occurring primarily during combustion processes. For high-temperature combustion, most of the nitrogen originates from the atmosphere, but some can also be derived from organic nitrogen compounds in fuels such as wood. Smaller amounts of nitric oxide are released as a by-product of microbial nitrification in soil, a process that is enhanced in the high-temperature tropical environment. Lightning, which is also most frequent in the tropics, adds a further small input to the global nitric oxide budget.

### Daytime nitrogen oxide chemistry

Nitric oxide is oxidized by $O_2$, $O_3$, $\cdot OH$ or $ROO\cdot$ (where R is an alkyl group). For example

$$\cdot NO + O_3 \rightarrow \cdot NO_2 + O_2 \tag{5.1}$$

The nitrogen dioxide produced in this way subsequently contributes to ozone and hydroxyl radical production and therefore is responsible (in part) for the initiation of a photochemical smog sequence. In the process, nitric oxide is regenerated and is therefore available to once again contribute to additional ozone and smog production. Fortunately, the $NO_x$ compounds have only a limited atmospheric lifetime, or otherwise smog events would persist and increase. But, unfortunately, the principal mechanism for their removal from the atmosphere is through conversion to nitric acid via oxidation of nitrogen dioxide by the hydroxyl radical

$$\cdot NO_2 + \cdot OH + M \rightarrow HNO_3 + M \tag{5.2}$$

where M is a third body. The second-order rate constant for the reaction has a value of $1.2 \times 10^{-11}(T/298)^{-1.6}$ cm$^3$ molecule$^{-1}$ s$^{-1}$. Because reaction 5.2 involves starting materials formed in part through photochemical processes, it is largely a daytime reaction.

---

**Example 5.2 Conversion rate for nitrogen dioxide to nitric acid**

Assuming a constant hydroxyl radical concentration of $2 \times 10^6$ molecules cm$^{-3}$ and a temperature of 20°C, what is the half-life of $NO_2$ according to this reaction?

$$\text{Rate} = k_2[\cdot NO_2][\cdot OH], \text{ where } k_2 \text{ is the second-order rate constant}$$

$$= k_1[\cdot NO_2], \text{ where } k_1 = k_2[\cdot OH] = 1.2 \times 10^{-11}(293/298)^{-1.6} \times 2 \times 10^6$$

$$k_1 = 2.5 \times 10^{-5} \, s^{-1}$$

$$\text{Half-life} = \ln 2 / (2.5 \times 10^{-5}) = 2.8 \times 10^4 \, s = 7.8 \, h$$

---

## Night-time chemistry

After sunset, an alternative sequence becomes more important for producing nitric acid. It involves the nitrate radical, which is formed during both day and night but accumulates only at night-time because it is destroyed by photolysis. The reaction for nitrate-radical formation is

$$\bullet NO_2 + O_3 \rightarrow \bullet NO_3 + O_2 \tag{5.3}$$

The radical can take part in a number of reactions. To some extent, it is destroyed by reactions with the $NO_x$ compounds.

$$\bullet NO_3 + \bullet NO_2 \rightarrow \bullet NO + \bullet NO_2 + O_2 \tag{5.4}$$

$$\bullet NO_3 + \bullet NO \rightarrow 2\ NO_2 \tag{5.5}$$

In the same manner as hydroxyl, nitrate radical is able to add to the double bond of olefins.

$$\bullet NO_3 + C_nH_{2n} \rightarrow \bullet C_nH_{2n}NO_3 \tag{5.6}$$

Because the product is also a radical, it is susceptible to further reactions. Typically, addition of the nitrate radical is followed by rapid addition of oxygen.

Also like hydroxyl, the nitrate radical can initiate reaction sequences by first abstracting a hydrogen. In these cases, nitric acid is formed:

with aldehydes

$$\bullet NO_3 + RCHO \rightarrow R\overset{\bullet}{C}O + HNO_3 \tag{5.7}$$

with alkanes

$$\bullet NO_3 + RH \rightarrow \bullet R + HNO_3 \tag{5.8}$$

In both reactions 5.7 and 5.8, R is an alkyl group. The alkyl radicals that are produced take part in further reactions such as the addition of dioxygen.

A kinetically rapid pair of reactions involving the nitrate radical and nitrogen dioxide results in the formation of dinitrogen pentoxide and then nitric acid.

$$\bullet NO_3 + \bullet NO_2 \rightleftharpoons N_2O_5 \tag{5.9}$$

$$N_2O_5 + H_2O_2 \rightarrow 2HNO_3 \tag{5.10}$$

In many environmental situations, most of the $NO_x$ present in the atmosphere at night-time proceeds through reactions 5.3, 5.9, and 5.10 and this is an additional means of removing these smog-precursor species by producing nitric acid. A smaller fraction of $NO_x$ is finally converted to PAN by the reaction

$$RC(O)OO\bullet + \bullet NO_2 \rightarrow RC(O)O_2NO_2 \text{ (PAN)} \tag{5.11}$$

## Atmospheric processes for removal of nitric acid

Nitric acid is removed from the atmosphere by either wet or dry deposition, and is one of the main contributors to precipitation acidity. To some degree it also reacts with ammonia, whose principal source is volatilization from nitrogen fertilizers, urea in animal urine, and other organic reduced nitrogen sources.

$$NH_3 + HNO_3 \rightarrow NH_4NO_3 \tag{5.12}$$

The ammonium nitrate can act as a condensation nucleus, a site on which a water droplet can form, or it is deposited as part of the solid aerosol. This issue will be described more fully in Chapter 6.

---

**Main point 5.2** Nitric acid is one of the important acidifying components of precipitation. Much of it is produced by oxidation of the nitrogen oxides with ozone and hydroxyl radical being the principal oxidants.

---

## 5.3 **Atmospheric production and removal of sulphuric acid**

### Oxidation of reduced sulphur species

#### Initial oxidation reactions

The production of sulphuric acid is a more complex process than that of nitric acid. For one thing, the reaction sequence can begin with a wide range of reduced as well as partially oxidized sulphur compounds, mostly of natural origin. Hydrogen sulphide, carbonyl sulphide, carbon disulphide, methyl mercaptan, dimethyl sulphide, and dimethyl disulphide all contain sulphur in its lowest ($-2$) oxidation state. These compounds are released from the oceans and from soils under reducing conditions as a consequence of various microbiological processes. Some features of their production will be discussed in Chapter 15. Warm temperatures favour microbial activity and so the release of reduced sulphur compounds is especially significant in the tropics.

Once in the atmosphere, a sequence of reactions begins. Hydrogen sulphide, carbon disulphide, and carbonyl sulphide are oxidized via hydroxyl giving the thionyl radical ($\cdot$SH) as an initial product.

$$H_2S + \cdot OH \rightarrow H_2O + \cdot SH \tag{5.13}$$

$$CS_2 + \cdot OH \rightarrow COS + \cdot SH \tag{5.14}$$

$$COS + \cdot OH \rightarrow CO_2 + \cdot SH \tag{5.15}$$

Of the three sulphide compounds, hydrogen sulphide and carbon disulphide react quickly but carbonyl sulphide (COS), released directly from the oceans or produced by oxidation of carbon disulphide, is in kinetic terms relatively stable with respect to the oxidation reaction (5.15) shown.

Further oxidation of thionyl leads to production of sulphur dioxide.

$$\cdot SH + O_2 \rightarrow SO + \cdot OH \tag{5.16}$$

$$\cdot SH + O_3 \rightarrow SHO\cdot + O_2 \tag{5.17}$$

$$SHO\cdot + O_2 \rightarrow SO + HOO\cdot \tag{5.18}$$

$$SO + \begin{array}{c} O_2 \\ O_3 \\ NO_2 \end{array} \bigg| \rightarrow SO_2 + \text{other products} \tag{5.19}$$

Some prominent phytoplankton groups living in surface waters of the oceans produce a large amount of dimethylsulphoniopropionate (DMSP). Through algal or bacterial enzyme-mediated processes, the DMSP is subsequently degraded, with dimethyl sulphide (DMS) being one of the products. Only a small portion of the dimethyl sulphide is released into the atmosphere, but in quantity, this is one of the most important sources of reduced sulphur compounds. In the future, the amount may increase as algal blooms proliferate due to greater nutrient pollution in the oceans and warming of the global climate. The hydroxyl radical reacts with dimethyl sulphide by hydrogen abstraction or by addition.

$$(CH_3)_2S + \cdot OH \quad \begin{array}{c} \rightarrow \\ \\ \rightarrow \end{array} \bigg| \begin{array}{l} CH_3\dot{S}CH_2 + H_2O \\ \\ CH_3\dot{S}(OH)CH_3 \end{array} \rightarrow \text{further oxidation products} \tag{5.20}$$

The further oxidation products include dimethyl sulphoxide and methane sulphonic acid, both of which have been found in the marine atmospheric aerosol. The pathways and mechanisms for the complex reaction sequences have not been clearly mapped out.

By means of these processes, any reduced sulphur compounds can be oxidized, with one of the principal products being sulphur dioxide. The sulphur dioxide is ultimately converted to sulphuric acid. In areas of the world remote from human activity the emitted reduced sulphur compounds, dimethyl sulphide in particular, are a major source of sulphur dioxide that produces excess acidity in atmospheric aerosols and precipitation.

Sulphur dioxide is also released in large quantities directly into the atmosphere as a by-product of sulphide ore smelting and fossil-fuel combustion. It is this additional anthropogenic source that gives rise to even greater acidity in some continental regions of the globe that are influenced by intense industrial activity.

### Oxidation of sulphur dioxide by homogeneous reactions

Sulphuric acid production from sulphur dioxide takes place by at least two distinct sets of processes. The first sequence occurs homogeneously in the gas phase and most frequently begins with reaction 5.21 as the rate-determining step.

$$SO_2 + \bullet OH + M \rightarrow HOSO_2^\bullet + M \tag{5.21}$$

The reaction as shown is third order but in the lower troposphere where the concentration of the 'third body' M—mostly dinitrogen and dioxygen—is large, it becomes pseudo-second-order. The second-order rate constant therefore would decrease with decreasing pressure moving to higher altitudes in the troposphere and stratosphere. However, a second factor must be considered in determining the rate constant. A feature of this and many other radical–radical and ion–molecule reactions involving simple species is that the activation energy is negative. Therefore, the decrease in temperature with altitude in the troposphere by itself would result in a corresponding increase in the rate constant for the reaction. Combining both temperature and pressure factors, best estimates of the second-order rate constant are shown in Fig. 5.1.

The $HOSO_2^\bullet$ radical can then undergo a number of relatively rapid reactions, some of which result in sulphuric acid production. The simplest and most important acid-producing process is

$$HOSO_2^\bullet + O_2 + M \rightarrow HOO\bullet + SO_3 + M \tag{5.22}$$

This is followed by dissolution in water to form sulphuric acid.

$$SO_3 + H_2O \rightarrow H_2SO_4 \tag{5.23}$$

The hydroperoxyl radical produced in the sequence also reacts with nitric oxide

$$NO + HOO\bullet \rightarrow NO_2 + \bullet OH \tag{5.24}$$

and the nitrogen dioxide and hydroxyl radical take part in the nitric-acid-generating sequence

$$\bullet NO_2 + \bullet OH + M \rightarrow HNO_3 + M \tag{5.2}$$

described in the previous section. Some of the hydroxyl radicals react with additional sulphur dioxide and so the set of reactions 5.21–5.24 is a self-accelerating sequence.

Starting again with sulphur dioxide, there are other quantitatively less important homogeneous reaction series that produce sulphuric acid, including direct reaction with atomic oxygen. The rate constant for the reaction between sulphur dioxide and atomic oxygen is similar to that for the reaction with hydroxyl radical, but the atomic oxygen tropospheric mixing ratio is approximately two orders of magnitude below that of hydroxyl.

### Oxidation of sulphur dioxide by heterogeneous reactions

In Chapter 3, we encountered a sequence of heterogeneous reactions that leads to large-scale losses of stratospheric ozone during the polar spring. Sulphuric acid can also be produced in a heterogeneous process when the required reactants are available in cloud droplets.

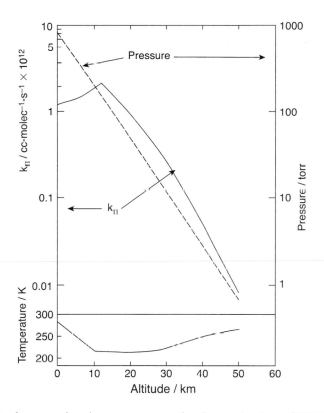

**Fig. 5.1** Variation in the second-order rate constant for the production of $HOSO_2^*$ as a function of altitude in the troposphere and stratosphere. (Redrawn from Calvert, J.G. and W.R. Stockwell, Mechanisms and rates of the gas-phase oxidations of sulphur dioxide and nitrogen oxides in the atmosphere, In *Acidic deposition: sulphur and nitrogen oxides* (ed. A.H. Legge and S.V. Krupa), Lewis Publishers Inc., Chelsea, Michigan; 1990.).

Beginning again with sulphur dioxide, the following reactions occur.

$$SO_2\,(g) \rightleftharpoons SO_2\,(aq) \qquad\qquad K_H = 1.81 \times 10^{-5}\ mol\ L^{-1}\ Pa^{-1} \qquad (5.25)$$

$$SO_2\,(aq) + 2H_2O \rightleftharpoons HSO_3^-\,(aq) + H_3O^+\,(aq) \qquad K_{a1} = 1.72 \times 10^{-2}\ mol\ L^{-1} \qquad (5.26)$$

$$HSO_3^-\,(aq) + H_2O \rightleftharpoons SO_3^{2-}\,(aq) + H_3O^+\,(aq) \qquad K_{a2} = 6.43 \times 10^{-8}\ mol\ L^{-1} \qquad (5.27)$$

Like that of carbon dioxide, the aqueous solubility of sulphur dioxide is pH dependent, but sulphur dioxide is much more soluble throughout the entire pH range. If the atmospheric mixing ratio of sulphur dioxide is 10 ppbv at $P^o$, the solubility is $2.2 \times 10^{-6}$ mol $L^{-1}$ when the aerosol has a pH of 4.0, and is $2.2 \times 10^{-3}$ mol $L^{-1}$ when the pH is 7.0. The solubility is one of the factors affecting the rate of the heterogeneous reaction. Oxidation of sulphur species takes place within the water droplets. The most important oxidant is hydrogen peroxide, a chemical whose atmospheric mixing ratio is around 1 or 2 ppbv and which is readily soluble in water ($K_H = 7.0 \times 10^{-1}$ mol $L^{-1}$ $Pa^{-1}$).

$$HSO_3^-\,(aq) + H_2O_2\,(aq) \underset{k_2}{\overset{k_1}{\rightleftharpoons}} HOOSO_2^-\,(aq) + H_2O \qquad (5.28)$$

Peroxymonosulphite, $HOOSO_2^-$, has a structure

and rapidly rearranges to form hydrogen sulphate, $HSO_4^-$, whose structure is

In protonated form, hydrogen sulphate is sulphuric acid. The combined rearrangement and protonation reaction is therefore

$$HOOSO_2^- \text{ (aq)} + H_3O^+ \text{ (aq)} \xrightarrow{k_3} H_2SO_4 \text{ (aq)} + H_2O \tag{5.29}$$

A simplified description of sulphuric-acid formation starting from dimethyl sulphide (DMS) is shown in Fig. 5.2.

In the steady-state condition, the rate of production of sulphuric acid by hydrogen peroxide oxidation of sulphur dioxide is calculated in the following way. Based on reaction 5.29, the rate of production of sulphuric acid is given by

$$\frac{d[H_2SO_4]}{dt} = k_3[HOOSO_2^-][H_3O^+] \tag{5.30}$$

Assuming a steady-state concentration of $HSO_4^-$

$$\frac{d[HOOSO_2^-]}{dt} = 0 = k_1[HSO_3^-][H_2O_2] - k_2'[HOOSO_2^-] - k_3[HOOSO_2^-][H_3O^+] \tag{5.31}$$

In eqn 5.31, $k_2' = k_2[H_2O]$ is a pseudo-first-order rate constant since $[H_2O] \gg [HOOSO_2^-]$.

$$k_1[HSO_3^-][H_2O_2] = [HOOSO_2^-](k_2' + k_3[H_3O^+]) \tag{5.32}$$

$$[HOOSO_2^-] = \frac{k_1[HSO_3^-][H_2O_2]}{k_2' + k_3[H_3O^+]} \tag{5.33}$$

Substituting eqn 5.33 into eqn 5.30,

$$\text{rate} = \frac{d[H_2SO_4]}{dt} = \frac{k_1k_3[HSO_3^-][H_2O_2][H_3O^+]}{k_2' + k_3[H_3O^+]} \tag{5.34}$$

The values of the rate constants are[1]

$$k_1 = 5.2 \times 10^6 \text{ L mol}^{-1}\text{ s}^{-1}$$

$$k_2' / k_3 = 10^{-1}$$

When the pH is greater than 2, $k_3[H_3O^+] \ll k_2'$ and the rate is given by

$$\text{rate} = \frac{k_1k_3}{k_2'} [HSO_3^-][H_2O_2][H_3O^+] \tag{5.35}$$

Using the equations for $K_H$ (see Chapter 11) and $K_{a1}$,

$$[H_3O^+][HSO_3^-] = K_H K_{a1} P_{SO_2} \tag{5.36}$$

---

[1] Martin, L.R., *Kinetic studies of sulphite oxidation in aqueous solution, in SO₂, NO and NO₂ oxidation mechanisms: atmospheric considerations* (ed. J.G. Calvert), Butterworth Publishers, Boston; 1984.

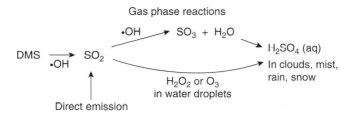

**Fig. 5.2** Summary of principal reactions that lead to production of sulphuric acid in atmospheric water droplets.

$$\text{rate} = \frac{k_1 k_3 K_H K_{al}}{k_2'} [H_2O_2]P_{SO_2} \tag{5.37}$$

$$= k'[H_2O_2]P_{SO_2} \tag{5.38}$$

This rate law applies between pH approximately 2 to 5 where hydrogen sulphite ($HSO_3^-$) is the principal aqueous sulphur (IV) species. Within this range, the oxidation of sulphur dioxide by hydrogen peroxide is the dominant mechanism and the reaction rate is approximately independent of pH. Below pH 2 the rate decreases as reflected by an increase in the denominator of eqn 5.34. At higher pH values, sulphite ($SO_3^{2-}$) becomes the dominant sulphur species and, because it does not react with hydrogen peroxide, the rate of oxidation via this oxidant again decreases.

A second heterogeneous pathway involves ozone as the oxidant. In this case both hydrogen sulphite and sulphite ions are oxidizable.

$$HSO_3^- (aq) + O_3 \rightarrow SO_4^{2-} (aq) + H_3O^+ (aq) + O_2 \tag{5.39}$$

$$SO_3^{2-} (aq) + O_3 \rightarrow SO_4^{2-} (aq) + O_2 \tag{5.40}$$

The rates of the ozone-based reactions are such that, taken together with the previous processes, below pH 5.5, hydrogen peroxide is the primary oxidant while, above that value, ozone becomes more important. Depending on the availability of oxidants, the heterogeneous reactions can make a greater contribution to sulphur dioxide oxidation than do the gas-phase processes.

### Catalytic enhancement of oxidation of sulphur dioxide

Ozone and hydrogen peroxide rapidly oxidize sulphur dioxide in water droplets. On the other hand, molecular oxygen is able to oxidize sulphite only very slowly in aqueous solution. However, the presence of small amounts of some metal ions catalyses the reaction. Metals that have been shown to increase the rate of the reaction include iron (II) and (III), manganese (II), copper (II), and cobalt (III). In acidified water, small concentrations of these metals are soluble and the soluble species are the catalytic agents. Furthermore, even in higher pH situations where metals such as iron (III) are very insoluble, surface-based catalysis still occurs. Other solids including carbon particles have also been found to increase the rate of oxidation by molecular oxygen. Even when catalysed, however, oxidation of sulphur dioxide by dioxygen in aqueous solution probably makes a relatively small contribution compared to other oxidation routes. Metal ions may also enhance the rate of heterogeneous reaction of sulphur dioxide by hydrogen peroxide and ozone.

## Atmospheric processes for removal of sulphuric acid and its precursors

After the various sulphur compounds have been oxidized and converted into sulphuric acid, the acid is dissolved in cloud water droplets and rained out or washed out of the atmosphere. But also, like its nitric acid counterpart, some sulphuric acid reacts with ammonia and forms two compounds that then become part of the atmospheric aerosol where they act as sites for the condensation of water vapour.

$$NH_3 + H_2SO_4 \rightarrow NH_4HSO_4 \tag{5.41}$$

$$2NH_3 + H_2SO_4 \rightarrow (NH_4)_2SO_4 \tag{5.42}$$

In this way, some of the acid in rainfall is neutralized. In most global situations, especially those in industrial areas of the world, there is an excess of acid over the ammonia base, so that the residual acidity remains high.

A second removal mechanism applies to long-lived reduced sulphur compounds. We noted earlier that carbonyl sulphide is quite stable to oxidation by the hydroxyl radical and has an atmospheric lifetime estimated to be between 0.5 and 7 years. As a consequence, diffusion of this particular sulphur-containing species into the stratosphere is an important removal process. In the stratosphere, it can undergo photochemical oxidation to produce sulphur dioxide and ultimately sulphate anion, which is an important component of the stratospheric aerosol.

### Volcanoes—the 1991 eruption of Mount Pinatubo

Volcanic activity can be a means by which sulphur dioxide is injected directly into the stratosphere where, in part through ozone-related reactions, it becomes sulphuric acid and increases the concentration of the aerosol there. This can lead to a significant depression of the global temperature by physically blocking solar radiation. The sulphur compounds in the stratosphere also take part in other chemical processes and this has diverse environmental consequences. For example, consider the case of the Mount Pinatubo volcanic eruption.

Mount Pinatubo is a volcanic mountain located about 100 km north-west of Manila in the Philippines. After being quiescent for 635 years, it erupted dramatically in June 1991, with peak releases on 14 and 15 June. Approximately 7 km$^3$ of magma was expelled as lava, and as ash into the atmosphere. There was heavy ashfall up to 40 km from the volcano, and the typhoon Yunya that occurred shortly after the volcanic eruption carried solids, depositing them as far away as Thailand and Singapore. The ash—a calc-alkaline pumice—contained phenocrysts of anhydrite (CaSO$_4$), indicating that there were high concentrations of sulphur in the magma.

Besides ash, there was a release of large quantities of gases with composition in the carbon–oxygen–hydrogen–sulphur family. Principal gases included water vapour and carbon dioxide along with about 20 Mt of sulphur dioxide (This amount [10 Mt expressed as sulphur] is about one-tenth of the annual global anthropogenic release of sulphur dioxide; see Table 5.5.) The gases were injected into the stratosphere at an altitude of 20 to 30 km, and the sulphur dioxide over time was converted to a sulphuric acid aqueous aerosol. The aerosol cloud drifted to the north-west and could be observed as far away as the Greenland–Iceland area in early January 1992. Various processes caused the cloud to disperse over a 1–3 year period.

There have been a number of environmental effects attributed to the aerosol. The dispersed cloud particles blocked solar radiation and a measurable, but not uniform, average global cooling was observed in the 2 years following the eruption. There is also evidence that ozone loss was accelerated, especially within the polar vortexes. This is attributed to enhanced conversion of dinitrogen pentoxide to nitric acid (reaction 5.10), a reaction that occurs readily on the surface of the sulphuric acid–ice crystals. By removing NO$_x$ from the stratosphere there was reduced tendency for nitrogen dioxide to react with chlorine monoxide (reaction 3.54). This, in turn, augmented the chlorine catalytic cycle for ozone destruction and caused a reduction of ozone concentrations.

The amount of sulphur dioxide released from Pinatubo was too large to have occurred only by exsolution (release out of solution) from the magma at the time of the eruption; it is believed that pre-eruption releases of large amounts of vapour also occurred. Such emissions occur regularly at other terrestrial and marine sites, and are not always associated with catastrophic volcanic eruptions. As we noted at the beginning of the book, this type of release of gases has occurred throughout the entire expanse of Earth history. This led to the formation of our planet's unique atmosphere. Much of the water on Earth was derived in this way.

Main point 5.3 The sulphuric acid in precipitation originates from a number of chemical precursors including sulphur dioxide emitted directly into the atmosphere, usually from anthropogenic sources, and natural-origin reduced sulphur compounds. The latter are oxidized to sulphur dioxide with hydroxyl radical as the primary oxidizing agent. The sulphur dioxide dissolves in water droplets and is then further oxidized to sulphuric acid via several homogeneous and heterogeneous processes.

## 5.4 **Acidifying agents in precipitation**

The major ions present in precipitation are well-known and are listed in Table 5.1. As noted, sodium, potassium, calcium, and magnesium ions are cations of strong bases, and chloride, nitrate, and sulphate are anions of strong acids. As such, all these species are themselves neutral and therefore the only major ions that perturb the acid–base balance of the water are ammonium and hydronium ion itself. The activity of hydronium ion, of course, directly determines the solution pH. Since ammonium is a very weak acid ($pK_a = 9.25$), in the presence of even a small excess of hydronium ion, it is unable to donate protons and has a negligible effect on the precipitation pH. This is not to say that ammonium lacks acid-producing capability. On the contrary, when deposited in soil or water under aerobic conditions, microbial oxidation of ammonium takes place. This produces nitrate and in the process generates two hydronium ions for each ammonium molecule.

$$NH_4^+ \text{ (aq)} + 2O_2 + H_2O \xrightarrow{\text{microorganisms}} NO_3^- \text{ (aq)} + 2H_3O^+ \text{ (aq)} \qquad (5.43)$$

In this indirect way, ammonium in precipitation is a potent contributor to acidification. More will be said about this very important reaction called nitrification in Chapter 18 when we deal with the chemistry of soil processes.

We have indicated that the hydronium ion in rain is associated with either nitric or sulphuric acids produced by mechanisms described above. If these two components were the only sources, there should be a good correlation between hydronium ion concentration and that of sulphate and / or nitrate. Of the numerous studies designed to examine the issue, many do show excellent correlations (correlation coefficient > 0.8) for one or other of the relations. But there are a number of exceptions indicating that additional factors, including meteorological anomalies, other emissions, and proximity to source, overrule any simplified relation. A good example is in the prairie regions of western North America where hydronium ion is best correlated—in an inverse fashion—to calcium ion, indicating terrestrial control of precipitation acid–base balance through calcium carbonate and other alkaline soil minerals.

The relative importance of nitric and sulphuric acids depends on distance from source because the rate of conversion of $NO_x$ to nitric acid, and its deposition velocity is greater than corresponding rates for sulphuric acid. This can be shown by the following example that considers the rate-determining step for the two principal homogeneous reaction processes.

### Example 5.3 **Rates of oxidation of NO$_2$ and SO$_2$**

Consider a situation where the atmospheric concentrations of nitrogen dioxide and sulphur dioxide are, respectively, 50 and 25 µg m$^{-3}$. These are typical concentrations observed in heavily industrialized areas such as in western Europe and eastern North America. A reasonable average 24-h value for the concentration of hydroxyl radical during the summer months is $1.7 \times 10^6$ molec cm$^{-3}$. The relevant pseudo-second-order rate constants at the Earth's surface for reactions 5.2 and 5.21 are $1.2 \times 10^{-11}$ cm$^3$ molec$^{-1}$ s$^{-1}$ and $1.2 \times 10^{-12}$ cm$^3$ molec$^{-1}$ s$^{-1}$, both estimates made at $P°$ and 25°C.

The atmospheric concentration of nitrogen dioxide is first converted into units of molec cm$^{-3}$ to be consistent with the units of the rate law

$$50 \text{ µg m}^{-3} = \frac{50 \times 10^{-6} \text{ g m}^{-3}}{46 \text{ g mol}^{-1}} \times 6.02 \times 10^{23} \times \text{molec mol}^{-1} \times 10^{-6} \text{ m}^3 \text{ cm}^{-3}$$

$$[NO_2] = 6.5 \times 10^{11} \text{ molec cm}^{-3}$$

Likewise, for sulphur dioxide the concentration is

$$25 \ \mu g \ m^{-3} = \frac{25 \times 10^{-6} \ g \ m^{-3}}{64 \ g \ mol^{-1}} \times 6.02 \times 10^{23} \ molec \ mol^{-1} \times 10^{-6} \ m^3 \ cm^{-3}$$

$$[SO_2] = 2.4 \times 10^{11} \ molec \ cm^{-3}$$

For nitrogen dioxide, the rate of oxidation is

$$\frac{-d[NO_2]}{dt} = k_{NO_2} \ [\cdot NO_2][\cdot OH]$$

$$= 1.2 \times 10^{-11} \ cm^3 \ molec^{-1} \ s^{-1}$$

$$\times 6.5 \times 10^{11} \ molec \ cm^{-3} \times 1.7 \times 10^6 \ molec \ cm^{-3}$$

$$= 1.3 \times 10^7 \ molec \ cm^{-3} \ s^{-1}$$

$$= 4.8 \times 10^{10} \ molec \ cm^{-3} \ h^{-1}$$

This last figure represents a loss rate of approximately 7% of the original concentration of nitrogen dioxide in one hour.

For sulphur dioxide, a similar calculation is as follows:

$$\text{rate of oxidation} - \frac{-d[SO_2]}{dt} - k_{SO_2}[SO_2][\cdot OH]$$

$$= 1.2 \times 10^{-12} \ cm^3 \ molec^{-1} \ s^{-1} \times 2.4$$

$$\times 10^{11} \ molec \ cm^{-3} \times 1.7 \times 10^6 \ molec \ cm^{-3}$$

$$= 4.9 \times 10^5 \ mole \ cm^{-3} \ s^{-1}$$

$$= 1.8 \times 10^9 \ molec \ cm^{-3} \ h^{-1}$$

The initial loss rate of sulphur dioxide is therefore approximately 0.7% of the original concentration in one hour.

Note that we have only considered the homogeneous oxidation processes in this calculation.

---

Taking into account these and other oxidation pathways, it is found that, when both sulphur and nitrogen oxides are emitted from an industrial region, and the emissions along with transformation products move downwind together, the molar ratio of sulphate to nitrate increases away from the source. For example, the ratio is approximately 1:1 to 1.5:1 in the Netherlands, a level indicative of the industrial heartland of western Europe. Moving north-east with prevailing winds, the ratio becomes 2:1 in southern Scandinavia and as high as 5:1 in northern Scandinavia. The ratio increase with time and distance is clear evidence that sulphuric acid is formed more slowly than nitric acid.

> **Main point 5.4** Rainfall composition at any place on the Earth depends on the nature of emissions from various sources, distance from the sources, patterns of air movement, and availability of oxidizing agents and other reactants.

## 5.5 Rain, fog, and snow chemistry—similarities and differences

### Rain

The chemical composition of rain is highly variable depending on the geographic location and the influence of natural and anthropogenic chemical processes on the atmosphere in that region. We have emphasized the role of nitrogen and sulphur compounds in determining the acidity of rain and other precipitation forms. Table 5.1 reported the major element composition of rain at several sites. Additional elements (including metals) are present in rain in trace amounts, again depending on location. To some extent the trace components are derived from soil and other dust particles that act as nuclei around which water condenses to form cloud droplets. The solubility of metals from such sources depends on the nature of the metal and the original form in which it was present. Metals associated with a silicate mineral matrix are almost completely insoluble. Where insoluble iron (III) and aluminium hydrous oxide particles (also commonly of terrestrial origin) are found in water droplets, they act as scavengers by adsorbing various chemical species on their surface. Elevated amounts of these solids therefore can suppress the solubility of other metals. The rainwater pH is another factor that controls metal solubility, with most metals becoming more soluble under more acid conditions.

## Fog

As in rain, water droplets in fog and mist also contain chemical species accumulated from the atmosphere. The composition is similar to that of rain, but concentrations tend to be higher in fog because of its location near the Earth's surface where levels of contaminating gases and other species are usually greater. One of the foggiest areas in the world is the Bay of Fundy on the Atlantic coast of Eastern Canada between Nova Scotia and New Brunswick. During the summer season from April to October, warm air is drawn into the bay from the south. Passing over the cold ocean water, it is chilled, causing condensation and frequent heavy fog conditions. During this season, areas adjacent to the Bay of Fundy are subjected to fog for 12–30% of the time, sometimes for 3 to 5 days continuously at a stretch.

There is concern that birch trees growing in the forests adjacent to the bay would be adversely affected by prolonged exposure to the acid-containing fog. In one study, analysis of fog composition along a 37.5-km transect inland from the coast has been carried out. The volume-weighted mean concentrations (over the 1987 growing season) of major constituents at all five sites are given in Table 5.3.

These average concentrations are greater than those that have been measured in many rainfall samples (Table 5.1). By following the trend of values beginning at the coast and moving inland, it was found that the concentrations of most of the dissolved ions increased steadily and this was attributed to evaporation of the aqueous solvent. Increases were largest for hydrogen ion and for sulphate, probably an indication of an additional cause—rapid heterogeneous oxidation of dissolved sulphur dioxide in the aqueous aerosol during the time taken for the fog to drift away from the ocean. The oxidant may have been ozone or alkyl peroxides, both produced by reactions involving the hydrocarbon emissions from the forest.

## Snow

The chemistry of snow must be considered in two aspects. First is the nature of the *snowfall* as precipitation—i.e. its composition at the time it is deposited on the surface of the Earth. Secondly, because snow frequently remains on the ground for extended periods of time, it is subject to further inputs from the atmosphere by wet and dry deposition processes. Therefore, we must also consider the chemistry of the *snowpack* as an accumulated deposit.

**Table 5.3** Concentrations of major constituents of fog near the Bay of Fundy, Canada[a].

| Species | $H^+$ | $Na^+$ | $K^+$ | $Ca^{2+}$ | $Mg^{2+}$ | $NH_4^+$ | $Cl^-$ | $NO_3^-$ | $SO_4^{2-}$ |
|---------|-------|--------|-------|-----------|-----------|----------|--------|----------|-------------|
| Conc / $\mu mol\ L^{-1}$ | 330 (pH = 3.5) | 78 | 31 | 13 | 11 | 50 | 61 | 160 | 245 |

[a] Volume-weighted mean values from five sites, taken between April and October, 1987. (Cox, R.M., J. Spavold-Tims, and R.N. Hughes, Acid fog and ozone: their possible role in birch deterioration around the Bay of Fundy, Canada, *Water, Air, Soil Pollution*, **48** (1989), 263–76.)

**Table 5.4** Composition of two fresh snow samples (bracketed values are standard deviations).

| Location | Concentration / μmol L$^{-1}$ | | | | | | | |
|---|---|---|---|---|---|---|---|---|
| | $H_3O^+$ | $Na^+$ | $K^+, Ca^{2+}$ | $NH_4^+$ | $Mg^{2+}$ | $Cl^-$ | $NO_3^-$ | $SO_4^{2-}$ |
| Antarctica surface snow[a] | 1.52 (0.60) pH, 5.82 | 0.64 (0.26) | Not detected | 0.11 (0.04) | 0.073 (0.030) | 0.84 (0.31) | 0.82 (0.35) | 0.26 (0.09) |
| Ciste Mhearad Scotland[b] | 279 (31) pH, 3.55 | | | | | 13 (5) | 23 (9) | 86 (25) |

[a] The Antarctic results are from 14 samples of surface snow taken along a transect moving inland from 100 to 430 km in Terre Adelie. From Legrand M. and R.J. Delmas, Spatial and temporal variations of snow chemistry in Terre Adelie (East Antarctica), *Ann. Glaciol.*, **7** (1985), 20–5.

[b] The limited data from Scotland are based on 15 samples taken along a 700-m transect. From Brimblecombe P., M. Tranter, P.W. Abrahams, I. Blackwood, T.D. Davies, and C.E. Vincent, Relocation and preferential elution of acidic solute through the snowpack of a small, remote, high-altitude Scottish catchment, *Ann. Glaciol.*, **7** (1985), 141–7.

Table 5.4 shows concentrations of ionic species in freshly fallen snow in the Antarctic and in Scotland. The extremely low values found in the Antarctic samples are indicative of a location remote from anthropogenic sources of these ions. In both cases, for samples relatively close to one another, considerable variability in concentrations was observed in spite of there being no obvious local influences that should affect the results. The inhomogeneity indicates different rates of atmospheric scavenging of ions over space and time and / or the effects of lateral movement of snow due to wind action.

At other locations around the Earth, there is even more variability in the chemical composition of snow samples.

Snow remaining on the ground over the winter season is subject to chemical alteration due to inputs from dry and wet deposition influenced by urban or industrial sources as well as natural organic debris—the latter is especially important in forested areas. The alterations of composition while snow is on the ground contribute to spatial variability in physical and chemical composition and, perhaps more importantly, can influence the melting processes and the nature of the meltwater.

Aside from the additional deposits, during the winter season, snow undergoes *metamorphosis* with individual particles coalescing and recrystallizing into larger grains.[2] As part of this process, the solute ions are partially excluded from the ice crystal lattice and tend to migrate to the crystal surfaces. In winter it is also normal that there are periods of partial melting of the snowpack due to higher temperatures and / or to exposure to intense sunlight. During these periods, the early melt fractions encounter the surface impurities and dissolve them in the meltwater, leaving reduced concentrations in the remaining snow. Such events may occur several times before complete melting in the *spring run-off*. The total amount of solute available for dissolution declines as winter proceeds but, during each event, an initial flux of high-concentration solution is produced. For the major precipitation anions, preferential elution occurs in the order: sulphate > nitrate > chloride. This means that early meltwater is enriched in sulphate. The snowpack is generally depleted of ions but in relative terms is enriched in chloride—which will then be carried away in the final meltwater.

The combination of additions of organic and inorganic species to the snowpack by wet and dry deposition, and removal of chemical species by midwinter melting means that snowpack

[2] Jeffries, D.S., Snowpack storage of pollutants, release during melting, and impact on receiving waters, In *Acidic precipitation*, Vol. 4 (ed. Norton, S.A., S.E. Lindberg, and A.L. Page), Springer-Verlag, New York; 1989.

chemistry inevitably changes over the season. For any particular species, depending on location and winter climate, the snow concentration may show an increase or a decrease.

In later parts of the book (Chapters 11 and 18), we will discuss the ability of water and soil to neutralize inputs of acid. Where acid is provided continuously and slowly over extended periods of time, depending on the nature of the soil, it may be neutralized. However, a sudden release of acids into the aqueous or terrestrial environment can lead to a phenomenon known as *acid shock*. This term is used to refer to the large flux of water that passes over and through the soil when the final spring thaw occurs. Because the rapid release means there is limited opportunity for contact with the soil, a large proportion of the dissolved ions remain unreacted and end up directly in surface or groundwater. Although meltwater released at that time may not have as high a concentration of some species as was present in earlier releases, the great volume of water ensures that the soil, rivers, and lakes that receive the water are also receiving large amounts of cations (including hydronium ion), anions, and organic species. The sudden influx can have a major effect on water, soil, and biota.

> **Main point 5.5** Various chemical species are present in small concentrations in rain, snow, and fog. Most of these species are derived from natural sources; anthropogenic influences in certain local environments may, however, add significantly to the occurrence of similar species.

## 5.6 **The global picture**

### Sources and sinks

It is very difficult to make quantitative estimates of all the natural and anthropogenic sources and sinks of nitrogen and sulphur compounds found in the atmosphere, and amounts measured and calculated by different researchers show large variability. The data reported in Table 5.5 are estimates made for the Earth as a whole over the past two decades and give some sense of the major processes affecting atmospheric concentrations. The ultimate sinks for these compounds are through deposition onto water and soil; we will see some of the effects of this later.

We have noted that there are a number of reduced sulphur species emitted from the oceans that contribute to aerosol and precipitation acidity. Sea-salt particles also contain sulphur in the form of alkali and alkaline-earth metal sulphates. These particles act as condensation nuclei for water and play an important role in cloud and fog formation. For the most part, however, the sulphate is a chemically unreactive species.

Of the compounds that are precursors of sulphuric and nitric acids, about half are emitted as a result of human activities, mostly related to energy production, while the other half arise from naturally occurring geochemical and biogenic processes. There are two major anthropogenic sulphur dioxide sources—emissions from the smelting of sulphide-based ores and the combustion of fossil fuels. The former sources include production of copper, nickel, lead, and zinc, which are frequently found as metal sulphide minerals.

One well-known case is that of the high-grade nickel–copper ores at Sudbury, Ontario, Canada. In

---

**Sources and sinks of tropospheric gases**

- Sources: Primary sources—the chemical is released directly into the atmosphere.

   Secondary sources—atmospheric reactions involving primary emission products produce the chemical of interest.

- Sinks: Chemicals can be removed from the troposphere by deposition onto land or water or by leaking into the stratosphere. They may also be 'removed' by reactions to form other species.

**Table 5.5** Major sources of nitrogen and sulphur compounds in the atmosphere.

| Nitrogen compounds | N / g × $10^{-12}$ $y^{-1}$ | | | | Sulphur compounds | S / g × $10^{-12}$ $y^{-1}$ | | |
|---|---|---|---|---|---|---|---|---|
| | a | b | c | d | | e | f | g |
| **$NH_3$** | | | | | **Solid species, mostly $SO_4^{2-}$** | | | |
| Biogenic Volatilization | 122 | | | | Sea salt | 44 | | |
| **$NO_x$** | | | | | Dust | 20 | | |
| From stratosphere | 1 | | | 0.6 | **Reduced sulphur** | | | |
| Atmospheric oxidation of $NH_3$ | 1 | | | 0.9 | Biogenic (oceans and land) | 98 | | |
| Lightning | 5 | 10 | | 5 | DMS | | 19 | 15.5 |
| Biogenic | 8 | | | 7 | **Partially oxidized sulphur** | | | |
| Biomass combustion | 12 | | | 8 | Volcanoes (average) | 5 | 13.4 | |
| * Fossil-fuel combustion | 20 | 21 | 34* | 22 | Fossil-fuel combustion / smelting | 104 | 78 | 67 |
| **Total $NO_x$** | | | 52 | 44 | | | | |

[a] Jaffe, D.A., The nitrogen cycle. In *Global biogeochemical cycles*, (eds. Butcher, S.S., R.J. Charlson, G.H. Orians, and G.V. Wolfe), Academic Press, London; 1991.
[b] Mosier, R., Bleken, M.A., Chaiwanakupt, P., Ellis, E.C., Freney, J.R., Howarth, R.B., Matson, P.A., Minami, K., Naylor, R., Weeks, K.N., Zhu, Z.L. Policy implications of human-accelerated nitrogen cycling *Biogeochemistry*, **57** (2002), 477–516.
[c] Moomaw, W.R. Energy, industry and nitrogen: Strategies for decreasing reactive nitrogen emissions. *Ambio* 31 (2002), 184–189.
[d] Lee, D. S., Köhler, I., Grobler, E., Rohrer, F., Sausen, R., Gallardo-Klenner, L., Olivier, J.G.J., Dentener, F.J., Bouwman A.F. Estimations of global no, emissions and their uncertainties. *Atmo. Environ.*, **31** (1997), 1735–1749.
[e] Scriven, R., What are the sources of acid rain? In Scottish Wildlife Trust, *Report of the acid rain inquiry*, Edinburgh; 1985.
[f] Gondwe, M., Krol, M., Gieskes, W., Klaassen, W., de Baar, H. The contribution of ocean-leaving DMS to the global atmospheric burdens of DMS, MSA, SO₂, and NSS $SO_4^=$ *Global Biogeochem. Cycles*, 17 (2003), 1056.
[g] Barth M.C., Rasch, P.J., Kiehl, J.T., Sulphur chemistry in the National Center for Atmospheric Research Community Climate Model: Description, evaluation, features, and sensitivity to aqueous chemistry. *J. Geophys. Res*, **105** (2000), 1387–1415.

the nineteenth century and the early part of the twentieth century, refining of the ore was done in open roasting beds, using much of the timber from forests in the local region, releasing huge quantities of sulphur dioxide at ground level. The ambient sulphur dioxide and the acid generated from it destroyed much of the remaining vegetation in the Sudbury area and the barren, shallow soils were eroded from the underlying bedrock. In recent years, stringent controls have reduced the emissions substantially. At present, there are smaller levels of emissions and the reduced amounts of sulphur dioxide are released through a 'superstack' 400 m above ground level. Of course, while the emissions that are now released have minimal effect in the local area, they still contribute to the regional and global budget of sulphur dioxide.

All fossil fuels contain some sulphur. Coal is considered to be the major source, with sulphur contents ranging from fractions of a per cent to 10% in some cases. Smaller amounts are present in liquid fuels. Gasoline may contain 10 to 500 ppm sulphur depending on its origin and the refining process. Concentrations in diesel fuel typically fall between 1000 and 5000 ppm. There can be large amounts of sulphur compounds in some natural gas (methane) supplies, but these are substantially removed in the refining process.

You will recall that combustion-produced nitrogen oxides always originate from atmospheric dinitrogen whenever the burning temperature is very high, as in internal combustion engines or in large-scale industrial units that burn fossil fuels. This is called *thermal* $NO_x$. When biomass is burned as in forest fires or for domestic heating and cooking purposes, however, the temperature is usually too low to oxidize substantial amounts of atmospheric nitrogen, yet nitric oxide is still released. In these cases, the nitrogen oxides are derived almost exclusively from the fuel itself and is called *fuel* $NO_x$. The amount released is then equal to the amount of nitrogen in the biomass multiplied by the conversion efficiency. This latter term is often around 10% and depends on combustion conditions; it can only be determined experimentally for each situation.

$$NO_x \text{ emitted} = \text{mass of biomass} \times \text{fraction of N in biomass} \times \text{conversion efficiency} \quad (5.44)$$

Table 5.6 gives values estimated for annual nitrogen oxide emissions associated with agricultural practices in various regions in the tropics. The total nitrogen oxides produced are calculated to be between 3.2 and 6.1 Tg y$^{-1}$, a significant part of the 12 Tg y$^{-1}$ attributed to biomass combustion (Table 5.5).

Most atmospheric ammonia is derived from biogenic sources (see Chapter 6); nevertheless some of these sources have origins closely related to humans. The cattle industry—both in terms of milk and meat production—has grown with the human population and ammonia produced from manure makes a large contribution to the atmospheric budget. A small portion of the volatilized ammonia is oxidized in the atmosphere and the rest (about 8 Tmol y$^{-1}$) is neutralized by atmospheric nitric and sulphuric acids of which 4 and 5.5 Tmol y$^{-1}$, respectively, are potentially produced from the atmospheric nitrogen and sulphur compounds. These acids are sufficient to completely neutralize 4 (via nitric acid) + 2 × 5.5 (via sulphuric acid) = 15 Tmol y$^{-1}$ of ammonia, and therefore, taking the Earth as a whole, there is excess acid that contributes to precipitation acidity.

If all anthropogenic releases of acidic and basic substances were eliminated, natural production of both might be in closer balance and the global average pH could be near the 5.7 value.

## Fermi question

China is rapidly expanding its production of electricity (see below). If it were to double its current production using coal-fired power plants, how would this affect the global budget of sulphur dioxide that is released annually to the atmosphere?

**Table 5.6** Annual emissions of nitrogen oxides from biomass burning associated with agricultural production in various tropical areas, based on fuel N and conversion efficiencies[a].

| | Total area / $10^{12}$ m$^2$ | Biomass burned / $10^{15}$ g dry matter y$^{-1}$ | Fuel N content / % | Conversion efficiency / % | Nitrogen oxide emission / $10^{12}$ g N y$^{-1}$ | Nitrogen oxide flux from total area / g N m$^{-2}$ y$^{-1}$ |
|---|---|---|---|---|---|---|
| Tropical forest | 15.9 | 0.8–2.0 | 1 | 13 | 1.0–2.6 | ~0.1 |
| Tropical woodland, shrubland, and grassland | 22.1 | 2.0–3.7 | 0.6 | 10 | 1.2–2.2 | ~0.1 |
| Agricultural land | 17.6 | 1.7–2.1 | 0.6 | 10 | 1.0–1.3 | ~0.1 |

[a] From Galbally, I.E. and R.W. Gillett, Processes regulating nitrogen compounds in the tropical atmosphere. In *Acidification in tropical countries* (ed. Rodhe, H. and R. Herrera), John Wiley and Sons, Chichester, 1988.

## Recent trends in release of acid-producing gases

The release into the atmosphere of sulphur and nitrogen compounds varies from country to country around the Earth and as a consequence, rainfall chemistry also varies. Over the previous century, about 60 or 70% of the anthropogenic emissions of both elements have come from sources in Europe and North America and some of the most severe acid precipitation problems occur in these continents. The average pH in parts of the highly industrialized world has been as low as 4.0, and individual events containing even more acidic precipitation have been observed. Deposition rates of sulphate and nitrate are correspondingly high in and near the same industrialized regions. Figure 5.3 shows turn-of-the-century sulphur emissions in Europe expressed in two ways—total mass of sulphur and mass per unit area of land. Sulphur emissions, in the form of sulphur dioxide, are documented in part because of the importance of this gas as a precursor of acid rain. In addition, sulphur depositions from rainfall, mostly as sulphate, can have adverse effects on soils by leaching out important metal ions like calcium and magnesium. Other cations and especially anions can also adversely affect the properties of the receiving water and soil. Some of these effects will be discussed in later chapters.

An examination of global sulphur emissions from 1850 to the present shows that maximum releases to the atmosphere occurred around 1990. Since that time, to a large extent as a result of initiatives taken in the major emissions regions of western Europe and North America, there was a significant decline up to near the turn of the century. In the new millennium, release rates again began to rise with the location of greater emissions shifting from North America and western Europe to eastern Europe and the (former) Soviet Union. The European maps in Fig. 5.3 show that the greatest sulphur emissions were occurring in the eastern part of the continent, but there were also large releases in other heavily industrialized areas as well.

Global nitrogen emissions likewise stabilized for a time but since 2000 have been increasing once again.

Beginning in about 1990 and especially in the new millennium, parts of Asia have become the centre of attention. In both China and India there has been a dramatic increase in fuel use and

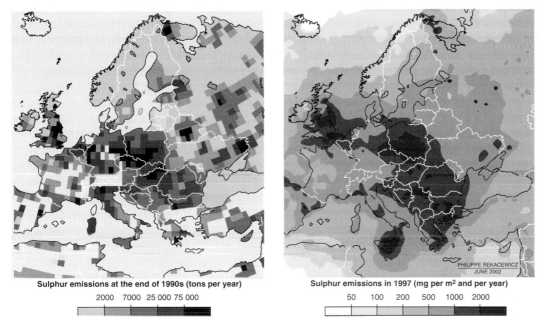

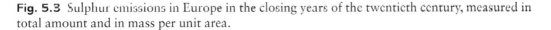

**Fig. 5.3** Sulphur emissions in Europe in the closing years of the twentieth century, measured in total amount and in mass per unit area.

in resultant emissions of sulphur and nitrogen oxides. This is largely associated with industrial development, including the commissioning of many new electric power plants, most of which are fuelled by coal. Another major source of the increase is the rise in availability and use of personal automobiles. Total sulphur dioxide emissions in that region increased from 23 to 51 Mt y$^{-1}$ over the period 1980 to 2003 while NO$_x$ emissions went from 11 to 29 Mt y$^{-1}$. China was responsible for about 70% of the increase in both cases.

It is undeniable that development activities in many low-income countries in heavily populated regions of the world are increasing their consumption of fossil fuels. As long as this trend continues, it is critical that emphasis be put on technologies that minimize release of acid- and smog-generating chemicals into the atmosphere. Fortunately, such technologies exist and improvements are being made yearly. Evidence of this is that, since the introduction of the Clean Air Act Amendments in the United States in 1990, while coal use increased by more than 30%, SO$_2$ emissions decreased by 20% and NO$_x$ emissions remained stable over the next 10 years. Similar improvements have been made both for sulphur and nitrogen compounds in other places around the world.

---

**Main point 5.6** Areas where acid precipitation is endemic are usually in regions of intense industrialization—Europe and North America being prime examples. In recent years, the problem has spread to parts of Asia. Anthropogenic sources of both nitrogen and sulphur oxide emissions are associated with combustion processes. Sulphur dioxide is also released during the smelting of sulphide minerals.

**Literature link**  Long-range atmospheric transport of acidifying substances

Regarding emissions of acid- and smog-generating gases, we have seen that there is a growing focus on Asia and this is reflected in research on various aspects of the subject—chemistry, physics, engineering and general environmental science. A paper entitled *Long-range transport of acidifying substances in East Asia—Part II: Source–receptor relationships* by M. Lin, T. Oki, M. Bengtsson, S. Kanae, T. Holloway, D.G. Streets (Atmospheric Environment **42**, 5956–5967 (2008)) considers the subject of *source–receptor (S / R)* relationships associated with the primary and secondary emission products in that region.

In the discussion that describes the chemical transformations that the emitted sulphur and nitrogen compounds undergo, you will recognize much of the chemistry as having been presented in this and earlier chapters. Ammonia is another player in the gas-phase reactions that go on in the atmosphere as the primary products move with the global air currents (see Chapter 6 of this book). We have already indicated that ammonia reacts with both sulphate and nitrate to form airborne aerosol particles. Because sulphate is the preferred reactant, it is consumed first and only if there is excess ammonia is a significant amount of ammonium nitrate produced. Interestingly, ammonium nitrate aerosol has a longer atmospheric lifetime than its precursors, ammonia and nitric acid. For this reason, a reduction in ammonia emissions in the source region reduces the longer range transport of ammonium nitrate to downwind receptor areas and increases dry deposition of nitric acid in the local area. If there is a reduction in nitrogen oxide emissions, there is a small increase in deposition of sulphate species in the source area. The authors attribute this to more hydroxyl radical being available to react with sulphur dioxide (reaction 5.21) rather than with NO$_x$ (reaction 5.2). On the other hand, reduced emissions of sulphur dioxide allow for more complete conversion of NO$_x$ to the longer-lived ammonium nitrate aerosol with a resulting increase in nitrate depositions in remote receptor areas.

The paper uses knowledge of sulphur and nitrogen atmospheric chemistry along with carefully measured analytical data and sophisticated modelling to understand what factors—regional sources or long-range transport from remote sources—are most important in determining levels of sulphate and nitrate deposition in various parts of Eastern Asia. For example, in Japan, most of the nitrate deposition was from local sources, much of anthropogenic origin, and associated with the transportation sector. A smaller amount originated from Korea. At the time of the study, the Miyakejima volcano[3] was active and accounted for about 80% of the sulphate depositions on the Japanese islands. There was a predictable seasonal effect in that most of the deposition occurred during the wet months from April to October. Throughout the drier winter, winds carried the sulphate offshore to the north-west Pacific Ocean. In terms of anthropogenic emissions of sulphur compounds, local sources were most important, but there were also significant amounts from sources in south-east and central China. Clearly, this type of information is important in its own right, but it also plays a central part in negotiations between nations regarding regulations of emissions.

## 5.7 Control of anthropogenic nitrogen and sulphur emissions

We have seen that the anthropogenic sources of nitrogen oxides and gaseous sulphur compounds centre around energy-related activities. Four approaches for reducing these emissions are possible—decreasing energy use by various efficiency measures, producing energy via non-combustion processes, preventing emissions of the problem gases, or removing the gases after they have been generated. All of the approaches are technically possible and there are philosophical, political, and economic arguments to be made for and against each one. In Chapter 4, we

---

[3] The Miyakejima volcano is located about 180 km south of Tokyo and began erupting in June, 2000. At its peak it was releasing 80 000 t of sulphur dioxide each day, but by 2006 emissions were reduced to about 2000 t per day.

considered catalytic methods of reducing nitric oxide emissions from vehicles. Here, using coal combustion as an example, we will look briefly at technology related to methods of minimizing emissions of both nitrogen and sulphur compounds. Simultaneously, other environmental effects of the modified technology must be considered.

## Fluidized-bed combustion

New types of combustion chambers have been designed so as to enhance the efficiency of coal combustion and of heat transfer and therefore to minimize fuel use. Fluidized-bed combustion (FBC) for burning coal is one such technique (Fig. 5.4).

In the FBC chamber, pre-heated air is forced upward through a bed of powdered coal. The passage of air, as well as convection generated by the hot gases from the burning finely divided particles of coal, creates a fluid-like suspension. In this configuration, uniform and complete combustion occurs with reduced emissions of carbon monoxide. The combustion temperature is somewhat lower than that in a static bed, thus reducing the amount of nitric oxide produced. A modification of the method allows for the simultaneous injection of powdered limestone into the bed so that there is a reaction between sulphur dioxide and lime (calcium oxide) to produce solid calcium sulphate.

$$CaCO_3 \text{ (s)} \rightarrow CaO \text{ (s)} + CO_2 \text{ (g)} \tag{5.45}$$

$$CaO \text{ (s)} + SO_2 \text{ (g)} + \tfrac{1}{2}O_2 \text{ (g)} \rightarrow CaSO_4 \text{ (s)} \tag{5.46}$$

At the high bed temperature, unburned coal particles are mechanically separated from the flue gas by a centrifugal cyclone device and fed back into the combustion bed. Heavy ash, including the calcium sulphate, settles through a grid under the fluidized bed and finer particles, carried upward in the flue gases, are trapped by an electrostatic precipitator or a fabric filter. Although conventional precipitators are capable of removing well over 99% (by mass) of the airborne particulates emitted during coal combustion, the efficiency may be much less (about 30%) when calculated on the basis of number of particles. In particular, these systems are much less effective in controlling particles with diameters less than approximately 5 μm and it is these colloids that are potentially most hazardous to human health. More will be said about atmospheric particulates and their control in the next chapter.

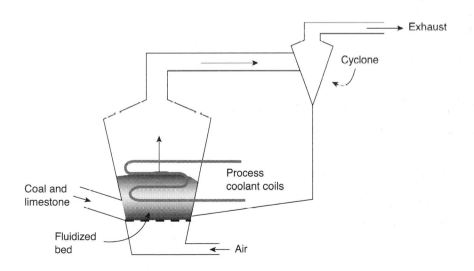

**Fig. 5.4** A fluidized-bed combustion unit with cyclone for removal of particulate material in the flue gases.

As a means of removing sulphur dioxide, this is a highly efficient process, effecting around 90% recovery. Nitrogen oxides emission is reduced by about 50% due to the controlled combustion conditions, which also enhance conversion of carbon monoxide to carbon dioxide.

### Retrofitted flue-gas desulphurization

Existing conventional coal-fired plants can be retrofitted with devices that remove sulphur dioxide from the flue gases. Among the many processes used the most common are lime and limestone slurry scrubbers in which the combustion gas passes through an aqueous slurry where reactions to form calcium sulphite take place:

With hydrated lime slurry

$$Ca(OH)_2 + SO_2 \text{ (g)} \rightarrow CaSO_3 \text{ (g)} + H_2O \tag{5.47}$$

With limestone slurry

$$CaCO_3 + SO_2 \text{ (g)} \rightarrow CaSO_3 \text{ (s)} + CO_2 \text{ (g)} \tag{5.48}$$

The insoluble calcium sulphite may be oxidized downstream to produce $CaSO_4 \cdot 2H_2O$, which is the mineral gypsum.

$$CaSO_3 \text{ (s)} + \tfrac{1}{2}O_2 \text{ (g)} + 2H_2O \rightarrow CaSO_4 \cdot 2H_2O \text{ (s)} \tag{5.49}$$

which settles in disposal ponds or may be recovered for use in applications such as manufacturing plaster or plasterboard.

This technology efficiently desulphurizes the gas stream (~ 90%) but vast quantities of water and lime (or limestone) are required and the resulting wastes are correspondingly large. For a 1000-MW coal-fired electricity-generating plant supplied with coal containing 10% ash and 2% sulphur, about 10 000 t of coal are burned daily. The solid wastes include 1000 t of ash. The daily volume of water required for the scrubber slurry is around 7000 m³ and the limestone requirement is about 600 t. Over 900 t of gypsum are produced in the process for each day of operation.

An obvious alternative to sulphur removal by these processes is the use of low-sulphur coal. The decision about reduction of sulphur emissions then becomes one of economics—whether it is less expensive to transport low-sulphur coal to the plant site or to equip the facility with the requisite control system.

### The SONOX process for removal of both sulphur and nitrogen precursors

A recently developed process for control of acid gas emission from power plants is called SONOX. This process, developed in Canada by Ontario Hydro at its small 640 MJ h$^{-1}$ Combustion Research Facility, is soon to be tested on full-sized boilers.

The control system involves in-furnace injection of an aqueous slurry of a calcium-based sorbent, usually powdered limestone, and a nitrogen-containing additive, usually urea, at temperatures ranging between 900 and 1350°C. The following reactions occur in the high-temperature atmosphere of the furnace reactor

$$CaCO_3 \text{ (s)} \xrightarrow{\text{heat}} CaO \text{ (s)} + CO_2 \text{ (g)} \tag{5.45}$$

$$CaO \text{ (s)} + SO_2 \text{ (g)} + \tfrac{1}{2}O_2 \text{ (g)} \rightarrow CaSO_4 \text{ (s)} \tag{5.46}$$

$$NH_2CONH_2 \text{ (s)} + 2NO \text{ (g)} + \tfrac{1}{2}O_2 \text{ (g)} \rightarrow 2N_2 \text{ (g)} + CO_2 \text{ (g)} + 2H_2O \tag{5.50}$$

Spraying the additives into the furnace via a high-pressure nebulizer ensures rapid evaporation of the solvent and efficient 'cracking' of the calcium carbonate and urea to produce calcium oxide and the reactive amidogen radical ($NH_2C\overset{\bullet}{O}NH$) respectively. A schematic of the process is shown in Fig. 5.5.

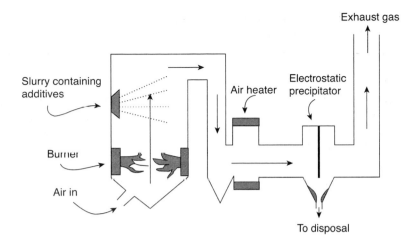

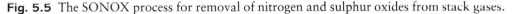

**Fig. 5.5** The SONOX process for removal of nitrogen and sulphur oxides from stack gases.

Capture efficiency of sulphur dioxide and nitrogen oxides depends on the nature of the additive, the rate of addition, the spray characteristics, the reactor temperature, and the sulphur content of the coal. Optimum conditions include:

- furnace temperature ~ 1150°C;
- concurrent injection;
- mean droplet diameter ~ 6.6 μm.

The sulphur dioxide sorbent is ~ 90% porous limestone and 10% dolomite or hydrated lime, and the solids are present at a concentration of 40% in an aqueous slurry concentration, calcium:sulphur ratio 2.5 to 3:1. The nitrogen oxides sorbent is urea or ammonium carbonate in aqueous solution at a stoichiometric molar ratio, additive:nitrogen oxide = 1.7 to 2.0:1. Under these conditions up to 85% sulphur dioxide and 85–95% nitrogen oxides removal can be effected. When urea is used as an additive, the concentration of nitrous oxide in the flue gases is augmented from a value of 10–25 ppmv (without additive) to 50–150 ppmv. More studies may result in methods for reducing these concentrations.

The solid waste obtained after slurry injection is about double in mass of that which results from combustion of coal without emission control. (The actual amount depends on the ash content of the coal as well as the amount of sulphur dioxide sorbent used. This latter amount depends, in turn, on the sulphur content of the coal.) The additional material in the solid waste consists of unreacted calcium oxide and calcium sulphate and consideration must be given to disposal of these mixed residues.

## Conversion of coal to gaseous and liquid forms

Finally, the environmental consequences of using coal to manufacture synthetic gaseous and liquid fuels, often called *synfuels*, must be considered. The conversion of coal to gaseous or liquid forms is carried out in order to create energy commodities that can be transported via pipelines, are readily stored in containers, are clean, and are suitable for use in small-scale facilities, particularly in vehicular engines. Conversion, especially of the poorer grades of coal, to liquid or gaseous forms also allows for the upgrading of energy content of the fuel. The fuel components of coal are principally carbon and hydrogen, and the basic principle for conversion is to increase the relative proportion of hydrogen compared to carbon. We will look at these technologies in a later section.

A comprehensive description of many nitrogen oxide removal technologies is given in Additional Resources 3.

---

**Main point 5.7** Technologies for control of nitrogen and sulphur oxides in industrial emissions have been developed. Economics of the control processes and problems associated with disposal of the waste materials are two factors limiting application of these technologies.

---

## ADDITIONAL RESOURCES

1. Brimblecombe, P., H. Hara, D. Houle, M. Novak, (eds.), *Acid rain—deposition to recovery* Springer, Dordrecht, the Netherlands, 2007.

2. Stern, D.I., Global sulphur emissions from 1850 to 2000, *Chemosphere*, 58 (2005), 163–175.

3. Lani, B.W., T.J. Feeley III, J. Murphy, L. Green, A review of DOE / NETL's advanced $NO_x$ control technology R&D program for coal-fired power plants, 2005 available at http://www.netl.doe.gov/technologies/coalpower/ewr/nox/index.html, accessed October, 2009.

4. Maps showing trends in concentration and deposition of S and N in the United States: *National atmospheric deposition project,* http://nadp.sws.uiuc.edu/data/animaps.aspx, accessed October, 2009.

5. World Resources Institute—EarthTrends Environmental Information, searchable database for information on emissions of acidifying gases and many other environmental topics, http://earthtrends.wri.org/, accessed October, 2009.

---

## PROBLEMS

1. The tropospheric processes discussed in this chapter cannot be considered as independent reactions. Discuss the relationship between the production of nitrogen-oxide species and the formation of 'acid rain' and the role played by carbon monoxide, methane, and the hydroxyl radical.

2. In a particular 3000 km² region of southern Sweden, the annual rainfall averages 850 mm, its mean pH is 4.27, and 66% of the hydrogen ion is associated with sulphuric acid with the remaining 34% derived from nitric acid. Calculate whether soils of this region are subject to excessive sulphate loading if the only source of sulphate is rainfall and if the recommended maximum is set at 20 kg $SO_4^{2-}$ ha$^{-1}$.

3. The mean monthly pH values of rainfall in Guiyang city in Guizhou province in southern China in 1984 were as follows:

    Jan   3.9, 4.0, 3.8, 4.1, 4.0, 4.5, 4.5, 4.1, 3.7, 3.8, 3.7, 3.4   Dec

    For the same year, the measurements at Luizhang, an adjacent rural area were:

    Jan   4.3, 4.4, 4.4, 4.2, 4.5, 4.9, 4.9, 4.6, 4.8, 4.3, 5.4, 5.4   Dec

    (a) Calculate the mean monthly pH of the rain at the two locations.

    (b) What is the ratio of the mean hydrogen ion activity at the two sites?

    (c) What is likely to be the most important anion in the rainfall at the urban site?

    Refer to Dianwu, Z. and X. Jiling, Acidification in southwestern China, In *Acidification in tropical countries* (ed. Rodhe, H. and R. Herrera), John Wiley and Sons, Chichester; 1988.

4. The ionic composition (in units of mg m$^{-3}$) of an atmospheric aerosol in a tropical rain forest is

$$SO_4^{2-}, 207; \; NO_3^-, 18; \; NH_4^+, 385; \; K^+, 180; \; Na^+, 247.$$

The pH of the aerosol is 5.22.

Use these data to calculate the total positive and negative charge 'concentration' (mol m$^{-3}$) in the aerosol and suggest reasons that might account for any discrepancy in anionic and cationic charge.

5. A report on rainfall composition in Egypt* gives results obtained at several sites. One of these is for average rainfall in Cairo where the pH was 7.24:

$$SO_4^{2-}, 12.23; \; NO_3^-, 5.99; \; NH_4^|, 1.85; \; Cl^-, 2.9; \; Ca^{2+}, 5.3.$$

Concentrations are in mg L$^{-1}$. After doing a charge-balance calculation, explain how you would interpret the results.

*Daifullah, A.A.M. and A.A. Shakour, Chemical composition of rainwater in Egypt, AJEAM-RAGEE 6, (2003) 32–43.

6. The tropospheric mixing ratios of carbon monoxide are higher in the Northern Hemisphere than in the Southern Hemisphere. However, it has been observed that there was a general global decline in carbon monoxide concentration everywhere in the 1990s. Two reasons have been suggested for this—one is the eruption of Mount Pinatubo and the other is the occurrence of several relatively dry years in the tropics. Comment on these two possibilities in terms of tropospheric and stratospheric processes.

7. The name *reactive nitrogen* has been recommended as a new short term for an extensive group of atmospheric nitrogen compounds, and is given by:

$$NO_y = NO + NO_2 + NO_3 + 2N_2O_5 + HONO + HNO_4$$
$$+ HNO_3 + aerosol \; nitrate + PAN + organic \; nitrates.$$

Discuss situations of atmospheric chemistry that we have described in this and earlier chapters where use of this term could be employed.

8. Nitrogen oxide is formed at night by dissociation of dinitrogen pentoxide (the reverse of reaction 5.9). For the first-order reaction

$$N_2O_5 \rightarrow NO_2 + NO_3$$

the rate constant is $3.14 \times 10^{-2}$ s$^{-1}$ at 25°C, and $6.88 \times 10^{-1}$ s$^{-1}$ at 55°C. Calculate the half-life of this molecule at the two temperatures. For a concentration of N$_2$O$_5$ of 3.6 ppbv, calculate the length of time it could take for the concentration to be reduced to 1.0 ppbv at a constant temperature of 25°C. Calculate the Arrhenius parameters, $A$ and $E_a$, for the reaction.

9. The United States Department of Energy has set a goal within their Innovation for Existing Plants program to limit the amount of NO$_x$ released from their current fleet of fossil fuel-fired electricity generating plants. These now have a capacity of 320 GW. The goal is to emit less than 0.10 pounds of NO$_x$ per MMBtu of fuel consumed. Assume that these facilities operate at 30% efficiency, and over the year at 70% capacity, calculate the number of tonnes of NO$_x$ that will be emitted annually. Table 8.6 will provide useful information for solving this problem.

10. Using the SONOX process, how much limestone would be required each year in order to effect quantitative removal of sulphur dioxide from a power plant whose daily consumption of coal (1.5% sulphur) is 6000 t?

The following two problems use concepts that are developed in Chapter 11.

11. Assuming an atmospheric pressure of 83 kPa, an atmospheric mixing ratio of 1.5 ppbv for hydrogen peroxide, and a value of $K_H(H_2O_2) = 7.0 \times 10^{-1}$ mol L$^{-1}$ Pa$^{-1}$, calculate its solubility in the cloud water droplets. Will this concentration depend on pH over the range 5 to 8?

12. Using the values for constants provided in reactions 5.25–5.27, calculate the solubility of sulphur dioxide in water at pH = 9.0.

13. In the United States and other parts of the world, control of both sulphur and nitrogen oxide emissions in the past decades have had a significant impact on their concentrations in precipitation and the amounts deposited into water and soil. Consult the maps found through Additional Resources 4. Note the regions of major depositions of these compounds. What are the most important sources? Explain what legislative initiatives are likely to have had the greatest effect on the reductions observed. Comparing nitrate and ammonia, what are the reasons for the geographical difference in their distribution? What is the likely reason for the increase in ammonia levels during the time when amounts of the other compounds are decreasing?

14. Consult the EarthTrends database (Additional Resources 5) and find the sulphur dioxide and nitrogen oxides emissions for the three years 1990, 1995 and 2000 for your country. Determine the annual emissions in mass of $SO_2$ *per capita* and mass of $NO_2$ *per capita*, and compare this national data with the same ratios for the world as a whole. Do these results lead to any conclusions and / or policy initiatives related to your country? Note that in this database, amounts of emissions are given based on mass of $SO_2$ and $NO_2$, whereas many other sources (including Table 5.5 in this book) report the same emissions using mass in terms of S and N.

# Chapter 6
# Atmospheric aerosols

The focus of this chapter is on understanding the role played by atmospheric aerosols in the global environmental chemistry context. We will accomplish this by:

- developing an understanding of the nature of aerosols and their environmental significance
- identifying sources of aerosols, both natural and anthropogenic
- investigating the condensation chemistry processes for formation of aerosols
- focusing on their concentrations, lifetimes and other properties
- examining control technologies that are used to minimize industrial emissions of particulates.

An aerosol is a suspension of particles in a gas and an atmospheric aerosol consists of particles that remain aloft in the air.

When we refer to particles in an aerosol, in the definition we are including both solids and liquids. Particles are distinguished from smaller gas molecules or molecular clusters by their ability to cause incoherent scattering of visible light and therefore to interfere with light transmission. As a consequence, the presence of a high concentration of aerosol is indicated by a hazy appearance in the atmosphere. To scatter visible light, particles must have dimensions comparable with or greater than the wavelengths of that light, say, within an order of magnitude (for example, at least one-tenth of 400 nm, that is 40 nm or 0.04 µm).

There are a number of factors that determine how long particles can stay suspended in the air and we will examine these in detail later in this chapter. Large particles readily settle out. Except for those of very low density, most particles with dimensions greater than 10 µm require strong air currents to keep them aloft. On the other hand, very small particles have limited lifetimes as independent entities, because they come together and coagulate to form larger ones. Particles within the size range 0.01–1 µm are most likely to remain suspended for long periods of time, sometimes up to a month or even longer, thus allowing them to move with the air mass for long distances across the landscape.

Aerosol particles are known by a variety of names, depending on the source and nature of the particles. Many of the names are common; we are all familiar with terms like dust, smoke, fly ash, and pollen (solids in gas) and cloud, mist, fog, and smog (liquids in gas). Figure 6.1 documents some properties of various types of atmospheric aerosols.

Many aerosol particles—dust, pollen, smoke—are released or injected into the air as preformed entities. There is another class, however, that results from chemical processes involving gaseous species reacting together in the atmosphere to form liquid or solid particles. Clouds and mist are simple examples of these *condensation aerosols*. Another condensation process is the generation of photochemical smog as described earlier in Chapter 4.

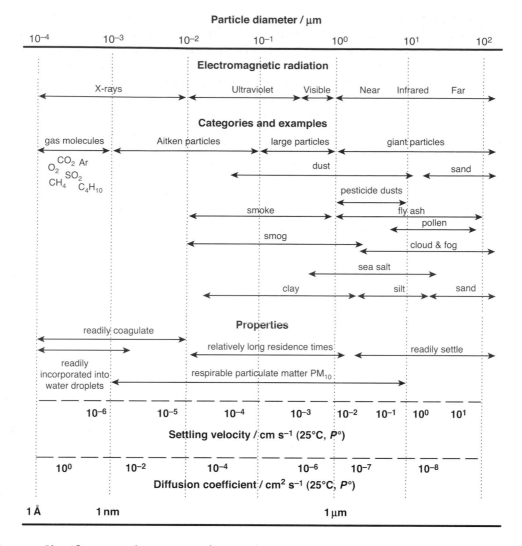

**Fig. 6.1** Classification and properties of atmospheric particulates. For the purposes of comparison, dimensions of gaseous molecules and wavelengths of electromagnetic radiation are also given.

## Aerosol measurement categories

While aerosols are present everywhere, excessive concentrations and / or the presence of particular chemical species may lead to human health problems, and therefore they are regularly monitored at many locations. Several parameters are determined to assess the concentration of particulates. Typical of these parameters are the following operationally defined aerosol measurements that are regularly made in Ontario, Canada.

- Coefficient of haze (COH) is determined by drawing 300 linear metres of air through a porous filter tape, after which the optical absorbance of the tape is compared with standards. The coefficient of haze is equal to the numerical value of 100 times the absorbance. A result of 6 or greater indicates air that may cause adverse symptoms to persons suffering from respiratory problems such as asthma. This method of sampling favours retention of solid aerosol particles in the size range 5–10 μm.

- Total suspended particulate (TSP), expressed in units of $\mu g\ m^{-3}$, is determined by gravimetrically measuring the quantity of particulates that have been obtained after filtering air at a rate of $1.4\ m^3\ h^{-1}$. An average concentration of individual measurements made over a 1-year period of greater than $60\ \mu g\ m^{-3}$ is considered excessive. Values of about $10–30\ \mu g\ m^{-3}$ are observed in many locations and individual readings greater than $500\ \mu g\ m^{-3}$ have been observed in the core of cities like Toronto.

- Inhalable particulates (IP), also expressed in units of $\mu g\ m^{-3}$, are those particulates that are smaller in dimension than $10\ \mu m$. These have been divided into two size groups, coarse and fine. Coarse particles cover the range between 2.5 and $10\ \mu m$ and tend to be derived from more physical processes. This size category of the aerosols is referred to as the $PM_{10}$ (PM, particulate matter) fraction. Fine particles ($PM_{2.5}$) cover the range less than $2.5\ \mu m$ and are mostly generated by combustion processes and condensation (chemical) reactions in the air. Both types of particles are considered important because they are the agents of many serious respiratory problems. The importance of the $PM_{10}$ and $PM_{2.5}$ categories is that the smaller the particulate, the more serious a health risk it poses. Fine particles can be deeply inhaled, interfering with lung's cellular activity and the reoxygenation of blood.
An observed relation that approximately relates total suspended particles and inhalable particles is

$$IP = 0.45\ TSP \qquad (6.1)$$

- Using a passive technique, total dustfall (TDF) is measured in $g\ m^{-2}\ month^{-1}$ by weighing dust that settles into an open-topped container over a 30-day period. A value greater than $7.0\ g\ m^{-2}\ month^{-1}$ (mass per unit area per time) of settled dust is considered excessive, and this value is frequently found to be exceeded in industrial cities such as Hamilton, Ontario.

As would be expected, it is very difficult to determine quantitatively the input rate of particulates to the atmosphere even from point sources. When all natural (mostly diffuse) and anthropogenic sources around the world are considered together, estimates become exceedingly problematic. Table 6.1 lists the major sources and gives approximate ranges for the yearly input. For obvious reasons, the table does not include essentially aqueous aerosols such as fog and clouds.

The total annual global production of aerosol particles then appears to be between 2500 and $4000\ Tg\ y^{-1}$. Although this range is only a highly approximate estimate, it is consistent with many reported values. We will look at aspects of the formation and properties of some of the common aerosol materials.

**Table 6.1** Estimated ranges of yearly input fluxes of particles that make up atmospheric aerosols.

| Aerosol | Natural (N) or anthropogenic (A) | Annual flux[a] / $Tg\ y^{-1}$ |
|---|---|---|
| Sea spray | N | 1000–1500 |
| Dust | N, A | 100–750 |
| Forest fires | N, A | 35–100 |
| Volcanic emissions | N | 50 (highly variable) |
| Meteors | N | ~1 |
| Anthropogenic combustion | A | ~50 |
| Condensation | N, A | ~1500 |

[a] Values obtained from various sources. $1\ Tg = 10^{12}\ g$.

Also discussed later in this chapter, in the section titled 'Condensation aerosols', and in Chapter 8, dealing with the subject of global climate, are the topics of cloud formation and their influence on the radiation budget of the Earth, respectively. Despite not being included in Table 6.1 clouds are one of the most important environmental aerosols.

---

**Main point 6.1** Atmospheric aerosols consist of particles of solid or liquid in the air. These particles are present in all outdoor and indoor atmospheres and, depending on their type and amount, play important environmental roles or can adversely impact human and environmental health.

---

## 6.1 **Sources of aerosols—a closer look**

### Sea spray

At any given time, whitecaps (waves with broken and foaming crests) cover approximately 2% of the ocean surface. The vigorous wind-driven action of the waves generates numerous extremely small bubbles that form and collapse at a rate of greater than $10^6$ events $m^{-2}$ $s^{-1}$. These include not only the bubbles that foam and make the whitecaps white, but also many that are small (anywhere from 5 to 500 µm in diameter) and individually invisible. Figure 6.2 illustrates the initial processes of sea-spray aerosol formation. Within milliseconds after a bubble has formed, the pressure of the surrounding water causes it to collapse in on itself, contorting and then rupturing the surface film, producing typically 1–10 very small water droplets. A few larger drops are also formed when the collapsing bubble ejects a stream of water that breaks apart as it falls under the force of gravity. The film disintegration droplets are typically between 5 and 25 µm in diameter and contain a mass of sea salt between 2 and 300 pg, while the central jet droplets are about 25 to 500 µm in diameter with 300 pg to 2 µg of salt in each.

After forming, many of the larger droplets fall back into the ocean, but smaller ones are wafted into the atmosphere where the water rapidly evaporates creating a solid aerosol having a sea salt chemical matrix. The particle composition reflects that of the soluble matter in sea water but, because the bubbles are produced at the surface of the sea, the composition shows any anomalies associated with surface water. In fact, the surface microlayer of a natural water body is enriched in surface-active components. Many of these are amphiphilic organic macromolecules that can sequester both neutral organic solutes and metal ions. Therefore, the salt matrix is frequently enriched in species other than the sodium, chloride, and other major solutes in bulk sea water. The enrichment is reflected in the composition of the sea-salt aerosol and its extent is described by a chemical concentration factor (CCF) that is defined in relation to one of the major elements, usually sodium, of the ocean.

$$CCF = \frac{(C_X/C_{Na})_{aerosol}}{(C_X/C_{Na})_{sea}} \qquad (6.2)$$

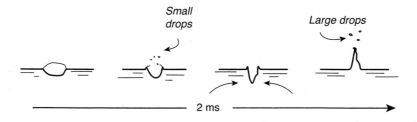

**Fig. 6.2** Progress of sea-spray aerosol formation shown from left to right, lasting 2 ms.

In eqn 6.2, $C_X$ and $C_{Na}$ in the numerator are the concentrations of the element of interest and of sodium in the aerosol, while the denominator terms are the respective concentrations in sea water itself.

Ocean-derived CCF values that are greater than 100 have been observed for some elements, especially ones like mercury, lead, and cadmium that reach the ocean surface by atmospheric transport. These elements also tend to form complexes with carboxylic acid- and nitrogen-containing ligands that are structural components of some organic macromolecules found in the oceans. Large CCF values have also been measured for certain organic species.

## Dust

Aerosol dust is produced by simple physical processes whereby solid materials having very small diameters, such as clay minerals and finely divided organic matter, are lifted into the atmosphere by wind currents. When air currents drive these particles along the surface, they collide with other solids on the ground causing fragmentation into still smaller particles that may then become airborne, adding to the aerosol. It is found that the chemical composition of dust reflects the general composition of the solid surface from which it was derived. For example, desert dust storms consist of mostly siliceous material, and sometimes this is carried for hundreds or even thousands of kilometres to remote locations. In West Africa, northerly winds, called the *Harmattan*, coming off the Sahara between December and February, carry dust at concentrations of up to 1000 $\mu$g m$^{-3}$ that settles over vast areas from Liberia to the Cameroons. The dust has a composition similar to that of the sands in the great desert to the north-west. Similarly, dust from the Gobi Desert in north-western China and Mongolia is carried far to the south and east, settling as far away as Japan and the Pacific Ocean or even further. In April 2001, a dust cloud, reliably measured as having originated in the Gobi Desert, produced a haze at an altitude of 7 to 9 km that was observed over wide areas of the Midwestern United States.

Chemical concentration factors analogous to those for the sea-spray aerosol have also been determined for dust. For the land-based material, a reference element that is a major component of the crust is used. Silicon is an obvious choice, but because of analytical difficulties it is rarely used. Aluminium, the most abundant metal in the geosphere, is easier to determine and makes a good reference element. Factors affecting species enrichment in dust are more varied than in the marine situation. They include anomalies in the surface soil composition, many a result of human activities such as agriculture and a wide range of urban influences.

Urban dust has been the subject of a good number of studies, but usually the investigations are done on settled dust, which favours larger particulates but also includes particles that are sufficiently large and / or heavy so that they were never a part of the aerosol. Besides soil components, city dust contains vegetative plant fragments, cement, tire and brake-lining particles, solid aerosols from vehicle exhaust, and many other synthetic and natural materials in smaller amounts. The urban aerosol contains the same substances in the form of fine particles that can be legitimately termed dust but it contains other non-dust components too. Smoke, pollen, and condensation products are intermixed with dust, and identification of individual sources is very difficult.

## Combustion products

Combustion products are formed as a consequence of a wide variety of natural and human-induced activities ranging from forest fires to fossil-fuel-fired power production plants. In most cases, the principal products are carbon dioxide and water vapour, and these are accompanied by other gases in smaller amounts. Depending on the fuel and the manner of burning, particulate matter is simultaneously emitted. Where combustion of carbon-based fuels is incomplete, elemental carbon is given off as a black smoke. This is the case when the ratio of fuel to oxidant is larger than the stoichiometric value and / or the combustion temperature is relatively low.

We have seen that diesel engines operate under these conditions, making them a major source of carbon emissions. On the other hand, a white plume from a factory stack is usually due to condensed water vapour (also a combustion product) and is then evidence of efficient and complete reaction.

Even when combustion of the carbonaceous material is complete, other particulates derived from minor constituents originally present in the fuel are released. Coal invariably contains non-combustible ash that is mostly siliceous in nature. Some of the minor elements remain in the combustion chamber along with the siliceous residue as *bottom ash*, while the finer particles are released into the atmosphere as *fly ash* unless otherwise removed by post-combustion methods. Table 6.2 gives ranges of fly ash composition resulting from coal combustion. Unburned carbon is not included in the listing.

Water extracts of fly ash are generally alkaline, giving pH values as high as 11. This suggests that ash collected in precipitators has potential value as a low-grade lime amendment for acidic soils. However, trace elements are frequently present in significantly large concentrations that disposal of ash collected in precipitators becomes a problem. Metals like lead and mercury appear to be deposited on the surface of the very fine material. This supports a theory that low-boiling metals present in fuels volatilize during combustion and then condense on to ash particle surfaces in cooler regions of the furnace or stack. Being surface deposits they are readily leachable and also available for biological uptake. We will say more in Chapter 19 about ash produced during combustion of urban solid waste materials.

**Table 6.2**  Range of composition of fly ash from coal combustion[a].

| Component | Range / % | Mean / % |
|---|---|---|
| Si | 9.0 to 28 | 21 |
| Al | 4.6 to 15 | 11 |
| Fe | 2.5 to 18 | 7.6 |
| Ca | 0.7 to 22 | 6.2 |
| K | 0.3 to 2.5 | 1.4 |
| S | 0.1 to 6.4 | 1.3 |
| Mg | 0.2 to 4.2 | 1.1 |
| Na | 0.1 to 6.3 | 0.9 |
| Ti | 0.1 to 1.0 | 0.7 |
| P | 0.1 to 1.0 | 0.3 |
|  | **Range / $\mu g\ g^{-1}$** | **Mean / $\mu g\ g^{-1}$** |
| Zn | 27 to 2900 | 450 |
| V | <95 to 650 | 270 |
| Cr | 37 to 650 | 250 |
| Pb | 21 to 2100 | 170 |
| As | 8 to 1400 | 160 |
| U | 11 to 30 | 19 |
| Cd | 6 to 17 | 12 |

[a] Most of these data are based on measurements of US fly ash as reported in Ainsworth, C.C. and D. Rai, *Chemical characterization of fossil-fuel combustion wastes*, EPRI EA-5321, Electric Power Research Institute, Palo Alto, California; 1987. Where sulphur emissions from coal combustion are controlled by a lime slurry scrubbing process, the ash has elevated calcium sulphite or sulphate levels.

### Polyaromatic hydrocarbons (PAHs)

One group of combustion products that has been widely studied and is of particular concern because of its carcinogenicity is the class of compounds called polyaromatic hydrocarbons (PAHs). These two- to eight-ring aromatic compounds are produced when wood, coal, and other carbon-based fuels are burned in the presence of limited amounts of oxygen. Along with elemental carbon, PAHs are products of incomplete burning under relatively low temperature conditions. Another significant point source release of PAHs is a result of carbon-anode baking associated with aluminium production.

Typical PAH compounds include angular, pericondensed, and linear compounds as shown from left to right below.

Chrysene (angular)    Pyrene (pericondensed)    Anthracene (linear)

Altogether, more than 150 PAH compounds containing two to eight rings, and some with alkyl or other substitution, have been identified as products present in smoke. Although these compounds are termed aromatic, application of the Hückel rule indicates that a species like pyrene does not meet the '$4n + 2\pi$-bonding electrons' criterion for aromaticity. The criterion is met, however, if one considers only the periphery of the compound. Nevertheless, we will treat all PAHs as aromatic and draw structures accordingly. Molar masses run from 128 to over 300. The smaller species are somewhat volatile, while those with larger molar mass are present as solids, usually as surface deposits on soot and other combustion product particles. Chrysene (molar mass = 228) has been observed to be distributed approximately equally between the vapour and solid phases.

Air measurements made in the industrial city, Hamilton, Ontario, have indicated about $10\,\text{ng}\,\text{m}^{-3}$ PAH in summer and $30\,\text{ng}\,\text{m}^{-3}$ in winter.[1] The PAH compounds were found to be associated with soot particles, 80% of which were less than 3.3 μm in diameter. While we would expect to identify PAH compounds in an industrial urban centre, we might not expect that they could also be found in regions remote from major sources of combustion. However, concentrations of $1\,\text{ng}\,\text{m}^{-3}$ or more have been detected in North American Arctic regions. This is evidence of their non-reactivity so that they persist while carried by wind currents from distant sources in Eurasia and North America. The time taken to travel these distances may be more than 1 year. The resultant behaviour has been described in terms of a *grasshopper effect*. During the summer season, southerly winds and high temperatures favour gas-phase transport in a northward direction, but in winter the solid phases are favoured so that there is less transport by the northerly air currents and the compounds remain essentially stationary. In a subsequent warm season they 'hop' further toward the pole. The PAHs are unreactive, long-lived compounds and have half-lives often estimated to be in the range of 5 to 9 years.

Because PAHs are toxic, persistent and semi-volatile, they have been included in the United Nations Economic Commission for Europe (UNECE) Convention on Long-Range Transboundary Air Pollution (LRTAP) 1998 Protocol on Persistent Organic Pollutants (commonly called POPs) that entered into force in 2003. It is encouraging that where appropriate measures are taken, emissions and resultant atmospheric levels of these compounds have decreased over the last decade. In the UK it is estimated that emissions of PAHs have declined from almost 8 million

[1] Isidorov, V.A., *Organic chemistry of the Earth's atmosphere*, Springer-Verlag, Berlin; 1990.

tonnes in 1990 to 1.2 million tonnes in 2005.[2] Most of this decline is associated with control on emissions from the aluminium and energy industries and restrictions on agricultural burning. Transport and domestic heating remain as the principal present-day sources of PAHs.

## Condensation aerosols

### Ammonium sulphate and nitrate

Primary emission products are those that are produced during combustion or other processes, and then released directly into the atmosphere. Secondary products can be produced by reactions that take place after their release. These products are called *condensation aerosols*. There are many and varied chemical processes involving gaseous reactants that result in formation of liquid or solid condensation particles in the atmosphere.

Three major condensation-formed components of both the continental and oceanic aerosol are ammonium hydrogen sulphate ($NH_4HSO_4$), ammonium sulphate (($NH_4)_2SO_4$) and ammonium nitrate ($NH_4NO_3$). Each of these solid species is produced as a result of parallel series of reactions.

The first series begins with ammonia that is released into the atmosphere from several natural and anthropogenic sources. During microbial degradation of decaying biomass and organic matter in soil and water, nitrogen compounds like proteins are ammonified to emit ammonia / ammonium ion into the surroundings. An important source of nitrogen-rich biomass that undergoes these reactions is animal excreta deposited on soil or maintained in manure piles. Especially problematic are locations where livestock are housed in large numbers in a limited amount of confined space. One of the nitrogen compounds in animal urine is urea and it hydrolyses to produce ammonia and carbon dioxide.

$$CO(NH_2)_2 + H_2O \rightarrow CO_2 + 2NH_3 \tag{6.3}$$

The actual amount of ammonia given off depends on temperature, moisture, soil texture and pH, and the nature of plants growing where the excreta are deposited. On a global scale, it may be that up to 20% of the nitrogen in animal waste is volatilized as ammonia.

Similarly, where synthetic fertilizers containing reduced nitrogen (ammonia, urea, ammonium nitrate or sulphate) are used, it is possible that ammonia species are released. Gaseous ammonia is given off in an alkaline environment, while ammonium ion is favoured under acidic or neutral conditions. The ammonium ion is soluble in the aqueous soil solution and, assuming that the soil does not have an excessively low pH, it becomes *fixed* onto cation-exchange sites (Chapter 18) and / or is oxidized to nitrate before being taken up by the growing crop. Nitrate is, of course, a very soluble ion and is not volatilized into the atmosphere.

There is little loss of volatile ammonia if the organic matter degradation occurs under well-aerated (meaning oxidizing) conditions or if the fertilizer is incorporated into a surface soil and efficiently used by plants. On the other hand, when application of these same fertilizers takes place in soils that are waterlogged due to flooding or deliberately submerged during rice culture, the reducing conditions maintain the nitrogen in the ammonia / ammonium form and, depending on soil solution pH, substantial loss of ammonia due to volatilization may occur.

Additional ammonia is also released from a range of industrial processes.

Ammonia is one of the gaseous compounds that reacts to form the atmospheric aerosol; the other compounds are sulphuric and nitric acid. As described in detail in Chapter 5, oxidation reactions where oxygen and the hydroxyl radical serve as oxidizing agents convert reduced sulphur-containing gases into sulphur dioxide. This sulphur dioxide is augmented by additional amounts released directly from other, mostly industrial, sources and is further oxidized to form

[2] Meijer, S., Sweetman, A.J., Halsall, C.J., and Jones, K.C. Temporal trends of polycyclic aromatic hydrocarbons in the U.K. atmosphere: 1991–2005. *Environ. Sci. Technol.* **42** (2008), 3213–3218.

sulphuric acid. The ammonia and sulphuric acid react together to produce ammonium hydrogen sulphate or ammonium sulphate (reactions 6.4 and 6.5) in the form of particles that are approximately 0.1 to 1 μm in diameter. Because sulphuric acid is usually in excess, the former compound is dominant in most situations.

$$NH_3 + H_2SO_4 \rightarrow NH_4HSO_4 \qquad (6.4)$$

$$2NH_3 + H_2SO_4 \rightarrow (NH_4)_2SO_4 \qquad (6.5)$$

Ammonium nitrate is formed in similar fashion via an analogous reaction between ammonia and nitric acid.

$$NH_3 + HNO_3 \rightarrow NH_4NO_3 \qquad (6.6)$$

All three hydrophilic ammonium salt particulates act as nuclei around which water condenses to form clouds.[3] Where there is a high concentration of particles, the clouds that form consist of a large number of very small droplets. Such clouds are whiter and more reflective than those containing fewer large drops. This is one way in which cloud reflectivity, an important regulator of global climate, is influenced by atmospheric aerosols.

### Organic condensation nuclei

Smog-forming reactions that were described in Chapter 4 are an example of complex condensation processes that lead to the formation of a liquid aerosol. There are also natural processes through which some types of organic condensation nuclei form, creating an atmospheric aerosol. A good example is the haze that develops over heavily forested areas during warm summer days. Terpenes and related compounds are low molar mass chemicals synthesized within the leaves and stems of a variety of plants. These compounds are relatively volatile and so vaporize into the atmosphere, producing the attractive odours that characterize forests. One of the terpenes, α-pinene, is produced by coniferous species including pine, spruce, and fir trees, and has been measured in forest atmospheres at concentration levels between 0.1 and 50 ppbv (0.5 and 300 μg m$^{-3}$).

α-pinene

For deciduous species such as willow, oak, poplar, and aspen, a more characteristic emission product is isoprene and concentrations between 1 and 10 ppbv (3 and 30 μg m$^{-3}$) of this compound have been observed. Emission rates of these and other compounds depend on the tree species and are maximal during the daytime and in warm temperatures.

$$CH_2{-}C{-}CH{=}CH_2$$
$$\mid$$
$$CH_3$$

isoprene

Both isoprene and the terpenes are highly reactive with respect to photochemical oxidation and undergo reactions that are very much analogous to the urban photochemical smog

---

[3] Cloud seeding to encourage rain / snowfall is sometimes done by injecting a hydrophilic solid (usually silver iodide, which has a crystal structure very similar to ice) into the moist atmosphere. The solid particles act as a locus where water vapour condenses to form a liquid droplet. In principle, this creates a positive feedback effect due to the release of heat during condensation that causes an updraft of warm moist air. This is then a source of additional water vapour capable of generating more cloud.

formation processes. Two oxidative pathways[4] have been proposed. The first involves the presence of $NO_x$ species as a source of the hydroxyl radical. Reaction sequence 6.7 illustrates the hydroxyl radical-initiated oxidation of isoprene. You will recognize that this sequence is similar to ones already described for hydrocarbons in Chapter 4 (reactions 4.13 to 4.17).

$$k_{6.7\text{overall}} \approx 9 \times 10^{-11} \text{ cm}^3 \text{ molecule}^{-1} \text{ s}^{-1} \tag{6.7}$$

There are alternative processes that employ ozone as the initiator of oxidation and these can follow several routes to produce a range of products. Some are shown in reaction 6.8.

$$k_{6.8\text{overall}} \approx 10^{-16} \text{ cm}^3 \text{ molecule}^{-1} \text{ s}^{-1} \tag{6.8}$$

While the oxidation of isoprene initiated by hydroxyl has an observed rate constant, $k_{6.7, \text{ overall}}$, of approximately $10^{-10}$ cm$^3$ molecule$^{-1}$ s$^{-1}$, the ozone-based sequence has a much smaller observed rate constant, $k_{6.8, \text{ overall}}$, of $10^{-16}$ cm$^3$ molecule$^{-1}$ s$^{-1}$. Yet both pathways may contribute comparably to the oxidation of isoprene. This is because the hydroxyl radical is present in smaller mixing ratios than ozone. The hydroxyl-radical concentration is near $10^{-5}$ ppbv compared to the 30 ppbv ozone concentration; these are typical levels of the two species in forested areas remote from urban influences.

The oxygenated products including aldehydes, ketones, and carboxylic acids make up some of the constituents of a photochemical haze that is characteristic of certain forests on otherwise

---

[4] Isidorov, V.A., *Organic chemistry of the Earth's atmosphere*, Springer-Verlag, Berlin; 1990, and Hanst, P.L., J.W. Spence, and E.O. Edney, Carbon monoxide production in photooxidation of organic molecules in the air, *Atmos. Environ.*, **14** (1980), 1077.

clear sunny days. The Great Smokey Mountains of North Carolina in the USA are often cited as an example of a location of where this natural phenomenon is evident. In recent years, the level of VOC emissions from US forests has actually increased at a rate of 6% per decade, with even greater increases in the southern states. Nevertheless, it is possible that human activities can affect the otherwise natural process since reactants like ozone and $NO_x$, required to create the haze, are augmented by combustion and other processes. The particles in the haze aerosol typically are less than 0.3 μm in diameter, putting them in the 'large' and 'Aitken particle' size range. Total annual global emissions of gaseous organic species from forests have been estimated to be around 20 Tg $y^{-1}$.

## Arctic haze—atmospheric pollution in a remote area

The term 'Arctic haze' was first used over 50 years ago by J. Murray Mitchell[5] to describe reduced visibility observed during weather reconnaissance flights in the Arctic. Today, the term conveys the notion of haziness due to atmospheric pollution in a part of the Earth that is perceived to be a pristine environment. Since the early 1970s scientific investigation has helped us gain a better understanding of the cause of this phenomenon.[6]

Arctic haze is the result of wind-blown dust and industrial emissions (predominantly sulphur dioxide, but also including hydrocarbons, soot, metals, several persistent organic pollutants (POPs) and other gases and particulates) that are carried to the Arctic from mostly Eurasian locations. There is a distinct seasonal variation in the haze. From December to April, when Arctic air covers areas far to the south including some heavily industrialized regions in Eurasia and North America, particulate and gaseous pollutants continue to be transported to the north. During the northern winter season, natural atmospheric cleansing processes are less efficient in the cold dry Arctic, and this allows the particulates to accumulate in the atmosphere. Another important factor in the production of the haze is the steep inversion layer that is present in the Arctic in late winter and early spring. A temperature differential of up to 30–40°C may exist between the lower surface temperature and the higher (albeit still quite low) air temperature several hundred metres above. These factors combine to allow a buildup of diverse aerosol components and the brown haze, in appearance not unlike that of a large southern city, becomes highly visible at the emergence of the spring sun.

Arctic haze consists mainly of particles of variable chemical and physical properties. The particle concentration ranges from 10 to 4000 particles $cm^{-3}$ with a geometric average of 200 to 350 particles $cm^{-3}$. The most numerous particles are those in the size range 0.005–0.2 μm. Particles in the range 0.1–1 μm consist mainly of sulphate-derived aerosols and are primarily responsible for the visible haze. There are also coarse particles in the range 1–10 μm, and giant particles (> 10 μm) that result mainly from soil and sea spray. While the larger particles contribute a significant fraction to the aerosol mass, there are relatively few of these particles and they have little influence on the actual appearance of haze.

The haze spreads over a vast area, stretching from Greenland through coastal regions of the Canadian Arctic and Alaska and sometimes into Eastern Siberia, and occupies altitudes below 9 km, with a maximum concentration at an altitude of 4 to 5 km. The January–April average Arctic haze has a reported composition consisting of 2 μg $m^{-3}$ of $SO_4^{2-}$, 1.0 μg $m^{-3}$ of organic compounds, 0.3–0.5 μg $m^{-3}$ of black carbon, and smaller concentrations of other substances as well as a few μg $m^{-3}$ of water.

The period of May to November is associated with conditions that are 20 to 40 times less hazy; at this time, the residual aerosols consist mostly of wind-blown dust and sea spray.

---

[5] Murray Mitchell, J., Visual range in the polar regions with particular reference to the Alaskan Arctic, *J. Atmos. Terr. Phys.* Special Supplement (1956), 195–211.
[6] Arctic Monitoring and Assessment Program (AMAP), *Arctic pollution 2009, a downloadable report from* www.amap.no, accessed October 2009.

During January to April the haze can remain suspended for considerable periods of time, but eventually some of the heavier particles settle. The black soot and other dark-coloured particles reduce the reflectivity (albedo) of the snow so that solar radiation is more readily absorbed. The heating effect is one reason why melting of ice and snow in this region is occurring at an unusually rapid rate and this raises considerable concern in terms of changing global climate and effects on sea levels.

> **Main point 6.2** Atmospheric aerosols are derived from natural sources—sea spray, dust, etc. and from anthropogenic sources—industrial, combustion, etc. Aerosols can be produced directly from the source or can be generated by condensation processes in the atmosphere.

## 6.2 **Aerosol concentrations and lifetimes**

Table 6.3 lists typical total aerosol concentration ranges. Values are invariably reported in units of number or mass of particles per volume of air. Concentrations involving moles (mixing ratios) cannot be used since the aerosol always consists of a mixture of ill-defined species.

While the aerosol particles consist of liquid or solid materials, on a mass basis their concentrations in the atmosphere may not be as large as that of some minor gaseous constituents. The atmospheric mean mixing ratio of methane gas is about 1.77 ppmv, which is equivalent to 1200 µg of the compound in a cubic metre of air. This is a much larger concentration than the 10 to 100 µg m$^{-3}$ that is typical of the total concentration for many aerosols in the atmosphere.

Two types of physical processes are most important in determining the lifetime of aerosol particles. The first of these is settling, which is the primary means of removal of larger particles from the atmosphere. Settling occurs due to the force of gravity and in a simple way is described by the Stokes relation. As given in eqn 6.9, Stokes' law determines the terminal velocity or settling velocity (Fig. 6.1) of spherical particles falling under the force of gravity in a fluid.

$$v_t = \frac{(\rho_p - \rho_a)Cgd_p^2}{18\eta} \tag{6.9}$$

where $v_t$ = terminal velocity of particles / m s$^{-1}$; $\rho_p$ = density of particle / g m$^{-3}$; $\rho_a$ = density of air = $1.2 \times 10^3$ g m$^{-3}$ at $P^o$ and 25°C; $C$ = the Stokes–Cunningham slip correction factor (see Table 6.4); $g$ = 9.8 m s$^{-2}$ = acceleration due to gravity; $d_p$ = particle diameter / m; and $\eta$ = viscosity of air = $1.9 \times 10^{-2}$ g m$^{-1}$ s$^{-1}$ at $P°$ and 25°C.

The dimensionless Stokes–Cunningham slip correction factor $C$ accounts for the discontinuous nature of fluid interactions when the particle size is small compared with the molecular mean free path in air.

**Table 6.3** Total concentration ranges of aerosols in various geographic settings.

| Geographic setting | Typical concentration / µg m$^{-3}$ |
|---|---|
| Open ocean | 10 to 150 |
| Sea shore | up to 500 |
| Vegetated rural areas | 10 to 50 |
| Arid areas | up to 500 |
| Urban atmosphere | up to 200 |

**Example 6.1 The terminal velocity of an aerosol particle**

Consider the case of a dust particle of geological origin whose particle diameter is 10 $\mu$m and density is 2.5 g mL$^{-1}$ ($= 2.5 \times 10^6$ g m$^{-3}$).

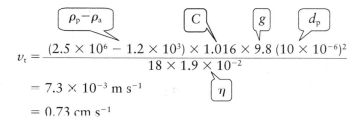

$$v_t = \frac{(2.5 \times 10^6 - 1.2 \times 10^3) \times 1.016 \times 9.8\,(10 \times 10^{-6})^2}{18 \times 1.9 \times 10^{-2}}$$

$$= 7.3 \times 10^{-3} \text{ m s}^{-1}$$

$$= 0.73 \text{ cm s}^{-1}$$

For most particles, the density of air could be neglected in these calculations.

Opposing the downward movement is the natural convective upward movement of air. As a broad generalization, particles greater than 10 $\mu$m in size are considered to be settleable, while those smaller than 10 $\mu$m remain suspended until removed by other processes such as washout in rain. Estimates of terminal or settling velocities are included in Table 6.4 and Fig. 6.1. A number of approximations, including assumptions about particle density and shape, have been made in producing these estimates.

The other physical process that determines particle lifetime is referred to as coagulation. This process involves the coming together by Brownian diffusion of small particles to form larger ones, a rate that is greater for smaller-size particles.

The rate of coagulation in a monodisperse system of particles (i.e. an aerosol made up of particles having uniform size) with a particular composition and density, is then given by

$$-\frac{dN}{dt} = 4\pi D C d_p N^2 \tag{6.10}$$

**Table 6.4** Aerosol transport properties assuming spherical particles, density 2.0 g cm$^{-3}$, in air at $P^\circ$ and 25°C.

| $d_p$ / $\mu$m | $C$ | $v_t$ / cm s$^{-1}$ | $D$ / m$^2$ s$^{-1}$ | $t_{1/2}$[a] |
|---|---|---|---|---|
| 0.001 | 216 | | $5.14 \times 10^{-6}$ | 1 min |
| 0.005 | 43.6 | | $2.07 \times 10^{-7}$ | 0.5 h |
| 0.01 | 22.2 | | $5.24 \times 10^{-8}$ | 2 h |
| 0.05 | 4.95 | | $2.35 \times 10^{-9}$ | 38 h |
| 0.1 | 2.85 | $1.7 \times 10^{-4}$ | $6.75 \times 10^{-10}$ | 110 h |
| 0.5 | 1.326 | $2.0 \times 10^{-3}$ | $6.32 \times 10^{-11}$ | 520 h |
| 1.0 | 1.164 | $6.8 \times 10^{-3}$ | $2.77 \times 10^{-11}$ | 690 h |
| 5.0 | 1.032 | $1.5 \times 10^{-1}$ | | |
| 10.0 | 1.016 | $6.0 \times 10^{-1}$ | | |
| 50.0 | 1.003 | $1.5 \times 10^{1}$ | | |
| 100.0 | 1.0016 | $5.8 \times 10^{1}$ | | |

[a] Half-life calculations are expressed in terms of coagulation; for this calculation the particle number density is taken to be $10^9$ m$^{-3}$.

where $N$ = particle concentration / m$^{-3}$; $D$ = diffusion coefficient of the particles in air / m$^2$ s$^{-1}$; $C$ = Stokes–Cunningham slip correction factor (Table 6.4); and $d_p$ = particle diameter / m. Assuming that $D$, $C$, and $d_p$ are constant, the rate of the coagulation process is second order.

$$-\frac{dN}{dt} = k_2 N^2 \qquad (6.11)$$

The half-life (in seconds) is then given by

$$t_{1/2} = \frac{1}{k_2 N} = \frac{1}{4\pi D C d_p N} \qquad (6.12)$$

---

**Example 6.2  The half-life (with respect to coagulation) of an aerosol particle**
An aerosol consisting of particles, all 0.01 μm (= $0.01 \times 10^{-6}$ m) in diameter, with a number concentration, $N$, of $10^9$ per m$^3$, the half-life is estimated to be

$$t_{1/2} = \frac{1}{4 \times 3.14 \times 5.24 \times 10^{-8} \times 22.2 \times 0.01 \times 10^{-6} \times 10^9}$$

$$= 6800 \text{ s}$$

$$\approx 2 \text{ h}$$

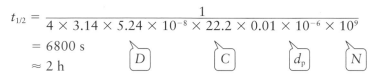

In this calculation, the particle diffusion coefficient was $5.24 \times 10^{-8}$ m$^2$ s$^{-1}$. The other values of $t_{1/2}$ reported in Table 6.4 were calculated in a similar manner.

---

Where particle sizes approach molecular dimensions, the above equations based on collisions limited by Brownian diffusion do not apply and the rate of coagulation is better described using relations derived from the kinetic theory of gases.

Because diffusion of particles is inversely proportional to the square of diameter, coagulation processes are more significant for very small particles but become negligible when the diameter is greater than approximately 0.01 μm. Table 6.4 lists diffusion coefficients as well as slip correction factors, terminal settling velocities, and half-lives for particles in the atmosphere at $P°$ and 25°C.

As a consequence of settling and coagulation processes, large and small particles, respectively, have relatively short residence times in the atmosphere. As shown in Fig. 6.3, the longest-lived particles are those whose size is in the intermediate range.

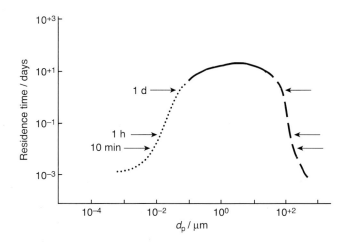

**Fig. 6.3** Residence times of particles in the atmosphere. The dotted line indicates that coagulation is the principal removal process, the dashed line indicates settling, and the solid line represents particles that are relatively long lived.

> **Main point 6.3** Aerosol particles cover a size range from 1 nm to 100 μm in diameter, but it is particles in the range 0.01–10 μm that are most stable in suspension. Particles smaller than 0.01 μm in diameter tend to coagulate into larger units, while those larger than 10 μm readily settle out. The aerosol fraction that consists of very fine particles is of special concern to human health because of its association with respiratory problems.

**Literature link** Characterizing urban soot

There are several reasons why it is important to identify the nature of carbonaceous soot from combustion processes. Epidemiological data have shown clearly that long-term breathing of air polluted with smoke derived from various sources can have serious adverse effects on human respiratory and cardiac systems. What is not clear, however, is which species within the aerosol are most responsible for the ensuing health problems. Carbonaceous material in smoke is also implicated as affecting regional or even global climates by its interaction with solar radiation. When present in an aerosol, black carbon particles absorb solar radiation as it passes through the troposphere, preventing it from reaching the Earth's surface. When this occurs, as has happened in South Asia due to widespread forest fires in Indonesia especially in 1997–8, there is a cooling of the regional climate. An opposite effect occurs when aerosol particles are deposited on the surface, especially on snow. Low-altitude warming then occurs as the particles are able to absorb radiation that otherwise might be reflected back into space as we noted in the case of Arctic haze.

These and other issues present an analytical challenge to develop methods that can characterize urban smoke in order to better understand such phenomena. The paper *Toward distinguishing woodsmoke and diesel exhaust in ambient particulate matter* by A. Braun, F.E. Huggins, A. Kubatova, S. Wirick, M.M. Maricq, B.S. Mun, J.D. McDonald, K.L. Kelly, N. Shah, and G.P. Huffman (Environ. Sci. Technol. **42**, 274–280 (2008)) addresses this problem. The authors point out that a standard method, gas chromatography with mass spectrometric detection, gives thousands of peaks on analysis of extracts of soot (collected smoke particles) samples, making meaningful characterization very difficult. What is needed is some kind of a characteristic marker that would identify a strong feature of samples, enabling the analyst to distinguish between sources.

To this end, the authors propose using near-edge X-ray absorption fine structure (NEXAFS) spectroscopy as a non-destructive probe of molecular structure of carbonaceous particulate matter. Specifically, they employ the method to distinguish between the carbonaceous material emitted from diesel engines and that from wood fires. The principal characteristic of diesel smoke is a NEXAFS peak at ~285 eV associated with aromatic C=C bonds. Wood smoke gives a smaller peak at that energy, but a well developed one at ~287 eV, this being assigned to C–OH that is characteristic of lignin combustion products. Using these and other less-intense peaks in the spectra of samples, the authors were able to assign the relative contribution of diesel emissions and wood burning in aerosol samples from several American cities. Further developments and application of this methodology will require examining other sources of soot. Aging or weathering of the soot sample will also alter its chemical structure and this must be taken into account in interpreting results.

What other analytical methods could be investigated for examining such samples—either as they are collected in solid forms or after extraction with various solvents?

## 6.3 Air pollution control for particulate emissions

It is possible to minimize emissions of aerosol particles from point sources such as thermal electrical generating stations or industrial smelting units. Obviously, other point-source emissions such as volcanic eruptions cannot be attenuated. Nor is it usually possible to control particulate emissions from non-point sources such as sea spray, wind-blown dust (e.g. the Harmattan; see earlier section on 'Dust'), or forest or grassland fires. In the latter case, steps can be taken to prevent some uncontrolled fires, and this would lessen the amount of aerosol released to the atmosphere.

Elimination of much of the solid particulate material from the industrial gaseous emission stream can also be achieved by prevention, through methods such as improving the efficiency of combustion or using a better grade of fuel. However, in many cases some production and release of particulates is inevitable and containment is therefore required before the waste gases are released into the atmosphere.

Containment is accomplished using devices that remove the aerosol particles from the fast-moving stack gas stream. Common collection methods include settling chambers, cyclones, fabric filters, scrubbers, and electrostatic precipitators, which are depicted in simple diagrammatic form in Fig. 6.4.

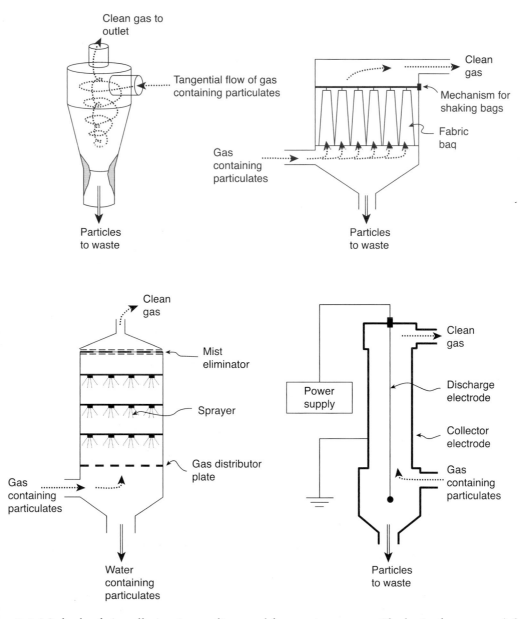

**Fig. 6.4** Methods of air pollution (aerosol) control from point sources. Clockwise from upper left: cyclone, fabric filter, electrostatic precipitator, scrubber.

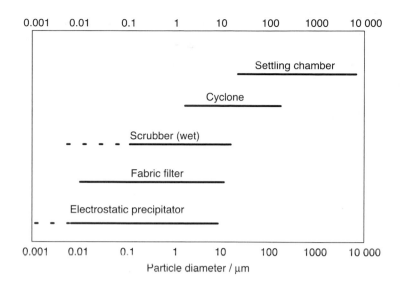

**Fig. 6.5** Particle sizes efficiently removed by various aerosol control devices used in industry.

Figure 6.5 compares the various methods in terms of the particle sizes that are most effectively removed using the different technologies.

Settling chambers are the simplest form of collection device and are one of the most widely applied methods of particulate control. The construction includes chambers with a variety of baffles and open spaces designed to allow the particles sufficient time to settle under the force of gravity. Because settling rates are limited by gravity, the method is most effective only for large particles (>10 μm).

Cyclones are cone-shaped devices that cause the waste gas stream to swirl rapidly in spiral fashion, causing larger particles to move towards the wall of the cone by centrifugal force. Once in contact with the wall, the particles slide down the inside of the cone to a collection container below. Stokes' law again determines the extent of removal of particles, but the 'settling' rate can be greatly enhanced by the increased force due to the cyclone action. The effective removal range therefore extends to much smaller particles.

A bag or fabric filter operates in a manner that is similar to that of a vacuum cleaner—that is, the air stream is made to pass through a porous fabric material—and is effective for particulates in the 0.01–10 μm range. Bags or fabric filters are sensitive to temperature and humidity and, because the fine pore filters become clogged during use, they must be periodically cleaned.

Scrubbers allow for contact between the gas stream and a fine mist or spray of water. By using water droplets to capture small particles, their size is effectively increased so that they are able to settle more rapidly. Various scrubber designs have been developed.

An electrostatic precipitator causes the particles in a gas stream to become charged by electrons produced through an electrical discharge between two electrodes. Once the particles are charged they migrate to the positive electrode and are collected and removed from the emission stream.

**Main point 6.4** Technology for capturing aerosol particles released from industrial processes is based on filtration, settling rates, incorporation into an aqueous phase, or electrostatic precipitation.

## ADDITIONAL RESOURCES

1. Kondratyev, K.Y., L.S. Ivlev, V.F. Krapivin and C.A. Varotsos, *Atmospheric aerosol properties—formation, processes and impacts,* Praxis Publishing Ltd., Chichester, U.K.; 2006.

2. Penner, J., D. Hegg and R. Leaitch, Unraveling the role of aerosols in climate change. *Environ. Sci. Tech.,* **35** (2001), 332A–40A.

3. Linak, W.P. and J.O.L. Wendt, Toxic metal emissions from incineration: mechanisms and control, *Prog. Energy Combust. Sci.,* **19** (1993), 145–85.

4. Pye, K., *Aeolian dust and dust deposits,* Academic Press, London; 1987.

5. NASA Total Ozone Mapping Spectrometer (TOMS), Aerosol Index, http://jwocky.gsfc.nasa.gov/, accessed October, 2009.

## PROBLEMS

1. Which of the two—cloud droplets or raindrops—would you expect to be more effective in scavenging gases from the atmosphere? Explain your reasoning.

2. A fog consists of 10 000 droplets of water per cm$^3$. The average diameter of the drops is 1.5 μm. Compare the mass of water in the liquid phase to that in the gaseous form if the temperature is 35°C and the relative humidity is 100%.

3. Compare the number of moles of nitrogen–oxygen species in air containing 300 ppbv nitric oxide, with fog consisting of 10 000 droplets per cm$^3$, having an average diameter of 2 μm, and containing nitrate ion at a concentration of $3 \times 10^5$ mol L$^{-1}$.

4. A fly ash ($\rho = 1.8$ g mL$^{-1}$) aerosol consists of particles averaging 13 μm in diameter and with a concentration of 800 μg m$^{-3}$. Use the average diameter to calculate the settling velocity (cm s$^{-1}$) and settling rate (g m$^{-2}$ s$^{-1}$) of the particles in still air.

5. Suggest a possible sequence of reactions (beginning with oxidation via the hydroxyl radical) by which dimethyl sulphide (($CH_3)_2S$) can be oxidized to produce a sulphuric acid aerosol.

6. Polyaromatic hydrocarbons are commonly associated with soot particles, as adsorbates on the surface. Calculate the relative and actual surface areas (actual values in m$^2$ g$^{-1}$) of soot particles ($\rho = 0.6$ g mL$^{-1}$) having diameters 20 μm and 2 μm, respectively, assuming that the particles are spheres. Speculate on the health implications with regard to human intake of PAH compounds.

7. In fly ash, the lead and chromium concentrations are measured for four particle-diameter ranges. Results are as follows: Suggest reasons for these trends.

| Particle diameter / μm | Concentration / μg g$^{-1}$ | |
|---|---|---|
| | Lead | Chromium |
| >10 | 870 | 330 |
| 6–10 | 990 | 760 |
| 2–6 | 1100 | 1800 |
| <2 | 1300 | 2700 |

8. Refer to the TOMS website given in Additional Resources 5. From the main page, select the 'Aerosol' link under 'Product' on the left-hand side of the page. At the bottom of the new page, choose the output 'global image'. Use the OMI satellite (default) and fill in the date. A dust storm that originated from the Gobi desert formed and dissipated quickly in early April, 2006. Access the data in the form of a series of global images for the period from April 5 to 10 and describe the progress of the dust storm. How is the aerosol index measured?

9. Refer to the TOMS website given in Additional Resources 5 and the instructions given in Problem 8. An unusual dusty haze appeared over Sydney, Australia on September 23, 2009. Follow the progress of this storm by observing the global images for September 20 to 25. Where did it originate and what was its final fate? Refer also to the website of the Department of Environment, Climate Change and Water of the Government of New South Wales, Australia— http://www.environment.nsw.gov.au/AQMS/aqidaily.htm, accessed October 2009.

   Using the link 'Search air quality data' on the left side of the page, search the archived data and observe the Air Quality Index value at Sydney on Sept 23, 2009. What is the AQI value for $PM_{10}$ aerosols on that day and how does this relate the actual concentration of these particles in the atmosphere? Note: A description of the Air Quality Index will be given in the next chapter.

10. A sample of solid aerosol is collected from a rural area and found to have as composition $Al = 2400$ mg kg$^{-1}$ and $Pb = 3.4$ mg kg$^{-1}$. Using data for the crustal abundance of elements (Appendix B.1), calculate the chemical concentration factor for lead. What might be the reason for a value such as you have obtained?

# Chapter 7
# Chemistry of urban and indoor atmospheres

**The focus of this chapter is on the air that surrounds us and the concerns we should be aware of as we continue with our work and recreation. We will accomplish this by:**

- describing the principal atmospheric pollutants in the urban environments where most of us spend the majority of our time
- reviewing the factors that affect indoor-air quality and discussing its common contaminants.

In previous chapters, we have examined some of the important chemical processes that occur in the troposphere. This has allowed us to consider specific types of atmospheric pollution including pollution due to sulphur dioxide, particulates, carbon monoxide, and a range of oxidants such as ozone and organic peroxides.

In the present section we will make use of this and additional information to describe the chemical composition of air in places where people live. We begin with an examination of air quality in major urban areas around the world. We will see that there are serious problems in some cities and that, to a very large extent, such problems are a direct consequence of energy use—energy in all its traditional and modern applications. In particular, the combustion of fossil fuels in motor vehicles, in space heating and cooling, in power generation and industrial processes, and the incineration of waste materials are all major contributors to atmospheric pollution. Use of petroleum products, especially in motor vehicles, results in ground-level emissions of carbon monoxide, volatile hydrocarbons, nitrogen oxides, and sometimes lead. Wherever these compounds are released, aldehydes and other secondary pollutants are also formed. The combustion of biomass and coal produces substantial concentrations of solid particulate matter along with nitrogen oxides, PAH compounds, and, in the case of much of the world's coal, sulphur dioxide as well. Open burning of refuse is an important cause of air pollution in many countries and this is a source of a wide variety of volatile organic compounds (VOCs) and solid particulate matter (SPM). Particulates are also provided from non-fuel sources—notably dust thrown up by movement of people and vehicles and wind-blown dust, especially in cities adjacent to deserts or dry, barren areas.

A second part of the chapter will examine related issues concerning the chemistry of atmospheres inside buildings. Depending on the building design, the same combustion products that are found in the outdoor urban atmosphere are again a source of concern indoors. Construction, furnishing, and cleaning materials are additional potential sources of other gaseous chemicals that are found in indoor atmospheres.

Epidemiologists have demonstrated several known and potential linkages between air pollution and human health. In the short term, very high levels of air pollution can produce acute respiratory distress. Long-term effects, however, are probably more common but the direct connections with air pollution more difficult to demonstrate. As an exception, the long-term direct relation between cigarette smoke and the incidence of lung cancer has now been well documented in numerous studies. Some PAH compounds, which are emitted where coal, wood and other solids are burned, are also known to be carcinogens and ozone, another common tropospheric pollutant, has been shown to damage lung tissue. In a more general sense, particulate matter interferes with normal respiratory functions and is associated with asthma, bronchitis and emphysema. In some cases it may also affect the cardiovascular system. Air pollution exacerbates the condition of people who already suffer from respiratory and cardiovascular diseases. One study[1] in Europe estimates that for the 36 million persons living in the cities studied, approximately 22 000 premature deaths occur annually; these are deaths that could have been prevented if strict guidelines to minimize air pollution had been followed. Unlike contaminated drinking water, bad air does not usually lead to rapid death and for this reason urban administrators often pay less attention to air quality than they do to the urban water supply. Nonetheless, it should be of the highest priority to ensure that the air we breathe is as fresh and healthy as possible.

---

**Fermi question**

Use the European statistics above (similar statistics have been provided in other countries) to estimate the number of premature deaths that could be attributed to air pollution in the city where you live. What factors would cause large deviations from these predictions.

---

## 7.1 Pollutants in the urban atmosphere

As more and more people move into and make their lives in cities, an emphasis on setting standards for air quality has been a preoccupation of jurisdictions around the world.

---

**The principal air pollutants**

- **Sulphur dioxide ($SO_2$).** Produced when fossil fuels high in sulfur (coal and oil) are burned, during smelting, or from other industrial processes.

- **Nitrogen oxides ($NO_x$).** Produced from both mobile sources like cars and stationary sources like power plants.

- **Particulate matter (PM).** Refers to fine particles suspended in air. PM is released from vehicles, power plants, and industrial process in nearly equal measure.

- **Carbon monoxide (CO).** An odorless gas emitted from vehicles, particularly those without a catalytic converter. Some emissions result from industrial fossil-fuel burning.

- **Lead.** Emitted by vehicles burning leaded gasoline, and from coal combustion, lead smelting and other metal processes.

- **Ozone.** Created under certain weather conditions (sunny, still days) by VOCs and nitrogen oxides.

- **Volatile organic compounds (VOCs).** Include hydrocarbons, alcohols, aldehydes, and ethers. VOCs play a role in ozone formation and are emitted by industrial processes and vehicles.

---

[1] Apheis Third Year Report, Air pollution and health: A European information system (2002–3).

**Table 7.1** Summary of European Union (EU) and World Health Organization (WHO) Air Quality Guidelines[a].

| Pollutant | Maximum time-weighted average concentration | | Averaging time[b] |
|---|---|---|---|
| | EU | WHO | |
| Sulphur dioxide | 350 $\mu$g m$^{-3}$ | 500 $\mu$g m$^{-3}$ | 1 h |
| | 125 $\mu$g m$^{-3}$ | 125 $\mu$g m$^{-3}$ | 24 h |
| | 20 $\mu$g m$^{-3}$ | 50 $\mu$g m$^{-3}$ | 1 y |
| Carbon monoxide | n.a. | 30 mg m$^{-3}$ | 1 h |
| | n.a. | 10 mg m$^{-3}$ | 8 h |
| Nitrogen dioxide | 30 $\mu$g m$^{-3}$ | 40 $\mu$g m$^{-3}$ | 1 y |
| | 200 $\mu$g m$^{-3}$ | 200 $\mu$g m$^{-3}$ | 1 h |
| Ozone | 180 $\mu$g m$^{-3}$ | 180 $\mu$g m$^{-3}$ | 1 h |
| | 120 $\mu$g m$^{-3}$ | 120 $\mu$g m$^{-3}$ | 8 h |
| Suspended particulate matter (SPM) | | | |
| Black smoke | n.a. | 100–150 $\mu$g m$^{-3}$ | 24 h |
| | n.a. | 40–60 $\mu$g m$^{-3}$ | 1 y |
| Total suspended particulates | n.a. | 150–230 $\mu$g m$^{-3}$ | 24 h |
| | n.a. | 60–90 $\mu$g m$^{-3}$ | 1 y |
| Respirable particulates (PM$_{10}$) | 50 $\mu$g m$^{-3}$ | 70 $\mu$g m$^{-3}$ | 24 h |
| Lead | n.a. | 0.5–1 $\mu$g m$^{-3}$ | 1 y |

[a] Values were obtained from footnotes 3 and 4. Regulations for other countries are also given in 4.
[b] The averaging time refers to the time period during which the weighted-average value should not exceed the specified guideline concentration.

A widely used set of guidelines is that recommended by the World Health Organization (WHO) of the United Nations (Additional Resources 5). Some of the WHO guidelines are summarized in Table 7.1 but many countries and groups of nations have established their own regulations.[2]

Guidelines always include a time period over which measurement is made, as human exposure and potential toxicity depend on both atmospheric concentration and duration of contact with the atmosphere.

$$\text{Exposure} = \text{concentration} \times \text{time}$$

Therefore, a safe mixing ratio over an 8-h period will always be smaller than the ratio allowed for a 1-h period. For example, the WHO guidelines specify that it is acceptable to be exposed to carbon monoxide at a level of 30 mg m$^{-3}$ for 1 h but, over an extended time of 8 h or more, the average level should not exceed 10 mg m$^{-3}$.

[2] Baldasano, J.M., E. Valera, and P. Jiménez, Air quality data from large cities. *The Science of the Total Environ.* **307** (2003) 141–165.

Considering the WHO guidelines, studies of air quality in 20 of the world's megacities have been carried out through the United Nations Environment Program (UNEP).[3] In this context, a megacity is defined as an urban agglomeration with a population of ten million or more. These cities are found on every continent except Australia and Antarctica and represent a variety of climatic, cultural, and technological situations. Since the publication of the UNEP report (1992), even more extensive monitoring of urban air has been carried out and Table 7.2 gives a selection of data based on comprehensive documentations reported in 2003[2] and 2008.[4]

## Suspended particulate matter

Most cities monitor values of solid particulate matter (SPM), usually as total solid particulates (TSP), but increasingly also as concentrations of small particulate matter ($PM_{10}$ or $PM_{2.5}$). It is clear that there is a severe problem of excessive concentrations of atmospheric particulates in many large cities around the world. The average levels in many of these, and other cities, is well in excess of the guideline level of 60 to 90 $\mu g\ m^{-3}$ set by the WHO. It is not uncommon that peak concentrations exceed 1000 $\mu g\ m^{-3}$. The human health consequences associated with such high values depends on the nature of the particulates; those derived from coal and those in the $PM_{10}$ and especially the $PM_{2.5}$ categories have been shown to be particularly hazardous. The yearly averages of these very fine particle fractions are less commonly measured, but seriously high averages have been observed in a number of Asian megacities as well as in other places.

In many urban areas, including Beijing, Karachi, and Dhaka, the elevated SPM levels are associated with a combination of sources—diesel vehicles, domestic fuel consumption, and / or nearby coal-fired power generation. In cities like Beijing, Cairo, Delhi, Tehran, and Mexico City this is supplemented at various times of the year by high natural loadings of wind-blown particulates. Of the cities listed here, it is mostly places in Europe and North America that consistently meet WHO guidelines with respect to particulates. This leads to a conclusion that there is a close relation between air quality in terms of atmospheric particulate concentrations and national wealth.

## Sulphur dioxide

Sulphur dioxide levels are affected by fuel types and are especially high where there is heavy use of diesel vehicles and where combustion of coal is common. As such, total solid particulates and sulphur dioxide levels often appear to be related. There are exceptions where we find cities that have high TSP but low sulphur dioxide levels. Delhi is an example and this may be due to the emphasis on compressed natural gas as the fuel to substitute for diesel in buses and other vehicles for public transportation. Chongqing, Tehran and a number of other cities in Asia have ambient levels of sulphur dioxide whose annual average concentration exceeds WHO guidelines by a factor greater than three. Peak daily concentrations are sometimes even greater than the higher value specified as the 10 minute limit.

In contrast, elsewhere around the world, urban sulphur dioxide concentrations are usually well within the defined range, although brief high-level excursions have been observed in London and Sao Paulo as well as in Kolkata during the dry season. Because coal use is now restricted in many cities, there has been a significant decline in sulphur dioxide concentrations over the past two decades and further improvements are predicted.

[3] World Health Organization and the United Nations Environment Programme, *Urban air pollution in megacities of the world*, Blackwell, Oxford; 1992.
[4] Gurjar, B.R., T.M. Butler, M.G. Lawrence, J. Lelieveld, Evaluation of emissions and air quality in megacities. *Atmos. Environ.*, **42** (2008), 1593–1606.

**Table 7.2** Summary of air quality in large cities from all continents around the globe[a].

| City | TSP MAC $\mu g\ m^{-3}$ | PM$_{10}$ MAC $\mu g\ m^{-3}$ | SO$_2$ MAC $\mu g\ m^{-3}$ | NO$_2$ MAC $\mu g\ m^{-3}$ | O$_3$ MAC $\mu g\ m^{-3}$ | O$_3$ Max 1 h $\mu g\ m^{-3}$ |
|---|---|---|---|---|---|---|
| Athens | 55 | | 25 | 58 | 49 | 228 |
| Bangkok | 100 | | 18 | 39 | | |
| Beijing | 377 | | 90 | 122 | | |
| Buenos Aires | 185 | | 20 | 20 | | |
| Cairo | 593 | | 69, 37 | 59 | | |
| Capetown | | | 21 | 57 | 37 | |
| Chongqing | 320 | | 340 | 70 | | |
| Dhaka | 516 | | 120 | 83 | | |
| Delhi | 415, 405 | | 24, 18 | 41, 36 | | |
| Karachi | 668 | | 13 | 30 | | |
| London | | 28 | 10 | 33 | 30 | 154 |
| Los Angeles | | 39 | 9 | 66 | | 225 |
| Madrid | 35 | | 19 | 63 | | 177 |
| Melbourne | — | 17 | 3 | 15 | | 191 |
| Mexico City | 201, 201 | 52 | 46, 47 | 55, 56 | 72 | 546 |
| Moscow | 100, 150 | | 15 | 80, 170 | | |
| Mumbai[b] | 240, 243 | | 33, 19 | 39, 43 | | |
| New York | 27 | 24 | 26, 22 | 70, 63 | | 272 |
| Paris | | 22 | 9 | 43 | 35 | 238 |
| Sao Paulo | 53 | | 18 | 47 | | 403 |
| Seoul | 84 | | 44 | 60 | | |
| Shanghai | 246 | | 53 | 73 | | |
| Tehran | 248 | | 209 | | | |
| Tokyo | 49, 40 | | 78, 19 | 68, 55 | | |

[a] Data from references in footnote 2 (and in bold) 3. MAC is the mean annual concentration, and Max 1 h is the greatest concentration observed over a one-hour period.

## Nitrogen dioxide

The previous data show that particulates and sulphur dioxide problems are associated with the use of what are often considered lower grades of fuel, and limited resources to control emissions. Nitrogen dioxide emissions, on the other hand, are excessive in many places including some of the wealthiest cities in the world. This is, of course, because the major source is from automobiles and the fuel quality has little to do with the amount released. There are, of course, increasingly stringent efforts to control automobile emissions of this gas, but at the same time, the number of automobiles and the extent of their use have both increased, more than making up for the improved emission controls. Approximately two thirds of the cities listed in Table 7.2 show annual averages that are greater than the WHO guidelines.

## Ozone

Ozone is a secondary pollutant, and as such its concentration gives an integrative measure of emissions both of gaseous hydrocarbons and nitrogen oxides emanating from fossil-fuel combustion. Not all jurisdictions, however, monitor ozone on a regular basis. Where it is measured, exceedences are seen to be common. Mexico City, for example, provides a particular set of conditions (see Chapter 4) that are near ideal for the formation of ozone and this is reflected in the high average and extremely high hourly values that are sometimes observed. The metropolitan area of Mexico City occupies an area of about 2500 km$^2$. Situated at an altitude of 2240 m, the region is surrounded by mountains. Within this contained territory, winds tend to be light and frequent temperature inversions occur, leading to persistent accumulations of air pollutants. There is an average of about seven hours of sunshine every day throughout the year. In 2008, the total population of the Mexico City agglomeration was estimated to be 22.8 million, giving a density of close to 9000 persons per square kilometre. Many and varied industries are located within the area and there is a vast and complex transport system; in sum, the total consumption of energy estimated to be greater than 500 PJ (1 PJ (petajoule) = $10^{15}$ J) each year. The combination of all these factors provides a 'perfect storm' situation in which ozone levels three times above guideline levels are sometimes observed.

## Lead

Airborne lead depends on the number density of motor vehicles, the concentration of lead additives in the fuel, and the availability of unleaded fuels. Concentrations in leaded gasoline vary between 0.1 and 2 g L$^{-1}$ although, increasingly, use of tetraethyl lead to augment the octane number is being abandoned in favour of other less toxic formulations. High-performance engines such as those in racing cars require the use of leaded fuel. Where tetraethyl lead has already been eliminated, atmospheric lead concentrations are measurably smaller. This includes Brazilian cities where low-lead gasohol produced from sugar cane is frequently used as a fuel. Residual high concentrations of lead persist, however, in roadside soils around the world.

## Air quality indices

We have seen that there are several specific issues that together are used to describe the quality of air. Each one of these issues is important, but to make matters simpler for the public, it is useful to develop a single index that can be used to give an overall picture of air quality. This has been done in many jurisdictions. An example is the Air Quality Index (AQI) in the United States.

The Environmental Protection Agency (EPA) in the United States measures air quality in all parts of the country and publishes a daily AQI based on the data obtained. The following six priority pollutants are measured regularly in order to generate the index:

- carbon monoxide (CO);
- nitrogen dioxide (NO$_2$);
- particulates (PM$_{10}$ and PM$_{2.5}$);
- sulphur dioxide (SO$_2$);
- ozone (O$_3$, 1 h and 8 h);
- lead (Pb).

For each parameter, the measured value is compared with a defined standard considered safe and a scaled value is determined using a range 0 to 500. Table 7.3 is an example of AQI values (and associated colours) that correspond to specific concentrations of individual pollutants. The index is calculated by:

$$I = \frac{I_{\text{high}} - I_{\text{low}}}{C_{\text{high}} - C_{\text{low}}}(C - C_{\text{low}}) + I_{\text{low}} \tag{7.1}$$

where $I$ = the Air Quality Index,

$C$ = the measured pollutant concentration,

$C_{low}$ = the concentration breakpoint that is $\leq C$,

$C_{high}$ = the concentration breakpoint that is $\geq C$,

$I_{low}$ = the index breakpoint corresponding to $C_{low}$,

$I_{high}$ = the index breakpoint corresponding to $C_{high}$.

'Breakpoints' refer to the defined high and low values of either concentration or corresponding index. Note that the concentrations of gaseous species are given as mixing ratios in ppmv, in contrast to the WHO guidelines where concentrations ($\mu$g m$^{-3}$) are used throughout.

The EPA-defined index and concentration breakpoint values are given in Table 7.3.

---

**Example 7.1  Calculation of an Air Quality Index (AQI) value**

Consider a day in which the 8-h average mixing ratio of carbon monoxide in an urban area is 7.5 ppmv. The AQI is calculated using eqn. 7.1

$$I = \frac{100 - 51}{9.4 - 4.5}(7.5 - 4.5) + 51 = 82$$

The AQI value is 82, within the moderate range.

---

Using these calculations, any value $<50$ is considered to be acceptable in terms of general human health, 50 to 100 levels may affect a small number of especially sensitive persons, air in the 100 to 150 range is unhealthy for sensitive persons, and $>150$ can have a widespread effect on the general population. Values greater than 150 are especially serious and call for an air pollution alert or warning.

**Table 7.3** Air Quality Index (AQI) values and colours compared to breakpoint concentration ranges for individual pollutant species[a].

| Categories | Breakpoint AQI values (colour) | Breakpoint concentration values (ppmv) | | | ($\mu$g m$^{-3}$) |
| --- | --- | --- | --- | --- | --- |
| | | $O_3$ (8 h) | CO (8 h) | $SO_2$ (24 h) | PM$_{2.5}$ |
| Good | 0–50 (green) | 0.000–0.064 | 0.0–4.4 | 0.000–0.034 | 0.0–15.4 |
| Moderate | 51–100 (yellow) | 0.065–0.084 | 4.5–9.4 | 0.035–0.144 | 15.5–40.4 |
| Unhealthy for sensitive groups | 101–150 (orange) | 0.085–0.104 | 9.5–12.4 | 0.145–0.224 | 40.5–65.4 |
| Unhealthy | 151–200 (red) | 0.105–0.124 | 12.5–15.4 | 0.225–0.304 | 65.5–150.4 |
| Very unhealthy | 201–300 (purple) | 0.125–0.374 | 15.5–30.4 | 0.305–0.604 | 150.5–250.4 |
| Hazardous | 301–500 (maroon) | >0.375 | >30.5 | >0.605 | >250.5 |

[a] The United States Environmental Protection Agency, *Technical assistance document for reporting of daily air quality—the Air-quality index*, EPA-454 / B-09-001, 2009. http://www.epa.gov/airnow/aqi_tech_assistance.pdf, accessed November 2009.

Importantly, in the USA when a number of air-quality parameters are measured and the AQI for each is calculated, the reported AQI is given, not by combining individual values, but rather by using the single value that gives the highest reading. For example if, on a given day, the highest individual AQI value is 106 and relates to particulates, the air quality is reported as 'unhealthy for sensitive persons' even if all of the other parameters fall within the 'good' category.

Many other countries use AQI methodologies based on that proposed by the USA EPA, although the details of calculation of individual values and the scales used differ from jurisdiction to jurisdiction. In order to make integrative comparisons of air quality in cities around the world, a combined Multi-Pollution Index (MPI) has been proposed, based on WHO guideline values.[4] The MPI involves averaging index values for individual pollutants. When cities are to be compared, multiple observations are made and included in the calculation.

$$\text{MPI} = (1/n)[\Sigma\{(AC_i - GC_i)/GC_i\}] \qquad (7.2)$$

where $n$ = the number of individual observations
$AC_i$ = a concentration value for an individual atmospheric pollutant
$GC_i$ = the guideline value for that pollutant.

The MPI calculation can be illustrated using a simple case of single measurements of three pollutants:

| Pollutant | $AC_i$ | $GC_i$ |
|-----------|--------|--------|
| $SO_2$ | 63 µg m$^{-3}$ | 125 µg m$^{-3}$ |
| CO | 12 mg m$^{-3}$ | 10 mg m$^{-3}$ |
| $NO_2$ | 222 µg m$^{-3}$ | 200 µg m$^{-3}$ |

---

**Example 7.2 Calculation of the Multi-Pollution Index (MPI) value**
Using eqn. 7.2 and the values provided in the text above, the MPI is calculated as follows

$$\text{MPI} = (1/3)[(63 - 125)/125 + (12 - 10)/10 + (222 - 200)/200]$$
$$= -0.50 + 0.20 + 0.11 = -0.19$$

The MPI value is −0.19, within the good air-quality range.

---

From the calculation, you will see that values of MPI are usually fractional, with negative numbers indicating good air quality and positive numbers showing poor air quality. Reference 4 (above) provides data for 18 cities involving 18 measurements on each of 3 pollutants over an 11-year period. As examples, the calculated MPI values included: Tokyo −0.27, Buenos Aires −0.01, Kolkata 0.59 and Dhaka 2.40.

**Main point 7.1** Urban air pollution problems are frequently associated with the products of combustion in industry, in vehicles, and for domestic purposes. Suspended particulate matter, carbon monoxide, sulphur dioxide, nitrogen oxides, and ozone are common pollutants. Many large cities around the globe exhibit excessive levels of one or more of these pollutants.

**Literature link** Volatile organic compounds in the Chinese atmosphere

The Pearl River Delta region of southern China has experienced dramatic growth in population and in industrialization during the past 30 years. With these developments has come a significant increase in emissions of airborne pollutants, and these are from many sources including vehicle exhaust, gasoline and diesel vapour, paint, asphalt, industrial and residential coal burning, biomass burning, and the petrochemical industry. Each one of the sources generates a characteristic suite of volatile organic compounds (VOCs) into the atmosphere. As we know, many of these chemicals can be harmful in themselves and they also act as precursors for smog formation. A research article by Liu, Y., M. Shao, L. Fu, S. Lu, L. Zeng, and D. Tang entitled *Source profiles of volatile organic compounds (VOCs) measured in China: Part I.* (Atmospheric Environment **42**, 6247–6260 (2008)) describes the methodology that was used to determine the chemical nature of emissions from individual sources and to assess the relative magnitude of each.

Samples of emissions from the sources noted above were collected, and two capillary gas chromatography procedures were used for analysis. One employed an alumina column with a flame-ionization detector and was used for low molar mass (C2 to C4) hydrocarbons. The other used a semi-polar column with quadrupole mass spectrometry detection to determine C4 to C12 straight- and branched-chain hydrocarbons including aromatics.

The various sources of VOCs each required different sampling strategies. Headspace analysis from sealed bottles was used to collect samples of fuel emissions due to evaporation. For different types of paint and asphalt, ambient-air collection was carried out in the open air adjacent to the actual painting or paving operations. Similarly, ambient-air samples in the region of a major petrochemical industry were obtained. Canisters were used to collect air 1 m downwind of chimneys during household burning of various biomass fuels and different types of coal. Samples from every situation generated their own characteristic emission signal, each described in detail in the paper. In total, ninety-two individual species were identified and quantified. Quantification was done by calculating the percentage associated with a single compound compared with the total amount of all compounds in the sample.

Vehicles were considered to be the major source of emitted VOCs in this region of China. Light and heavy duty vehicles as well as motorcycles were tested. Chassis dynamometer tests were carried out on both gasoline- and diesel-powered vehicles, and tailpipe exhaust samples collected in each case. The samples were diluted with purified air and a portion withdrawn using a constant-volume sample system. Ethylene, isopentane, benzene, and toluene were found to be principal species in all the exhaust emissions. Diesel engine exhaust also contained heavier C9 to C12 alkanes, compounds that were absent when gasoline was fuel. On the other hand, gasoline engines emitted a range of methyl-substituted benzene compounds not found in diesel exhaust. There were interesting differences in volatile emission products from automobile and motorcycle engines, both fuelled with gasoline. From cars, ethylene, propene, benzene and toluene were prominent products of fuel combustion. In contrast, motorcycle engine emissions contained a number of C5 and C6 branched-chain alkanes. These are characteristic of the original fuel and indicated that combustion was less efficient in the small engines. Additional data were provided and taken together, these show that for each individual situation, the emission profile depended on a number of factors such as engine design and emission control systems as well as, most importantly, fuel type.

By examining results from a large number of situations, possible tracer (marker) compounds were identified. Tracers are compounds that always made up a significant fraction of the sample from a given source, and therefore individually or as ratios of two or more compounds can be used to predict the origin of an unknown sample.

A number of important issues are addressed in this work and several additional research questions emerge:

- Are there alternatives to the analytical methodology used here?
- How do the results compare with those described in other situations? Recall, for example, the data in Chapter 4 that was obtained in an earlier study in Taipei, Taiwan.
- Is the goal of searching for unique marker compounds realistic?
- Are there obvious differences in the human health implications of emissions from different sources?

## 7.2 Indoor air quality

### Factors influencing indoor air quality

Many of us spend the greater portion of our lives indoors, and the atmospheres we encounter there can exhibit an extreme variety of conditions. They range from the highly purified, chemically and microbially sterile atmosphere in a hospital operating room to the dusty environment of a poorly ventilated feed mill. Space does not permit us to deal with such specific and individual cases, but we will consider the nature of indoor atmospheres under 'normal' conditions such as in homes. Here too there are a number of possibilities. In some parts of the world where the climate is warm year round, a sizeable fraction of the population lives in houses constructed from clay-rich soils or other fresh or baked-earth materials. Because of the warm temperatures, these homes are open much of the time and air exchange is very rapid. However, sometimes heating (in cooler seasons) or cooking is done over an open fire in a room without a chimney and a variety of fuels are used so that a complex suite of gaseous and particulate emissions is produced. The range of air properties in these houses will be much different from those built where the climate is temperate or cold. In the latter regions, construction materials are typically a combination of brick, stone, wood, various plastics, and metals. Often such homes are well insulated and sealed so that there is little exchange between outdoor air and that inside. Many activities take place within the home, including cleaning, cooking, heating by open or enclosed fires and, in some cases smoking, to name a few. The result is a different set of variable air conditions.

Four major factors determine the quality of indoor air.

- The first is the nature of the ambient air, outdoors around the building. Whether it is rapid or slow, exchange inevitably does occur so that the indoor atmosphere is always influenced by air outside. In the case of some pollutants, the only sources are from outdoors. We have seen that ozone production is a photochemical process requiring ultraviolet radiation. Therefore, it cannot normally be produced within a building (except by certain electrical devices such as copying machines and electrostatic precipitators used for dust control), yet indoor ambient levels of 5 to 15 $\mu$g m$^{-3}$ are typically measured. In most cases all of the ozone present in a building originated in the atmosphere outside. As would be expected therefore, the ratio $C_{ozone}$(indoors) / $C_{ozone}$(outdoors) is usually substantially smaller than unity.

- The second factor is the infiltration rate or air-exchange rate for a building. This relates to the site and design of the building. In the open design home of the tropics, there are many exchanges per hour. On the other hand, an insulated climate-controlled house in winter typically has an air-exchange rate of 1 h$^{-1}$, while tightly sealed 'energy-efficient' buildings exchange air only every 2 or even 10 h giving air-exchange rates of 0.5 to 0.1 h$^{-1}$, respectively. In these latter cases, the influence of the outside atmosphere on air quality inside is much less. An important corollary is that any atmospheric chemical that originates indoors accumulates to a greater degree when air exchange is limited.

- A third influence on indoor air is the material present in the building as construction material or in other forms. We shall see that many modern polymers used for construction are sources of formaldehyde and their presence in the building provides a passive but continuous supply of this chemical. Likewise, building materials derived from the Earth—clays, laterite, concrete—contain varying trace amounts of radioactive elements that generate radon gas, ultimately a source of alpha-radiation. Every material, just by being present, is a potential source for emission of a variety of chemicals.

- Finally, and perhaps most importantly, the activities that go on inside a building determine the nature of the indoor atmosphere. Combustion for heating or cooking, whether isolated from the atmosphere in the living space, as in a high-efficiency natural-gas furnace, or in an open fire, releases gaseous and / or particulate emissions. Smoking is another combustion process that has a major effect on the atmosphere of an enclosed area. Any cleaning activity, whether by mechanical means that can raise dust, or with cleaning aids that contain volatile solvents, etc., alters the indoor atmosphere. In fact, every activity inside a building inevitably has a small or large effect on the atmospheric composition within the enclosed space.

A general equation describing the steady-state behaviour of a stable compound with respect to indoor and adjacent outdoor concentrations is

$$R_i = k_e C_i - k_e C_o \tag{7.3}$$

where $R_i$ (units of concentration per time) is the net rate of production of the compound inside, $C_i$ and $C_o$ are the indoor and outdoor concentrations, respectively, and $k_e$ is the first-order rate constant for atmosphere exchange (defined above as the air-exchange rate, with units of time$^{-1}$). In the steady state the interior concentration is given by

$$C_i = C_o + R_i / k_e \tag{7.4}$$

For cases where the outdoor contribution is negligible,

$$C_i = R_i / k_e \tag{7.5}$$

Where none of the chemical is produced indoors,

$$C_i = C_o \tag{7.6}$$

unless the chemical is lost by reaction as it infiltrates the building. Furthermore, each of these relations assumes a stable chemical, meaning that it is not lost either by reaction or deposition inside the building.

---

### Example 7.3 Concentration of indoor levels of VOCs

Consider a situation where an open fire in a well-ventilated home produces VOCs at a rate of 30 mg m$^{-3}$ h$^{-1}$. A complete exchange of air takes place every 5 min (1/12 h). The ambient outdoor concentration of VOCs is 75 $\mu$g m$^{-3}$. Calculate the expected indoor concentration of these compounds.

In terms of hours, air exchange is 1/12 h. Therefore $k_e = 1/(1/12) = 12$ h$^{-1}$. Using eqn 7.4.

$$C_i = C_o + R_i / k_e$$

$$C_i = 75 \ \mu g \ m^{-3} + \frac{30\,000 \ \mu g \ m^{-3} \ h^{-1}}{12 \ h^{-1}}$$

$$= 2500 \ \mu g \ m^{-3}$$

---

**Main point 7.2** Air quality indoors depends primarily on the nature of the surrounding outdoor air, the air-exchange rate, the materials used in construction and furnishing the house, and activities taking place in the enclosed space.

## 7.3 Common indoor air contaminants

In this section we will consider three types of air-quality issues that are commonly encountered in indoor situations. The nature and concentration of the chemicals relate to all four of the factors affecting air quality, including the types of structural materials present in the building and the activities that take place indoors.

## Radioactivity

Everywhere, we are exposed to a constant low level of radiation of various sorts including radioactive emissions from elements that are present, usually in trace quantities, in the Earth's crust. Inside a building, most radioactivity above the outdoor background level is associated with radon. Radon (Rn, element 86) is a dense, radioactive, noble gas that is itself the product of radioactive decay processes beginning with uranium-238 and thorium-232. The two parent isotopes are present at low concentration in many geological materials and have half-lives of 4.5 and 14 billion years, respectively. Because of their long half-lives, trace quantities persist to the present day in soil that surrounds structures, in building materials derived from rock or other geological sources, and also in water that is in contact with the soil and rocks. Both uranium and thorium decay through complex radiochemical reaction series, emitting alpha, beta, and gamma radiation, as shown in Fig. 7.1. Intermediate daughter products in the sequences include radon-222 and radon-220.

When there are dangerous levels of radiation in indoor air they are not usually associated *directly* with the emissions from uranium and thorium. Rather, the dangerous radiation emanates from daughter nuclei formed in the initial decay process. When alpha particles are given off by uranium or thorium in solid materials or water, the heavy matrix absorbs most of the alphas before they reach the surrounding air. Only a small number of particles are able to escape out of the material surface and this forms part of the background radiation within a building. The heavy alpha particles are also readily absorbed by air and very few move more than 30 or 40 cm away from the solid or liquid. Where they do encounter biological tissue they are even more strongly absorbed and do not penetrate much beyond the surface. As a consequence, there is little concern about such low-level direct alpha emissions from uranium and thorium.

The situation is different with the radon gas produced in the decay sequence of the two heavy elements. Being a gas, radon escapes from the construction materials and the surrounding soil and water; it penetrates through cracks in a building and is released into the indoor atmosphere where it can be inhaled. Decay of radon-222 (half-life = 3.8 days) and radon-220 (half-life = 55.6 s) via alpha emission then takes place directly inside the lungs. The radioactive decay products include isotopes of metals such as polonium-218, -216, -214, and -212 that are deposited on the internal tissues. Further decay of these radioactive isotopes releases more alpha particles that are then available to interact immediately with cell molecules. Of the two isotopes of radon, radon-220 has a very short half-life giving it little time to escape into and accumulate in the ambient atmosphere. It is therefore a less abundant and less important form of the element than is radon-222.

It is instructive to consider a semi-quantitative relation between production and loss of radon in an enclosed air space—a relation that is analogous to eqn 7.1. We can assume a steady-state indoor rate of radioactive decay, $A_i$, over time—measured as activity in a 1 L volume of air, Bq $L^{-1}$, using the SI unit for radioactivity.[5] Note that activity acts as a surrogate for measuring the concentration of a radioactive element. Outside the building, the decay rate is also constant at $A_o$ (Bq $L^{-1}$). In the present situation, two first-order rate constants are required to describe the interchanges in radioactivity concentration: $k_e$ ($h^{-1}$) is the air-exchange rate as defined earlier and $k_d$ ($h^{-1}$) is the radioactive decay constant, in this case for radon-222. Equation 7.1 is therefore modified for this new situation

$$R_i + k_e A_o = k_e A_i + k_d A_i \qquad (7.7)$$

---

[5] The becquerel (Bq) is defined as one disintegration per second. Another basic (non-SI) unit for radiation is the curie (Ci), which is equal to $3.7 \times 10^{10}$ Bq.

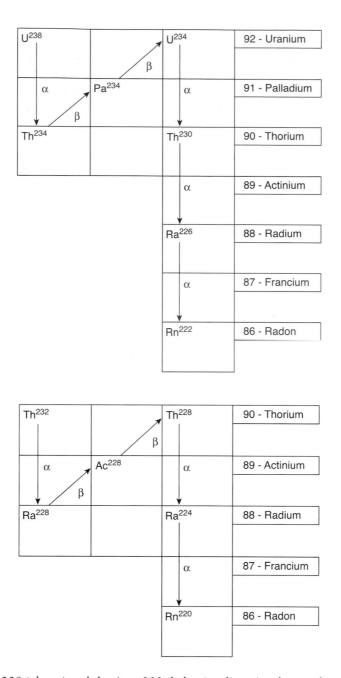

**Fig. 7.1** Uranium-238 (above) and thorium-232 (below) radioactive decay schemes to produce radon-222 and radon-220, respectively. The product isotopes are gases and are themselves radioactive, decaying further to stable products.

The left-hand terms in eqn 7.7 include internal and external sources of radon, and the right-hand terms are for loss by exchange and radioactive decay, respectively. Rearranging eqn 7.7, the interior radioactivity concentration is given by

$$A_i = (R_i + k_e A_o)/(k_d + k_e) \tag{7.8}$$

## Example 7.4 **Indoor radioactivity**

Consider a situation where the rate of production of radioactivity inside the building, $R_i$ is 10 Bq m$^{-3}$ h$^{-1}$, and the external radioactivity concentration, $A_o$, is 4.0 Bq m$^{-3}$. Calculate the steady-state interior radioactivity concentration, $A_i$, in two situations: (a) an open building with excellent air-exchange rate = 20 h$^{-1}$ and (b) a tightly sealed 'energy-efficient' building with air-exchange rate = 0.10 h$^{-1}$.

Radon-222 has a half-life of 3.8 days, equivalent to a decay constant, $k_d = \ln 2/t_{1/2}$, of 0.00754 h$^{-1}$. Using eqn 7.8.

$$A_i = (R_i + k_e A_o)/(k_d + k_e)$$

For the open building, with rapid air exchange,

$$A_i = \frac{10 + (20 \times 4)}{0.00754 + 20}$$

$$= \frac{90}{20}$$

$$= 4.5 \text{ Bq m}^{-3}$$

For the tightly sealed, energy-efficient building.

$$A_i = \frac{10 + (0.10 \times 4)}{0.00754 + 0.10}$$

$$= \frac{10.4}{0.107}$$

$$= 97 \text{ Bq m}^{-3}$$

Clearly, the rate of production and the exterior activity of radon would also be important determinants affecting activity inside. In Example 7.2, the value used for outdoor radioactivity associated with radon is the global average of about 4.0 Bq m$^{-3}$. Levels of radioactivity less than 10 Bq m$^{-3}$ are considered low, levels around 100 Bq m$^{-3}$ are normal, and levels greater than 4000 Bq m$^{-3}$ are high. The World Health Organization recommends 100 Bq m$^{-3}$ as the optimum reference level to minimize health hazards but, if this is not achievable under country-specific conditions, it should not in any case exceed 300 Bq m$^{-3}$.

The relations describing decay in the radioactive chemical example serve as a good example for other indoor-air contaminants that undergo degradation by normal chemical processes. A more general version of eqn 7.7 that applies to these situations is

$$R_i + k_e A_o = k_e A_i + k_r A_i \tag{7.9}$$

where the terms are defined as before, except that $k_r$ is the rate constant for any reaction that causes degradation or loss of the contaminant of interest. In indoor air situations, loss refers most commonly to sorption by surfaces such as concrete, plaster, paints, vinyl, and various fabrics used for carpets and upholstery. This important issue is very complex since sorption may be an equilibrium surface phenomenon or it may involve diffusion deep into pores of the materials and be essentially irreversible.

## Volatile organic compounds

Volatile organic compounds (VOCs) that are present in indoor atmospheres arise from a number of sources. Many organic compounds are emitted from construction materials, furnishings, and consumer products in the building. Combustion processes generate an additional suite of VOCs.

Construction materials, including wood composites made from chips or sawdust, insulating foams, floor tiles, carpet, and the adhesives used for installation, all contribute to the VOC burden in an indoor atmosphere. Emissions from freshly manufactured and installed material are particularly high in a new building; they decline rapidly over the first few months and then more slowly over extended periods of time. There are many aromatic and aliphatic compounds that contribute to VOC concentrations, with chloroform, acetone, chlorinated compounds, and formaldehyde being predominant in many locations.

Consumer products used in homes contribute other VOCs to the atmosphere. For example, latex paint contains toluene, ethylbenzene, 2-propanol, and butanone. Cleaning agents, household solvents, detergents, and waxes all contain various small molar mass volatile organic species.

Small concentrations (typical values are between 10 and 200 $\mu$g m$^{-3}$ in a new construction) of formaldehyde are observed in the air of many buildings and its behaviour has been extensively studied. Formaldehyde or its precursors are present in a wide variety of modern materials found in homes and other buildings. They are found in resins that are used to make plywood and various types of particle board. Anti-shrink, anti-wrinkle, colour-retention, and fire-retardant chemicals modify many fabrics; all of these may also be sources of the chemical. Some papers such as pre-pasted wallpaper contain formaldehyde resins. Urea formaldehyde (UF) polymers, in several forms, are perhaps the largest source of formaldehyde. Where the polymers have been made into high-density moulded or extruded plastics, the surface area is relatively small and chemical reactions to release formaldehyde in gaseous form are slow. However, the UF polymers are also widely used in the form of foams that produce a sponge-like porous solid with an enormous surface area. Such UF foams are convenient and efficient insulating materials for structures built in areas of cold climate. However, UF foams have been shown to be sources of significant quantities of formaldehyde, with emission occurring by both rapid- and slow-release processes depending on whether the formaldehyde is in free or combined forms.

There is initial rapid evolution of free formaldehyde, and $N$-methylol end-groups present in the resin can react rapidly, releasing additional product (reaction 7.10).

$$R-\underset{\underset{H}{|}}{N}-\underset{\underset{}{\overset{O}{\|}}}{C}-\underset{\underset{H}{|}}{N}-CH_2-OH \longrightarrow R-\underset{\underset{H}{|}}{N}-\underset{\underset{}{\overset{O}{\|}}}{C}-NH_2 + CH_2O \qquad (7.10)$$

UF resin end group

When these sources are expended, continuous nearly steady-state slow release occurs, mostly due to hydrolysis reactions of methylene bridge groups in the polymer backbone (reaction 7.11).[6] Release rates are enhanced at high temperatures and, because the reactions are due to hydrolysis, humid conditions also increase the rate during the steady-state period.

$$R-\underset{\underset{H}{|}}{N}-\underset{\overset{O}{\|}}{C}-\underset{\underset{H}{|}}{N}-CH_2-\underset{\underset{H}{|}}{N}-\underset{\overset{O}{\|}}{C}-\underset{\underset{H}{|}}{N}-R + H_2O \longrightarrow 2R-\underset{\underset{H}{|}}{N}-\underset{\overset{O}{\|}}{C}-NH_2 + CH_2O \qquad (7.11)$$

UF resin backbone

A qualitative picture of the pattern of predicted formaldehyde release to the indoor air of a newly constructed building is shown in Fig. 7.2.

Loss of atmospheric formaldehyde occurs by reactions that were discussed in Chapter 4. The first step (reaction 7.12) of its decomposition is a photochemical process that requires energetic radiation having wavelengths at the lower end of the UV-A and in the UV-B regions of the

[6] Allan, G.G., J. Dutkiewicz, and E.J. Gilmartin, Long-term stability of urea-formaldehyde foam insulation, *Environ. Sci. Technol.*, **14** (1980), 1235–41.

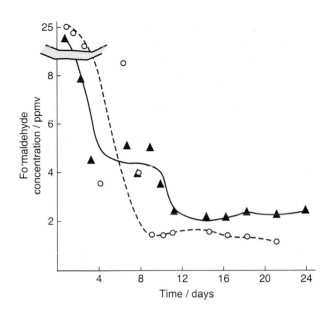

**Fig. 7.2** Time sequence of formaldehyde release in a building containing formaldehyde-based polymers. The plot shows the rapid initial release of formaldehyde, declining to give a lower steady-state concentration in the air. Room maintained at 33°C; measurements taken over a 30-min sampling period. Filled triangles, high-humidity conditions; open circles, low humidity conditions. (Figure redrawn from data provided in Gammage, R.B. and K.C. Gupta, *Formaldehyde*, in Additional Resources 3, p. 113.)

spectrum. Fluorescent lamps emit significant amounts of UV-A radiation, while radiation from incandescent lights is mostly in the visible and near-infra-red regions.

$$\text{HCHO} \xrightarrow{\quad hv, \lambda < 330 \text{ nm} \quad} \text{HCO·} + \text{·H} \tag{7.12}$$

The two radicals react further with oxygen, producing the hydroperoxy radical

$$\text{·H} + \text{O}_2 \rightarrow \text{HOO·} \tag{7.13}$$

$$\text{HCO·} + \text{O}_2 \rightarrow \text{HOO·} + \text{CO}_2 \tag{7.14}$$

The hydroperoxy radical is a highly reactive species and is consumed by a number of possible reactions, one of which is the oxidation of nitric oxide.

$$\text{HOO·} + \text{NO} \rightarrow \text{·OH} + \text{NO}_2 \tag{7.15}$$

Outdoor clean-air concentrations of formaldehyde are typically less than 10 ppbv but, in buildings that contain substantial formaldehyde-producing materials, concentrations between 100 and 500 ppbv have frequently been observed. Values that are even several times higher than these are not unusual. A mixing ratio of 100 ppbv has been proposed as a level of concern.

In recent decades, in Europe and North America there has been an increasing number of complaints about the indoor atmosphere from residents of some private homes and workers in offices and other commercial buildings. It is claimed that the climate in certain buildings produces unhealthy symptoms, commonly including mucosal and skin irritations, tiredness, dizziness, headaches, and general unease. Where this is a common occurrence, the building is often characterized as *sick* and the collection of symptoms are together referred to as *sick-building syndrome.*

While the syndrome is not usually associated with a well-defined source, there are common features related to materials used in construction and furnishings, inadequate ventilation, and features of the building's operation. Most probably, formaldehyde and other organic chemicals as have been discussed contribute to the unhealthy symptoms.

### Polybrominated diphenyl ether (PBDE) compounds

One class of chemicals that has recently come under considerable scrutiny is the polybrominated diphenyl ethers (PBDE), whose general structure is shown here.

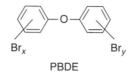

PBDE

These compounds with varying degrees of bromine substitution are commonly used as fire retardants by incorporation in many commercial and household products, including plastic casings for appliances and fabrics used for clothing, carpets, and other materials. (See Chapter 3 for a brief discussion of the action of fire retardants.) Like formaldehyde, the PBDEs slowly volatilize into the atmosphere where they are persistent, bioaccumulative, and toxic. Effects on brain development in the young as well as liver toxicity, disturbance of thyroid hormone levels, and endocrine system disruption are some of the consequences of excessive exposure. The widely used pentabrominated congener (penta-BDE) is associated with the greatest number of health effects. There is special concern that increasing levels have been found in the breast milk of women who live in countries and situations where there is significant exposure to treated products. Levels in breast milk are measured in relation to the lipid mass. These concentrations would historically be zero, in the absence of anthropogenic inputs, but they have been increasing exponentially in the last two or three decades. Typical values in Sweden have been found to be about 4 ng g$^{-1}$, while in North America they are much higher. Samples collected in New York State in 1997 averaged about 140 ng g$^{-1}$, but are highly variable between regions and even within regions. Rapidly rising levels (up to 72 ng g$^{-1}$ wet weight) are also observed in fish, some in areas remote from obvious sources of the compound. In the early part of the present century, several countries are taking steps to limit the use of PBDEs.

The effects of the various volatile chemicals used in modern manufacturing are enhanced in situations where they are able to accumulate by adsorption on surfaces within the building. This occurs when the building contains a large 'concentration' of high surface area materials such as books and papers on open shelves, carpets, draperies, and other textiles. Chemicals are sorbed by these high-surface-area materials, and then can be re-emitted especially when the temperature is elevated—often when people are present in the building. Many volatile organic compounds have been measured in the air of such buildings. Formaldehyde and phenol are two commonly encountered chemicals, but other aliphatic and aromatic hydrocarbons, aldehydes, ketones, acids, alcohols, and esters are frequently found in measurable concentrations in the atmosphere. Careful choice of building and furnishing materials and efficient ventilation are therefore design considerations required in the construction of a healthy building. Note that this can present challenges in building design. A well-insulated building with limited air exchange is energy-efficient but we have shown that it may be 'unhealthy' because of the accumulation of harmful chemicals in the indoor air. Engineering solutions usually specify bringing in a controlled flux of outdoor air that is pre-heated (or pre-cooled) by means of an efficient heat-exchange system.

### Gaseous products of indoor combustion

Combustion inside a building contributes to the concentration of VOCs and it also is a source of stable inorganic gases and sometimes particulates. Burning fuel indoors is a common practice—usually for purposes of heating or cooking. The nature of the combustion products depends on the fuel used as well as on the fuel:oxidant ratio and other combustion conditions. There are many types of fuel, but oxygen in air is invariably the oxidant. The extent to which combustion products are released into the building depends on the design of the heat production system. Such designs range from high-efficiency furnaces that draw in air and release the emission products outside the building, to heating or cooking devices that generate and release both gases and particulates inside a poorly ventilated room. Tobacco smoking is another combustion issue.

Combustion of carbon-based fuels always produces carbon dioxide leading to increased atmospheric mixing ratios of this gas. Even where ventilation is poor, however, carbon-dioxide levels rarely exceed 1000 ppmv (recall that outdoor ambient levels are 387 ppmv), a mixing ratio that is usually considered to be acceptable. Carbon-monoxide emissions are also associated with all burning activities but enhancement of indoor levels may be minimized by ensuring that combustion conditions favour complete oxidation to carbon dioxide. A reasonable air-exchange rate is also important so that levels of carbon monoxide will not be higher than 10 ppmv. Nitric oxide (NO) is derived from the combined nitrogen that may be present in the fuel. Wood or other forms of biomass contain substantial amounts of this element, but there is very little in petroleum products or natural gas. An additional amount of nitric oxide is produced from the dinitrogen in air during high-temperature combustion. Because the rate of generation depends on flame temperature, much more is produced by burning of natural gas ($T_{flame} > 2000°C$) than of biomass such as wood ($T_{flame} < 1000°C$). The post-combustion oxidation reaction NO $\rightarrow$ NO$_2$ requires photochemically generated species such as peroxy radicals in order to progress rapidly. One of the routes for producing the hydroperoxy species was shown in reactions 7.12–7.14. Mixing ratios greater than 100 ppbv of either NO$_x$ species are considered to be excessive. Except in the unusual case of indoor burning of high-sulphur coal, sulphur dioxide emissions are not a serious problem.

Volatile organic carbon compounds are other products of combustion, their nature and amount again depending on the fuel type and burning conditions. Coal, wood, and other forms of biomass tend to release more gaseous hydrocarbons and other partially oxidized VOCs than do petroleum and natural gas products. When the former fuels are burned in unvented appliances, elevated levels of these gases are found in the indoor air. Proper venting of exhaust gases minimizes this problem. Formaldehyde is a minor product of combustion of all types of fuel.

Tobacco smoking is a source of many VOCs including aldehydes, ketones, organic bases like nicotine, organic acids, and hydrocarbons. Cigarette smoke is yet another source of formaldehyde with an estimated amount of 2.4 mg being emitted by each cigarette.[7] The directly inhaled air drawn through a cigarette may contain formaldehyde concentrations more than 400 times greater than the level of concern. Mainstream cigarette smoke (and associated vapour) carries more than 4000 chemicals with it. A comprehensive survey of VOCs that are found in smoke from common American brands of cigarettes is presented in Additional Resources 1. More will be said about the products of tobacco combustion in the next section.

### Indoor particulates

Much of the solid aerosol in a building is due to dust lifted into the air by human activities and air circulation. Release of particulate matter also accompanies the combustion of coal and biomass materials. The particle size of the aerosol derived from combustion of these solid fuels

[7] Olander, L., J. Johansson, and R. Johansson, Tobacco smoke removal with room air cleaners, *Scand. J. Work Environ. Health*, **14** (1998), 390–7.

is largely within the $PM_{10}$ range, with many particles having diameters less than 2 μm making them able to penetrate deep into the respiratory passages. A concentration of 100 μg m$^{-3}$ total solid particulates is a typical guideline value for an acceptable upper limit inside a building. (Refer to Table 7.1 for recommended outdoor values.)

In homes in rural Mexico, it has been found that very high levels of particulates were present during the (typically 2–3 h) period required for the preparation of tortillas. In one study,[8] four preparation methods were considered and the average particulate concentrations within the $PM_{10}$ range were:

- 1140 μg m$^{-3}$, for biomass fuel in an unvented environment;
- 330 μg m$^{-3}$, for unvented liquefied petroleum gas (LPG);
- 540 μg m$^{-3}$, for a mixture of unvented biomass and LPG;
- 430 μg m$^{-3}$, for vented biomass.

All of these values, but especially those associated with biomass fuels, are higher than the recommended acceptable levels.

Smoking can be a significant contributor to respirable particulate matter inside a building. Figure 7.3 illustrates the estimated range of exposure to particulates for persons inside a building where smokers are present. The respirable suspended particulates (RSP) concentration is plotted against the volume density of persons in the enclosed area. In this theoretical calculation, it is assumed that at any time one-third of the persons in the room are smokers and, at any time

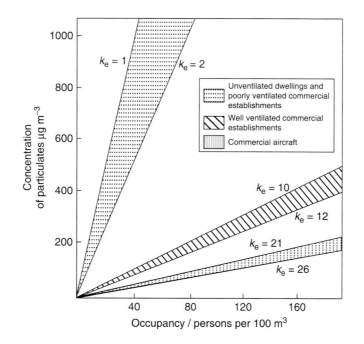

**Fig. 7.3** Theoretical steady-state density of respirable suspended particulates from environmental tobacco smoke in an enclosed indoor space. Air-exchange rate is $k_e$ / h$^{-1}$. (Redrawn from Repace, J.L. and A.H. Lowrey, Indoor air pollution, tobacco smoke, and public health, *Science*, **208** (1980), 464–71.)

[8] Brauer, M., K. Bartlett, J. Regalado-Pineda, and R. Perez-Padilla, Assessment of particulate concentrations from domestic biomass combustion in rural Mexico, *Environ. Sci. Technol.*, **30** (1996), 104–9.

during the day, for every three smokers one cigarette burns constantly. Typical occupancy densities are 4 per 100 $m^{-3}$ for office space, 25 per 100 $m^{-3}$ for restaurants, and up to 50 per 100 $m^{-3}$ for crowded places like auditoriums, trains, and airplanes. Values of the air-exchange rate, $k_e$, often range from 1 $h^{-1}$ for an unventilated building through 10 $h^{-1}$, for a well-ventilated commercial establishment, to 26 $h^{-1}$ for a commercial aircraft with excellent air exchange. It is clear that high occupancy by smokers and poor air exchange, or a combination of the two factors, can lead to unacceptable levels of RSP within the confined areas. It is fortunate that, within the past two decades, many jurisdictions have taken strong steps to limit smoking in enclosed public areas.

In Chapter 6 we saw that polyaromatic hydrocarbons (PAHs) are associated with particulate and gaseous emissions from coal and biomass. While outdoor concentrations of PAHs are often below 1 ng $m^{-3}$, levels may be several times higher in communities where extensive biomass combustion takes place. Concentrations are further enhanced in geographic situations such as valley communities or during inversion events. There are cases of wood-burning communities in the USA where, under unfavourable meteorological conditions, outdoor levels of benzo($a$)pyrene have been measured around 10 ng $m^{-3}$ and occasional levels above 100 ng $m^{-3}$ have been reported. (For analytical reasons, levels of benzo($a$)pyrene are often reported as a surrogate for total PAH. This may be misleading, since other particular species may be present in much higher concentrations and, in any case, proportions of all species vary in each situation.)

Indoor concentrations of PAH compounds are greatly affected by the amount and conditions of combustion in the building. Where outdoor levels of benzo($a$)pyrene are in the usual low range, residences with enclosed wood stoves can have concentrations of PAHs in a range around 5 ng $m^{-3}$ and those with open fireplaces (but a well-vented chimney) often have levels above 10 ng $m^{-3}$. On the other hand, where cooking is done with open fires in unvented traditional homes, concentrations can be much greater—for example, levels of 70 ng $m^{-3}$ for benzo($a$)pyrene and other PAHs have been measured in houses in Burundi. In addition to benzo($a$)pyrene, wood and other biomass smoke emissions are dominated by the lighter molar mass PAHs including phenanthrene, pyrene, fluoranthene, and anthracene.

Tobacco smoking is a source of gaseous and solid PAH compounds as well as of particulates in general. It has been shown that one unfiltered cigarette provides about 25 ng of benzo($a$)pyrene to the smoker.[9] Using this figure and a breathing rate of 23 $m^3$ $d^{-1}$, each cigarette provides as much of the compound as does continuous breathing over 1 day of an atmosphere containing 1 ng $m^{-3}$. A 20-cigarette-a-day smoker is then exposed to the equivalent of an atmosphere with a constant level of benzo($a$)pyrene of about 20 ng $m^{-3}$.

> **Main point 7.3** Radioactivity associated with radon gas, volatile organic compounds, gaseous combustion products and particulate matter of diverse origins are important sources of air quality problems inside buildings.

## ADDITIONAL RESOURCES

1. Polzin, G.M., R. Kosa-Maines, D. Ashley, and C.H. Watson, Analysis of volatile organic compounds in mainstream cigarette smoke. *Environ. Sci. Technol.* 41 (2007), 1297–1302.

2. Otson, R. and P. Fellin, Volatile organics in the indoor environment: sources and occurrence, In *Gaseous pollutants: characterization and cycling* (ed. J.O. Nriagu). John Wiley and Sons, Inc., New York; 1992.

[9] National Research Council, *Particulate Polycyclic Organic Matter*, Committee on Biological Effects of Atmospheric Pollutants, National Academy of Sciences, Washington, D.C., 1972 as reported in Additional Resources 2.

3. Walsh, P.J., C.S. Dudney, and E.D. Copenhaver (eds.), *Indoor air quality*. CRC Press, Inc., Boca Raton, Florida; 1984.

4. Air Now, *International Air Quality*. http://www.airnow.gov/, accessed October 2009. Note: Click on 'international' at the bottom of the home page. This takes you to a listing of many national websites where air quality information is available; some countries report data as on-going real-time information as well as archiving previous data.

5. World Health Organization, *Air quality guidelines—global update, 2005*, and *Air quality guidelines for Europe 2nd Edn 2000*. Both available online from http://www.euro.who.int/, accessed under *publications*, *air*, October 2009.

## PROBLEMS

1. Atmospheric pollutants are sometimes classified into two categories:
   - primary pollutants—those that are emitted directly from the source;
   - secondary pollutants—those that are produced by reactions in the open atmosphere.

   Which category(s) would each of the following fall into: carbon monoxide; carbon dioxide; sulphur dioxide; nitric oxide; nitrogen dioxide; ozone; PAH compounds; formaldehyde?

2. The average concentrations ($\mu g\ m^{-3}$) of gaseous pollutants in Krakow, Poland have been measured to be: ozone 43, $PM_{10}$ 39, sulphur dioxide 23 and nitrogen dioxide 44. The maximum one-hour level of ozone was 201 $\mu g\ m^{-3}$. What are the likely sources of these pollutants and which ones appear to be problems or potential problems in terms of air quality?

3. Access the *Air Now* website (Additional Resources 4) and go to the international listings (under the quick links near the bottom of the page). Choose the United Kingdom, London Air Quality Network link and then on the next page from the main menu bar select Download Data. From the sites menu, fetch *Windsor Roadside* and on the next screen choose oxides of nitrogen. Follow the trend in oxides of nitrogen using both 15-min and 1-h data for the one-day period July 1 to July 2, 2008. Explain the observed trend and compare the peak values with the WHO guideline for this pollutant.

4. Access the *Air Now* website as above and choose the Federal Environmental Agency Germany—Air Quality Division. The information can be accessed in either German or English. For the current day, examine the maps of Germany that depict ozone, carbon monoxide, nitrogen dioxide and sulphur dioxide, following the time period from 6 am to 12 noon. Comment on the changing pattern over the 6-h period, taking into account the locations of different values and the time of the year during which the information is provided. Also comment on the daily mean $PM_{10}$ value provided. How do the German data compare with WHO guidelines?

5. The indoor / outdoor concentration ratios for carbon monoxide are usually near 1, for carbon dioxide they are usually greater than 1, and for sulphur dioxide they are usually less than 1. Give reasons for these normal situations and cite instances that might be exceptions to the values.

6. In the uranium-238 decay series, radium-226 is an immediate radioactive parent of radon-222. Consider its position in the periodic table and indicate its usual oxidation state, its probable mobility in water, and factors affecting its solubility in soil / water systems.

7. As an experiment, gas cooking units were activated until high levels of carbon monoxide and $NO_x$ compounds were measured in a house and then the gas burners were turned off. The carbon dioxide levels returned to background values in 1.6 h, but the $NO_x$ levels only required 0.7 h before they had reached the 'normal' concentration. Suggest an explanation.

8. In Chinese village homes, heating is often done by an open wood or coal fire. Where outdoor PAH concentration is 0.60 ng $m^{-3}$, the rate of emission of PAH from indoor combustion is

3.5 ng m$^{-3}$ h$^{-1}$, and the air-exchange rate is 2 h$^{-1}$, estimate the indoor-air concentration of PAH compounds. Assume that the only loss mechanism is by air exchange.

9. At a gathering of 30 people in a room 6 m × 9 m × 3 m, half are smoking, on average, 3 cigarettes per hour. Each cigarette when smoked releases 2.4 mg of formaldehyde. The air exchanges five times every hour. Assuming that the outdoor formaldehyde concentration is negligible, calculate the steady-state concentration in the air inside the room. Does this concentration exceed the concentration (100 ppbv) considered to be the level of concern?

10. A beaker of benzene is left open in a closed lab of dimensions 4.2 m × 5.0 m × 2.6 m. It evaporates at a rate of 7 mL per hour. The air-exchange rate in this unventilated room is 0.3 h$^{-1}$. Room temperature is 20°C and pressure is $P^o$. What is the steady-state mixing ratio of benzene in the air of this room. Suppose the room is now a properly ventilated lab with an air-exchange rate of 2.5 h$^{-1}$. Recalculate the steady state-mixing ratio of benzene.

11. A basement apartment is situated in an area where the outdoor radioactivity is 3 Bq m$^{-3}$. The construction materials and surrounding soil release radon into the basement atmosphere at a rate of 72 Bq m$^{-3}$ h$^{-1}$. With open windows, etc., the summer exchange rate is 8 h$^{-1}$ while in the winter it is 0.4 h$^{-1}$. Calculate the steady-state radon activity in both seasons and compare to the WHO reference levels.

# Chapter 8
## The chemistry of global climate

The focus of this chapter is to examine the ways in which global climate is influenced by the composition and chemistry of the atmosphere. This will be done by:

- discussing the composition of the atmosphere and investigating the energy balance of incoming solar radiation and outgoing infra-red (black body) radiation from the Earth
- identifying greenhouse gases and their infra-red radiation absorption properties as well as examining the radiative influence of aerosols
- introducing the relative instantaneous radiative forcing (RIRF) and global-warming potential (GWP) indices for measuring the impact of greenhouse gases
- providing an overview of energy resources, the relationships between carbon-based fuels and carbon dioxide emissions, and mitigation strategies.

In the simplest manner, climate may be defined as average weather. At a particular geographical location, the local climate describes the average of day to day weather variations—temperature, wind, amounts and patterns of precipitation—over a period of many years. Global climate is a spatial average of all local climates around the world.

Climate is not a static phenomenon but rather shows changes—some unidirectional, some cyclical—over periods of geological time. Figure 8.1 shows how the Earth's temperature has varied over the past quarter million years. During most of this period, the temperature has usually been significantly lower than at present.

The global climate and the changes it undergoes are determined by many factors, one very significant one being the chemical nature of the Earth's atmosphere. Gases and aerosol species determine the degree to which the Sun's radiation is absorbed or reflected, and they do the same for radiation emitted back into space from the Earth itself. In a complex manner then, temperature, along with precipitation, wind patterns, and other climatic features, is affected by the chemical composition of the atmosphere. At the same time, the reverse effect is true as well. Changes in climate bring about changes in atmospheric composition and so it is frequently difficult to identify cause and effect.

In this chapter, we will consider how the chemistry of the Earth's atmosphere plays a role in determining its average temperature. We will also examine some of the human activities that are affecting the chemical composition in ways that can cause changes in global climate patterns.

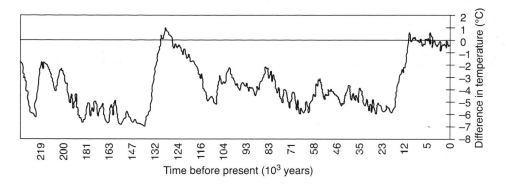

**Fig. 8.1** Temperature deviation from current temperatures at Vostok Antarctica. Difference in temperature (°C) from present-day temperatures in the past (displayed in thousands of years before the present), derived from deuterium. Source: J. Jouzel, et al. (Webpage: http://cdiac.ornl.gov/ftp/trends/temp/vostok/vostok.1999.temp.dat, accessed October 2009.)

## 8.1 Composition of the Earth's atmosphere

The average mixing ratio of the nine principal gases in the dry troposphere is given in Table 8.1. Because all these gases have lengthy atmospheric residence times, the values are relatively constant at all locations around the Earth.

The most highly variable major gaseous component in the troposphere is water. Its residence time (calculated from the values in Fig. 1.4) is about 10 days—a period that is much smaller than the time required for complete mixing of the troposphere. It is for this reason that the mixing ratio of water vapour varies from day to day and place to place.

The maximum mixing ratio of water that can be present in the atmosphere is related to temperature through the vapour pressure ($P_V$) curve given in Fig. 8.2. Relative humidity ($H_R$) is an expression of the percentage of this maximum (equilibrium) value that obtains in a given situation. In the dry tropics, a temperature of 30°C and relative humidity 40% could be typical. This corresponds to a non-equilibrium $P_{actual (H_2O)}$ of

$$H_R / 100 \times P_{V(H_2O)} = P_{actual (H_2O)}$$

$$0.40 \times 4.24 \text{ kPa} = 1.7 \text{ kPa}$$

**Table 8.1** Average composition of the dry troposphere.

| Component | Mixing ratio |
| --- | --- |
| Nitrogen | 78.08% |
| Oxygen | 20.95% |
| Argon | 0.93% |
| Carbon dioxide | 387 ppmv |
| Neon | 18 ppmv |
| Helium | 5 ppmv |
| Methane | 1.77 ppmv |
| Hydrogen | 0.53 ppmv |
| Nitrous oxide | 0.32 ppmv |

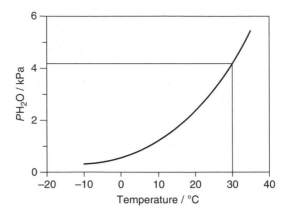

**Fig. 8.2** The vapour pressure of water $P_{V(H_2O)}$ as a function of temperature. The value of $P_{V(H_2O)}$ at 30°C is determined from the plot, or calculated to be 4.24 kPa.

At $P°$, the mixing ratio of $H_2O$ is then equal to the ratio of its partial pressure to total pressure.

$$\text{mixing ratio (\%)} = \frac{P_{actual(H_2O)}}{P°} \times 100 = \frac{1.7 \text{ kPa}}{101 \text{ kPa}} \times 100 = 1.7\%$$

An equivalent calculation for a situation in a cooler region where the temperature is −10°C and the humidity 100% gives a value of $P_{V(H_2O)} = P_{actual\ (H_2O)} = 0.26$ kPa and a mixing ratio of $0.0026 = 0.26\%$. These two examples illustrate that:

- water is a quantitatively important component (ranked third or fourth according to mixing ratio) of the atmosphere;
- the atmospheric concentration of water in time and space is highly variable;
- the mixing ratio of water depends on both temperature and the extent of non-equilibrium as expressed by the relative humidity.

> **Main point 8.1** The Earth's atmosphere consists of a number of gases present in relatively constant (over time and space) mixing ratios. The only quantitatively important species whose concentration is highly variable is water vapour.

## 8.2 Energy balance

### Energy from the Sun

The ultimate source of most energy available on Earth is the Sun. The energy[1] that the Sun (or any other object, including the Earth) emits is described in terms of *black-body radiation*. According to the Planck relation[2] for black-body radiation the emissive energy per unit wavelength, of any material with a finite temperature, is given by

$$M_\lambda = \frac{2\pi hc^2}{\lambda^5}\left(\frac{1}{e^{(hc/kT\lambda)} - 1}\right) \tag{8.1}$$

---

[1] The use of the word energy in the context of eqn 8.1 and subsequent equations is common but not quite correct. The unit, watt (W), measures power, which is energy per unit time (J s⁻¹).

[2] Grum, F. and R.J. Becherer, *Optical radiation measurements*, Volume 1, *Radiometry*, Academic Press, New York; 1979.

where $M_\lambda$ = the emissive energy W m$^{-2}$ (the units signify watts per square metre of surface per metre of wavelength), $h$ = Planck's constant = $6.63 \times 10^{-34}$ J s, $c$ = velocity of light = $3.00 \times 10^8$ m s$^{-1}$, $\lambda$ = wavelength / m, $k$ = Boltzmann's constant = $1.38 \times 10^{-23}$ J K$^{-1}$, and $T$ = temperature / K.

Figure 8.3 shows a plot of $M_\lambda$ versus $\lambda$ for different temperatures. The area under the curves between any two wavelengths equals the total power per area (W m$^{-2}$) emitted between these two wavelengths. The spectrum at 5000 K corresponds closely to the situation for the Sun (whose surface temperature is about 5800 K). As is evident, solar radiation is given off as a broad continuum that spans part of the near-ultraviolet, all of the visible, and part of the infra-red regions of the electromagnetic spectrum. A good approximation of the wavelength (in m) of maximum emission (for objects with temperatures greater than 100 K) is given by

$$\lambda_{max} = \frac{2.88 \times 10^{-3}}{T} \tag{8.2}$$

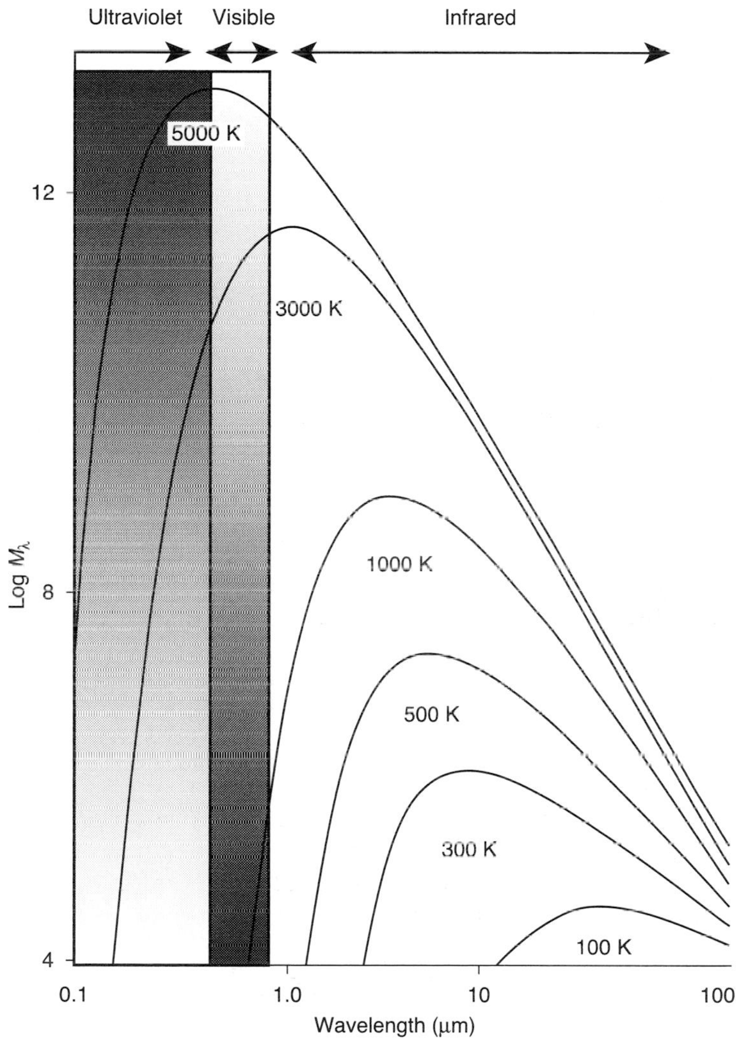

**Fig. 8.3** Spectral distribution of a black-body emitter. Units of $E_\lambda$ are W m$^{-2}$ μm$^{-1}$. (Redrawn from Grum, F. and R.J. Becherer, *Optical radiation measurements*, Volume 1, *Radiometry*. Academic Press, New York; 1979.)

At 5800 K, the $\lambda_{max}$ is then predicted to be about 500 nm, in the green region of the visible spectrum.

### Solar energy and the Earth

For reasons we have discussed in earlier chapters, much of the solar radiation, especially high-energy radiation, is absorbed by molecules in the upper atmosphere. The total energy reaching the part of space occupied by the Earth (that is, the energy that would be received by a satellite orbiting above the Earth's atmosphere) is called the solar flux, $F_s$, and is 1368 W m$^{-2}$ averaged over a one-year period and over the entire Earth's surface. However, only a portion of this energy can actually reach the Earth's surface and only part of this is absorbed. Figure 8.4 diagrammatically shows what happens to this solar energy as it flows into the global system. The numerical values are normalized to a value of 100 units, corresponding to the total flux of incoming solar radiation.

Of the total solar flux represented as 100 units, a portion is reflected back into space and does not contribute to the energy budget of the Earth. The reflected portion includes 6 units from the Earth's surface, 17 units due to clouds, and 8 units contributed by aerosols including dust, salt from sea spray, smoke from fires, and volcanic ash. Thus the total reflectivity of the Earth is 31 units. This is commonly referred to in percentage terms as the Earth's *albedo* being 31% (or $A = 0.31$). Of the non-reflected 69 units of solar energy, 4 units are absorbed and thus retained within the Earth's atmosphere by water droplets in the clouds, and 19 units are absorbed by other aerosol particles and gaseous species such as ozone. Thus, 23% of the solar energy reaching the Earth is absorbed in the atmosphere and only 46% is actually absorbed by land or water. Once absorbed, the relatively high-energy, short-wavelength (UV and visible) solar radiation is available to be used as an energy source for biomass growth, or to contribute to warming of the Earth's surface and to phase transitions of water. These are ways in which the energy is stored, but the situation is a steady-state one and, sooner or later, an amount of energy equivalent to that received must be re-emitted. This release to the atmosphere occurs as lower-energy, longer-wavelength (infra-red (IR)) radiation.

If the re-emitted energy were completely lost in space, we could predict the temperature of the Earth's surface in the following way. At any given time, solar radiation strikes half of the globe, but not directly—not at right angles to the entire surface (Fig. 8.5). The average value of the solar flux, $F_s$, over space and time is, as previously stated, 1368 W m$^{-2}$

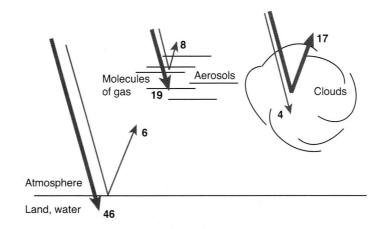

**Fig. 8.4** Relative energy flows of solar radiation into the Earth's environment (based on 100 units). Arrows show solar energy that is absorbed or reflected by components of the Earth's environment.

---

**Definitions of important global climate terms**

- **Greenhouse effect**—the heating of the Earth's near-surface atmosphere by the trapping of out-going infra-red radiation by atmospheric gases
- **Global warming**—the increase of the near-surface average global temperature
- **Radiative forcing**—describes the net energy in watts per square metre made available to the Earth associated with the increase in the concentration of each greenhouse gas, since 1750 (see Table 8.3)
- **Relative instantaneous radiative forcing (RIRF)**—a measure of the ability of an incremental addition of a greenhouse gas, into the present atmosphere, to increase the absorption of infra-red radiation. $CO_2$ has a defined RIRF value of 1.0.
- **Global-warming potential (GWP)**—a calculated value that allows for comparing the potential of individual greenhouse gases to contribute to overall warming over a defined period of time. $CO_2$ has a defined GWP value of 1.0.

---

The integrated value of $F_s$ at all angles over the spherical Earth is equivalent to the full flux continuously and directly striking an area normal to the solar beam $= \pi r^2$ where $r$ is the Earth's radius. The total solar energy reaching the Earth is then given by $F_s \pi r^2$.

The total amount of solar energy absorbed by the Earth is equal to this integrated flux ($F_s = 1368$ W m$^{-2}$) minus the portion that is reflected back into space. This portion is the *albedo* (A).

$$E_s = F_s (1 - A)\pi r^2 \tag{8.3}$$

where $E_s$ = total solar energy absorbed by the Earth / W and $r$ = radius of the Earth / m.

In the steady state, this absorbed energy is exactly balanced by the average energy emitted from the Earth. To estimate emitted energy, we make use of Wien's law (eqn 8.4), which gives the total energy radiated from any object, a black body having temperature T / K, at all wavelengths from 1 square metre of the body's surface. Equation 8.4 is obtained by integration of eqn 8.1.

$$F = \sigma T^4 \tag{8.4}$$

In referring to radiative emission from the Earth we will write eqn 8.4 as

$$F_e = \sigma T_e^4 \tag{8.5}$$

where $F_e$ = radiant flux from the Earth / W m$^{-2}$, $\sigma$ = the Stefan–Boltzmann constant,

$$\sigma = \frac{2\pi^5 k^4}{15h^3 c^2} = 5.67 \times 10^{-8} \text{ W m}^{-2} \text{K}^{-4}$$

and $T_e$ = effective temperature of the Earth / K. The terms $h$, $k$ and $c$ are defined in eqn. 8.1.

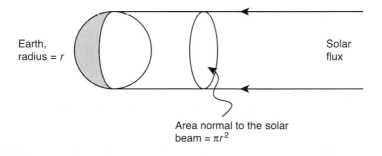

**Fig. 8.5** Solar flux ($F_s$) and the Earth.

The total radiative energy, $E_e$, emitted from the entire area of the Earth's surface (radius $= r$, area $= 4\pi r^2$) is then

$$E_e = 4\pi r^2\, \sigma T_e^4 \tag{8.6}$$

where $E_e$ = total energy emitted by the Earth / W.

Over any substantial time span, in the steady state, total energy absorbed from the Sun ($E_s$) = total energy emitted by the Earth ($E_e$). Therefore,

$$F_s(1 - A)\pi r^2 = 4\pi r^2\, \sigma T_e^4 \tag{8.7}$$

Knowing the values of $F_s$, $A$, and $\sigma$, the temperature of the Earth is then predicted by

$$T_e = \left(\frac{(1-A)\,F_s}{4\sigma}\right)^{1/4} \tag{8.8}$$

Inserting these values,

$$T_e = \left(\frac{(1 - 0.31) \times 1368\ \mathrm{W\ m^{-2}}}{4 \times 5.67 \times 10^{-8}\ \mathrm{W\ m^{-2}K^{-4}}}\right)^{1/4}$$

$$= 254\ \mathrm{K}$$

$$= -19°\mathrm{C}$$

The calculation therefore leads us to predict an average global surface temperature of $-19°$C. In fact, the measured average surface temperature is known to be $+17°$C, 36°C higher than the calculated value (the atmosphere just above the Earth's surface has an average temperature about 3°C less than this value). The same kind of calculation can be made for other planets within the solar system. It is instructive to note their calculated and actual temperatures along with those of the Earth (Table 8.2).

Why are actual temperatures on these planets greater than the ones predicted in this simple way? In each of the cases, the positive value of $\Delta (= T_{actual} - T_{calculated})$ can be attributed to a 'greenhouse effect'. Recall that our simple calculation included an assumption that all of the radiation emitted from the Earth escaped into space. It is this assumption that is incorrect and leads to the error in our calculation. In fact, a significant portion of the radiation that is re-emitted from the planet's surface is not immediately radiated into space. Rather, it is absorbed by gases in the planetary atmosphere, particularly the lower troposphere, and in this way it contributes to enhanced warming. Life as we know it on Earth is compatible with the climate that results from these conditions. We call the enhanced warming the 'greenhouse effect'.

On both Venus and Mars the principal gas responsible for greenhouse warming is carbon dioxide, but the amount of this gas in the atmospheres of these two planets is greatly different. On Mars, the total pressure of carbon dioxide is only about 0.6 kPa; therefore the greenhouse effect is small, about 6°C. In contrast, the Venusian atmosphere contains about 95% carbon dioxide and its partial pressure is greater than 9000 kPa generating an enormous greenhouse warming effect of about 505°C.

Coming back to the situation on Earth, as we move upwards in the troposphere, we reach an altitude above which emitted IR radiation is absorbed only to a small extent and the radiation does go directly into space. This altitude is about 5.5 km, where the temperature is actually near

**Table 8.2** Temperatures at the surface of three planets where $\Delta = T_{actual} - T_{calculated}$.

| Planet | Calculated $T$ / K | Actual $T$ / K | $\Delta$ / K |
|--------|--------------------|----------------|--------------|
| Earth  | 254                | 290            | $+ 36$       |
| Mars   | 17                 | 223            | $+ 6$        |
| Venus  | 227                | 732            | $+ 505$      |

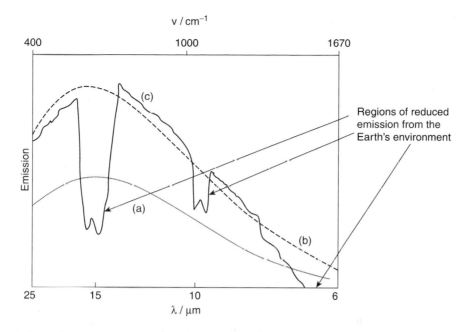

**Fig. 8.6** Infra-red emission spectra for black-body radiators at (a) 240 K, (b) 280 K, and (c) the Earth. (Redrawn from Wayne, R.P., *Chemistry of atmospheres*. Clarendon Press, Oxford; 1991.)

the predicted value of 254 K (see Fig. 2.1). Observed from space, the Earth behaves as a black body having this temperature.

The radiation that is emitted from the Earth's surface is infra-red radiation in the 'thermal IR' region from about 3 to 40 μm. To understand what gases are responsible for absorbing this radiation and preventing some of it from going into space, we can look at the Earth's IR emission spectrum as viewed from satellites orbiting above the farthest reaches of the atmosphere. Figure 8.6 shows three such emission spectra—theoretical spectra for black-body radiators at (a) 280 K and (b) 240 K and (c) the actual emission spectrum from the Earth.

In the figure, we see that the actual IR emission spectrum from the Earth (solid line) appears to track closely to the 280 K spectrum (dashed line) with the predicted (eqn 8.2) maximum emission at 10.3 μm (970 cm$^{-1}$), near that actually observed. However, there is a major difference in the Earth's emission spectrum compared with the theoretical black-body spectrum in that the Earth's radiation is reduced in certain regions, especially those around $15 \pm 2$ μm (770–590 cm$^{-1}$), $9.5 \pm 0.5$ μm (110–1000 cm$^{-1}$), and $< 8$ μm (1250 cm$^{-1}$). This indicates that species capable of absorbing and attenuating radiation in these ranges, *radiative forcing* must be present in the Earth's atmosphere. Shortly, we will examine what these species are.

The greenhouse effect is a measure of the extent to which this absorption retains heat near the Earth's surface. The magnitude of this greenhouse effect can be quantitatively estimated, and a value in good agreement with the observed value is obtained, but we will not estimate it here. The calculations are difficult and involve several simplifying assumptions.

**Main point 8.2** Solar energy in the ultraviolet and visible regions of the electromagnetic spectrum is the primary source of energy available on the Earth's surface. When it is absorbed, it is converted into longer-wavelength (infra-red) radiation that is emitted outward from the Earth. Some of this energy is absorbed in the lower atmosphere.

## 8.3 The greenhouse gases and aerosols

### Infra-red absorption properties of gaseous species

The species that interact with thermal radiation emitted from the Earth's surface are now well known. It is these 'greenhouse gases' that give rise to the absorption peaks that are observed in the three regions of the IR spectrum as described above. Their concentrations, and contribution to warming are listed in Table 8.3. We will look at each of the species in turn.

### Water

For the Earth, water vapour is actually the most important of all greenhouse gases, and it absorbs IR in the ranges 4000–3300 $cm^{-1}$ (2.5–3.0 $\mu m$), 2000 to 1250 $cm^{-1}$ (5–8 $\mu m$), as well as over a broad range below 700 $cm^{-1}$ (above 14 $\mu m$) (Fig. 8.7). While the mixing ratio of water vapour is highly variable in time and space (Chapter 2), the global average relative humidity is constant at about 1% and there are no anthropogenic activities that directly cause it to increase to a significant extent. Nevertheless, gaseous water is involved in feedback processes.

**Table 8.3** Past and present greenhouse-gas concentrations in the troposphere, and their contribution to radiative forcing[a].

| Gaseous compound | Tropospheric concentration | | Contribution to radiative forcing / $W\ m^{-2}$ |
| --- | --- | --- | --- |
| | Before 1750 | At present | |
| Carbon dioxide | 280 ppmv | 387 ppmv | 1.66 |
| Methane | 0.70 ppmv | 1.774 ppmv | 0.48 |
| Nitrous oxide | 0.27 ppmv | 0.319 ppmv | 0.16 |
| Ozone | 0.025 ppmv | 0.034 ppmv | 0.30 (net) |
| CFC-11 | 0 pptv | 257 pptv | |
| CFC-12 | 0 pptv | 544 pptv | |
| CFC-113 | 0 pptv | 80 pptv | |
| Carbon tetrachloride | 0 pptv | 94 pptv | |
| Methyl chloroform | 0 pptv | 34 pptv | 0.34 total for all halocarbons |
| HCFC-22 | 0 pptv | 146 pptv | |
| HFC-23 | 0 pptv | 14 pptv | |
| Perfluoroethane | 0 pptv | 3 pptv | |
| Sulphur hexafluoride | 0 pptv | 4.8 pptv | 0.002 |
| Aerosols | | | −1.2 |

[a] The contribution to greenhouse warming is usually referred to as increased radiative forcing; this describes the average additional energy in watts per square metre made available to the Earth associated with the increase in concentration of each gas. Radiative forcing values used in this book are for changes relative to pre-industrial conditions defined at 1750.

Most of these and other values given in the present chapter are taken from the Intergovernmental Panel on Climate Change (IPCC) *Climate change 2007: Synthesis report*. The 12-month average carbon dioxide value (ending August 2009) was obtained from the National Oceanic and Atmospheric Administration, Earth Systems Research Laboratory, Global Monitoring Division website, http://www.esrl.noaa.gov/gmd/trends/co2_data_mlo.html, accessed September 2009. For regularly updated gas-mixing ratios from specific sites around the world see Additional Resources 5.

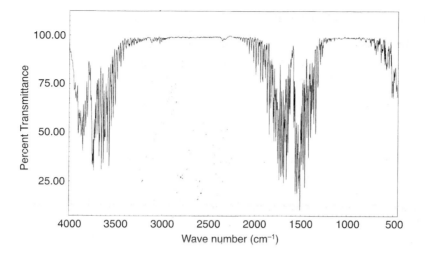

**Fig. 8.7** Infra-red absorption spectrum of water. The relation between wave number and wavelength is: wave number (cm$^{-1}$) = 10000/wavelength (m). (© BIO-RAD Laboratories, Sadtler Division, 2000.)

- Positive feedback occurs in that increased global warming means increased evaporation from ocean and land surfaces leading to higher atmospheric mixing ratios for water—therefore enhanced warming.
- Negative feedback results from the troposphere becoming more cloudy leading to increased reflection and absorption of the Sun's radiation. Because of this, the solar flux reaching the solid / liquid surface of the Earth is reduced.

It is not clear, however, which of these two phenomena will be most important in the future. At present, greenhouse warming associated with water vapour is estimated to be about 110 W m$^{-2}$; this is also the historic value.

### Carbon dioxide

Like water, carbon dioxide is a major contributor to greenhouse warming. It absorbs in the 710–530 cm$^{-1}$ (14–19 μm) range, and completely blocks the radiative flux between 670 and 630 cm$^{-1}$ (15 and 16 μm); it also absorbs very strongly between 2500 and 2300 cm$^{-1}$ (4.0 and 4.3 μm) (Fig. 8.8).

The many natural processes that are sources of carbon dioxide in the atmosphere include animal, plant, and microbial respiration and decay, and combustion of biomass[3] ({CH$_2$O}) through forest and grassland fires that are often started by lightning. Equation 8.9 in the left to right direction describes the overall reaction for all of these processes, recognizing of course that the biological reactions are very complex.

$$\{CH_2O\} + O_2 \rightleftharpoons CO_2 + H_2O \tag{8.9}$$

Regions in the oceans too are important sources of carbon dioxide release to the atmosphere. In the midlatitudes of the Pacific, in particular, upwelling of carbon-dioxide-rich waters causes release of large quantities of the gas.

There are also Earth processes that act as natural sinks. Photosynthesis is one of these and takes place when plants and some microorganisms grow on land and within both oceans and

---

[3] The symbol {CH$_2$O} is used here and elsewhere as a representation of the simplest formula for plant biomass, much of which consists of carbohydrate and related materials.

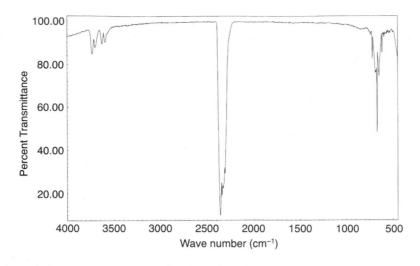

**Fig. 8.8** Infra-red absorption spectrum of carbon dioxide. The relation between wave number and wavelength is: wave number (cm$^{-1}$) = 10 000/wavelength ($\mu$m). (© BIO-RAD Laboratories, Sadtler Division, 2000.)

fresh water. The overall reaction, again a summary of very complex steps, is also given by eqn 8.9, this time reading from right to left. You can see then that synthesis and degradation of organic matter describe a cycle in which carbon dioxide is taken up (sink) and then released (source). Another uptake mechanism is dissolution in sea water where it then circulates in the great ocean currents. An 'ultimate' sink for the oceanic dissolved carbonate is its precipitation as calcite (limestone, $CaCO_3$) to form part of the sedimentary material.

Added to the natural sources and sinks, human activities have a significant impact on the global carbon cycle. The anthropogenic contributions are what we hear so much about in news reports. These sources of atmospheric carbon dioxide include carbon release via combustion of fossil fuels and forest destruction and burning. Besides releasing carbon dioxide into the atmosphere, the burning of growing trees eliminates their future contribution to carbon dioxide removal by the photosynthesis reaction.

It is estimated that about 8.0 Gt (as C)[4] of anthropogenic carbon dioxide are released to the atmosphere each year. Three-quarters is from fossil-fuel combustion and the rest from changing land use, primarily in the tropics. About 2.3 Gt (as C) of this are assimilated by dissolution in oceans and the same amount by increased plant growth rates. The remaining 3.4 Gt remain in the atmosphere. While the complex relations between sources and sinks are only partially understood, the net consequence of all the processes involving carbon dioxide is a steady annual increase of about 1.4 ppmv (about 0.5% of the 2009 concentration of 387 ppmv) in the atmosphere. The greenhouse warming effect due to carbon dioxide is estimated to be approximately 50 W m$^{-2}$, including an increase since pre-industrial times of 1.66 W m$^{-2}$; this latter figure is referred to as the radiative forcing.[5]

Water and carbon dioxide are the two most important greenhouse gases. Together, they absorb much of the radiation in the thermal infra-red region above 1300 cm$^{-1}$ (below 7.7 $\mu$m) and below 770 cm$^{-1}$ (above 13 $\mu$m). This leaves the region between 7.7 and 13 $\mu$m as a 'window'

---

[4] Releases of carbon dioxide may be presented in terms of carbon or as carbon dioxide. 8.0 t of carbon is equivalent to 29 t of carbon dioxide.
[5] Definition of radiative forcing is given in Table 8.3. All values given here are relative to the situation in 1750, the preindustrial age.

through which thermal energy escapes into space. Other gases, however, do absorb in that region partially closing the window; these greenhouse gases can have a major effect on heat retention in the atmosphere of the Earth.

### Methane

The present average tropospheric concentration of methane is 1.774 ppmv and its tropospheric lifetime is about 12 years. The small concentration of methane would have little effect on the Earth's energy balance if it absorbed radiation within the major absorption bands of water or carbon dioxide. Methane, however, absorbs in the ranges from 3300 to 2800 cm$^{-1}$ (3.0–3.6 $\mu$m) and 1400 to 1200 cm$^{-1}$ (7.1–8.3 $\mu$m), the latter lying in the window region noted above (Fig. 8.9).

Methane is produced where organic matter is found in an oxygen-depleted highly reducing aqueous or terrestrial environment (Chapter 15). For example, it is released from wetlands, including both natural and constructed wetlands as well as cultivated paddy (rice) fields. The amount released is positively correlated with temperature, and is related to vegetation and soil type. Methane is also produced during extraction, transport, and inefficient combustion of fossil fuels. In particular, significant losses occur from leakages in natural-gas pipelines. A third major source is from the digestive tracts of ruminants (cattle, sheep, goats) and termites. Claims have sometimes been made that methane release occurs mostly in low-income countries in the tropics—where the ruminant and termite populations are high and rice production is common. Recent estimates,[6] however, show that these sources together produce only about 30–40% of released methane. Other sources are not centred in the tropics and some, such as landfills and emissions associated with fossil fuels, are actually in greater abundance in highly industrialized societies. It is interesting that the rate of increase in atmospheric methane levels was about 20 ppbv per year up to 1998, but has since declined to about 8 ppbv per year. To a large degree this is attributed to improved maintenance on gas wells and pipelines in the countries of the Commonwealth of Independent States (CIS, States of the former Soviet Union).

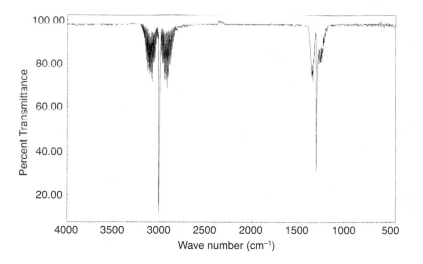

**Fig. 8.9** Infra-red absorption spectrum of methane. The relation between wave number and wavelength is: wave number (cm$^{-1}$) = 10 000 / wavelength ($\mu$m). (© BIO-RAD Laboratories, Sadtler Division, 2000.)

[6] $CO_2$ / *Climate report*, Canadian Climate Centre, Atmosphere Environment Service, Downsview, Ontario, Issue 98–1; 1998.

The principal sink for methane decomposition is oxidation via hydroxyl radicals in the troposphere, as has been discussed in Chapter 2.

$$CH_4 + \cdot OH \rightarrow \cdot CH_3 + H_2O \tag{8.10}$$

$$\longrightarrow \text{other reactions}$$

Smaller amounts of methane are removed from tropospheric air when taken up by soils and by leakage into the stratosphere.

The extent to which oxidation occurs depends on the availability of the hydroxyl radical, and this is determined to a large extent by the availability of carbon monoxide, which also reacts readily with hydroxyl radicals. Emissions of carbon monoxide are high in industrialized countries; in this way additional atmospheric methane build up is indirectly caused by excessive use of fossil fuels with consequent increased carbon-monoxide release.

Of considerable concern is a possible positive feedback process by which additional enormous quantities of methane could be released to the atmosphere. It is known that great reserves of clathrate hydrates such as $CH_4 \cdot 6H_2O$ are trapped in ice crystals in Arctic sediments and soils below the permafrost. The quantities could amount to $10^{17}$ kg or even more of carbon—much more than the present carbon content of the atmosphere ($\sim 10^{15}$ kg). A somewhat warmer climate could favour the release of a significant portion of these hydrocarbons greatly multiplying the warming trend.

At present, methane contributes 0.48 W m$^{-2}$ to radiative forcing.

### Ozone

Ozone absorbs between 1100 and 1000 cm$^{-1}$ (9 and 10 $\mu$m) in the IR spectrum and therefore acts as a highly efficient greenhouse gas. Greater production of NO$_x$ by fossil-fuel burning and forest and grassland fires has resulted in net low-altitude ('bad') ozone concentrations increasing by about 1.6% per year in the Northern Hemisphere. The increased absorption of IR radiation contributes to warming, but this is partly offset by the decrease in concentration of ozone in the stratosphere, discussed earlier. Ozone mixing ratios are highly variable in space and time, as would be expected from our discussion of urban air quality; its globally averaged mixing ratio is estimated to be 34 ppbv. The ozone radiative-forcing contribution in the troposphere is approximately 0.35 W m$^{-2}$; that due to stratospheric ozone is $-0.05$ W m$^{-2}$ resulting in net radiative forcing of 0.30 W m$^{-2}$.

### Nitrous oxide

Nitrous oxide absorbs IR radiation with several peaks above 2000 cm$^{-1}$ (below 5 $\mu$m) and exhibits a strong peak in the infra-red window between 1350 and 1150 cm$^{-1}$ (7.4 to 8.7 $\mu$m) (Fig. 8.10). The 2002 concentration of nitrous oxide was 318 ppbv and it is increasing at a rate of 0.3% per year, giving a present-day radiative forcing of 0.16 W m$^{-2}$.

Some nitrous oxide is released from industrial processes such as the production of adipic acid and nitric acid. The major sources, however, are from microbial denitrification in soils, lakes, and oceans. Denitrification is a term describing a group of microbiological reactions that convert nitrate to nitrous oxide, along with other nitrogen species. This will be discussed in Chapter 15. While denitrification is usually termed a natural process, it can be enhanced by human activities. With increased application of nitrogenous fertilizer, including animal manure, the supply of the nitrate substrate required for denitrification is augmented, leading to the production of more nitrous oxide. The amount released from soils is also greater where temperature and soil-moisture levels are high and where oxygen has been depleted. Additional emissions of nitrous oxide are produced from urban waste-landfill sites, and where there is direct sewage disposal into large bodies of water. The influx of this and other types of organic matter leads to emissions from the oceans especially in coastal regions and estuaries. There are no important

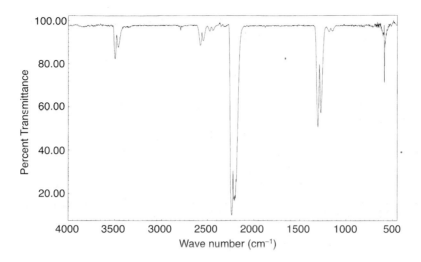

**Fig. 8.10** Infra-red absorption spectrum of nitrous oxide. The relation between wave number and wavelength is: wave number ($cm^{-1}$) = 10 000 / wavelength ($\mu m$). (© BIO-RAD Laboratories, Sadtler Division, 2000.)

tropospheric sinks for this gas, so it is lost only by slow leakage into the stratosphere where it undergoes photolytic degradation as described in Chapter 3; it therefore has a substantial tropospheric residence time estimated to be about 120 years. It has approximately the same effect on greenhouse warming as does ozone.

### Chlorinated fluorocarbons (CFCs) and other halogenated gases

In addition to their role as agents for the catalytic decomposition of stratospheric ozone, CFCs are also important greenhouse gases. They absorb in the range 1250–830 $cm^{-1}$ (8–12 $\mu m$) with each CFC having specific absorption bands in this critical window region. For example, CFC-12 has very strong peaks centred around 1050 $cm^{-1}$ (9.5 $\mu m$) and 900 $cm^{-1}$ (11.1 $\mu m$) (Fig. 8.11). The recently developed hydrochlorofluorocarbons (HCFCs) also attenuate radiation within the same range, but their residence time in the troposphere is substantially shorter than the CFCs. Overall, the present concentration of all CFCs and HCFCs taken together is between 1 and 2 ppbv and, until the time of the Montreal Protocol, had been growing at a rate of nearly 5% per year. The rate of increase of the CFCs has declined by a factor greater than two in the past decade, but HCFC concentrations are rising at a much higher rate. Radiative forcing associated with chlorinated fluorocarbons and other related gases is estimated to be 0.34 W m$^{-2}$.

Three fully fluorinated gases of industrial origin have recently come to the fore as potentially important contributors to greenhouse warming. They are present in trace amounts, but have lifetimes of thousands or tens of thousands of years. Tetrafluoromethane ($CF_4$) and hexafluoroethane ($C_2F_6$) both arise during electrolysis of alumina ($Al_2O_3$) in cryolite ($Na_3AlF_6$) at carbon electrodes, and release of the gases is estimated to be about 0.77 and 0.1 kg, respectively, per tonne of aluminium produced. Together, their atmospheric concentration is approximately 0.08 ppbv. The other gas is sulphur hexafluoride ($SF_6$), which has no natural source and is formed during magnesium production. The atmospheric mixing ratio of sulphur hexafluoride is increasing at the rapid rate of about 5% per year. With such extraordinarily long atmospheric lifetimes, there is no practical means by which the amounts of these fluorinated compounds could be reduced within any reasonable time period.

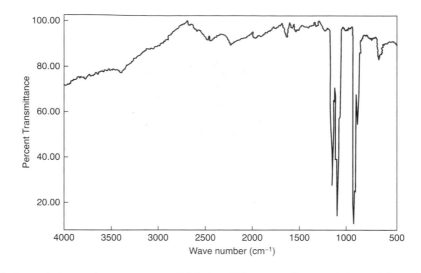

**Fig. 8.11** Infra-red absorption spectrum of dichlorodifluoromethane (CFC-12). The relation between wave number and wavelength is: wave number (cm$^{-1}$) = 10 000 / wavelength (μm). (©BIO-RAD Laboratories, Sadtler Division, 2000.)

## Aerosols

Clouds are the most important atmospheric aerosol in terms of reflecting and absorbing incoming radiation and emitted radiation from the Earth. The cooling effects of clouds on warm days, and their warming effects on cool nights are phenomena we all have recognized. Other aerosols, too, add to the complexities of the global energy balance situation. In particular, ammonium sulphate and other sulphate-based solid aerosols are becoming increasingly important. The sulphate aerosol derives from natural oceanic sulphide, particularly dimethyl sulphide emissions, as well as from anthropogenic sources of sulphur dioxide. In the Northern Hemisphere, about 90% results from human activities, while in the South most has a natural origin. Besides its direct role in backscattering incoming shortwave solar radiation, the presence of sulphate in the aerosol also affects processes of cloud formation. The net result of the direct and indirect processes is complex and varies from region to region but, overall, sulphate aerosols contribute to a negative global radiative forcing and therefore a measure of atmospheric cooling. As noted earlier, sulphate particulates that are periodically injected by volcanoes into the stratosphere also contribute to cooling of the troposphere.

Some biomass aerosols derive from combustion, with the release of fine smoke and soot, often called black carbon, into the atmosphere. Their extent varies around the globe and from year to year. Recent years, especially 1997–98, have seen increased incidents of widespread biomass burning in countries like Malaysia and Indonesia, as well as in parts of North America. The higher average global temperature in those years may have contributed to drying of forested areas and greater opportunities for fires to ignite and spread. In recent years, there is also clear indication of reduced industrial emissions emanating from Europe and Russia, but this appears to be offset by the increased emissions from countries in the South.

In contrast to the sulphate-based aqueous aerosols, the dark particles add to the positive radiative forcing of greenhouse gases. In some parts of South-east Asia, the local warming reduces daytime cloud cover, further enhancing the heating effects. Aerosols of industrial origin are also largely combustion based, and are usually found in the lower (<2 km) parts of the troposphere. Because they are readily washed out with precipitation, they have small atmospheric residence times, of the order of a few days, so their contribution to greenhouse warming is local and short-lived.

The radiative forcing associated with atmospheric aerosols is estimated to be −1.2 W m$^{-2}$.

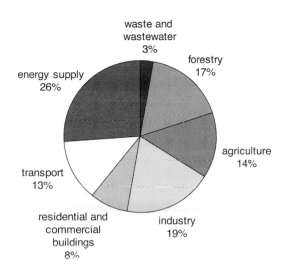

**Fig. 8.12** Share of different sectors in total anthropogenic GHG emissions in 2004 in terms of $CO_2$-equivalents. Redrawn from *Climate change 2007: Synthesis report*.

### Surface albedo changes

The surface *albedo* of the Earth is affected by two factors directly or indirectly related to the changes in global emissions. Where dark-coloured aerosol particles are deposited on snow-covered land, the reflective properties of snow are reduced, causing a net reduction of *albedo* in that area. The radiative forcing associated with this is estimated to be $0.1$ W m$^{-2}$. More significant is the opposite effect due to an increase in reflectivity caused by deforestation of land, such as the great rain forests in the mid latitude regions of the Earth. This causes a cooling effect equivalent to a radiative forcing of $-0.2$ W m$^{-2}$.

Figure 8.12 summarizes the $CO_2$ equivalent contribution of greenhouse gases associated with human activities. About 70% are related in some way to energy production and consumption. Because of the diversity of sources of carbon dioxide and other greenhouse gases, any meaningful initiative to control and reduce their release requires actions on several fronts.

---

**Main point 8.3** Water and carbon dioxide absorb large amounts of IR radiation, and have the potential to contribute to a warmer climate than would otherwise exist on the Earth. Several other trace gases—methane, ozone, nitrous oxide, and CFCs—also absorb IR radiation, most notably in the 'window' region. All of these are referred to as greenhouse gases. Both natural and anthropogenic aerosols contribute an additional warming effect.

---

## 8.4 Relative instantaneous radiative forcing (RIRF) and global warming potential (GWP)

Clearly, you can see that all of the species discussed above have a quantifiable ability to attenuate infra-red (IR) radiation. In the atmospheric chemistry literature, this attenuation or absorption process is referred to as *radiative forcing*, as defined in Table 8.3. To rank the relative importance of the various gases, at least three factors are involved:

- The present atmospheric concentration of the species: Increasing the concentration of gases that are present only in trace concentrations in the atmosphere can have a large effect on radiative forcing. On the other hand, if large concentrations are present so that essentially complete absorption of particular wavelengths of IR radiation already occurs, an additional amount of that species can only increase radiative absorption on the outer wings of the absorption lines. This may have only a limited effect on radiative forcing.

- The wavelengths at which the gas molecules absorb: If a particular species absorbs only in regions of the IR spectrum where absorption is nearly complete due to other species, then an increase in its concentration can add only minimally to warming. On the other hand, species that absorb in the 'window' region, have a much greater potential to increase retention of heat near the Earth's surface.

- The strength of absorption (absorptivity) per molecule: A small additional concentration of a strongly absorbing species will have a larger effect than the same concentration of a species that has limited intrinsic ability to absorb IR radiation.

The three factors relate to immediate effects, resulting from an incremental increase in the amount of the gas being present in the atmosphere. To estimate the long-term effects, the residence time of the gas must be considered as an additional factor.

In order to quantify the relative contribution of the various greenhouse gases to radiative forcing, it has been a goal to take into account the various factors and incorporate them into a single index. In fact, two such indices have been described and are widely reported and used.

The first index is the relative instantaneous radiative forcing (RIRF) index. As the name suggests, the RIRF value is a measure of the ability of an incremental addition of a gas, into the present atmosphere, to increase the absorption of infra-red radiation. The RIRF value for carbon dioxide is arbitrarily set at 1, and the RIRFs for other gases are relative to this value. For instance, the greater than four orders of magnitude ratio of RIRF for CFCs compared with carbon dioxide is due to the ability of CFCs to absorb strongly in the open window portion of the IR spectrum, the region where carbon dioxide and water do not absorb. This explains why an incremental addition of CFCs increases IR absorption to a very much larger extent than does the same mass increment of carbon dioxide. Figure 8.13 takes all three factors plus the actual amount of emissions and indicates the relative contribution of emissions of different greenhouse gases around the world in 2004.

A second index takes a longer view of the potential for contributing to greenhouse warming by taking into account the residence time of each species. To do this, Lashof and Ahuja[7] have

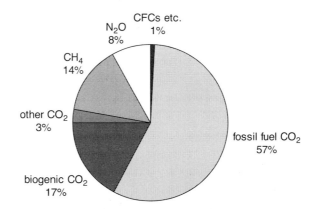

**Fig. 8.13** Share of different greenhouse gases of anthropogenic origin to global emissions in 2004. All values take into account quantity of emissions and relative instantaneous radiative forcing. Redrawn from *Climate change 2007: Synthesis report*.

[7] Lashof, D.A. and Ahuja, D.R., Relative contributions of greenhouse gas emissions to global warming, *Nature*, **344** (1990), 529–31.

developed an index of global-warming potential (GWP) applicable to any gas. The GWP value is now widely reported in documents such as those of the Intergovernmental Panel on Climate Change.

$$GWP = \frac{\int_0^t a_i(t)\, c_i(t)\, dt}{\int_0^t a_c(t)\, c_c(t)\, dt} \tag{8.11}$$

where $a_i(t)$ is the instantaneous radiative forcing due to a unit increase in the concentration of the gas $i$ (the ratio $a_i / a_c$ is equivalent to the RIRF value), $c_i(t)$ is the fraction of gas $i$ remaining at time $t$, and the integration limits 0 to $t$ represent the time span over which integration is carried out. The corresponding values with subscript c are for carbon dioxide and are in the denominator.

Defined in this way, GWP describes the long-term contribution of any gas by comparison to that of carbon dioxide. Any period can be selected for the time over which integration is done, but a common convention has developed that GWP values are calculated for a 100-y period. For carbon dioxide itself, both numerator and denominator terms are the same and so the GWP has a value of unity. Other greenhouse gases have GWP values ranging into the low tens of thousands (Table 8.4). From the table, for example, we see that addition of a particular amount of nitrous oxide to the atmosphere would contribute around 300 times more to greenhouse warming over a 100-y period, than would addition of the same amount of carbon dioxide.

In keeping with the three principles outlined above, $a_i$ depends on the concentration of the gas (e.g. raising the concentration of carbon dioxide from 387 to 397 ppmv will have a greater effect than raising it from 450 to 460 ppmv). The value of $a_i$ also depends on the absorptivity of the molecule and the wavelength(s) at which absorbance occurs. In particular, gases that absorb at wavelengths within the window region—covering a large portion of the frequency and wavelength range from 1250 to 830 cm$^{-1}$ (8–12 $\mu$m)—make an especially significant contribution to additional radiative forcing.

The term, $c_i$, depends on the residence time of the gas and can be calculated if the rates of all processes leading to loss or destruction of the gas are known. Because the GWP includes an accounting of the residence time, gases that rapidly dissipate from the atmosphere have a smaller

**Table 8.4** Greenhouse gas properties related to global warming.

| Gas | Atmospheric lifetime / y[a] | Relative instantaneous radiative forcing (RIRF) | Global-warming potential (GWP)[a] |
|---|---|---|---|
| $CO_2$ | 50–200[b] | 1 | 1 |
| $CH_4$ | 12 | 25 | 25 |
| $N_2O$ | 114 | 220 | 298 |
| CFC-11 | 45 | 17 000 | 4750 |
| CFC-12 | 100 | 23 000 | 10 900 |
| HCFC-22 | 12 | 14 000 | 1810 |
| $CCl_4$ | 26 | 9300 | 1400 |
| $C_2F_6$ | 10 000 | 37 000 | 12 200 |
| $SF_6$ | 3200 | 19 000 | 22 800 |

[a] Most atmospheric lifetime values and all RRIF and GWP values are taken from the Intergovernmental Panel on Climate Change (IPCC) *Fourth assessment report—the physical science basis*, 2007. GWP values are obtained by integration over a 100-y period.
[b] Reported residence-time values for carbon dioxide are highly variable. Differences are associated with the way in which oceanic uptake is measured, particularly whether the surface layer or the entire ocean is considered in the calculation.

long-term effect than do those with very long lifetimes. For example, CFC-11 and HCFC-22 have similar RIRF values, but the GWP for CFC-11 is almost three times greater than that of HCFC-22 as a consequence of the former compound's longer residence time.

While the physics and chemistry of the greenhouse effect are relatively well understood, there is great difficulty in predicting what climatic effects will result from increasing concentrations of the greenhouse gases. That the concentrations will continue to increase is beyond dispute. The difficulties in prediction are associated with the complex interactions between the various environmental processes and the resultant positive and negative feedbacks that result. For example, what happens to the additional carbon dioxide that each year is released into the atmosphere? We have indicated that there are several well-known sinks for this gas—the oceans being a major reservoir. But quantitative estimates describing carbon dioxide partitioning in water, the circulation patterns of the oceans, and the geochemical and biological processes that take up and release various carbonate species all require much more detailed knowledge than we presently have available. Adding to this are the equally complex terrestrial processes—both macro- and microbiological—and the interactions between them. A great deal of research effort is presently being devoted to modelling the fate of greenhouse gases. It is certainly worth noting that the 2007 report of the Intergovernmental Panel on Climate Change (IPCC), a document reviewed by more than 1500 scientists worldwide, repeats emphatically what the Panel had stated in earlier reports—that global warming is now occurring and that this is due in large part to the burning of fossil fuels. Further discussion of these important issues is beyond the scope of this book.

> **Main point 8.4**  The tropospheric concentration of all greenhouse gases that have anthropogenic origin is increasing. The degree to which this will affect the global climate depends on the extent of this increase as well as on complex positive and negative feedback mechanisms resulting from the chemical changes.

## 8.5 **Energy resources**

The greenhouse gases arise from many sources, but perhaps more than anything else the increasing release of several of the species into the atmosphere is associated with energy production and use. The life of all living species including humans depends on a supply of energy. Energy is involved in providing such basic necessities as the growing and preparation of food and the provision of warmth. It also fuels much of modern industry and is an aspect of all forms of transportation.

In considering sources and types of energy, certain ones such as fossil fuels and electricity readily come to mind. But there are many other forms as well, ranging from human and animal energy to energy obtained from the Sun and water or from nuclear fission. Whatever sources are used, the processes of extracting or manufacturing the energy and then putting it to use all have an effect on the atmospheric environment. As should be evident from previous chapters, some of the effects relate to smog, acid-precipitation precursors, and aerosol production. Here we are focusing on effects connected with release of greenhouse gases. One way of classifying the forms of energy is shown in Table 8.5.

The fundamental unit of energy is the joule ($J = N\,m = kg\,m^2\,s^{-2}$) but it has become customary to use units like barrels of oil, tonnes of coal, or kilowatt hours of electricity that relate to the particular energy form under consideration. Table 8.6 lists equivalents in joules for a variety of frequently encountered units. In this book, the joule is used as a common unit.

Some forms of energy such as fossil fuels are obtained as mineable deposits and these are non-renewable, but others, including the diverse forms of biomass, have the potential of being replenished and are available over an essentially limitless time period. There are, however, costs

**Table 8.5** Sources of usable energy.

| Primary sources | |
|---|---|
| Solar energy | Used directly or converted into electricity via photoelectric cells. Also, is the driving force for the water cycle, the ultimate energy source creating fossil fuels, and (through differential heating) causing wind and wave action. |
| Lunar energy | The cause of tides, which may be converted to useful forms including electricity. |
| Geoenergy | This is energy that is intrinsic to the structure and composition of the Earth. It includes nuclear and geothermal energy. These may be used as a source of heat or can be converted to mechanical or electrical energy. |
| **Derived sources** | |
| Fossil fuels | Includes coal, petroleum, and natural gas from various sources. These are primary combustion sources used as fuel for engines or to generate heat, which is often converted to electricity. |
| Biomass | Includes wood, straw, animal dung, sugar cane, corn, waste combustible products, etc. Used as fuels, or converted to other fuels or to electricity. |
| Hydro energy, wind energy, wave energy | Through the Sun's heating action on land and water these forms of energy are developed and the power can be used directly, but is most often converted to electricity. |
| Tidal energy | Results from the lunar gravitational pull on the ocean's water. May be used to generate electricity. |
| Electricity | Always a derived form of energy based on primary sources (solar, nuclear) or on other derived sources (fossils fuels, hydro power, etc.). |

**Table 8.6** Units of energy and equivalents in joules[a].

| Energy source | Unit | Abbreviation | Equivalent in joules |
|---|---|---|---|
| Natural gas | cubic metre | $m^3$ | $3.7 \times 10^7$ |
| | cubic foot | $ft^3$ | $1 \times 10^6$ |
| Petroleum | barrel | bbl | $5.8 \times 10^9$ |
| | tonne | t or TOE[b] | $3.9 \times 10^{10}$ |
| Tar sand oil | barrel | bbl | $6.1 \times 10^9$ |
| Shale oil | tonne | t | $4.1 \times 10^{10}$ |
| Coal | | | |
| Anthracite | tonne | t or TCE[c] | $3.0 \times 10^{10}$ |
| Bituminous | tonne | t or TCE[c] | $3.0 \times 10^{10}$ |
| Sub-bituminous | tonne | t or TCE[c] | $2.0 \times 10^{10}$ |
| Lignite | tonne | t or TCE[c] | $1.5 \times 10^{10}$ |
| Charcoal | tonne | t or TCE[c] | $2.8 \times 10^{10}$ |

(*Continued*)

**Table 8.6** (*Continued*).

| Energy source | Unit | Abbreviation | Equivalent in joules |
|---|---|---|---|
| Biomass (all on a dry weight basis) | | | |
| General | tonne | t | $1.5 \times 10^{10}$ |
| Miscellaneous farm wastes | tonne | t | $1.4 \times 10^{10}$ |
| Animal dung | tonne | t | $1.7 \times 10^{10}$ |
| Assorted garbage | tonne | t | $1.2 \times 10^{10}$ |
| Wood | tonne | t | $1.5 \times 10^{10}$ |
| | cubic metre | m$^3$ | $5 \times 10^{9}$ |
| | cord | 128 ft$^3$ | $2 \times 10^{10}$ |
| Fission | | | |
| Natural | tonne | t | $8 \times 10^{16}$ |
| Complete mass → energy conversion, $E = mc^2$ | tonne | t | $9 \times 10^{19}$ |
| Electricity | kilowatt hour | kWh | $3.6 \times 10^{6}$ |
| | terawatt year | TWy | $3.2 \times 10^{19}$ |
| General units | erg | erg | $1 \times 10^{7}$ |
| | calorie | cal | 4.18 |
| | British thermal unit | BTU | $1.05 \times 10^{3}$ |
| | ($10^5$ BTU) | therm | $1.05 \times 10^{8}$ |
| | ($10^{15}$ BTU) | quad | $1.05 \times 10^{18}$ |
| | ($10^{18}$ BTU) | Q | $1.05 \times 10^{21}$ |
| | horsepower hour | hp h | $3.6 \times 10^{6}$ |

[a] The data for individual commodities are obtained from several sources and many are estimates. Values for particular materials vary, especially for highly heterogeneous substances and substances with variable moisture content, such as different forms of biomass.
[b] TOE, tonnes of oil equivalent.
[c] TCE, tonnes of coal equivalent.

(including energy costs) required to sustain the renewability of such resources. For example, to maintain the productivity of a tree farm as an energy plantation, cultivation and fertilizers are needed. Both of these inputs have a significant energy component. We will look at these and other questions related to renewable resources in later sections.

The total amount of commercial energy used annually in the world is about 440 EJ[8] (1 EJ = $10^{18}$ J) but the patterns of commercial energy consumption are highly variable as shown in Table 8.7. Commercial energy is made up of forms that are purchased through regular channels, and excludes energy sources collected and used by individuals.

Many factors contribute to the great variability in energy consumption per capita. Some relate to the size of the country and its effect on transportation needs, others to climate, especially cold climates where heating of buildings is a necessity. Still others relate to the degree of

[8] BP, *Statistical review of world energy—June 2009*. Data are for 2008.

**Table 8.7** Annual commercial energy consumption in the regions of the world[a].

| Region | Energy consumption | | Population / $10^6$ | Annual commercial energy consumption per capita / GJ $y^{-1}$ |
|---|---|---|---|---|
| | EJ $y^{-1}$ | % of total | | |
| Africa | 13.9 | 3.1 | 967 | 14 |
| Asia Pacific | 156 | 35.2 | 3690 | 42 |
| Mid-East | 24.5 | 5.5 | 266 | 92 |
| Europe and Eurasia | 116 | 26.2 | 877 | 132 |
| Mexico, Central and South America | 29.3 | 6.6 | 681 | 43 |
| United States and Canada | 103 | 23.3 | 334 | 307 |
| World | 443 | 100 | 6705 | 65 |

[a] Energy data are for 2008 from British Petroleum, *Statistical review of world energy—June 2009*.

industrialization and the nature of industry. Perhaps most important is the population's expectation with respect to all these issues. It appears certain, however, that energy consumption will increase substantially in coming years, with much of the increase occurring in regions—Asia, Africa, South America—where energy consumption is much lower than the global average. The increase between 2007 and 2008 was approximately 1.4%, much of the growth occurring in China and India. This period included a major downturn in the global economy; the average annual increase in energy consumption over the previous decade was close to twice this value.

With this background, we can now look at some individual energy sources and the extent to which they contribute to global release of greenhouse gases. Earlier and later in the book we deal with some of the other environmental impacts associated with energy extraction and use.

> **Main point 8.5** Energy production and consumption is a major source of greenhouse gases. There are also environmental implications of many other types associated with each means of energy production.

## 8.6 Greenhouse gases associated with the use of fossil fuels

### Coal

The term 'coal' covers a range of sedimentary materials derived from the residues of plant materials that have been buried and subjected to high temperatures and pressures over extended periods of geological time. The degradation processes, which may initially take place under near-surface but anaerobic conditions, first lead to the formation of humic-like material (see Chapter 12) called peat. Further degradation under conditions of high heat and pressure is called coalification and produces coal (Fig. 8.14) with a spectrum of properties depending on the time and intensity of the reactions. Table 8.8 lists some chemical properties of representative American examples of four ranks of coal. The degree of coalification is lowest for lignite and increases from left to right in the table to anthracite.

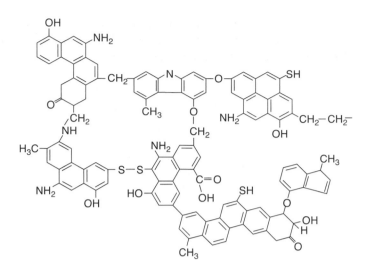

**Fig. 8.14** A hypothetical structure of coal. Note the similarities and differences between this structure and that of a hypothetical humate molecule as shown in Fig. 12.3.

Increasing coalification results in lower moisture and higher carbon percentages, factors that are reflected in a larger energy content per tonne of coal (Table 8.6). From this perspective, anthracite is most desirable but there is only a limited quantity of this type of coal and bituminous and sub-bituminous resources are most often exploited. The sulphur and ash contents vary depending on source, and some deposits contain up to 8 or 10% sulphur and / or up to 50% non-combustible ash.

Of the atmospheric pollutants given off during coal combustion, a number have been subjected to stringent controls by various political jurisdictions. We have earlier indicated problems with acid precursors, with atmospheric pollutants like mercury and PAHs and with ash. Coal varies in quality and its potential to generate pollutants depends on its source (Table 8.8). As an intrinsic by-product of carbon combustion, carbon dioxide is released and the release cannot presently be controlled in any practical sense. The same is true of any carbon-based fuel. There is current work, however, on mitigation strategies that are focused on ways of sequestering the released carbon dioxide.

**Table 8.8** Chemical properties of representative coal types[a].

| Coal rank[b] | Lignite | Sub-bituminous | Bituminous | Anthracite |
|---|---|---|---|---|
| Location | McLean, North Dakota | Sheridan, Wyoming | Muhlenberg, Kentucky | Lackawanna Pennsylvania |
| Moisture / % | 37 | 22 | 9 | 4 |
| Carbon / % | 41 | 54 | 65 | 80 |
| Ash[c] / % | 6 | 4 | 11 | 10 |
| Sulphur[c] / % | 0.9 | 0.5 | 2.8 | 0.8 |

[a] US Department of Energy, *Coal data: cost and quality of fuels*; US Department of Energy, Washington, DC; 1979.
[b] The ranking, from left to right, represents increasing coalification.
[c] Ash and sulphur concentrations depend on the geological setting and are independent of the coalification processes.

One can compare the combustion of the carbon component of coal (C) with that of natural gas ($CH_4$) and a heavy oil ($C_xH_{2x+2}$, represented as $C_{20}H_{42}$). The combustion reactions and the magnitude of enthalpy change for the reactions are as follows:

| | | | |
|---|---|---|---|
| Coal | $C + O_2 \rightarrow CO_2$ | $\Delta H_{comb} = -393.5 \text{kJ mol}^{-1}$ | (8.12) |
| Natural gas | $CH_4 + 2O_2 \rightarrow CO_2 + 2H_2O$ | $\Delta H_{comb} = -890.3 \text{kJ mol}^{-1}$ | (8.13) |
| Heavy oil | $C_{20}H_{42} + 30\frac{1}{2}O_2 \rightarrow 20CO_2 + 21H_2O$ | $\Delta H_{comb} = -13315 \text{kJ mol}^{-1}$ | (8.14) |

During the combustion of coal (approximated as carbon), for every GJ (1 GJ = $10^9$ J) of heat produced the amount of carbon dioxide released to the atmosphere is

$$\underbrace{\frac{10^9 \text{ J GJ}^{-1}}{393.5 \times 10^3 \text{ J}}}_{\Delta H_{comb}} \times 44 \times 10^{-3} \text{ kg} = 112 \text{ kg GJ}^{-1}$$

(molar mass of $CO_2$)

Corresponding amounts for the two other fuels are 66 kg ($CO_2$) $GJ^{-1}$ for oil and 49 kg ($CO_2$) $GJ^{-1}$ for natural gas. Of the fossil fuels, then, coal makes the greatest greenhouse-gas contribution in terms of release of carbon dioxide per unit of energy produced. This is one of the reasons why coal is a less-desirable fuel, aside from it being a major contributor to atmospheric burdens of particulates, acid rain precursors and other atmospheric pollutants.

The newer types of combustion processes such as fluidized-bed combustion, as described in Chapter 6, enhance efficiency of coal combustion and of heat transfer, and therefore serve to maximize energy output for a given amount of fuel.

Other traditional and emerging technologies, developed in order to produce a cleaner and more convenient fuel, convert the coal to gaseous or liquid products. A number of processes are possible as routes to coal gasification (Table 8.9). Several of these occur sequentially in commercial coal gasifiers some of which have been in use for decades, especially where natural gas has not been available. The product mixtures generally have low heat content, and widespread commercial applications await further technological developments.

Similarly, liquefaction of coal is used to produce a substitute for various types of petroleum products. The liquefaction can be done indirectly by first producing gaseous products and following this by a separate process to convert the gases to a liquid form. This procedure has been

**Table 8.9** Reaction sequences used in coal gasification processes.

| | | |
|---|---|---|
| Partial gasification | Coal $\xrightarrow{500-700°C}$ C + CH$_4$ + H$_2$ | Produces a relatively small amount of high energy content gases |
| Carbon—oxide | $C + O_2 \rightarrow CO_2$<br>$2C + O_2 \rightarrow 2CO$ | The carbon monoxide is a combustible product |
| Steam—carbon | $C + H_2O \xrightarrow{\text{heat, air}} CO + H_2$ | The two product gases are both combustible, but are diluted by nitrogen, giving a low energy content fuel |
| Catalytic methanation | $3H_2 + CO \rightarrow CH_4 + H_2O$ | Catalysts such as nickel oxide can enhance conversion of carbon monoxide into methane that has a higher energy content |

the basis of the *SASOL* method, developed and used in South Africa to produce more than 10 million litres per day of synthetic liquid petroleum. The process begins by heating the coal to temperatures between 600 and 800°C to partially volatilize it, producing a mixture of methane, hydrogen, and carbon. The mixture then moves through a heated region containing air and steam. The product gas mixture resulting from the complex reactions is a made up of about 10% methane, 20% carbon monoxide, 30% carbon dioxide, and 40% hydrogen. The mixture is treated by a process called the Fischer–Tropsch synthesis where the gas reacts under a pressure of $2 \times 10^6$ Pa in the presence of an iron catalyst, releasing heat and producing a liquid product. The hydrocarbon product contains many of the typical low, medium, and high molar mass species found in crude oil and may be refined into various commercial products.

Potentially more efficient are direct methods of liquefaction in which the coal is 'cracked' into large subunits that then undergo hydrogenation in a solution or slurry. Several direct approaches have been developed. *Solvent refined* coal is produced by mixing coal with a solvent (actually a product of the process itself), distilling off the solvent, and then heating to 450°C in the presence of hydrogen gas at a pressure of about $1 \times 10^7$ Pa. The products include both solid and liquid hydrocarbons. The use of a cobalt molybdenate catalyst can increase the yield of useful liquid products.

While the various conversion methodologies produce fuels that are more convenient to store, transport, and use and are cleaner, the production processes themselves create negative environmental impacts. The ash and gaseous by-products like hydrogen sulphide must be disposed of, and large amounts of energy are required to effect the conversion. Typically, for every unit of energy equivalent in the product fuel, 1.5 units of energy input are required. The energy is usually supplied in the form of coal itself so that the total quantity of some waste materials, like ash, are 2.5 times the amount that would result from the direct use of coal for producing the equivalent amount of heat. And importantly, the energy consumed in every step adds to the total release of carbon dioxide to the atmosphere.

## Petroleum

Petroleum is the fuel of choice in many situations; this is especially true in mobile applications. Petroleum is relatively easily and safely pumped from oil-bearing reservoirs. Crude petroleum is essentially a mixture of hydrocarbons, straight and branched-chain alkanes, aromatics and others with a wide range of molar masses. By separating the fractions during refining it becomes the source of a range of useful products, all of which are readily transported and stored. As a fuel, the products like gasoline are fairly low in sulphur and are suitable for burning to produce heat in domestic or industrial units or for producing mechanical power in internal combustion engines. Consumption patterns are highly tilted toward high-income nations—about two-thirds is used by the wealthiest one-quarter of the world's population.

Being a mixture of hydrocarbons many of which are saturated alkanes, the combustion takes a form like that shown in eqn 8.15.

$$C_xH_{2y} + \left(x + \tfrac{1}{2}y\right)O_2 \rightarrow xCO_2 + yH_2O \tag{8.15}$$

At present, about 37% of commercial energy consumed throughout the world is produced from crude oil. The use of petroleum-based fuels is therefore a major source of carbon-dioxide emissions even though it is more efficient than coal on the basis of carbon dioxide released per heat generated. Furthermore, although refined petroleum products contain a smaller concentration of sulphur than most sources of coal, it does contribute to sulphur-dioxide release to the atmosphere. Diesel, for example, contains significant concentrations of sulphur. Some non-conventional precursors of oil products, particularly shale oil and tar sands, do have much higher concentrations of the element (the Athabasca tar sand product contains about 4% sulphur) and this must be removed during purification.

**Table 8.10** The principal components of natural gas.

| | |
|---|---|
| Methane (75–100%) | Used directly or in the form of compressed natural gas (CNG) as an industrial and domestic fuel, and increasingly for mobile sources |
| Ethane (6–10%) | Used as a fuel or as a feedstock for petrochemical plants manufacturing ethylene |
| Propane and butane (5–8%) | Liquefied petroleum gases (LPGs)—used as fuels or as petrochemical feedstocks |
| Pentane and heavier hydrocarbons (1–4%) | Condensate, used as petrochemical feedstock |
| Nitrogen, carbon dioxide, hydrogen sulphide, helium (variable) | Components other than hydrocarbons |

While carbon-dioxide and sulphur-dioxide emissions contribute to *global atmospheric pollution* problems, burning of oil products in the transportation sector is especially important in contributing to *urban atmospheric pollution*. We have already noted the problems associated with carbon-monoxide release and with photochemical smog caused in part by nitrogen oxides and volatile organic carbon. All these compounds are present in emissions from internal combustion engines and are released directly at ground level in highly populated areas, although improved design has significantly lessened the quantities of these gases.

## Natural gas

Natural gas is found either in association with or independent of oil deposits. There is a considerable range in the composition of gas at the various locations but most deposits contain the components listed in Table 8.10.

In many ways, natural gas is the most desirable of the fossil fuels. On a mass basis it produces more heat and less carbon dioxide than either coal or petroleum products. In most cases, it has fewer undesirable impurities, such as sulphur, than other fuels and therefore burns more cleanly. Applications range from heat production for electric power generation, domestic and industrial space heating, and fuel for transport vehicles, to providing a feedstock in the petrochemical and nitrogen fertilizer industry. Besides being (inevitably) a source of atmospheric carbon dioxide, leakage of natural gas at gas wells along with losses during processing and transport may be the source of as much as 20% of the world release of methane—another important greenhouse gas.

Another environmental concern arises from transport of the gaseous products by pipeline where leakage and subsequent explosions may result. The liquid petroleum gases are gases under normal conditions (for butane the boiling temperature $T_b = -0.5°C$ and for propane $T_b = -42°C$ at $P°$) and are carried in pressurized and / or refrigerated containers. In 1979 a train derailment at Mississauga, Ontario exposed the community to a potential dangerous mix of propane and other chemicals and it was necessary to evacuate more than 200 000 persons from their homes. Nevertheless, the environmental dangers associated with the use of natural gas are somewhat less acute than those for other fossil fuels.

**Fermi question**

The total amount of carbon dioxide released worldwide each year from combustion of fossil fuels is approximately 23 Gt. In a one-year period, how much carbon dioxide is released into the atmosphere by all the world's humans, just by being alive and breathing?

## 8.7 **Sequestration of carbon dioxide**

Carbon dioxide is inevitably given off during the combustion of every carbon-based fuel and the amount is much larger than that of other gases like nitric oxide or sulphur dioxide. Attempts to find simple and cost-effective ways of capturing the large quantities in industrial stack gases or of compensating for the additional carbon dioxide after it has been released into the atmosphere have led to two types of development. In a general sense, such processes are referred to as carbon-dioxide sequestration or mitigation (Fig. 8.15).

- geological sequestration—capture of carbon dioxide from the stack gases, and pumping it into 'safe' reservoirs on land or in the oceans;
- biological sequestration—growing vegetation in quantities large enough to consume amounts of carbon dioxide equivalent to that released during energy consumption. The vegetation (biomass) that is produced can itself be used as a fuel, creating a closed production / consumption cycle.

### **Geological sequestration**

The goal of all methods of geological sequestration is to place the carbon dioxide deep in the Earth in a stable form where it will remain indefinitely.

As a first step, the geological-sequestration approach requires concentrating the carbon dioxide that is emitted during combustion. Recall that, on a molar basis, oxygen makes up only about 20% of the components of air. Therefore, if oxygen reacts completely when the fuel is burned, the stack gas would contain 20% carbon dioxide, with the rest being mostly nitrogen. In real situations, an excess of air is supplied so that typical stack gas composition is only about 10% carbon dioxide. There is a large economic and energy cost involved in pumping fluids deep into the Earth, so it is important that most of the harmless nitrogen and excess oxygen be removed before the carbon dioxide is transported into storage. Of the many approaches being considered to separate carbon dioxide from the other gases, the most common uses aqueous amine solutions such as methanolamine.

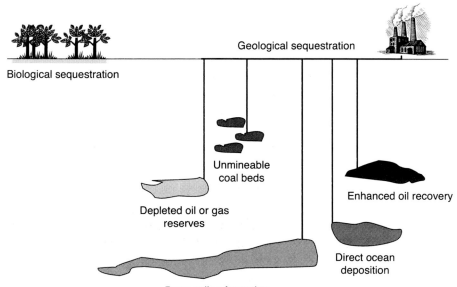

**Fig. 8.15** Carbon-dioxide sequestration possibilities.

$$CO_2 + 2RNH_2 \,(aq) \longrightarrow RNHCOO^- \,(aq) + RNH_3^+ \,(aq) \qquad (8.16)$$

To release the 'concentrated' carbon dioxide and recover the amine solvent, the solution is heated.

Another possible capture method uses chilled ammonia solution to scrub the gases, forming ammonium carbonate. The carbon dioxide can be driven off by heating the solution under pressure, and then the ammonia is recycled. Once the relatively pure stream of carbon dioxide has been created, it is transported by pipeline to the reservoir area. Before pumping it into a chosen reservoir it is usually compressed to form a supercritical fluid.

A variety of sedimentary formations can be employed for geological storage. In some cases, operating or depleted gas and oil wells are appropriate reservoirs. One project uses an oil field based on fractured carbonate rocks in Weyburn, Saskatchewan, Canada. Some of the carbon dioxide is generated on site, but most is piped in from an energy plant 325 km to the south in North Dakota, USA where coal is gasified to make synthetic gas (methane). In the gasification, a relatively pure stream of carbon dioxide is coproduced as a by-product. The Weyburn site sequesters about 1000 t of carbon dioxide per day, with a capability of increasing this amount five-fold. The benefits of sequestration are multiplied by the fact that the injected carbon dioxide releases additional petroleum from the site—an incremental production of 1600 m$^3$ per day is achieved.

Some permeable abandoned coal beds can also be used for storage of carbon dioxide. When injected into the cavities, the carbon dioxide is adsorbed onto the surface of the coal. Once again this process can provide a potential gain in energy production as methane is simultaneously released and becomes available for capture and use.

Other types of storage reservoirs are deep (>800 m) saline formations that can sequester and retain large quantities of carbon dioxide. Such formations exist both on land and deep under the ocean bed. The first commercial-scale project of this type is operated by a Norwegian company about 250 km off the Scandinavian coast in the North Sea. The brine-saturated unconsolidated sandstone deposit is thought to have a capacity for somewhere between 1 and 10 Gt of carbon dioxide, and in the first decade of the twenty-first century is being supplied with about 2700 t of the gas every day.

Ocean storage can also be done in two additional ways: one is by injecting carbon dioxide into the water column at depths below 1000 m, where it dissolves and becomes part of the ocean carbonate cycle (Chapter 15). A second way pumps the supercritical fluid to a depth greater than 3000 m. Under the great pressures there, the carbon-dioxide fluid is denser than water and forms an immiscible 'lake' that delays its dissolution into the surrounding ocean environment. Pumping can be done using a fixed pipeline from shore or from ships that carry the compressed gas.

It is clear that all of the geological-sequestration methods mentioned here involve major engineering operations. Furthermore, these methods can be applied only for large point sources; they are not appropriate for the many small sources like domestic heating or for carbon dioxide released by transportation. The capture, release and transport of carbon dioxide typically require an additional energy expenditure of up to 40% over that where no sequestration technology is applied.

Additional Resources 4 is a detailed report that documents currently used methods of sequestration technology.

## Biological sequestration and biomass based fuels

Biological sequestration is based on the idea that plants remove carbon dioxide from the atmosphere or from water as they are growing. Biomass refers to organic material produced as the solid product when photosynthesis occurs in growing plants (reaction 8.17).

$$CO_2 + H_2O \rightarrow \{CH_2O\} + O_2 \qquad (8.17)$$

Various forms of dry biomass can be and are used as fuel to provide energy for industrial and space heating and for cooking. Because biomass is synthesized using the solar flux as the primary source that drives its production, it can be called a form of derived energy. We think of biomass as a renewable resource because the average energy of the Sun reaching Earth remains constant, and crops can be grown and harvested year after year. However, solar energy is not the only required input. Water is obviously essential and soil also plays a role as a source of nutrients. To sustain the ability to produce biomass, the chemical and physical integrity of these two resources must be maintained. Besides being renewable, there is another unique and important feature associated with biomass as a fuel source. In principle, it is possible to sequester approximately the same amount of carbon dioxide during the growth phase as that released by combustion.

The various forms of biomass are converted to energy in two different ways. One is to burn them directly as a fuel source. If the biomass is considered to be a complex form of carbohydrate, the combustion reaction is expressed in simplest form as

$$\{CH_2O\} + O_2 \rightarrow CO_2 + H_2O \qquad \Delta H = -440 \text{ kJ mol}^{-1} \qquad (8.18)$$

As you can see, reaction 8.18 is the reverse of the photosynthesis reaction 8.17. It is for this reason that we can think of biomass growth and its use as a fuel as having little or no net impact on atmospheric carbon-dioxide levels. Based on reaction 8.18, one tonne of dry biomass would release $1.5 \times 10^{10}$ J of energy during combustion but actual values can be greater or (usually) less than this due to the fact that biomass is not exclusively carbohydrate. Included in the listing in Table 8.6 are several sources of biomass and their estimated energy-conversion factors.

There is a wide variety of forms of biomass that are used to produce energy. Wood from trees is the most common form but other crops such as sugar cane or maize (corn) are also sometimes grown for the specific purpose of producing a fuel. In some cases biomass energy sources are a by-product of other agricultural operations. Straw and animal dung are two examples. In the latter case, the by-product is a tertiary energy form being obtained from animals that feed on plants whose growth depends on the Sun. Although the by-products are often referred to as waste materials, this is not strictly true as they can serve many other useful purposes—for example, besides being fuel, they may also be used as animal feed, soil conditioners, or fertilizers.

Biomass is used throughout the world as a source of energy for heating, cooking, and other purposes. In many countries it is the principal source but, because its use may not be part of the commercial energy network, quantitative details about its production and consumption are difficult to obtain. Even tables of national energy use are usually incomplete with respect to information on biomass. In low-income countries, biomass has been estimated to contribute, on average, about 40% to energy consumption. The figure for higher-income countries is 1% and the global average is 14%. This means that, at present throughout the world, about 50 EJ of energy are supplied annually by biomass. Note that the energy data given in Table 8.7 is for commercial energy, that which is commercially traded and therefore accounted for in national records.

For the future, biomass is likely to continue to be an important energy source in spite of the fact that obtaining and using it is less convenient than for many of the commercial energy forms. In both high- and low-income countries, there is considerable interest and research directed toward developing sustainable and productive systems of *energy plantations* as sources of domestic and industrial fuel. Clearly, there are environmental considerations with respect to the present situation and the future prospects.

### Conversion of solar energy to biomass

The efficiency of conversion of solar energy into biomass is a simultaneous function of many factors. Of the total solar flux striking the Earth, on average only 46% is available to be absorbed at the Earth's surface. Of this, only about 43% can be used for photosynthesis by the green parts of growing plants. The 43% represents the portion of the solar spectrum between 400 and 700 nm that can be absorbed by chlorophyll in the chloroplasts. This is called photosynthetically active radiation (PAR) and it provides energy for the reaction involving carbon dioxide and water to produce carbohydrate and oxygen.

In some plants the conversion process takes place via a reductive pentose phosphate (RPP) cycle that produces, as its primary carboxylation product, a three-carbon acid. Such species are called C3 plants. The rest of the plant kingdom uses mechanisms in addition to the RPP cycle and incorporates carbon dioxide into a four-carbon acid. Naturally, these are termed C4 plants.

Amongst the C3 species are wheat, rice, soy beans, tomatoes, potatoes, and sugar beets. The C4 species include sorghum, maize, sugarcane, and desert grasses—all species that have the potential to produce large yields of biomass. In general, C3 species are common in temperate regions, while C4 types dominate in the tropics and subtropics especially in more arid regions.

Differences between C3 and C4 plants are important in terms of photosynthetic and water-use efficiency. The C4 species have higher rates of net photosynthetic production and transpire about 500 mol of water for each mol of carbon dioxide incorporated. In this way they are efficient biomass producers. This is in contrast to C3 species from which 1000 mol or more of water are lost per mol of carbon dioxide fixed. The lower rate of net photosynthesis in C3 plants is due to the fact that, in a warm sunny environment, a part of the photosynthesized material is lost via its reoxidation to carbon dioxide. This reverse process does not occur in C4 plants.

Of interest in the present context is the question as to whether photosynthetic rates increase due to increasing concentrations of carbon dioxide in the atmosphere. Because C4 plants efficiently carry out photosynthesis, they respond only slightly to carbon dioxide 'fertilization'. However, at least in the short term, C3 plants may grow much faster in an atmosphere containing more carbon dioxide. Laboratory growth experiments under an atmosphere with double the present carbon dioxide concentration have shown that photosynthesis increases substantially, leading to between 20 and 40% greater biomass production. More efficient photosynthesis is largely due to reduced rates of photorespiration. The increased growth rate is one of the sinks counteracting increasing releases of carbon dioxide to the atmosphere.

---

### Example 8.1 **Maximum efficiency of photosynthesis**

For any plant, the maximum theoretical efficiency obtainable in the photosynthesis process may be calculated with some precision. Eight quanta of PAR are required to fix one molecule of carbon dioxide.

Choosing a mean PAR wavelength of 575 nm, the energy required to fix one mol of carbon dioxide is

$$E = \frac{nNhc}{\lambda}$$

$$= \frac{8 \text{ mol} \times 6.02 \times 10^{23} \text{ mol}^{-1} \times 6.62 \times 10^{-34} \text{ J s} \times 3.00 \times 10^8 \text{ m s}^{-1}}{575 \times 10^{-9} \text{ m}}$$

$$= 1660 \text{ kJ}$$

Comparing the amount of solar energy required to generate one mole of biomass to the enthalpy value shown in reaction 8.17 (the reverse of the synthesis process), we can calculate the efficiency of energy conversion.

$$440/1660 \times 100\% = 27\%$$

Therefore, photosynthesis could be 27% efficient in terms of PAR or $0.46 \times 27\% = 12\%$ efficient in terms of the portion of the total solar spectrum that is absorbed by the Earth's surface.

---

While this is a theoretically achievable efficiency, there are practical reasons that lead to much lower rates of productivity than are estimated using the 12% figure. Respiration (chemically, the reverse of the photosynthesis reaction) reduces productivity by 20–80% or more. Microbial decomposition of synthesized material is another source of loss. As a result, the seasonal maximum growth rates measured for C4 plants are in the range of 22 g m$^{-2}$ d$^{-1}$(equivalent to 3.8 GJ ha$^{-1}$ d$^{-1}$) and for C3 species 13 g m$^{-2}$ d$^{-1}$ (equivalent to 2.2 GJ ha$^{-1}$ d$^{-1}$). A reasonable average total quantity of energy absorbed at the Earth's surface is 160 GJ ha$^{-1}$ d$^{-1}$. Therefore, the actual maximum conversion efficiency is about 1.4% and 2.4% for C3 and C4 plants, respectively. Even these percentages are high compared with what is attainable in practice using good agronomic practices. The actual average for well-managed production of maize grain plus stalk is 0.6% and for wheat grain plus straw is 0.3%. A large-scale global average efficiency might be close to 0.25%.

---

### Example 8.2 **Potential for energy recovery from biomass**

It is interesting to calculate how much energy could actually be produced annually in a given area of land. India has a land mass covering about 3 300 000 km$^2$. Of this, approximately 2 200 000 km$^2$ is arable or forested. Suppose all of this were devoted to generation of energy. A 0.25% photosynthetic energy conversion efficiency means that

$$0.25 \,/\, 100 \times 160 \text{ GJ ha}^{-1}\text{ d}^{-1} \times 365 \text{ d y}^{-1} \times 10^2 \text{ ha km}^{-2} = 1.46 \times 10^4 \text{ GJ km}^{-2}\text{ y}^{-1}$$

of photosynthetic energy would be produced, for a total energy production of

$$2\,200\,000 \text{ km}^2 \times 1.46 \times 10^4 \text{ GJ km}^{-2}\text{ y}^{-1} = 3.2 \times 10^{19} \text{ J y}^{-1} = 32 \text{ EJ y}^{-1}$$

This is equivalent to about 30 GJ per capita in that country.

---

The conclusion arising from this type of calculation is that planting the *entire* arable and forested land mass of India with energy-producing crops could generate a biomass supply sufficient to meet a modest energy demand. It would, however, come nowhere near providing energy at the level presently consumed in high-income countries. And this unlikely scenario leaves no land for food production or other essentials. In other words, biomass alone is not a solution to the world's energy needs. This conclusion is supported by an investigation[9] of the potential productivity of degraded and abandoned agricultural lands around the world. Using an estimated biomass production rate of 4.3 t ha$^{-1}$y$^{-1}$, this marginal and underutilized area would supply less than 10% of the current energy consumption for most countries, but it represents many times the energy demand of some nations in Africa. The biomass potential productivity of various ecosystem types is given in Appendix A.2. The values given there are approximate values, and assume a healthy ecosystem.

Although biomass alone can never satisfy the global energy needs, it must be recognized that biomass is now, and will remain, the basic energy source for domestic use by vast numbers of persons—especially those in rural communities in much of Africa, Asia, and Latin America. It has been estimated that 40–50% of the world's inhabitants depend entirely or in major part on wood fuel as their source of energy. A frequently cited average annual wood consumption figure (for persons relying on this energy source) is 1.0 m$^3$ per capita representing about 5 GJ of energy.

In the context of the present chapter, there is also the question of how much carbon dioxide is released when biomass is burned. We have already compared carbon dioxide release from various fossil fuels. For coal, 112 kg $CO_2$ is released for every GJ of energy produced. This carbon dioxide is considered as new to the atmosphere, due to the length of time it has been buried in the Earth and therefore not been part of the carbon cycle. The corresponding figure for biomass (as {$CH_2O$}) is approximately 100 kg, but is considered as already being part of the carbon cycle.

---

[9] Campbell, J.E., Lobell, D.B., Genova, R.C. and Field, C.B., *The global potential of bioenergy on abandoned agriculture lands*. Environ. Sci. Technol., 42 (2008), 5791–4.

In that sense, there is a benefit in using biomass rather than coal. More importantly, another question has been raised as to whether growing crops (often trees) as biomass sources can counteract carbon dioxide increases in the atmosphere by their photosynthetic activity. The answer to this question is complex and depends on the number of years of growth and the end use of the biomass produced. It has been shown[10] that the optimum situation is to grow trees for short periods of time—usually up to 10 years. During this growth period there are high rates of net carbon accumulation, but such rates cannot be sustained as the tree ages. Therefore at this stage the tree should be harvested, dried, and used as a fuel in replacement of fossil fuels. In this way the accumulated and released carbon dioxide would roughly balance while, at the same time, producing usable energy.

Another way of using biomass is to convert it by microbiological and / or chemical processes into other types of fuel that may then be burned. We will see in Chapter 16 that sewage sludge from the aerator of a secondary wastewater treatment plant is frequently digested anaerobically to produce methane. This kind of process is carried out in large-scale industrial operations as well as in small domestic units, many of which have been installed in low-income countries. Depending on the feedstock and the process, a variety of products or product mixtures are obtained. Table 8.11 lists some such feedstocks, processes, and products.

The chemistry and microbiology of some of these processes will be discussed in later chapters.

### Synthetic fuels from biomass-based resources

In a search for clean burning fuels that are based on renewable resources, a number of new developments have taken place in recent years. Two examples of such fuels are ethanol produced from carbohydrate sources like corn or sugarcane and biodiesel whose feedstock for production is oil of various types. We have discussed some of the issues surrounding these fuels in Chapter 4. Recall the calculation given there for the energy budget for ethanol production from corn. Clearly, there is an energy gain in producing the ethanol, but an energy investment is required to achieve that gain. Furthermore, other environmental resources are required in order to produce the corn or other crop used as feedstock. In summary, no biomass-based fuel can be called a truly renewable resource.

**Table 8.11** Biomass conversion processes and products.

| Feedstock | Process | Products |
|---|---|---|
| Dry biomass—wood, straw, husks, etc. | Gasification | Liquids—methanol |
| | | Gases—hydrogen, ammonia |
| | Pyrolysis | Solids—charcoal |
| | | Liquids |
| | | Gases |
| | Fermentation and distillation | Ethanol |
| Wet biomass—domestic and animal wastes, aquatic plants, etc. | Anaerobic digestion | Methane |
| Sugars—from juices and hydrolysed cellulose | Fermentation and distillation | Ethanol |

[10] Vitousek, P.M., Can planted forests counteract increasing atmospheric carbon dioxide? *J. Environ. Qual.*, **20** (1991), 348–54.

**Fermi Question**

How much land would be required to produce enough energy to cook three meals a day for your family for a year by using maize as a source of biomass? Consider two situations: (a) the entire plant is dried and burned directly as fuel, and (b) the maize grain is converted into ethanol.

### Literature link  Life-cycle analysis of fuel use

Conducting a life-cycle analysis is an increasingly important activity in many situations. Environmental life-cycle analyses involve estimating, as accurately as possible, the environmental effects of producing a product or carrying out a process, taking into account all the steps needed to produce that product or complete the process. Where a product is manufactured, the life cycle also includes considering the environmental effects of using the product and those related to what happens to the product when its useful life is over.

In the case of a fuel like coal, the emissions involved in producing energy include more than those that are released during combustion but also emissions from mining, transport and the disposing of wastes. If the coal is transformed into some other possibly cleaner fuel, there are further emissions in the conversion process. The paper *Comparative life-cycle air emissions of coal, domestic natural gas, LNG, and SNG for electricity generation* by Jaramillo, P, W.M. Griffin, and H.S. Matthews, (Environ. Sci. Technol., 41, (2007), 6290–6 considers these issues. The amount of carbon dioxide released during electricity generation is compared for the cases of domestic and imported natural gas, coal, and coal converted into gas. Emissions of carbon dioxide, sulphur dioxide and nitrogen oxides are accounted for. Figure 8.16 shows results for carbon dioxide using current technology. The figure depicts the amount of carbon dioxide released during combustion as well as the total amount released after including emissions during life cycle issues such as transportation and conversion. It is clear that taking the entire life cycle into account can add significantly to the total emissions of carbon dioxide and the other smog-generating gases. The large range of values is due to varying efficiencies of the processes involved.

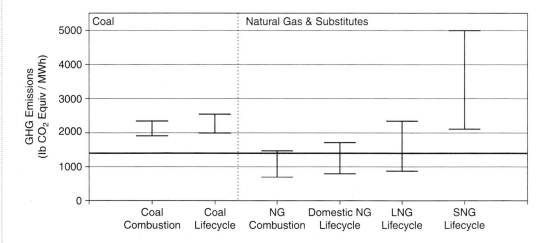

**Fig. 8.16** Fuel combustion and life-cycle GHG emissions for current power plants. The horizontal line represents the average greenhouse-gas emissions in carbon-dioxide equivalents released per MWh of electricity generated, for power plants in the USA. NG = natural gas (domestic), LNG = liquefied natural gas (imported), SNG = synthetic natural gas, produced from coal. The figure is reproduced from the above reference.

Looking ahead, if advanced technologies with carbon capture and sequestration were employed, life-cycle emissions from coal would be similar to those from the other sources. For sulphur and nitrogen oxides, the authors show that there are significant emissions in the upstream stages of the natural gas life cycles that contribute to a larger range in emissions of these compounds than is the case for coal. Considering all the emission products studied, the general conclusion is that a decision to import natural gas (as LNG) or to convert coal to synthetic natural gas should be based on more than just economic considerations.

---

**Main point 8.6** All fossil fuels release carbon dioxide when burned. The ratio, energy produced:carbon dioxide released is in the order: natural gas > petroleum > biomass > coal. Biomass, however, offers the possibility of sequestering atmospheric carbon dioxide during its growth.

---

## ADDITIONAL RESOURCES

1. IPCC Special Report, *Carbon dioxide capture and storage*, Cambridge University Press, New York, 2005.

2. Smil, V., *Energy at the crossroads: global perspectives and uncertainties*, MIT Press, Boston; 2003.

3. Houghton, J.T., L.G. Meira Filho, B.A. Callander, N. Harris, A. Kattenberg, and K. Maskell (eds.), *Climate change 1995: the science of climate change*, Cambridge University Press, Cambridge; 1996.

4. Miller, D.H., *Energy at the surface of the Earth*, Academic Press, New York; 1981.

5. The Carbon Dioxide Information Analysis Centre (CDIAC) under US Department of Environemnt, publishes trace gas concentrations from a number of specific sites around the world, http://cdiac.ornl.gov/. accessed November 2009.

6. The Intergovernmental Panel on Climate Change (IPCC), Assessment Reports, http://www.ipcc.ch/, accessed October 2009.

---

## PROBLEMS

1. What is the mass of water in 1 m$^3$ of air having a temperature of 32°C and a relative humidity of 83%?

2. Use eqn 8.4 to show why the total flux of solar radiant energy is about 10$^5$ times greater than that from the Earth.

3. The average albedo of the Earth is 0.31. Consider the following components of the Earth's surface  oceans, rainforests, deserts. How would you expect their albedo to differ from the average?

4. The current concentration of carbon dioxide in the atmosphere is 387 ppmv. It was indicated in the text that annual anthropogenic additions to the atmosphere are about 8 Gt (as C) of which about 4 Gt are removed into oceans and the terrestrial environment. Use these numbers to estimate the yearly net increase in atmospheric carbon dioxide mixing ratio in ppmv.

5. Express the amount (65 Mt) of carbon dioxide derived from the Kuwait fires in 1991 as a percentage of the total annual anthropogenic addition of the gas. Note, as indicated in the previous question, that the calculated increase in atmospheric carbon dioxide is only a fraction of what goes into the atmosphere.

6. There has been a steady decrease in the ratio of $^{14}$C to $^{12}$C in the atmosphere over the past decade. Explain how this is consistent with the view that the well-documented increase in

atmospheric carbon-dioxide concentrations is primarily due to emissions from the combustion of fossil fuels.

7. Estimates (Additional Resources 1) for emissions of methane to the atmosphere are given in the table below and the current atmospheric concentration is 1.774 ppmv. Calculate its residence time.

| Sources of atmospheric methane | Million tonnes per year |
| --- | --- |
| Wetlands and other natural sources | 160 |
| Fossil-fuel-related sources | 100 |
| Other anthropogenic sources of biological origin | 275 |

There may be $10^{14}$ t of methane hydrate ($CH_4 \cdot 6H_2O$) in the permafrost below the ocean floors. If 1% of this were to melt per year, what would be the increased concentration of methane (ppmv $y^{-1}$) in the atmosphere neglecting any removal processes. What sinks for methane would play a role in reducing this concentration?

8. Recent work[11] has shown that the flux of methane released from fens in the boreal forest area of Saskatchewan, Canada range from 176 to 2250 mmol $m^{-2}$ $y^{-1}$. Daily fluxes range from 1.08 to 13.8 mmol $m^{-2}$ $d^{-1}$. The data indicate that there are correlations between methane release and water depth (negative), water flow (negative), temperature (positive), and inorganic phosphorus in the sedimentary interstitial water (positive). Suggest reasons for these correlations.

9. The Arrhenius parameters for the reaction

$$N_2O \rightarrow N_2 + O$$

are $A = 7.94 \times 10^{11}$ $s^{-1}$ and $E_a = 250$ kJ $mol^{-1}$. The reaction is first order. Calculate the rate constant and half-life of nitrous oxide assuming a tropospheric mixing ratio of 319 ppbv $N_2O$ at 20°C and comment on the environmental significance of these results.

10. Discuss the possible effects of the following on greenhouse-gas chemistry:

    (a) the southern Pacific Ocean is seeded with the algal micronutrients zinc and iron;

    (b) CFCs cause further thinning of the ozone layer in the stratosphere;

    (c) urban air pollution leads to increased tropospheric ozone concentrations;

    (d) rice (paddy) is grown, under submerged conditions, on coarse sandy soils, rather than on fine clay-rich soils.

11. One 'climate-engineering' proposal for reducing the possibilities of global warming is to inject a sulphate aerosol into the stratosphere. Discuss the climatic and other atmospheric implications of this possible human intervention.

12. Tetrafluoromethane and hexafluoroethane produced during aluminium production are potent greenhouse gases, both having GWP values of approximately 10 000. Carbon dioxide is also released from the stoichiometric reduction of alumina by carbon. Use the figures given in the text to estimate the relative global warming impact over a 100-y period from these two sources.

13. To heat a modest-sized home during the winter in northern Europe may require approximately 2200 $m^3$ of natural gas per year. Calculate the amount of carbon dioxide released from the furnace over this period.

14. Estimate the amount (kg) of carbon-dioxide emissions associated with the use of 2000 kWh of electricity per month over a one-year period. Consider two cases: (a) electricity produced through burning coal and (b) electricity produced through natural-gas combustion.

[11] Rask, H., D.W. Anderson, and J. Schoenau, Methane fluxes from boreal forest wetlands in Saskatchewan, Canada, *Can. J. Soil Sci.*, **76** (1996), 230.

15. Estimate the area of solar collector surface required to produce all the world's electricity that is currently produced by nuclear power—The British Petroleum *Statistical review of world energy* indicates that the 2007 figure for nuclear power consumption was 622 million tonnes of oil equivalent (TOE).

16. In a life-cycle assessment of various power generation systems (*Life-cycle analysis of power generation systems*, Central Research Institute of Electric Power Industry, March 1995), the following greenhouse-gas emissions equivalents have been suggested. All values are in g of $CO_2$ $kWh^{-1}$.

    Hydro, 2 to 48; coal (modern plant), 790 to 1182; nuclear, 2 to 59; natural gas (cogeneration), 389 to 511; biomass (forestry waste combustion), 15 to 101; wind, 7 to 124; solar (photovoltaic), 13 to 731.

    (a) Recalculate the values for coal in terms of kg $CO_2$ $GJ^{-1}$, and compare these with the value (112 kg $GJ^{-1}$) given in the text.

    (b) In general terms, comment on possible sources of greenhouse-gas emissions associated with each of the generation options.

17. Use the information in Example 8.1 and following to predict the theoretical mass (g $m^{-2}$ $d^{-1}$) of carbohydrate that could be produced by photosynthesis in one square metre in one day. Compare this with the actual maximum values provided to show the overall efficiency of conversion.

18. How much $CO_2$ is released from the combustion of ethanol sufficient to generate 100 GJ of energy?

19. Although it took many years to be established, the Montreal Protocol regarding CFCs, was an international environmental agreement that has proved to achieve some measure of success. What additional challenges complicate negotiations surrounding an international agreement on climate-change issues?

20. One component of the global carbon dioxide emissions budget is the amount of this gas released by volcanoes. Data regarding amounts released are available from organizations like the United States Geological Survey (USGS website: http://volcanoes.usgs.gov/hazards/gas/index.php)

    (a) How much carbon dioxide was released by a recent volcano such as the April 2010 eruption of the Icelandic volcano at Eyjafjallajökull? What fraction of the global annual emissions does this represent?

    (b) In the case of the Icelandic volcano, air traffic over Europe was disrupted for the better part of one week. How much carbon dioxide was not emitted because of the reduced number of airplanes flying during this time? For assistance with this estimation, obtain information from the Aviation Environment Federation at http://www.aef.org.uk/

# Part B
# The hydrosphere

The shining water that moves in the streams and rivers is not just water but the blood of our ancestors. If we sell you the land, you must remember it is sacred and you must teach your children it is sacred, and that each ghostly reflection in the clear water of the lakes tells of events and memories in the life of my people. The water's murmur is the voice of my father's father.

*Chief Seattle of the Squamish Tribe, 1854*

# Chapter 9
# The hydrosphere

This chapter introduces the hydrosphere, with emphasis on the critical importance of water and its unique properties. Here, we will discuss:

- the distribution of water throughout the Earth
- the physical and chemical properties of water
- methods of expressing concentration of solutes in water.

Like air, water is one of the essentials that support all forms of plant and animal life. To many people, the quality of the environment is defined more than anything else by the quality of water that we see around us. Unsafe drinking water, turbid lakes and rivers, or ponds green with algae are obvious signs of a degraded water environment. On the other hand, we all appreciate seeing a clear mountain stream or a pristine lake, teeming with fish and a source of clean water for human consumption.

## 9.1 The global distribution of water

Water covers 73% of the Earth's surface, almost three times as much as the continents. But beyond this most visible form, water is an important component of the atmosphere and the terrestrial environment as well. Taken as a whole, we refer to the Earth's water as the hydrosphere or the aqueous environment. The distribution of the Earth's water is shown in Fig. 9.1.

The relatively small proportion of fresh water compared to that in the oceans is noteworthy. So too is the small fraction of the fresh water that is found in readily accessible surface liquid forms. Much larger amounts are present in glaciers, permafrost, ice and snow in the polar regions and in groundwater. Taken together, the most visible and most accessible forms of water represent less than 0.02% of the total water supply on Earth.

### The oceans

Making up some 97% of the total mass of the Earth's water, the oceans are quantitatively of primary importance in defining the world's water relations. Besides water, the oceans contain substantial quantities of many dissolved elements. Table 9.1 gives average values for the major constituents that found in sea water.

The constituents in Table 9.1 are 'conservative' ones having relatively constant composition through time and space, although many local variations, especially in coastal areas, have been documented. A more complete table of concentrations of these and also trace elements is given in Appendix B.1. Using highly sensitive methods of analysis, it has been shown that virtually

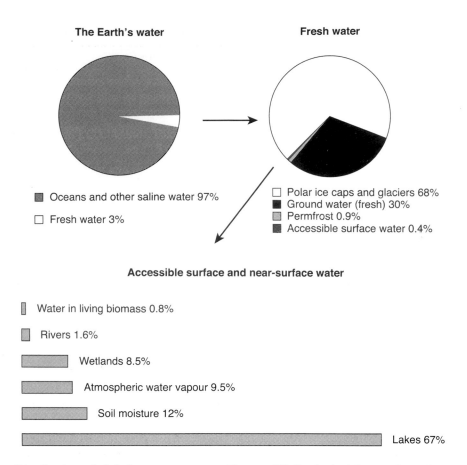

**Fig. 9.1** Distribution of global water resources. (Source: US Geological Survey (http://ga.water. usgs.gov/edu/waterdistribution.html, accessed November 2009) taken from Gleick, P. H., 1996: *Water resources*. In Encyclopedia of Climate and Weather, ed. by S. H. Schneider, Oxford University Press, New York, vol. 2, pp. 817–823.)

every stable element is present in sea water, in most cases at very low levels. For example, the gold concentration is estimated to be $2 \times 10^{-11}$ mol $L^{-1}$.

Oceanic residence times of the conservative elements are of the order of 10 000 to 100 000 000 years, with the shorter times for elements removed by precipitation from solution as solids. An example of this is the element silicon, whose residence time is approximately 20 000 years. The main source of silicon is as species dissolved from bedrock and sediments and carried into the oceans via streams and from groundwater. The principal removal process involves a biogenic reaction to form the siliceous skeletal material of diatoms, an abundant class of marine algae.

Another important species dissolved in sea water, whose equilibrium concentration ranges from 6 to just over 14 mg $kg^{-1}$ is molecular oxygen. The range of concentrations is influenced by both temperature and salinity. A detailed mathematical explanation is provided in Chapter 11.

Sometimes it is helpful to think of the chemistry of oceans in terms of processes that describe the vertical distribution of species, similar to the way in which we viewed the atmosphere. The ocean depth profile can be summarized as follows.

# Chapter 9
# The hydrosphere

This chapter introduces the hydrosphere, with emphasis on the critical importance of water and its unique properties. Here, we will discuss:

- the distribution of water throughout the Earth
- the physical and chemical properties of water
- methods of expressing concentration of solutes in water.

Like air, water is one of the essentials that support all forms of plant and animal life. To many people, the quality of the environment is defined more than anything else by the quality of water that we see around us. Unsafe drinking water, turbid lakes and rivers, or ponds green with algae are obvious signs of a degraded water environment. On the other hand, we all appreciate seeing a clear mountain stream or a pristine lake, teeming with fish and a source of clean water for human consumption.

## 9.1 The global distribution of water

Water covers 73% of the Earth's surface, almost three times as much as the continents. But beyond this most visible form, water is an important component of the atmosphere and the terrestrial environment as well. Taken as a whole, we refer to the Earth's water as the hydrosphere or the aqueous environment. The distribution of the Earth's water is shown in Fig. 9.1.

The relatively small proportion of fresh water compared to that in the oceans is noteworthy. So too is the small fraction of the fresh water that is found in readily accessible surface liquid forms. Much larger amounts are present in glaciers, permafrost, ice and snow in the polar regions and in groundwater. Taken together, the most visible and most accessible forms of water represent less than 0.02% of the total water supply on Earth.

### The oceans

Making up some 97% of the total mass of the Earth's water, the oceans are quantitatively of primary importance in defining the world's water relations. Besides water, the oceans contain substantial quantities of many dissolved elements. Table 9.1 gives average values for the major constituents that found in sea water.

The constituents in Table 9.1 are 'conservative' ones having relatively constant composition through time and space, although many local variations, especially in coastal areas, have been documented. A more complete table of concentrations of these and also trace elements is given in Appendix B.1. Using highly sensitive methods of analysis, it has been shown that virtually

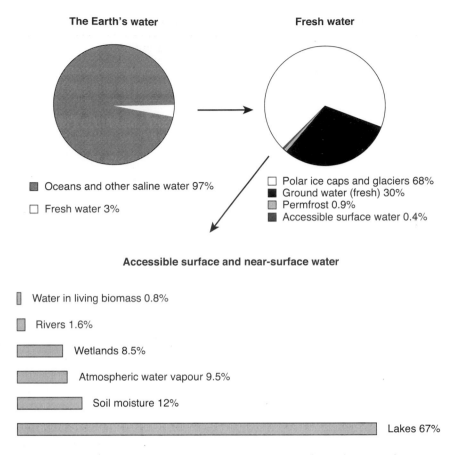

**Fig. 9.1** Distribution of global water resources. (Source: US Geological Survey (http://ga.water. usgs.gov/edu/waterdistribution.html, accessed November 2009) taken from Gleick, P. H., 1996: *Water resources*. In Encyclopedia of Climate and Weather, ed. by S. H. Schneider, Oxford University Press, New York, vol. 2, pp. 817–823.)

every stable element is present in sea water, in most cases at very low levels. For example, the gold concentration is estimated to be $2 \times 10^{-11}$ mol L$^{-1}$.

Oceanic residence times of the conservative elements are of the order of 10 000 to 100 000 000 years, with the shorter times for elements removed by precipitation from solution as solids. An example of this is the element silicon, whose residence time is approximately 20 000 years. The main source of silicon is as species dissolved from bedrock and sediments and carried into the oceans via streams and from groundwater. The principal removal process involves a biogenic reaction to form the siliceous skeletal material of diatoms, an abundant class of marine algae.

Another important species dissolved in sea water, whose equilibrium concentration ranges from 6 to just over 14 mg kg$^{-1}$ is molecular oxygen. The range of concentrations is influenced by both temperature and salinity. A detailed mathematical explanation is provided in Chapter 11.

Sometimes it is helpful to think of the chemistry of oceans in terms of processes that describe the vertical distribution of species, similar to the way in which we viewed the atmosphere. The ocean depth profile can be summarized as follows.

**Table 9.1** Composition of sea water—major inorganic chemical constituents[a].

| Component | Concentration / mg kg$^{-1}$ |
|---|---|
| Sodium | 10760 |
| Magnesium | 1294 |
| Calcium | 413 |
| Potassium | 387 |
| Strontium | 8 |
| Chloride | 19353 |
| Sulphate | 2712 |
| Hydrogen carbonate | 142 |
| Bromide | 67 |
| Boron | 4 |
| Fluoride | 1 |

[a] Data taken from Martin, D., *Marine chemistry*, Vol. 1 *Analytical methods*, Marcel Dekker, Inc., New York; 1968.

- The surface microlayer, a layer only micrometres in thickness, is highly enriched in some chemicals as a consequence of the elevated levels of surface-active organic materials, which have considerable ability to form complexes with many metals as well as non-metals and organic compounds. In Chapter 6, we saw that the enrichment influences the composition of the sea-spray aerosol.

- Below the surface microlayer, down to a depth of about 300 m, is a relatively well-mixed volume (the 'mixed layer'). For each element, a plot of concentration versus depth in the mixed layer has a unique shape depending on the properties of the element. Long residence time conservative elements such as rubidium and caesium have fairly uniform concentration versus depth profiles, influenced mainly by temperature. Near the biologically active regions where algae proliferate at the ocean surface, major and minor nutrient elements (nitrogen, copper, etc.) are depleted but their concentrations increase with depth. Non-nutrient elements, whose presence in the oceans is influenced by atmospheric sources (lead and vanadium are good examples), show decreasing concentration with depth through the mixed layer.

- Below the mixed layer is the metalimnion (also called the thermocline), a region of steadily decreasing temperature, extending over several hundred metres.

- Still deeper is the great mass of ocean water extending to depths of several km (the deepest point in the oceans is over 11000 m, the Mariana Trench, near the island of Guam in the North Pacific) that is only slowly mixed with that on the surface. In calculating the residence times noted above, the total ocean has been taken into account.

## Fresh water

Fresh water makes up only about 3% of the total global water resource, yet its importance far outweighs its quantitative contribution. Two-thirds of the fresh water is located in polar ice caps and alpine glaciers with 90% of that being tied up in the Antarctic continental ice sheet. Besides being an actual and potential source of high-purity water (you have probably heard of proposals to use ships to tow large icebergs from polar regions to areas where there is a shortage of potable water such as the Arabian Gulf—proposals that have never been implemented), the ancient

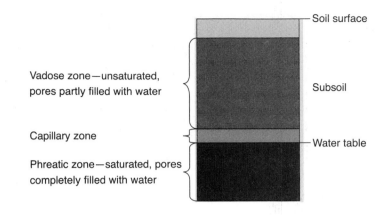

**Fig. 9.2** Nomenclature for zones in soil / permeable rock depth profiles.

glacial ice of the Antarctic, Greenland, and other places serves a useful environmental purpose in that trapped air bubbles and the water itself capture a record of atmospheric conditions through geological time. We have already cited examples of this in considering atmospheric carbon-dioxide concentrations and rainfall pH.

Most of the remaining fresh water is found within the ground. Where water is located in soil pores but is subject to periodic evaporation and replacement by air, it is referred to as soil pore water or soil moisture. Where the soil or rock pores are perennially filled—a saturation condition, below the water table—the water is termed groundwater. Some terminology frequently used by Earth scientists to describe terrestrial water is shown in Fig. 9.2. Groundwater makes up about 30% of the world resource of fresh water and is widely used by industry, for irrigation in agriculture, and as a source for municipal or individual domestic purposes.

Only about 0.4% of all fresh water is in easily accessible surface forms—rivers, ponds and lakes. Nevertheless, this 0.02% of Earth's total water supply dominates much of our study in environmental chemistry because it is visible, readily available and an essential requirement for the survival and growth of many forms of animal and plant life on the planet. Both quantity and quality of accessible water are of concern.

In terms of quantity, the fresh water resource is unevenly distributed. Figure 9.3 shows water distribution around the globe; quantities are measured as cubic metres per year per capita ($m^3 y^{-1}$ pc). There is a wide range in water distribution from over 600 000 $m^3 y^{-1}$ being available for each person in Iceland, to 20 000 in Sweden and Argentina, to less than 2000 in Germany and India, to 250 in Israel and 100 in Saudi Arabia. Whether the entire population has convenient access to the water is another question. In one assessment, countries that have available less than 2000 $m^3 y^{-1}$ pc of fresh water are considered to be in chronic water deficit.

The quality of water is also highly variable and there are many chemical and microbiological aspects that must be taken into account in determining its acceptability. This is a subject we will examine in Chapter 16. Background concentrations of elements in fresh water are also given in Appendix B.1.

A box model of the well-known water cycle has been provided in Fig. 1.4. The cycle includes not only the liquid and solid forms but also water vapour in the atmosphere. The total gaseous mass is not insignificant—equivalent to about 15% of that found in lakes and rivers. As we have seen, the nature of rain is strongly affected by anthropogenic processes that emit gaseous chemicals into the atmosphere. These gases dissolve in water droplets in the clouds, sometimes undergoing further reactions, and then fall to Earth in various forms of precipitation.

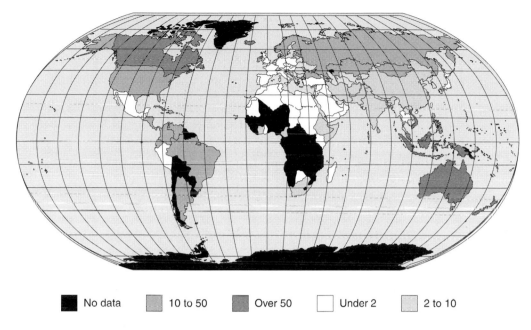

| | | | | | |
|---|---|---|---|---|---|
| ■ No data | ▨ 10 to 50 | ▨ Over 50 | □ Under 2 | ▨ 2 to 10 | |

**Fig. 9.3** Availability of internal renewable fresh water throughout the world. Map shows internal renewable water resources in thousand cubic metres per year per capita. (Data from Lean, G., D. Hinrichsen, and A. Markham, *Atlas of the environment*, Prentice Hall, New York; 1990.)

---

**Fermi question**

Water harvesting in both urban and rural areas is an important means of collecting otherwise wasted run-off from rainfall. Using a roof-top collection system, how much water could be collected each year by a family living in a modest single-family dwelling in Nairobi, Kenya? Is this a significant contribution to their yearly requirements?

---

**Literature link**  Vulnerability of water resources in South Asia

South Asia is home to about one quarter of the world's population, but has less than one twentieth of the world's renewable water resources. A report on the state of surface water in that part of the world focuses on the three major transboundary river basins—the Helmand, the Indus and the Ganges-Brahmaputra-Meghna—that are the source of water for much of this heavily populated region (Babel, M.S. and S.M. Wahid, *Fresh water under threat: South Asia,* A Report of the United Nations Environment Program, Nairobi, Kenya, 2008, available online at http://www.unep.org/Themes/Freshwater/PDF/Freshwater_under_threat_SAsia.pdf, accessed October 2009).

In an attempt to understand and quantify the threats to long-term sustainable water availability, Babel and Wahid developed a suite of indicators that measure various factors that can contribute to sustainability. The indicators are designed to assess four components of water vulnerability. *Resource stress* indicators measure *per capita* availability and variability of accessible freshwater in the three regions. *Development pressure* refers principally to infrastructure development and the indicators consider how much of the potentially available water has been made accessible in a sustainable manner. Water for agricultural, industry and domestic purposes is included in this category with the emphasis on provision of safe drinking water for the population. Indicators for *ecological insecurity* focus on the protection of water-supply sources by measuring vegetative cover in the source regions and the level of treatment of wastewater discharged back into the watersheds. *Management challenges* indicators investigate the human well-being and planning dimensions of water availability. As such they consider the functioning of institutions and communication mechanisms, efficiency of water use in contributing to the economic welfare of the regions, and access to proper sanitation facilities.

Indicators are carefully designed quantitative measures that provide a number or other descriptor that is able to give relevant information about the state of a system. They are often used for comparative purposes to measure progress over time or to compare systems in different locations. For each of the four components of water resource vulnerability, assessment was attempted using a small number of quantifiable indicators. Each one was scaled from 0 to 1, with 0 being the best value and 1 indicating a high level of vulnerability. For example, to measure ecological insecurity, two indicators were proposed.

The *ecosystem deterioration* ($EH_e$) *parameter* was defined as:

$$EH_e = \frac{A_d}{A}$$

where $A_d$ is the area of the river basin that is not covered by permanent vegetation, forests and wetlands and $A$ is the total area of the basin.

To measure water quality, a *water pollution parameter* ($EH_p$) was proposed to determine the extent to which untreated wastewater was discharged in the basin area.

$$EH_p = \frac{WW/WR}{0.15}$$

where WW is the total volume of untreated wastewater and WR is the total availability of water resources. Using this metric, any situation where 15% or more of the available water resource was used but returned untreated was assessed an $EH_p$ value of 1.

The indicators within each component were weighted and combined and all four combined values were further aggregated to give an overall vulnerability index. An especially high level of vulnerability was indicated for the Helmand and Indus watersheds and there was also considerable stress shown in the GBM region.

Indicators are widely used to provide a 'snapshot' assessment of the ecological or other status of a particular situation. After reading this chapter, do you think that the indicators, as chosen and designed, adequately describe the water vulnerability situation in South Asia? Do you think that the scaling and aggregation systems used are appropriate? You might wish to reconsider these questions after taking into account later material within the section on the hydrosphere, especially that in Chapter 16.

---

**Main point 9.1** Water is essential to all forms of life on the Earth. Of the vast amount available on Earth, only a very small percentage is in the form of readily available fresh water. This limited supply of fresh water, important for supporting development and quality of human life in rural and urban regions, is unevenly distributed among the countries of the world.

---

## 9.2 Physical and chemical properties of water

### Water, a unique substance

In order to develop a good understanding of the environmental relations in the hydrosphere, we begin with an examination of some basic principles of aqueous physical chemistry. Water is such a common substance that it is easy to forget or neglect the fact that it has very special characteristics. Its chemical nature determines the way the Earth is, and the forms of life that have developed. Many of its properties are unusual, even unique, in comparison with other compounds related by the position of their atoms on the periodic table. We will look at some of these properties in all three phases of water.[1]

---

[1] Additional properties of water are given in Appendix A.3.

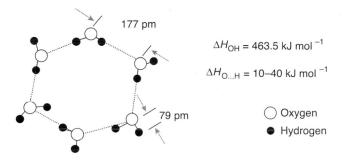

**Fig. 9.4** Structure of ice.

## Ice

The structure of ice takes the form of hexagonal puckered rings with bond lengths and bond enthalpies as shown in Fig. 9.4. Individual molecules are held together (actually they are held apart, as described below) by rather strong hydrogen bonds ($\Delta H_{O...H} = 10$–$40$ kJ mol$^{-1}$).

The density of ice is 0.917 kg L$^{-1}$, lower than that of liquid water, thus enabling it to float on the surface of the lake or pond. This important property affects the environment in many ways in regions with cold climates. The enthalpy of fusion (melting) is 6.02 kJ mol$^{-1}$, a value that is considerably higher than that of most other solids. In part, it is because of this high value that winter-time temperature fluctuations are reduced in areas adjacent to major water bodies.

## Liquid water—density changes with temperature

When ice melts, only about 12% of the hydrogen bonds are broken, indicating that liquid water at 0°C retains a considerable component of the ice structure. As noted above, the hydrogen bonds hold the water molecules apart from one another and the limited bond breaking that occurs on melting is sufficient to allow the individual molecules to come closer to each other. This is the reason why the density of liquid water at 0°C is greater than that of ice. As water is heated through the liquid temperature range, two competing factors affect its density. One is that there is further breaking of hydrogen bonds (an estimated 8% additional breakage occurs between 0 and 100°C, leaving 80% of the hydrogen bonds still intact in liquid water at its boiling point), and this leads to an increase in density. The second is that higher temperature results in greater kinetic energy of the molecules, causing thermal expansion and a density decrease. Of these factors the first dominates between 0 and 4°C and the second from 4 to 100°C. The combined effect is that water achieves its greatest density at approximately 4°C.

These density relations lead to a particular pattern of behaviour in lakes and other water bodies that are subject to major seasonal changes in temperature (Fig. 9.5).

- During the warm summer season, lakes develop a stable structure, with a three-layer profile (epilimnion, metalimnion, hypolimnion) similar to that described above for oceans. In the water near the surface, called the epilimnion, the temperature declines with depth. Of course, the depth of the epilimnion is much less than that in the marine situation. The magnitude of temperature differences and the resulting scale of the layers depend on the air temperature and other climatic factors as well as on the size and depth of the water body.

- When the warm season ends and a period of prolonged cooling occurs, surface water becomes more dense than the underlying warmer water. This causes it to sink, carrying with it a fresh supply of oxygen and nutrients and lifting the deep water upwards. The mixing process, called overturn, brings about a period of enhanced chemical and microbiological activity.

- As cold weather continues and a period of sustained subfreezing temperatures sets in, the lake develops a new type of stable (but inverted) structure. In the winter, ice floats on

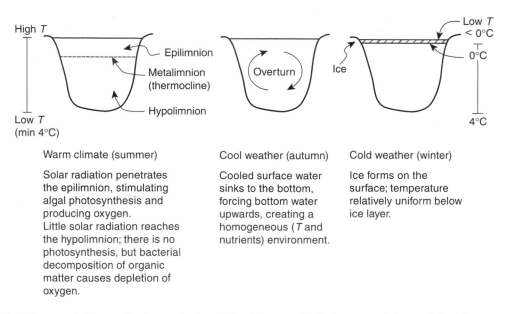

Warm climate (summer)

Solar radiation penetrates the epilimnion, stimulating algal photosynthesis and producing oxygen.
Little solar radiation reaches the hypolimnion; there is no photosynthesis, but bacterial decomposition of organic matter causes depletion of oxygen.

Cool weather (autumn)

Cooled surface water sinks to the bottom, forcing bottom water upwards, creating a homogeneous (*T* and nutrients) environment.

Cold weather (winter)

Ice forms on the surface; temperature relatively uniform below ice layer.

**Fig. 9.5** Seasonal changes in the vertical profile of a water body in parts of the world with temperate climate.

the surface, underlain by water with a temperature at or near 0°C, and the temperature increases downwards to a maximum of 4°C.

- As temperature increases in the spring, a second period of intense mixing occurs and the cycle repeats in subsequent years.

## Liquid water as a solvent

Water has a large dipole moment and dielectric constant—among the highest of any common liquid. These properties determine that it is a good solvent for many substances. In fact, it is sometimes called the 'universal solvent' since it is able to dissolve more substances than almost any other liquid. At the same time, we know that there are many non-polar, often organic substances that are essentially insoluble in water. Because of the high value of the dipole moment ($6.1 \times 10^{-30}$ C m), the water molecules orient themselves around ionic or polar species forming X . . . $H_2O$ bonds (Fig. 9.6). Being a negative free-energy process, bond formation to produce a hydrated species is a favourable reaction. Where the energy gain through hydration is greater than that required to break the original bonds in the solid, the substance is soluble. Many common ionic and polar substances are readily hydrated and are said to be *hydrophilic*, while non-polar, neutral substances have limited ability to form bonds with polar water molecules and are termed *hydrophobic*.

The number of water molecules that can surround an ion, as represented by the hydration number depends on several factors, the most important of which is the charge to radius ratio, as is illustrated by the data for alkali metals in Table 9.2.

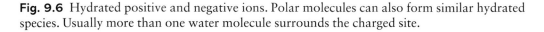

**Fig. 9.6** Hydrated positive and negative ions. Polar molecules can also form similar hydrated species. Usually more than one water molecule surrounds the charged site.

**Table 9.2** Charge and radius properties of the alkali metals in aqueous solution.

|  | Li⁺ | Na⁺ | K⁺ | Rb⁺ | Cs⁺ |
|---|---|---|---|---|---|
| Ionic radius / pm | 60 | 95 | 133 | 148 | 169 |
| Charge density / C pm$^{-1}$ | 0.0167 | 0.0105 | 0.0075 | 0.0068 | 0.0059 |
| Hydrated radius / pm | 340 | 276 | 232 | 228 | 228 |
| Hydration number | 23.3 | 16.6 | 10.5 | 10 | 9.9 |

Ions that have a large charge to radius ratio, that is, ones with a high concentration of charge, attract more water molecules around themselves. This leads to the interesting inverse relation between ionic radius and hydrated radius, as you can see in Table 9.2. This has implications with respect to ion-exchange reactions involving hydrated ions that take place on solid surfaces in the water column. The charge to radius ratio of the hydrated ions for the alkali metals, for example, is such that the rubidium ion is more strongly attracted to negative sites on suspended solids or sediments than is the sodium ion.

## Liquid water—complexation

'Free ions' that are dissolved in water, both cations and anions, are therefore really hydrated ions and these are referred to as 'aquo' complexes. The coordinated water is held in position by a combination of electrostatic and covalent forces, depending on the properties of the species with which it is associated.

In most natural water situations, there are many dissolved substances that can act as *ligands*, displacing the water and forming a new *complex* with the ion. This implies that the bond between the ion and the different ligand is more stable than the original one involving water. In many cases, the complex involves a central metal ion coordinated to one or more inorganic or organic ligands. Available in natural waters as potential complexing agents are a whole range of possible substances; examples include chloride or sulphate, small organic molecules containing an amino group, or larger organic molecules with several sites for forming a bond. The formation of complexes is a very important feature of aqueous chemistry of metal ions.

Stability of complexes is expressed using *stability constants*, also called *formation constants*. There are two types of such constants, as can be illustrated using a generic example. In the example, M is a metal ion and L is the ligand of interest. For simplicity, the charges on the metal ion and on the ligand (if it has a charge) are omitted. We will assume that it is possible for four ligand molecules to be coordinated with the metal. The complex formation reactions take place in a stepwise fashion, and each step has a corresponding formation constant, $K_f$.

$$M + L \rightleftharpoons ML \quad K_{f1} = [ML] / [M][L] \tag{9.1}$$

$$ML + L \rightleftharpoons ML_2 \quad K_{f2} = [ML_2] / [ML][L] \tag{9.2}$$

$$ML_2 + L \rightleftharpoons ML_3 \quad K_{f3} = [ML_3] / [ML_2][L] \tag{9.3}$$

$$ML_3 + L \rightleftharpoons ML_4 \quad K_{f4} = [ML_4] / [ML_3][L] \tag{9.4}$$

Each of the $K_f$ values is referred to as a stepwise formation constant. The sum of the four reaction steps describes the total reaction, and the overall formation constant for this process is designated as $\beta_4$.

$$M + 4L \rightleftharpoons ML_4 \quad \beta_4 = [ML_4] / [M][L]^4 \tag{9.5}$$

In the present case, the overall formation constant is related to the stepwise constants

$$\beta_4 = K_{f1} \times K_{f2} \times K_{f3} \times K_{f4} \tag{9.6}$$

In the general case, where the number of ligands bound to the metal ion is $n$,

$$\beta_n = K_{f1} \times K_{f2} \times K_{f3} \cdots \times K_{fn} \tag{9.7}$$

One further point to reiterate is that the reactions shown imply that the ligand is forming a series of complexes with a *free* or *uncomplexed* metal ion. In fact, as noted above, the so-called uncomplexed metal is invariably present as an aquo complex, and ligand addition to the metal actually involves displacement of a water molecule and replacement by the new ligand.

In the oceans, one ligand obviously present in large concentration is the chloride ion. As a result, some metals exist in the ocean primarily as chloro complexes. Mercury, for example, is found in clean ocean water at a concentration of approximately $5 \times 10^{-12}$ mol $L^{-1}$. The complexation reaction between mercury(II) and the chloride ligand is then described as

$$Hg(H_2O)_6^{2+} \text{ (aq)} + 4Cl^- \text{ (aq)} \rightleftharpoons Hg(H_2O)_2Cl_4^{2-} \text{ (aq)} + 4H_2O \tag{9.8}$$

The equilibrium constant for the reaction is $1.3 \times 10^{15}$ and the tetrachloro species is known to be a principal form of mercury (II) in sea water.

Many other ligands may be present depending on the water under consideration. Soluble organic substances like citrate ion, or macromolecules produced by decomposition of plant and animal tissue, act as coordinating ligands. In other instances, organic substances (such as nitrilotriacetic acid (NTA), which is sometimes used as a water-softening agent in detergents) or inorganic substances (like phosphate, also used in detergents or applied to soil as a fertilizer) of anthropogenic origin form complexes with dissolved substances. The nature and extent of complex formation depends on the properties of the central atom, and the availability and concentration of potential ligands. We demonstrate the quantitative (equilibrium) aspect of these complexes in Chapter 13.

## Liquid water—acid–base properties

Water is an amphiprotic substance and it undergoes autoprotolysis to form the hydronium ion and the hydroxyl ion. The equilibrium constant for autoprotolysis ($K_w$ or $K_{auto}$) depends on temperature

$$2H_2O \rightleftharpoons H_3O^+ \text{ (aq)} + OH^- \text{ (aq)} \quad K_w = 1.01 \times 10^{-14} \text{ at } 25°C \tag{9.9}$$

Because the products, $H_3O^+$ and $OH^-$, are themselves hydrated, a better representation of $H_3O^+$ would be $H_3O(H_2O)_3^+$, but we will use the simpler formula.

Neutral water, where the concentrations of hydronium and hydroxide ion are equal at a value of $10^{-7}$ mol $L^{-1}$, has a pH of 7.0 at 25°C. Most hospitable aqueous environments are ones whose pH is within one or sometimes two units of this value. Because autoprotolysis is an endothermic process, the extent to which it occurs depends on temperature and therefore the pH of neutral water is not always 7.

---

### Example 9.1 **pH of neutral water at the freezing point**
The value of $K_w$ at 0°C is $1.148 \times 10^{-15}$.

$$K_w = [H_3O^+][OH^-]$$

In neutral water,

$$[H_3O^+] = [OH^-] = x$$
$$x = (1.148 \times 10^{-15})^{1/2}$$
$$x = [H_3O^+] = 3.39 \times 10^{-8}$$

And

$$pH = -\log[H_3O^+] = 7.47$$

---

Being amphiprotic, water has the ability to both accept and donate protons from other substances and therefore acts as a Brønsted base or acid in the presence of proton donors or acceptors respectively. For example, oxalic acid and other carboxylic acids that are degradation products of natural organic matter (NOM) are a source of acidification of water by reactions such as

$$(COOH)_2 + H_2O \rightleftharpoons HOOCCOO^- \text{ (aq)} + H_3O^+ \text{ (aq)} \quad K_{a1} = 5.6 \times 10^{-2} \tag{9.10}$$

$$HOOCCOO^- \text{ (aq)} + H_2O \rightleftharpoons (COO^-)_2 \text{ (aq)} + H_3O^+ \text{ (aq)} \quad K_{a2} = 5.2 \times 10^{-5} \tag{9.11}$$

The doubly charged oxalate anion is the principal form of this molecule in most natural water situations. Other proton donors from natural or anthropogenic sources can also contribute to making water slightly acidic.

Where the pH of water is greater than 7, it is usually due to soluble carbonate species of natural geological origin. Sea water, for example, has a pH of 8.2; additional alkaline nature (higher pH) in water bodies is less common.

## Liquid water—redox properties

Oxidation and reduction reactions also play an important role in aqueous chemistry. Water has redox properties that define upper and lower potential limits for redox reactions of other substances that are present in the aqueous solution. The oxidation of water leads to evolution of $O_2$.

$$2H_2O \rightarrow O_2 \text{ (g)} + 4H^+ \text{ (aq)} + 4e \tag{9.12}$$

Based on the Nernst equation, the potential for this half-reaction as a function of pH is

$$E = 1.23 - 00591 \text{ pH at } 25°C \tag{9.13}$$

Under extreme reducing conditions, water decomposes producing hydrogen

$$2H_2O + 2e^- \rightarrow H_2 \text{ (g)} + 2OH^- \text{ (aq)} \tag{9.14}$$

The potential relationship to pH for this reaction is

$$E = -0.0591 \text{ pH at } 25°C \tag{9.15}$$

When plotted on a graph showing potential versus pH, the linear relationships in eqns 9.13 and 9.15 define the area of stability of water (Fig. 9.7).

Only species whose potential / pH characteristics locate them within the boundaries of stable water can exist in an aqueous system. Much more will be discussed about this important topic in Chapter 10.

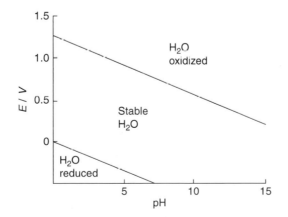

**Fig. 9.7** Stability of water as depicted by an $E$ / pH diagram.

Table 9.3 Vapour pressure of water as a function of temperature.

| $T$ / °C | $P_{vap}(H_2O)$ / Pa |
|---|---|
| 0 | $6.1 \times 10^2$ |
| 5 | $8.7 \times 10^2$ |
| 10 | $1.2 \times 10^3$ |
| 15 | $1.7 \times 10^3$ |
| 20 | $2.3 \times 10^3$ |
| 25 | $3.2 \times 10^3$ |
| 30 | $5.6 \times 10^3$ |

### Water vapour

The enthalpy of vaporization of water as shown in eqn 9.16 is the highest value for all common liquids, meaning that an unusually large amount of heat is consumed in causing the evaporation of water.

$$H_2O \text{ (l)} \rightleftharpoons H_2O \text{ (g)} \quad \Delta H = 40.6 \text{ kJ mol}^{-1} \tag{9.16}$$

As vaporization occurs, the remaining 80% of intact hydrogen bonds are broken, so that there is no structure remaining in the vapour state. The large enthalpy value contributes to temperature stability in areas over or adjacent to large water bodies.

The variation of vapour pressure of water is defined by eqn 9.16 and its relationship with temperature, and values are given in Table 9.3. The relation is shown graphically in Fig. 8.2. Therefore, at 20°C, 101 kPa pressure, and 100% humidity, the mixing ratio of water in the atmosphere would be

$$\frac{2.3 \text{ kPa}}{101 \text{ kPa}} \times 100\% = 2.3\%$$

Because of the movement of air masses, both horizontally and vertically, eqn 9.16 is rarely in equilibrium. Therefore, actual water vapour concentrations are usually less than the equilibrium value. The global range averages about 1%, but values are highly variable both temporally and spatially.

---

**Main point 9.2** In its physical and chemical properties, water is a unique chemical. Its physical properties are a major factor responsible for controlling climate. The acid–base, redox, and solvent properties determine its reactivity and the way in which elements are transported and made available to interact with other components of the living and non-living environment.

---

## 9.3 Concentration units used for aqueous solutions

One of the most important and fundamental ways used by chemists to express concentrations of solutes in aqueous solutions employs units based on molarity. Sea water contains approximately 1.97% w/v chloride (% weight / volume = mass (g) of chloride per volume (100 mL) of solution), and so it is a 0.556 mol $L^{-1}$ solution of the ion.

For elements or compounds present at lower levels, it is convenient to give concentrations in micromolar or nanomolar units. Typical would be zinc values ranging from 15 nmol $L^{-1}$ to 6.7 µmol $L^{-1}$ in the Toyohira River in Japan,[2] with the higher values found in samples within

---

[2] Sakai H., Y. Kojima, and K. Saito, Distribution of heavy metals in water and sieved sediments in the Toyohira river, *Wat. Res.*, **20** (1986), 559.

the Sapporo city boundaries. Average values for elements in fresh and sea water are given in Appendix B.1.

It is also common practice to express low concentrations such as the above value of zinc in units of parts per million (ppm), parts per billion (ppb), or parts per trillion (ppt). While widely used, there are several problems associated with the use of such units and these should be kept in mind. It is, first of all, very important to recognize that there is a significant difference between units such as ppm applied to water solutions compared to ppmv as used in atmospheric chemistry.

A unit like ppm when applied to either water or solids is fundamentally a mass ratio. If we say that the concentration of vanadium in a water sample is 0.15 ppm, we mean that there are 0.15 $\mu g$ of the element in 1 g of water or 1 mg kg$^{-1}$. Often, however, the unit ppm is thought of in terms of mass of solute per volume of aqueous solution. In this case 0.15 ppm is taken to mean 0.15 $\mu g$ mL$^{-1}$ or mg L$^{-1}$. These mass / volume units assume that the density of water is 1.00 kg L$^{-1}$. This is usually a good assumption for lakes, rivers, and other sources of fresh water, but it is not true for the oceans where the density is 1.025 kg L$^{-1}$ at 15°C. It is therefore preferable to specify the concentration units in unambiguous terms of mass per volume or mass per mass. A study of metals in the Huanghe (Yellow) River in China[3] found zinc concentrations ranging from 60 ng kg$^{-1}$ far inland to 352 ng kg$^{-1}$ near the estuary in Bohai. By using such units there is no uncertainty related to changes in density when moving from fresh to salt water.

A second complication with the *parts per . . .* type of units arises from the uncertainty associated with choosing a particular species for mass calculations. An example will illustrate this point. A 10 ppm ammonium ion solution corresponds to a molar concentration of $5.56 \times 10^{-4}$ mol NH$_4^+$ L$^{-1}$, calculated in the following way.

$$10 \text{ ppm} = 10 \text{ mg NH}_4^+ \text{ L}^{-1} = 10 \times 10^{-3} \text{ g NH}_4^+ \text{ L}^{-1}$$

$$= 10 \times 10^{-3} \div 18.0 \text{ mol NH}_4^+ \text{ L}^{-1}$$

$$= 5.56 \times 10^{-4} \text{ mol NH}_4^+ \text{ L}^{-1}$$

By an important reaction called nitrification that we will study in detail later, in Chapter 15, the ammonium can be oxidized to form nitrate. The stoichiometry is such that one ammonium ion becomes one nitrate ion. In units of ppm, this becomes

$$5.56 \times 10^{-4} \text{ mol NH}_4^+ \text{ L}^{-1} = 5.56 \times 10^{-4} \text{ mol NO}_3^- \text{ L}^{-1}$$

$$= 5.56 \times 10^{-4} \times 62 \times 10^3 \text{ mg NO}_3^- \text{ L}^{-1} = 34 \text{ ppm NO}_3^-$$

Obviously, the larger ppm value for nitrate compared to ammonium is not because any additional nitrogen-containing species has been created, but arises simply because of the differences in molar mass of the two nitrogen species. Creating confusion due to such apparent anomalies may be avoided by expressing concentrations in both cases in terms of nitrogen alone. The two values are then both reported in terms of N as 7.7 ppm.

---

### Example 9.2  Nitrate concentration expressed as NO$_3^-$ and as N
Consider water containing 34 ppm NO$_3^-$

$$34 \text{ ppm NO}_3^- = 34 \text{ mg NO}_3^- \text{ L}^{-1}$$

$$= 34 \times 14/62 \text{ mg N L}^{-1}$$

$$= 7.7 \text{ mg N L}^{-1}$$

$$= 7.7 \text{ ppm N}$$

---

[3] Zhang, J. and W.W. Huang, Dissolved trace metals in the Huanghe: the most turbid large river in the world, *Wat. Res.*, **27** (1993), 1.

Of course, there will never be any confusion of this kind if molar units are used—the concentration of nitrogen would remain at 0.56 mmol L$^{-1}$ whether as ammonium or as nitrate.

Yet another limitation with the use of units such as parts per million is that they give no indication of the concentration of reactive groups. For example, the concentration of total organic carbon (TOC) in a forest stream may be 9.0 ppm (= 9.0 mg L$^{-1}$). If the organic material being measured is largely dissolved humic material from the soil, then it will contain functional groups such as carboxylic acids that are capable of forming complexes with metals and therefore enhancing metal solubilization (a subject discussed in Chapters 12 and 13). One study concludes that the concentration of carboxylate functional groups per gram of humic material dissolved in the water is about 4 mmol g$^{-1}$ of humic material. Using a carbon percentage in humic material of 50%, the 9.0 ppm TOC could be expressed in the following way:

$$9.0 \text{ mg C L}^{-1} = 9.0 \times 10^{-3} \times 100/50 \text{ g humic material L}^{-1} \text{ water}$$

$$= 9.0 \times 10^{-3} \times 100/50 \text{ g humic material L}^{-1} \times 4 \times 10^{-3} \text{ mol carboxylate}$$

$$\text{groups g}^{-1} \text{ humic material}$$

$$= 7.2 \times 10^{-5} \text{ mol carboxylate groups L}^{-1} \text{ water}$$

This way of describing the concentration of available carboxylate groups could be used to conveniently describe aspects of the acid–base and complexing behaviour of the water.

---

### Example 9.3 **Concentration units for dissolved organic matter**

A natural water sample contains 2.2 mg L$^{-1}$ dissolved organic matter (DOM). The OM consists of many macromolecular species with average composition $C_{38}H_{43}O_{25}N_2S$. Recalculate the concentration as mol L$^{-1}$ and ppm (C).

Calculated from the average formula, the molar mass of the OM is 955 g mol$^{-1}$. Therefore, the molar concentration is

$$2.2 \times 10^{-3} \text{ g L}^{-1}/955 \text{ g mol}^{-1} = 2.3 \times 10^{-6} \text{ mol L}^{-1}$$

Within the formula given, the fractional mass of carbon is

$$38 \times 12/955 = 0.48$$

Therefore, 2.2 mg L$^{-1}$ of OM is equal to $0.48 \times 2.2 = 1.1$ mg L$^{-1}$ of carbon in the OM.

Note that ppm and mg L$^{-1}$ are equivalent, and so the concentration can be expressed as 1.1 ppm C.

---

**Fermi question**

Depending on the season, the Amazon River discharges over 100 000 m$^3$ of water per second into the Atlantic Ocean. After a one-second discharge is distributed evenly *via* the global hydrological cycle over an extended period of time, how many molecules of this water would be present in a glass of water obtained from a tap in Tokyo?

This question is quite realistic. Sometimes however, a similar question is asked, but instead of considering water, molecules of a specific contaminant in water are considered. What is much more problematic about this modified question?

---

**Main point 9.3** A wide variety of concentration units are employed to express concentrations of solutes in water. For chemical studies, molar units are usually most appropriate. Units for trace components of the type ppm, ppb, etc. are based on mass ratios and should be used with care.

## ADDITIONAL RESOURCES

1. Black, M. and J. King, *The atlas of water: mapping the world's most critical resource*, 2nd edn., Earthscan, London, UK; 2009.

2. Benjamin, M.M., *Water chemistry*, McGraw-Hill, New York; 2003.

3. Stumm, W. and J.J. Morgan, *Aquatic chemistry: chemical equilibria and rates in natural waters*, 3d edn., John Wiley and Sons, Inc., New York; 1996.

   There are several web-based downloadable resources available from international organizations. All were accessed in October 2009.

4. United Nations Environment Program, The Global Environment Monitoring System Water Program. *Water quality for ecosystem and human health*, 2006, http://www.gemswater.org/digital_atlas/digital_atlas.pdf.

5. United Nations Environment Program, *Global Environment Outlook Report 4*, GEO$_4$, Chapter 4, 2006. http://www.unep.org/geo/geo4/report/04_Water.pdf.

6. United Nations Environment Program, *Vital water graphics—an overview of the state of the world's fresh and marine waters.* http://www.grida.no/publications/vg/water2/.

7. The World Water Council deals with a broad range of water issues, with emphasis on political, social and environmental goals in the Millennium Development Goals context, *World water vision*, Earthscan Publications Ltd., London, UK, 2000. http://www.worldwatercouncil.org/index.php?id=961.

## PROBLEMS

1. Use Fig. 1.4 to estimate the residence time of water in the oceans and on the continents. In both cases, indicate limitations in interpreting these results. Keeping in mind the limitations, comment on the significance of residence time values in the three compartments of the environment.

2. Determine the water availability (in $m^3 \ y^{-1}$ pc) in your country using websites such as http://www.nationmaster.com/graph/hea_wat_ava-health-water-availability, and compare it with values for other countries as are shown in Fig. 9.3.

3. The concentration of titanium in the South Pacific ocean, near the island nation of Fiji, has been found to be approximately $3.0 \times 10^{-9}$ mol $L^{-1}$. Calculate the concentration in ppm, ppb, or ppt as appropriate.

4. The concentration of gold in the oceans averages approximately $2 \times 10^{-11}$ mol $L^{-1}$. Calculate the total mass in tonnes.

5. The principal anion and cation in Lake Huron, one of the *Great Lakes* in North America, are, respectively, hydrogen carbonate and calcium. The concentration of the former is approximately 1.05 mmol $L^{-1}$. Calculate the mass of solid calcium carbonate that would remain if 250 mL of Lake Huron water is evaporated to dryness.

6. The value of $K_w$ is $0.67 \times 10^{-14}$ at 20°C, $1.01 \times 10^{-14}$ at 25°C, and $1.45 \times 10^{-14}$ at 30°C. Calculate the value of $K_w$ at 10°C, and determine the pH of pure water at that temperature.

7. In a particular fresh water sample, the concentrations of cations and anions are (in μmol $L^{-1}$):

| | | | |
|---|---|---|---|
| $Na^+$ | 33 | $Cl^-$ | 120 |
| $K^+$ | 4 | $NO_3^-$ | 13 |
| $Mg^{2+}$ | 31 | $HCO_3^-$ | 270 |
| $Ca^{2+}$ | 160 | $CO_3^{2-}$ | 0.67 |
| | | $SO_4^{2-}$ | 11 |

Compare the concentration of total positive and negative charge in the solution. Assume that the difference is due to hydronium or hydroxyl ion, and calculate the pH.

8. The major ions and their concentrations (mmol $L^{-1}$) in sea water are:

| | | | |
|---|---|---|---|
| $Na^+$ | 470 | $K^+$ | 10 |
| $Mg^{2+}$ | 53 | $Ca^{2+}$ | 10 |
| $Cl^-$ | 547 | $SO_4^{2-}$ | 28 |
| $Br^-$ | 1 | $HCO_3^- + CO_3^{2-}$ | X |

Assume that charges of these species balance and calculate the total concentration of negative charge associated with the two carbonate species. With a pH of 8.2, calculate the concentrations of the two individual carbonate species.

9. In the open oceans the concentration of iron is approximately $1 \times 10^{-4}$ ppm in the surface water and $4 \times 10^{-4}$ ppm in the deep ocean. Corresponding values for concentration of aluminium are $9.7 \times 10^{-4}$ ppm and $5.2 \times 10^{-4}$ ppm. Why are the concentrations so low? Why is the ratio surface concentration: deep concentration < 1 for iron and > 1 for aluminium?

10. Manganese may precipitate as $MnCO_3$ from aqueous solution according to the reaction

$$Mn^{2+} (aq) + CO_2 (aq) + 3H_2O \rightarrow MnCO_3 (s) + 2H_3O^+ (aq)$$

The solubility product, $K_{sp}$, for manganese (II) carbonate is $5.0 \times 10^{-10}$. Use this value and the equilibrium constant values for the carbonate system to determine the minimum pH required to precipitate manganese (II) carbonate from a solution that contains $1.0 \times 10^{-3}$ mol $L^{-1}$ manganese (II) ion. Assume that aqueous carbon dioxide is in equilibrium with atmospheric carbon dioxide.

11. The following are controlling factors for the 'availability' of different elements:

- oxygen availability for iron;
- sulphide concentration for zinc;
- solution pH for chromium and silicon;
- carbonate concentration for calcium;
- sorption factors for copper.

Explain the chemical and environmental significance of these factors.

# Chapter 10
## Distribution of species in aquatic systems

The focus of this chapter is on the distribution of species in aquatic systems, and how to calculate and diagrammatically represent the distribution based on empirically determined constants. We will accomplish this by:

- illustrating methodology for generating a single variable species distribution diagram
- introducing two variable species distribution diagrams that are relevant to natural aquatic environments
- discussing redox chemistry in water and introducing the concept of *electron activity*
- introducing $pE$ / pH diagrams, with a detailed example of the sulphur system.

At the outset of the book, we noted that an accurate description of chemical composition in an environmental compartment requires knowledge of the forms or species of a particular chemical. Aluminium is a good example of an element for which information about species types is very important. In water whose pH is less than about 5, aluminium exists primarily as the free $Al^{3+}$ (hydrated) ion and this form can be taken up by plants and animals, sometimes with toxic effects. In slightly alkaline water, the common forms are partially deprotonated hydroxy species or complexes with organic ligands. The complexed forms are not as readily available for biological uptake and are considered to be less toxic. Similar kinds of pictures exist for other elements and compounds and the distribution of species in each case can have a major influence on environmental behaviour.

The species distribution (or 'speciation') for an element or compound depends on the nature of the chemical and also on the particular environmental conditions in which it is found. It is often possible to calculate or predict the forms that will be present, assuming one has available the appropriate analytical data and the required thermodynamic constants. The additional assumptions that are usually made include an idea that distribution is not affected by reaction rate—in other words, it is at thermodynamic equilibrium. For rigorous calculations, activities rather than concentrations should be used, but this requires knowledge of the total ionic composition of the solution. While recognizing that the simplifying assumption that activity and concentration are equivalent leads to errors in the calculation, we choose to use concentrations in most cases in the book. This is done partly to avoid making complex relations appear even more complex. Furthermore, in many situations, the errors generated by neglecting activity coefficients are smaller than those arising from uncertainties in the analytical data available for multicomponent environmental materials.

Two calculation situations are encountered:

- Detailed individual calculations related to specific sets of conditions—for example, a calculation to show what fraction of an acidic species is in protonated and what fraction is in deprotonated form at a given solution pH.
- Diagrams that show how species distributions can vary as conditions change. In the process of constructing such diagrams, methods for calculating individual situations become clear. There are many kinds of distribution diagrams. We will look into the construction of several types and we will examine how to interpret these and others. We will do this by way of considering some important environmental examples.

> **Main point 10.1** The environmental behaviour of an element or compound depends on the particular form of the species that is present. In the aqueous environment; the species distribution depends on a number of factors including the nature of the chemical and environmental conditions. In order to determine the species distribution of a particular substance, it is usual to make an assumption that all forms are in equilibrium with the surroundings. Furthermore, it is frequently assumed that concentration and activity of species are the same.

## 10.1 Single-variable diagrams

### Phosphate species

A single-variable diagram is a plot of some measure of species concentration ($y$-axis) versus a particular variable like pH, redox status, or concentration of an important complexing ligand ($x$-axis). A well-known case, useful for describing the chemistry of species that exhibit acid–base behaviour, is a plot of alpha ($\alpha$), the fractional concentration, of individual species against pH. A good example of this application is the phosphate system. Phosphorus exists in water almost exclusively as P (V) species, particularly in forms of orthophosphate.

$$H_3PO_4 \rightleftharpoons H_2PO_4^- \rightleftharpoons HPO_4^{2-} \rightleftharpoons PO_4^{3-} \tag{10.1}$$

To calculate the distribution of these four species as a function of pH we need only know the values of the acid dissociation constants that are given in Table 10.1 and also in Appendix B.4. These three equilibrium constants apply to the successive dissociation from the acid of its three protons, and apply to an aqueous solution at 25°C. For example, the first dissociation is approximated by the relation

$$K_{a1} = \frac{[H_2PO_4^-][H_3O^+]}{[H_3PO_4]} \tag{10.2}$$

The fraction, $\alpha_{H_3PO_4}$, of undissociated $H_3PO_4$ in a solution containing phosphate species is

$$\alpha_{H_3PO_4} = \frac{[H_3PO_4]}{[H_3PO_4] + [H_2PO_4^-] + [HPO_4^{2-}] + [PO_4^{3-}]} = \frac{[H_3PO_4]}{C_p} \tag{10.3}$$

**Table 10.1** Acid dissociation constants for phosphoric acid.

|  | $K_a$ | $pK_a$ |
|---|---|---|
| First dissociation | $7.1 \times 10^{-3}$ | 2.15 |
| Second dissociation | $6.3 \times 10^{-8}$ | 7.20 |
| Third dissociation | $4.2 \times 10^{-13}$ | 12.38 |

where $C_p$ = the total concentration of all four orthophosphate species. For the other phosphate species, similar fractions are given by

$$\alpha_{H_2PO_4^-} = \frac{[H_2PO_4^-]}{C_p} \tag{10.4}$$

$$\alpha_{HPO_4^{2-}} = \frac{[HPO_4^{2-}]}{C_p} \tag{10.5}$$

$$\alpha_{PO_4^{3-}} = \frac{[PO_4^{3-}]}{C_p} \tag{10.6}$$

The three dissociation constant expressions can be rearranged to give the concentration of each individual species in terms of $[H_3PO_4]$ and $[H_3O^+]$.

$$[H_2PO_4^-] = \frac{K_{a1} \times [H_3PO_4]}{[H_3O^+]} \tag{10.7}$$

$$[HPO_4^{2-}] = \frac{K_{a1} \times K_{a2} \times [H_3PO_4]}{[H_3O^+]^2} \tag{10.8}$$

$$[PO_4^{3-}] = \frac{K_{a1} \times K_{a2} \times K_{a3} \times [H_3PO_4]}{[H_3O^+]^3} \tag{10.9}$$

$$C_p = [H_3PO_4] + [H_2PO_4^-] + [HPO_4^{2-}] + [PO_4^{3-}]$$

$$= [H_3PO_4]\left(1 + \frac{K_{a1}}{[H_3O^+]} + \frac{K_{a1} \times K_{a2}}{[H_3O^+]^2} + \frac{K_{a1} \times K_{a2} \times K_{a3}}{[H_3O^+]^3}\right) \tag{10.10}$$

From eqn 10.3

$$\alpha_{H_3PO_4} = \frac{[H_3PO_4]}{[H_3PO_4]\left(1 + \dfrac{K_{a1}}{[H_3O^+]} + \dfrac{K_{a1} \times K_{a2}}{[H_3O^+]^2} + \dfrac{K_{a1} \times K_{a2} \times K_{a3}}{[H_3O^+]^3}\right)} \tag{10.11}$$

We then multiply the top and bottom of the right-hand side of the equation by $[H_3O^+]^3$

$$\alpha_{H_3PO_4} = \frac{[H_3O^+]^3}{[H_3O^+]^3 + [H_3O^+]^2 \times K_{a1} + [H_3O^+] \times K_{a1} \times K_{a2} + K_{a1} \times K_{a2} \times K_{a3}} \tag{10.12}$$

Using similar calculations, we find that

$$\alpha_{H_2PO_4^-} = \frac{[H_3O^+]^2 \times K_{a1}}{[H_3O^+]^3 + [H_3O^+]^2 \times K_{a1} + [H_3O^+] \times K_{a1} \times K_{a2} + K_{a1} \times K_{a2} \times K_{a3}} \tag{10.13}$$

$$\alpha_{HPO_4^{2-}} = \frac{[H_3O^+] \times K_{a1} \times K_{a2}}{[H_3O^+]^3 + [H_3O^+]^2 \times K_{a1} + [H_3O^+] \times K_{a1} \times K_{a2} + K_{a1} \times K_{a2} \times K_{a3}} \tag{10.14}$$

$$\alpha_{PO_4^{3-}} = \frac{K_{a1} \times K_{a2} \times K_{a3}}{[H_3O^+]^3 + [H_3O^+]^2 \times K_{a1} + [H_3O^+] \times K_{a1} \times K_{a2} + K_{a1} \times K_{a2} \times K_{a3}} \tag{10.15}$$

Note that each of eqns 10.12–10.15 can be used to calculate the fraction of an individual species in the phosphate system at a given pH. When the $\alpha$ values are plotted over a range of pH values, we have a distribution diagram. Calculating and plotting are most conveniently done using a standard spreadsheet-type computer program as shown in the example later in this section. Figure 10.1 is the plot for phosphate species distribution over the pH range from 0 to 14.

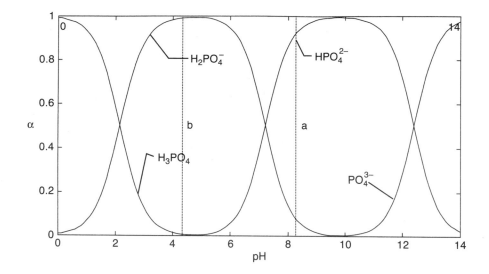

**Fig. 10.1** Distribution of phosphorus species expressed as the fraction, $\alpha$, as a function of aqueous solution pH.

Using this plot, estimates of the phosphorus species distribution are readily made. For example, in open water of the Kenyan (eastern) part of Lake Victoria, phosphorus levels are typically about 12 $\mu$g L$^{-1}$ (as P) and the pH of the water is 8.2. Under these conditions, the fractional values of the principal species are given by points where line 'a' intersects the species curves on Fig. 10.1 and it is estimated graphically that $\alpha_{H_2PO_4^-} = 0.08$ and $\alpha_{HPO_4^{2-}} = 0.92$, giving concentrations $[H_2PO_4^-] = 1$ $\mu$g L$^{-1}$ and $[HPO_4^{2-}] = 11$ $\mu$g L$^{-1}$. In another setting, a sample of pore water from a forest soil in the Canadian shield, total soluble phosphorus is 62 $\mu$g L$^{-1}$ and the pH 4.3. In this situation (points on line 'b' on Fig. 10.1) the principal species is almost exclusively $[H_2PO_4^-] = 62$ $\mu$g L$^{-1}$.

We have mentioned that there are assumptions in these calculations. One assumption not previously noted is that there are no interactions with other species in the system. Furthermore, as noted earlier, no account was taken of the solution ionic strength. Soil solutions typically have an ionic strength of 0.002, which leads to an activity coefficient of about 0.95 for a singly charged ion. Incorporating the activity coefficient into the calculation would make only a small difference and the correction would be even smaller for the lake water. While assuming zero ionic strength in this situation leads to a small error, the error could be much larger for a medium like sea water that contains a high ionic concentration. Apparent acid dissociation constants (which are affected by large changes in activity coefficient due to the high ionic strength) have been calculated for phosphoric acid in sea water and they are:

$$K_{a1} = 2.4 \times 10^{-2} \; pK_{a1} = 1.62$$
$$K_{a2} = 8.8 \times 10^{-7} \; pK_{a2} = 6.06$$
$$K_{a3} = 1.4 \times 10^{-9} \; pK_{a3} = 8.85$$

A substantially different distribution diagram therefore would apply to sea water compared with that calculated above for a 'fresh-water' system.

## Cadmium complexes with chloride

A second version of a single-variable distribution diagram is a plot of the fractional concentration of a particular species versus a chosen variable, usually the concentration of an important ligand.

As an example, we will consider the distribution of aqueous cadmium chloro complexes as a function of chloride ion concentration. Again, we will use concentrations rather than activities in the calculation. We start by assuming that, in the absence of chloride or any other complexing ligand, cadmium exists in aqueous solution as an aquo complex, perhaps $Cd(H_2O)_4^{2+}$. (We will omit the complexed waters and refer to the aquo species as $Cd^{2+}$ in further discussions.) This aquo complex remains without significant deprotonation (see Section 13.1) as long as the pH is less than about 8.5. Chloride ion forms complexes with cadmium in a stepwise fashion, with displacement of one water molecule each time a chloride is added. We apply the general relations for complex formation (eqns 9.1–9.4) to the cadmium / chloride situation.

$$Cd^{2+} + Cl^- \rightleftharpoons CdCl^+ \quad K_{f1} = \frac{[CdCl^+]}{[Cd^{2+}][Cl^-]} = 7.9 \times 10^1 \tag{10.16}$$

$$CdCl^+ + Cl^- \rightleftharpoons CdCl_2 \quad K_{f2} = \frac{[CdCl_2]}{[CdCl^+][Cl^-]} = 4.0 \tag{10.17}$$

$$CdCl_2 + Cl^- \rightleftharpoons CdCl_3^- \quad K_{f3} = \frac{[CdCl_3^-]}{[CdCl_2][Cl^-]} = 2.0 \tag{10.18}$$

$$CdCl_3^- + Cl^- \rightleftharpoons CdCl_4^{2-} \quad K_{f4} = \frac{[CdCl_4^{2-}]}{[CdCl_3^-][Cl^-]} = 0.6 \tag{10.19}$$

The reactions may also be described using 'overall' steps and the overall stability constants are symbolized as $\beta_f$. It is readily seen that

$$\beta_{fn} = K_{f1} \times K_{f2} \times \cdots K_{fn} \tag{10.20}$$

$$Cd^{2+} + Cl^- \rightleftharpoons CdCl^+ \quad \beta_{f1} = K_{f1} = 7.9 \times 10^1 \tag{10.21}$$

$$Cd^{2+} + 2Cl^- \rightleftharpoons CdCl_2 \quad \beta_{f2} = K_{f1} \times K_{f2} = 3.2 \times 10^2 \tag{10.22}$$

$$Cd^{2+} + 3Cl^- \rightleftharpoons CdCl_3^- \quad \beta_{f3} = K_{f1} \times K_{f2} \times K_{f3} = 6.4 \times 10^2 \tag{10.23}$$

$$Cd^{2+} + 4Cl^- \rightleftharpoons CdCl_4^{2-} \quad \beta_{f4} = K_{f1} \times K_{f2} \times K_{f3} \times K_{f4} = 3.8 \times 10^2 \tag{10.24}$$

Note that there can be significantly different values for the stability constants, $K_f$ or $\beta_f$, in the literature, depending on how the measurements were made. One reliable and widely used compilation is that in Hogfeldt, E., *Stability constants of metal-ion complexes*, IUPAC Chemical Data Series, No. 21, Pergamon Press, Oxford; 1982.

The total concentration of cadmium in an aqueous solution containing chloride is then

$$C_{Cd} = [Cd^{2+}] + [CdCl^+] + [CdCl_2] + [CdCl_3^-] + [CdCl_4^{2-}] \tag{10.25}$$

We can derive expressions for the concentration of each of the five cadmium species in the following way. We begin by dividing eqn 10.25 by $[Cd^{2+}]$

$$\frac{C_{Cd}}{[Cd^{2+}]} = 1 + \frac{[CdCl^+]}{[Cd^{2+}]} + \frac{[CdCl_2]}{[Cd^{2+}]} + \frac{[CdCl_3^-]}{[Cd^{2+}]} + \frac{[CdCl_4^{2-}]}{[Cd^{2+}]} \tag{10.26}$$

Substituting the expressions for the $\beta$ functions,

$$\frac{C_{Cd}}{Cd^{2+}} = 1 + \beta_{f1}[Cl^-] + \beta_{f2}[Cl^-]^2 + \beta_{f3}[Cl^-]^3 + \beta_{f4}[Cl^-]^4 \tag{10.27}$$

Rearranging eqn 10.27,

$$[Cd^{2+}] = \frac{C_{Cd}}{1 + \beta_{f1}[Cl^-] + \beta_{f2}[Cl^-]^2 + \beta_{f3}[Cl^-]^3 + \beta_{f4}[Cl^-]^4} \tag{10.28}$$

Similarly, the concentrations of other cadmium chloro species are given by

$$[CdCl^+] = \frac{\beta_{f1}[Cl^-]C_{Cd}}{1 + \beta_{f1}[Cl^-] + \beta_{f2}[Cl^-]^2 + \beta_{f3}[Cl^-]^3 + \beta_{f4}[Cl^-]^4} \qquad (10.29)$$

$$[CdCl_2] = \frac{\beta_{f2}[Cl^-]^2C_{Cd}}{1 + \beta_{f1}[Cl^-] + \beta_{f2}[Cl^-]^2 + \beta_{f3}[Cl^-]^3 + \beta_{f4}[Cl^-]^4} \qquad (10.30)$$

$$[CdCl_3^-] = \frac{\beta_{f3}[Cl^-]^3C_{Cd}}{1 + \beta_{f1}[Cl^-] + \beta_{f2}[Cl^-]^2 + \beta_{f3}[Cl^-]^3 + \beta_{f4}[Cl^-]^4} \qquad (10.31)$$

$$[CdCl_4^{2-}] = \frac{\beta_{f4}[Cl^-]^4C_{Cd}}{1 + \beta_{f1}[Cl^-] + \beta_{f2}[Cl^-]^2 + \beta_{f3}[Cl^-]^3 + \beta_{f4}[Cl^-]^4} \qquad (10.32)$$

Again, using an appropriate computer program, the fraction, $\alpha$, of each cadmium species can be calculated as a function of chloride ion concentration. For the chloride ion, we choose a range of values from 0 to 0.56 M, the latter being the chloride concentration in sea water. The plot then forms another type of distribution diagram and is shown in Fig. 10.2.

## The estuarine environment—Chesapeake Bay as an example

Figure 10.2 can be useful in several situations. As an example, we may use it to describe the changing cadmium species distribution in an estuary. An estuary has been defined[1] as 'a semi-enclosed coastal body of water that has a free connection with the open sea and within which sea water is measurably diluted with fresh water derived from land drainage'. A principal feature of any estuary is a regular variation from low to high in dissolved salt concentration as one moves from the inflowing river to the mouth where it opens out into the ocean. In the vertical dimension salt concentration changes as well. Usually, there is a steady but sometimes irregular increase in

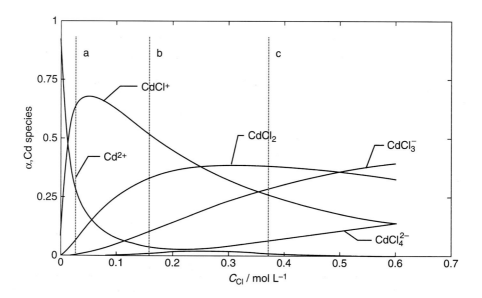

**Fig. 10.2** Distribution of cadmium chloro complexes as a function of the concentration of chloride ion in water. The range of chloride concentrations is from zero to 0.56 mol L$^{-1}$; the latter value is the approximate concentration in sea water.

[1] Pritchard, D.W., What is an estuary: physical viewpoint, In *Estuaries*, American Association for the Advancement of Science, Publication No. 83, Washington, DC 1967.

salt concentration with depth, in part due to the greater density of the high salt solution. There are also marked seasonal fluctuations in concentration, with reduced levels being characteristic of periods of heavy precipitation or surface run-off. The Chesapeake Bay in Maryland, USA is at the mouth of the Susquehanna River and is the largest estuary on the Atlantic coast of the USA. Figure 10.3 shows the springtime variation in *salinity* in a vertical section along the axis of Chesapeake Bay extending 300 km from river (left) to open ocean (right).

For sea water, salinity is defined as the mass in grams of the solids that can be obtained from 1 kg of sea water after all the carbonate is converted to oxide, the bromine and iodine replaced by chlorine, all organic matter oxidized, and the residue dried at 480°C to constant weight. The symbol commonly used for salinity is ‰, which is equivalent to per cent, but with a base of one thousand. Sea water has an average salinity of 35‰ (meaning 3.5% salt content) representing a chloride concentration of approximately 0.56 mol L$^{-1}$. For the present discussion we will assume a direct linear relation between salinity and chloride molarity.

We can now estimate the fractional distribution of cadmium species throughout the estuary. At the outlet of the Susquehanna River (salinity ~1‰, [Cl$^-$] ~0.02 mol L$^{-1}$), as represented by line 'a' in Fig. 10.2, the approximate fractions of the three principal species are Cd$^{2+}$ = 0.29, CdCl$^+$ = 0.64, CdCl$_2$ = 0.07. It is interesting that, even at very low chloride ion concentrations, there is considerable tendency for formation of complexes. About half-way (144 km) down the bay the surface water salinity has increased to 10‰ corresponding to [Cl$^-$] = 0.16 mol L$^{-1}$ (line 'b' on Fig. 10.2). At this point the species have fractions as follows: Cd$^{2+}$ = 0.04; CdCl$^+$ = 0.52; CdCl$_2$ = 0.33; CdCl$_3^-$ = 0.10; and CdCl$_4^{2-}$ = 0.01. Near the mouth of Chesapeake Bay, the salinity of surface water is 23‰ or [Cl$^-$] = 0.37 mol L$^{-1}$. This corresponds to line 'c' on the distribution diagram. Here, the cadmium species fractional distribution is CdCl$^+$ = 0.26, CdCl$_2$ = 0.39, CdCl$_3$ = 0.29, and CdCl$_4^{2-}$ = 0.06. It should not be surprising that the cadmium present in estuarine water shows an increasing tendency to be complexed with chloride as one moves

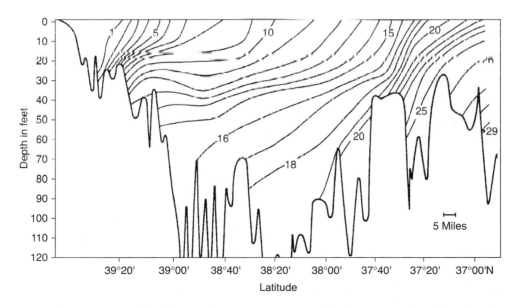

**Fig. 10.3** Salinity contours for Chesapeake Bay in springtime. The left (west) side of the diagram is the inland part of the estuary and salinity approaches that of fresh water. The right (east) side is where the estuary opens out into the Atlantic Ocean (Schubel, J.R., *The estuarine environment*, American Geological Institute; 1971). Note also the increase in salinity with depth, indicating that the fresh water from the river layers on to the surface of the saline matrix. Reprinted with permission.

to higher salinity water. The values we have calculated are fractional results. Using these along with appropriate measurements of total cadmium concentration, we could estimate the amount of each species at any point in the estuary.

The information about cadmium speciation helps us to understand some observed issues about the toxicity of this element. Cadmium is toxic to humans and other organisms including marine organisms, the mechanism of toxicity being connected with its ability to compete with calcium for binding to specific proteins. In the marine environment its toxicity is closely related to the abundance of free $Cd^{2+}$ with chloro complexes being much less available for biological uptake. As a consequence, there is an inverse relation between cadmium uptake and salinity. In the Chesapeake Bay, we could therefore expect that organisms located at the river outlet would accumulate more cadmium than similar ones further toward the open ocean.

## Equilibrium calculations done on spreadsheets

Species-distribution diagrams and equilibrium calculations that define them, like the two examples we have just looked at, have become quite accessible through the use of spreadsheet programs or even more sophisticated software packages or shareware (see *PHREEQC for windows* in Chapter 11 and Appendix D.1). The spreadsheet calculation example shown below will illustrate how the calculations for the carbonic acid ($CO_2$ (aq)) system are entered into a spreadsheet to generate the data required to plot the distribution diagram for a simple aqueous environment. In Chapter 11, we will expand upon this particular equilibrium system to include both atmospheric $CO_2$ and solid $CaCO_3$.

In generating the species distribution diagram for carbonic acid, the fraction of the individual species ($\alpha$) will be determined and plotted as the y-coordinate against the variable pH. In the calculations, the concentration of hydronium ions is needed, but we will calculate this value from the simple numerical value of pH entered. It is typical to use pH increments of 0.10, ranging from 0.00 to 14.00, for a total of 141 data points.

In the spreadsheet, choose five consecutive columns, e.g. B–F, and place a label for the value that will appear in the cells, as shown.

| Cell | A | B | C | D | E | F | G | H | I | J |
|------|---|---|---|---|---|---|---|---|---|---|
| 1 | | | | | | | | | | |
| 2 | | pH | $H_3O^+$ | $CO_2$ | $HCO_3^-$ | $CO_3^{2-}$ | | $K_{a1}$ | $4.5 \times 10^{-7}$ | |
| 3 | | 0.00 | 1.0 | 1.0 | 0.0 | 0.0 | | $K_{a2}$ | $4.7 \times 10^{-11}$ | |
| 4 | | 0.10 | $7.9 \times 10^{-1}$ | | | | | | | |
| 5 | | 0.20 | $6.3 \times 10^{-1}$ | | | | | | | |

For example, cell B2 contains the label pH and below it, the cells can be filled with the pH (variable) values 0.00 through to 14.00, as partly shown above. The two $K_a$ values for this diprotic acid are also needed for this calculation and can be added to the spreadsheet, off to the side and out of the way of the active columns. The formulae are now ready to be entered into the spreadsheet in the row below the labels.

The pH can be converted into the $H_3O^+$ concentration by the calculation $10^{-pH}$. This is done here in this example spreadsheet by entering the following formula in cell C3.

$$= 10^\wedge -B3$$

In this case the returned value in cell C3 will be 1.0.

Copy this formula in cell C3 down the rest of the column (to cell C143) and the hydronium ion concentrations will now appear in these cells. They would be best viewed with scientific notation showing one decimal.

Now the formulae for each of the carbonate species can be entered into the cells D3, E3 and F3, making use of both the new C column (hydronium ion) values and the two $K_a$ values in cells I2 and I3.

The fraction of $CO_2$ can be calculated by

$$\alpha CO_2 = (H_3O^+)^2 / ((H_3O^+)^2 + H_3O^+ \times K_{a1} + K_{a1} \times K_{a2})$$

and in the spreadsheet (cell D3) enter the formula:

$$= +(+C3)^2 / ((C3)^2 + (C3) * \$I\$2 + \$I\$2 * \$I\$3)$$

Note: the use of the $ signs anchor the cells I2 and I3, so they remain the same cell locations when the formula is copied, since we will always need just these two values. However, we do not anchor the C (hydronium ion) value since it changes when moving down the column.

The fraction of $HCO_3^-$ can be calculated by

$$\alpha HCO_3^- = (H_3O^+) \times K_{a1} / ((H_3O^+)^2 + H_3O^+ \times K_{a1} + K_{a1} \times K_{a2})$$

and in the spreadsheet (cell E3) enter the formula:

$$= +(+C3) * \$I\$2 / ((C3)^2 + (C3) * \$I\$2 + \$I\$2 * \$I\$3)$$

The fraction of $CO_3^{2-}$ can be calculated by

$$\alpha CO_3^{2-} = K_{a1} \times K_{a2} / ((H_3O^+)^2 + H_3O^+ \times K_{a1} + K_{a1} \times K_{a2})$$

and in the spreadsheet (cell F3) enter the formula:

$$= +(\$I\$2 * \$I\$3) / ((C3)^2 + (C3) * \$I\$2 + \$I\$2 * \$I\$3)$$

The formulae entered will be shown in the function line, but the values returned (in row 3) will appear as in the example portion of the spreadsheet above.

The formulae are now copied in columns D, E and F down to row 143; the complete set of $\alpha$ values then appears.

Plotting packages (e.g., scatter plot for Excel) within spreadsheet programs can then be used to produce a graphical output. In this case the carbonate species-distribution diagram is shown in Fig. 1.2.

> **Main point 10.2** Single-variable distribution diagrams show how the concentrations of different species change as a function of the one defined variable—typically pH or the concentration of a particular complexing ligand. Equations developed for each species can be used to calculate individual species concentrations in specified conditions.

## 10.2 Two-variable diagrams—p*E* / pH diagrams

Both the phosphorus and cadmium distribution diagrams share an obvious limitation in that they describe the behaviour of a particular chemical in terms of only a single environmental variable, pH or chloride concentration, respectively. Clearly, in complex real systems, there are many variables operating simultaneously. A further step towards accurately describing natural systems in graphical form is then to create a two-variable diagram. In such diagrams, the dominant species are plotted in two dimensions as a function of two independent variables. There are obvious merits in creating diagrams that take into account the simultaneous involvement of two factors, but we shall see that some information (a detailed description of concentrations) is lost in order to depict the results on a flat surface.

### Electron activity measured as p*E*

A very widely used type of two-variable diagram for describing chemical behaviour in the hydrosphere is the p*E* / pH diagram, also called a Pourbaix diagram. We shall discuss how to construct and interpret such plots, but before doing this it is necessary to introduce the concept of p*E*.

Analogously to pH, the measure of acidity in aqueous solutions, pE is defined as the negative logarithm of the electron activity[2]

$$pE = -\log a_e \tag{10.33}$$

A large negative value of pE indicates a large value for the electron activity in solution, which implies that reducing conditions obtain. This is the situation in anoxic water bodies such as swamps. Conversely, a large positive value of pE implies low electron activity in solution and oxidizing conditions, as would be the case in well-aerated surface water. In practice, pE values in water range from approximately $-12$ to 25. We shall see the reason for this range shortly.

While the definition of pE is simple and understandable to chemists, direct measurements of electron activity are not easily made in the way that measurements of pH are done. To show how pE is calculated and measured it is helpful to look at some examples.

Consider the simple half-reaction.

$$Fe^{3+}\ (aq) + e^- \rightleftharpoons Fe^{2+}\ (aq) \tag{10.34}$$

$$K_{eq} = \frac{a_{Fe^{2+}}}{a_{Fe^{3+}} \times a_e} \tag{10.35}$$

$$\frac{1}{a_{e^-}} = \frac{K_{eq} \times a_{Fe^{3+}}}{a_{Fe^{2+}}} \tag{10.36}$$

Using the definition of pE and taking logs of both sides of eqn 10.36, we have

$$pE = -\log a_e = \log K_{eq} + \log \frac{a_{Fe^{3+}}}{a_{Fe^{2+}}} \tag{10.37}$$

since

$$\Delta G^\circ = -2.303\ RT \log K_{eq} \tag{10.38}$$

$$= -nFE^\circ \tag{10.39}$$

where $n$ has the usual electrochemical meaning, i.e. the number of electrons transferred in the half-reaction.

Therefore, at 298 K ($R = 8.314\ \text{J K}^{-1}\ \text{mol}^{-1}$ and $F = 96\ 485\ \text{C mol}^{-1}$), we have

$$\log K_{eq} = \frac{nFE^\circ}{2.303\ RT} = \frac{nE^\circ}{0.0591} \tag{10.40}$$

In this case $n = 1$. Therefore,

$$\log K_{eq} = \frac{E^\circ}{0.0591} \tag{10.41}$$

and

$$pE = \frac{E^\circ}{0.0591} + \log \frac{a_{Fe^{3+}}}{a_{Fe^{2+}}} \tag{10.42}$$

Under standard conditions, $a_{Fe^{3+}} = a_{Fe^{2+}} = 1$,

$$\log \frac{a_{Fe^{3+}}}{a_{Fe^{2+}}} = 0 \tag{10.43}$$

and

$$pE = pE^\circ = \frac{E^\circ}{0.0591} \tag{10.44}$$

For non-standard conditions,

$$pE = pE^\circ + \log \frac{a_{Fe^{3+}}}{a_{Fe^{2+}}} \tag{10.45}$$

---

[2] Both pH and pE are *defined* in terms of activity. In equilibrium derivations in this section, we will employ activities for all species. However, while recognizing the inconsistency, we will revert to use of concentrations for soluble species other than hydronium and hydroxyl ions and the electron in subsequent calculations.

When the standard $pE°$ value and the actual activities (usually approximated by concentrations) of $Fe^{3+}$ and $Fe^{2+}$ are substituted into this equation, the $pE$ of a particular environmental system can be calculated.

In the general case, for a reaction

$$aA + ne^- \rightleftharpoons bB \tag{10.46}$$

where A and B are the oxidized and reduced forms of a redox couple, the reaction quotient ($Q$) is defined as

$$Q = \frac{(a_B)^b}{(a_A)^a} \approx \frac{[B]^b}{[A]^a} \tag{10.47}$$

The reaction quotient takes the form of an equilibrium constant, but uses activities (or, as an approximation, concentrations) that obtain under any conditions, not just those at equilibrium.

The general form of eqn 10.45 is then

$$pE = pE° - \frac{1}{n}\log Q \tag{10.48}$$

## Methods of calculating p*E*°

Values of $pE°$ for a number of half-reactions of environmental interest are given in Appendix B.5. When additional values are required, there are several methods that make use of other readily obtainable information.

The first method involves using eqn 10.44. Where a tabulated $E°$ value for a half-reaction is available, the $pE°$ is readily calculated. For

$$Fe^{3+}\,(aq) + e^- \rightleftharpoons Fe^{2+}\,(aq) \tag{10.34}$$

$$E° = +0.771\ V$$

Therefore,

$$pE° = +\frac{0.771\ V}{0.0591\ V} = 13.0$$

Being a ratio of two potentials, the $pE°$ value is a dimensionless number.

A second method for calculating $pE°$ makes use of the relations in eqns 10.40 and 10.44.

$$\log K_{eq} = \frac{nE°}{0.0591} = npE° \tag{10.49}$$

$$pE° = \frac{\log K_{eq}}{n} \tag{10.50}$$

This relation is applicable when an $E°$ value is not available, but where the appropriate equilibrium constant is known.

In some cases, several reactions may be combined to produce an overall half-reaction. Consider the half-reaction for which no tabulated $E°$ value is easily found.

$$Fe(OH)_3 + 3H_3O^+\,(aq) + e^- \rightleftharpoons Fe^{2+}\,(aq) + 6H_2O \tag{10.51}$$

The reaction is the sum of

$$Fe(OH)_3 \rightleftharpoons Fe^{3+}\,(aq) + 3OH^-\,(aq) \tag{10.51a}$$

$$Fe^{3+}\,(aq) + e^- \rightleftharpoons Fe^{2+}\,(aq) \tag{10.51b}$$

$$3H_3O^+\,(aq) + 3OH^-\,(aq) \rightleftharpoons 6H_2O \tag{10.51c}$$

For reaction 10.51a, $K_a = K_{sp} = 9.1 \times 10^{-39}$ and $\log K_a = -38.0$. For reaction 10.51b, $pE_b^\circ = \log K_b = 0.771/0.0591$ and $\log K_b = +13.0$. For reaction 10.51c,

$$K_c = \frac{1}{(K_w)^3} = 10^{42}$$

and

$$\log K_c = +42.0$$

For the original, overall reaction

$$\log K_{rxn} = \log K_a + \log K_b + \log K_c$$

$$= -38.0 + 13.0 + 42.0$$

$$= +17.0$$

Using eqn 10.50 ($n = 1$)

$$pE^\circ \text{ overall} = +17.0$$

There is a third method for calculating $pE^\circ$ values and this requires combining eqns 10.38 and 10.49.

$$\Delta G^\circ = -2.303\ RTnpE^\circ \tag{10.52}$$

$$pE^\circ = \frac{-\Delta G^\circ}{2.303\ RTn} \tag{10.53}$$

Consider the redox reaction

$$SO_4^{2-}\ (aq) + 10H_3O^+\ (aq) + 8e^- \rightleftharpoons H_2S\ (aq) + 14H_2O \tag{10.54}$$

$$\Delta G^\circ = \Delta G_f^\circ\ (H_2S) + 14\Delta G_f^\circ\ (H_2O) - \Delta G_f^\circ\ (SO_4^{2-}) - 10\Delta G_f^\circ\ (H_3O^+) - 8\Delta G_f^\circ\ (e^-)$$

Using thermochemical tables and noting that $\Delta G_f^\circ$ for the aqueous electron is 0 and that, for the hydronium ion, has the same value as for water,

$$\Delta G^\circ = -27.86 + 14 \times (-237.18) - 10 \times (-237.18) - (-744.60)$$

$$= -231.98\ \text{kJ}$$

$$pE^\circ = \frac{-(-231.98)\ \text{kJ} \times 1000\ \text{J kJ}^{-1}}{2.303 \times 8.314\ \text{J mol}^{-1}\text{K}^{-1} \times 298.2\ \text{K} \times 8\ \text{mol}}$$

$$= 5.08$$

This final method for calculating $pE^\circ$ values is perhaps the most generally useful but, depending on circumstances, any one of the three methods may be employed. Once $pE^\circ$ values are available, it is possible to determine $pE$ for particular non-standard environmental conditions. A two-part example follows.

## Chromium in tannery wastes

Traditional leather-tanning processes involve treating the hides with an aqueous solution of chromium (III). Suppose the wastewater from a tannery contains 26 mg L$^{-1}$ chromium, in its original $Cr^{3+}$ state. As the effluent flows downstream, the dissolved oxygen can cause $Cr^{3+}$ to be oxidized to $Cr_2O_7^{2-}$.

We will calculate the extent of oxidation for a situation where oxygen in the stream water is in equilibrium with atmospheric oxygen, and has a pH of 6.5 ($a_{H_3O^+} = 10^{-6.5}$). The first step is to calculate the value of $pE$. We can then use this value to calculate the concentrations of $Cr^{3+}$ and $Cr_2O_7^{2-}$ assuming that the chromium species are also in equilibrium with the system.

Step 1 The relevant reaction for atmospheric $O_2$ in equilibrium with water (synonymous with calling this *a well-aerated system*) is

$$O_2 \text{ (g)} + 4H_3O^+ \text{ (aq)} + 4e^- \rightleftharpoons 6H_2O \tag{10.55}$$

$$E° = 1.23 \text{ V}$$

$$pE° = \frac{1.23 \text{ V}}{0.0591 \text{ V}} = 20.8$$

Using eqn 10.48,

$$pE = pE° - \frac{1}{n} \log \frac{1}{P_{O_2}/P° \times (a_{H_3O^+})^4}$$

$$= 20.8 - \frac{1}{4} \log \frac{1}{0.209 \times (10^{-6.5})^4}$$

$$= 14.1$$

Note that, in these calculations, the pressure is given as a ratio of $P_{O_2}/P°$, which is numerically identical to pressure in atmospheres. For oxygen, which makes up 20.9% of the atmosphere, $P_{O_2} = 21\,200$ Pa; $P° = 101\,325$ Pa.

Step 2 For the Cr system

$$Cr_2O_7^{2-} \text{ (aq)} + 14H_3O^+ \text{ (aq)} + 6e^- \rightleftharpoons 2Cr^{3+} \text{ (aq)} + 21H_2O \tag{10.56}$$

$$E° - 1.36 \text{ V} \quad \text{and} \quad pE° = 23.0$$

$$pE = pE° - \frac{1}{6} \log \frac{[Cr^{3+}]^2}{[Cr_2O_7^{2-}](a_{H_3O^+})^{14}}$$

Since the chromium and oxygen systems are in equilibrium, $pE$ is the same for both.

$$14.1 = 23.0 - \frac{1}{6} \log \frac{[Cr^{3+}]^2}{[Cr_2O_7^{2-}](10^{-6.5})^{14}}$$

$$= 23.0 - \frac{1}{6} \log \frac{1}{(10^{-6.5})^{14}} - \frac{1}{6} \log \frac{[Cr^{3+}]^2}{[Cr_2O_7^{2-}]}$$

$$= 7.8 - \frac{1}{6} \log \frac{[Cr^{3+}]^2}{[Cr_2O_7^{2-}]}$$

$$\log \frac{[Cr^{3+}]^2}{[Cr_2O_7^{2-}]} = -37.8$$

$$\frac{[Cr^{3+}]^2}{[Cr_2O_7^{2-}]} = 1.6 \times 10^{-38}$$

This very small ratio indicates that virtually all of the $Cr^{3+}$ would be oxidized to $Cr_2O_7^{2-}$ and this, in fact, has serious environmental consequences. A widely used method of leather tanning involves a two-bath process in which the hides are soaked for several hours in a tank containing chromic acid in order to effect cross-linking between proline and hydroxyproline residues in protein molecules in the animal skin as a way of toughening the material. The hides are then transferred into another tank containing reducing agents such as sucrose in order to reduce the excess chromium (VI) to chromium (III). The excess is then discharged into the waste stream.

The leather industry is important throughout the globe, with India the world's major producer of leather goods. In that country there are large industrial establishments manufacturing shoes, luggage, garments, etc., but much of the industry is small-scale and scattered in numerous villages in rural areas of the country. The northern part of Tamil Nadu state in south India is one of the major centres of the leather industry.

The quantity of chromium (III) wasted in effluents of a small-scale chrome tannery in India is estimated to be approximately 0.4 kg per 100 kg of raw hides. This is about one-half of that added in the first bath. As our calculation shows, thermodynamics predicts that the element will be oxidized to $Cr_2O_7^{2-}$. While chromium (III) is a required trace element for mammals, chromium (VI) can be highly toxic and is also carcinogenic to humans and other mammals. It is also toxic to plants, and large areas of the leather-producing areas of Tamil Nadu are devoid of vegetation. The consequences to humans in that region are not documented. In any case, the release of such large amounts of chromium with its subsequent conversion to the thermodynamically stable $Cr_2O_7^{2-}$ species is a major environmental issue.

A proposed method[3] for the removal of chromium (III) from tannery wastes involves treating the wastes with successive batches of *Sargassum* seaweed. This material is dried and pre-treated with solutions of sulphuric acid and calcium and magnesium chloride before being exposed to acidic, chromium-containing solutions. Removal of up to 35 mg of chromium per g of seaweed was achieved; this was attributed to cation exchange at carboxylic acid sites on the *Sargassum*. A more detailed discussion of the nature of organic matter and its role in ion-exchange reactions is given in Chapter 12.

> **Main point 10.3** Electron activity is measured by the p$E$ function. High p$E$ values are characteristic of oxidizing conditions such as are found in well-aerated water. Low p$E$ is measured under reducing conditions, as in water where oxygen has been excluded or consumed.

## p$E$ / pH diagrams

We are now in a position to construct a p$E$ / pH diagram. The diagram will take the form of a two-dimensional plot of p$E$ ($y$-axis) versus pH ($x$-axis) in which areas in the diagram define the regions where particular species are dominant. In saying a particular species dominates, we must define conditions at the boundary between domains. This requires that we calculate and draw lines on the diagram corresponding to these boundaries.

An important issue, then, is how to define the boundary conditions in different situations. For the case of a reaction between two species where one is dissolved in the water and the other is a gas, the boundary condition is set as $P° = 101\ 325$ Pa $= 1$ atm. This is the minimum pressure required for gas evolution from the aqueous solution. In other words, when $P_{gas}$ is greater than 101 325 Pa, the gas phase itself is said to pre-dominate over the dissolved gas. As an example, this type of boundary condition is used in the case of the evolution of hydrogen gas from an acidic solution

$$2H_3O^+ \text{ (aq)} + 2e^- \rightarrow H_2 \text{ (g)} + 2H_2O \tag{10.57}$$

Another type of condition is required for the situation where only soluble species are involved, or where there is a soluble species reacting to form one that is insoluble. Examples of these two cases are

$$Sn^{4+} \text{ (aq)} + 2e^- \rightarrow Sn^{2+} \text{ (aq)} \tag{10.58}$$

and

$$MnO_2 \text{ (s)} + 4H_3O^+ \text{ (aq)} + 2e^- \rightarrow Mn^{2+} \text{ (aq)} + 6H_2O \tag{10.59}$$

In these cases, we arbitrarily define a concentration (as an estimate of activity) below which a particular species is considered to be soluble. When the concentration is higher than this arbitrary value, we are saying that the solid form is more important. Any concentration may be chosen but it is common to use values between $10^{-5}$ and $10^{-2}$ mol $L^{-1}$. Of course, in constructing a diagram one should consistently use the same concentration for all boundaries.

[3] Aravindhan, R., B. Madhan, J.R. Rao, B.U. Nair, and T. Ramasami, Bioaccumulation of chromium from tannery wastewater: an approach for chrome recovery and reuse, *Environ. Sci. Technol.*, **38** (2004), 300.

We will illustrate how boundary conditions are chosen and used in the construction of a $pE$ / pH diagram by considering the aqueous sulphur system where the species of interest are $SO_4^{2-}$ (aq), $HSO_4^-$ aq, S (s), $HS^-$ (aq), and $H_2S$ (aq). Before beginning calculations involving these species, however, there are two preliminary calculations common to all diagrams pertaining to the hydrosphere. These are calculations of the boundaries defining water stability.

## Water-stability boundaries

Water itself has a limited range of $pE$ and pH values within which it is stable. Under highly reducing conditions (low $pE$), water is reduced.

$$2H_2O + 2e^- \rightleftharpoons H_2 \text{ (g)} + 2OH^- \text{ (aq)} \tag{10.60}$$

For this reaction

$$E^\circ = -0.828V$$

$$pE^\circ = -14.0$$

Equation 10.48 for this reaction is written as

$$pE = pE^\circ - \tfrac{1}{2}\log(P_{H_2} / P^\circ \times (a_{OH^-})^2) \tag{10.61}$$

(Note again that here and in subsequent calculations, we will use concentrations rather than activities and pressures rather than fugacities.)

For the boundary involving a gas, we choose the condition that

$$P_{H_2} = P^\circ = 101\ 325 \text{ Pa}$$

$$pE = -14.0 - \log(a_{OH^-})$$

$$= -14.0 + pOH$$

Since

$$pH + pOH = 14$$

then

$$pE = -pH$$

This line then defines the boundary for water stability with respect to reduction and is shown as the lower line on Fig. 10.4. Where the $pE$ value is less than the pH value, water is unstable. It is possible also to calculate the same line by taking the reduction reaction to be

$$2H_3O^+ \text{ (aq)} + 2e^- \rightleftharpoons H_2 \text{ (g)} + 2H_2O \tag{10.62}$$

Considering the other extreme, highly oxidizing conditions, water is unstable with respect to $O_2$ evolution and the reaction is written as

$$6H_2O \rightleftharpoons 4H_3O^+ \text{ (aq)} + O_2 \text{ (g)} + 4e^- \tag{10.63}$$

For this reaction

$$E^\circ = 1.229 \text{ V}$$

$$pE^\circ = E^\circ / 0.0591 = 20.80$$

The production of oxygen from water is an oxidation reaction. It is important to note, however, that we use the IUPAC convention of defining $E^\circ$ and $E$, and $pE^\circ$ and $pE$, in terms of the reverse, reduction process.

$$pE = pE^\circ - \tfrac{1}{4}\log(1 / ((P_{O_2} / P^\circ) \times (a_{H_3O^+})^4)) \tag{10.64}$$

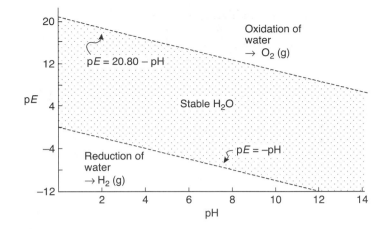

**Fig. 10.4** Region of stability and stability boundaries for water on a pE/pH diagram.

Once again, the boundary condition requires that the pressure of the gas equal atmospheric pressure

$$P_{O_2} = P^\circ = 101\ 325\ \text{Pa}$$
$$pE = 20.80 - \log(1/a_{H_3O^+})$$
$$= 20.80 - pH$$

This then defines the upper line on Fig. 10.4 and the region between the two lines is the stability region for water on a pE / pH diagram. Above the top boundary, water is oxidized with the evolution of oxygen, while below the lower boundary it is reduced and releases hydrogen.

## The sulphur system
Now we can deal with species in the sulphur system. To create the pE / pH diagram, we will be defining boundaries between species and superimposing these on the water stability diagram. We may consider the various pairs of species in any order but it is helpful to have an idea of what to expect before beginning calculations. For example, we expect that $HSO_4^-$ will be important at low pH and $SO_4^{2-}$ at high pH. Similarly, $SO_4^{2-}$ would exist under oxidizing, high-pE conditions, while $HS^-$ is an important reduced, low-pE form. For all soluble species, we will choose $10^{-2}$ mol L$^{-1}$ as our arbitrary definition for the boundary.

## The $SO_4^{2-}$ / $HSO_4^-$ boundary
The equation describing this boundary is an acid–base reaction and requires hydronium ion, but there is no oxidation or reduction involved.

$$SO_4^{2-}\ (\text{aq}) + H_3O^+\ (\text{aq}) \rightleftharpoons HSO_4^-\ (\text{aq}) + H_2O \qquad (10.65)$$

To make things easier, we use $H^+$ as an abbreviation for $H_3O^+$, the hydronium ion, but, of course, the result would be the same if the latter species were used in the equations and calculations. $\Delta G$ values are found in Appendix B.2.

$$\Delta G^\circ = \Delta G_f^\circ(HSO_4^-) - \Delta G_f^\circ(SO_4^{2-}) - \Delta G_f^\circ(H^+)$$
$$= -755.90 - (-744.60) - 0$$
$$= -11.39\ \text{kJ} = -11\ 390\ \text{J}$$

$$\log K = \frac{-\Delta G°}{2.303\,RT}$$

$$= \frac{1.995/11\,390}{2.303 \times 8.314 \times 298.2} = 1.995$$

$$K = \frac{[HSO_4^-]}{[SO_4^{2-}]a_{H^+}}$$

At the boundary, $[HSO_4^-] = [SO_4^{2-}] = 10^{-2}$, and

$$K = \frac{1}{a_{H^+}}$$

$$\log K = pH = 1.995$$

Therefore, the boundary between $HSO_4^-$ and $SO_4^{2-}$ is a vertical line at pH 1.995. At less than this value, $HSO_4^-$ is the dominant form; at greater than it $SO_4^{2-}$ is most important. See line 'a' on Fig. 10.5.

### The $HSO_4^-$ / $S°$ boundary

$$HSO_4^- \text{ (aq)} + 7H^+ \text{ (aq)} + 6e^- \rightleftharpoons S° \text{ (s)} + 4H_2O \qquad (10.66)$$

$$\Delta G° = \Delta G_f° (S) + 4\Delta G_f° (H_2O) - \Delta G_f° (HSO_4^-) - 7\Delta G_f° (H^+) - 6\Delta G_f° (e^-)$$

$$= 0 + 4(-237.18) - (-755.99) - 0 - 0$$

$$= -192.73 \text{ kJ} = -192\,730 \text{ J}$$

$$pE° = \frac{-\Delta G°}{2.303nRT} = \frac{+192\,730}{2.303 \times 6 \times 8.314 \times 298.2} = 5.626$$

$$pE = pE° - \frac{1}{6} \log \frac{1}{[HSO_4^-][a_{H^+}]^7}$$

At the boundary, $[HSO_4^-] = 10^{-2} \text{ mol L}^{-1}$ so

$$pE = 5.626 - \frac{7}{6} \log \frac{1}{a_{H^+}} - \frac{1}{6} \log \frac{1}{10^{-2}}$$

$$= 5.626 - \frac{7}{6} pH - 0.333$$

$$= 5.293 - 1.167\,pH$$

This is line 'b' on Fig. 10.5. Above it is the domain of $HSO_4^-$; below it is $S°$.

### The $SO_4^{2-}$ / $S°$ boundary

$$SO_4^{2-} \text{ (aq)} + 8H^+ \text{ (aq)} + 6e^- \rightleftharpoons S° \text{ (s)} + 4H_2O \qquad (10.67)$$

$$\Delta G° = \Delta G_f° (S) + 4\Delta G_f° (H_2O) - \Delta G_f° (SO_4^{2-}) - 8\Delta G_f° (H^+) - 6\Delta G_f° (e^-)$$

$$= 0 + 4(-237.18) - (-744.60) - 0 - 0$$

$$= -204.12 \text{ kJ} = -204\,120 \text{ J}$$

$$pE° = \frac{-\Delta G°}{2.303nRT} = \frac{+204\,120}{2.303 \times 6 \times 8.314 \times 298.2} = 5.958$$

$$pE = pE° - \frac{1}{6} \log \frac{1}{[SO_4^{2-}](a_{H^+})^8}$$

At the boundary, $[SO_4^{2-}] = 10^{-2}$ mol $L^{-1}$ so

$$pE = 5.958 - \frac{8}{6} \log \frac{1}{a_{H^+}} - \frac{1}{6} \log \frac{1}{10^{-2}}$$

$$= 5.958 - \frac{8}{6} pH - 0.333$$

$$= 5.625 - 1.333 \, pH$$

This is line 'c' on Fig. 10.5. Above it is the domain of $SO_4^{2-}$; below it $S°$.

## The $S°$ / $H_2S$ boundary

$$S \, (s) + 2H^+ \, (aq) + 2e^- \rightleftharpoons H_2S \, (aq) \tag{10.68}$$

$$\Delta G° = \Delta G_f° (H_2S) - \Delta G_f° (S) - 2\Delta G_f° (H^+) - 2\Delta G_f° (e^-)$$

$$= -27.86 - 0 - 0 - 0$$

$$= -27.86 \text{ kJ} = -27 \, 860 \text{ J}$$

$$pE° = \frac{-\Delta G°}{2.303nRT} = \frac{+27 \, 860}{2.303 \times 2 \times 8.314 \times 298.2} = 2.440$$

$$pE = pE° - \frac{1}{2} \log \frac{[H_2S]}{(a_{H^+})^2}$$

At the boundary, $[H_2S] = 10^{-2}$ mol $L^{-1}$ so

$$pE = 2.440 + 1 - pH$$

$$= 3.440 - pH$$

This is line 'd' on Fig. 10.5. Above it is the domain of $S°$; below it $H_2S$.

## The $SO_4^{2-}$ / $H_2S$ boundary

$$SO_4^{2-} \, (aq) + 10H^+ \, (aq) + 8e^- \rightleftharpoons H_2S \, (aq) + 4H_2O \tag{10.69}$$

$$\Delta G° = \Delta G_f° (H_2S) - 4\Delta G_f° (H_2O) - \Delta G_f° (SO_4^{2-}) - 10\Delta G_f° (H^+) - 8\Delta G_f° (e^-)$$

$$= -27.86 + 4(-237.18) - (-744.60) - 0 - 0$$

$$= -231.98 \text{ kJ} = -231 \, 980 \text{ J}$$

$$pE° = \frac{-\Delta G°}{2.303nRT} = \frac{+231 \, 980}{2.303 \times 8 \times 8.314 \times 298.2} = 5.079$$

$$pE = pE° - \frac{1}{8} \log \frac{[H_2S]}{[SO_4^{2-}](a_{H^+})^{10}}$$

At the boundary, $[H_2S] = [SO_4^{2-}] = 10^{-2}$ mol $L^{-1}$, so

$$pE = 5.079 - \frac{10}{8} \log \frac{1}{a_{H^+}}$$

$$= 5.079 - 1.25 \, pH$$

This is line 'e' on Fig. 10.5 (a very small segment, difficult to distinguish from line 'd'). Above it is the domain of $SO_4^{2-}$; below it $H_2S$.

## The H₂S / HS⁻ boundary

$$H_2S\ (aq) \rightleftharpoons HS^-\ (aq) + H^+\ (aq) \tag{10.70}$$

As for the $HSO_4^-/SO_4^{2-}$ boundary, this is not a redox reaction and therefore the line will be vertical with the protonated species on the left.

$$\Delta G^\circ = \Delta G_f^\circ (HS^-) + \Delta G_f^\circ (H^+) - \Delta G_f^\circ (H_2S)$$

$$= 12.08 + 0 - (-27.86)$$

$$= 39.94\ kJ = 39\ 940\ J$$

$$\log K = \frac{-\Delta G^\circ}{2.303\ RT} = \frac{-39\ 940}{2.303 \times 8.314 \times 298.2} = -6.995$$

$$K = \frac{[HS^-]a_{H^+}}{[H_2S]}$$

At the boundary, $[HS^-] = [H_2S] = 10^{-2}$ mol L⁻¹, so

$$K = a_{H^+}$$

$$\log K = \log a_{H^+} = -pH = -6.995$$

$$pH = 6.995$$

This is line 'f' on Fig. 10.5. To the left is the domain of $H_2S$ and to the right that of $HS^-$.

## The SO₄²⁻ / HS⁻ boundary

$$SO_4^{2-}\ (aq) + 9H^+\ (aq) + 8e^- \rightleftharpoons HS^-\ (aq) + 4H_2O \tag{10.71}$$

$$\Delta G^\circ = \Delta G_f^\circ (HS^-) + 4\Delta G_f^\circ (H_2O) - \Delta G_f^\circ (SO_4^{2-}) - 9\Delta G_f^\circ (H^+)\quad 8\Delta G_f^\circ(e^-)$$

$$= 12.08 + 4(-237.18) - (-744.60) - 0 - 0$$

$$= -192.04\ kJ = -192\ 040\ J$$

$$pE^\circ - \frac{-\Delta G^\circ}{2.303nRT} - \frac{+192\ 040}{2.303 \times 8 \times 8.314 \times 298.2} - 4.204$$

$$pE = pE^\circ = \frac{1}{8} \log \frac{[HS^-]}{[SO_4^{2-}](a_{H^+})^9}$$

At the boundary, $[HS^-] = [SO_4^{2-}] = 10^{-2}$ mol L⁻¹, so

$$pE = 4.202 - \frac{9}{8} \log \frac{1}{a_{H^+}}$$

$$= 4.204 - 1.25\ pH$$

This is line 'g' on Fig. 10.5. Above it is the domain of $SO_4^{2-}$; below it $HS^-$.

## The completed diagram

The completed diagram is shown in Fig. 10.5. To interpret the plot, it is helpful to make use of a template (Fig. 10.6) that contains the $H_2O$ stability boundaries and, within these, *approximate* p*E* / pH regions for a variety of environmental regimes.

It is evident from the template and also noteworthy that most environmental regimes are located in one of two well-separated clusters. One set are those associated with well-oxygenated environments and these run across the upper boundary (high p*E*) of water stability. Rainfall, fast-flowing rivers and lakes that are subject to vigorous wave action are in this category, with pH defining the individual environments within the cluster. Near the bottom of the stability region (low p*E*) are environments where oxygen is not present, usually as a result of having been consumed by

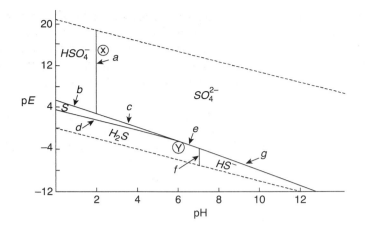

**Fig. 10.5** The $pE$ / pH diagram for the aqueous sulphur system.

materials in the water. Most of these anaerobic situations are associated with waters containing abundant decomposable organic matter, either of natural origin or added as effluent wastes. This would include swamps and bogs and submerged soils containing abundant organic matter. The broad region in the middle of the diagram is usually a region of non-equilibrium states that exist only for a short period of time. An example would be where sewage is discharged into a river. Initially, oxidation of the large flux of organic material uses up most of the dissolved oxygen. Once oxidation is complete, by contact with the atmosphere above, the river slowly accumulates oxygen again and reverts to its aerobic state. Only for a brief time, as it moved from anaerobic to aerobic conditions, would the $pE$ value pass through the intermediate region.

## Application of $pE$ / pH diagrams
Using the sulphur diagram, we can look at two specific environmental situations.

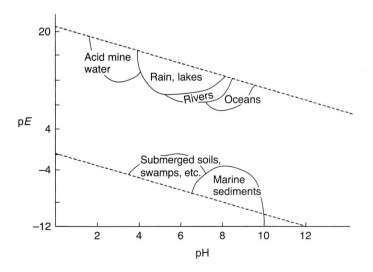

**Fig. 10.6** Template for use with $pE$ / pH plots. The indicated areas show typical $pE$ and pH values for commonly encountered aqueous, soil, and sediment environments.

- For mine wastes, at a pH of 2.5 and exposed to the atmosphere, so that they are well aerated (pE ~ 15), corresponding to point X on Fig. 10.5, the most important sulphur species in solution would be sulphate. Where the originally mined material was sulphide ore, such as from the copper–nickel ores of Sudbury, Ontario, Canada, sulphur might initially go into solution as sulphide but in oxygenated surface water would eventually (the kinetics are relatively fast) be oxidized to sulphate.

- For a swamp or paddy (rice) field where soil containing a high content of organic matter (OM) is submerged, the OM acts as a reducing agent and creates a low-pE condition. For example, a submerged soil might have pH = 6 and pE = −3, corresponding to point Y in the figure. This is near several boundaries, but it would not be surprising to detect the presence of $H_2S$ in the interstitial water of the sediment.

Additional important pE / pH diagrams will be given in the problems section of this chapter and others are discussed at later points in the book.

---

**Main point 10.4** Two variable diagrams indicate regions on a two-dimensional plot where individual species predominate. No detailed information about concentrations within the domains is given. For environmental situations, pH and pE are two key properties that define the nature of species in an aqueous environment. The pE / pH diagram is therefore a common form of the two-variable diagram.

---

## 10.3 Measurements of pE

In principle, it should be simple to measure pE in a real environment. The requirement for measurement is an indicator electrode that is inert and develops a potential in response to the ratio of redox couples that are in true equilibrium. A platinum electrode of small size can serve this purpose. In addition, a reference electrode is used so that the indicated potential may be measured against a known value. As with all potential measurements it is essential to ensure that negligible current be drawn during the measurement process. This is readily ensured by using an electronic voltmeter. Figure 10.7 shows an apparatus suitable for determining pE. The entire setup is similar to that required to take pH readings, except that the platinum redox indicator electrode replaces the glass pH electrode. The electrodes can be immersed in water or in wet soil or sediment. While the measurement process appears to be simple, in practice, non equilibrium conditions make stable and meaningful readings exceedingly difficult.

---

### Example 10.1 Calculation of pE from a measured potential

Suppose the potential in soil pore water is measured in the field and the value is found to be +713 mV versus the saturated calomel electrode (SCE).

The value of E is recalculated versus the normal hydrogen electrode (NHE), using the known potential, +0.242 V, of the SCE.

$$E \text{ versus NHE} = 0.713 + 0.242 \text{ V}$$
$$= 0.955 \text{ V}$$

The value of pE is then readily calculated using eqn 10.44

$$pE = \frac{0.955}{0.0591} = 16.2$$

A pE of 16.2 indicates an oxidizing regime.

---

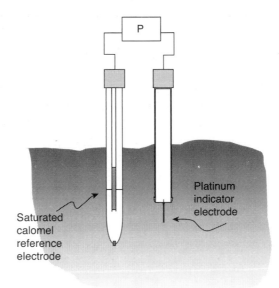

**Fig. 10.7** Apparatus for measuring p$E$ in water, soil, or sediment in the laboratory or in the field. The indicator electrode is often a platinum disc or wire; any appropriate reference electrode can be used. P, the potential measuring device, is usually an electronic voltmeter.

As has been suggested, non-equilibrium conditions frequently obtain in water, soil, and sediments. Furthermore, a typical environmental sample will be comprised of many species that are part of several different redox couples and that are themselves not in equilibrium. The result will be an unstable, drifting potential. For this and other reasons it is frequently not possible to make accurate and precise p$E$ readings as can be done for pH. It is, however, usually possible to obtain an approximate value and define a regime as generally in the oxidizing or reducing category. As the template indicates, an intermediate redox status is rare and usually transient as the system moves to a more stable higher or lower p$E$ value.

> **Main point 10.5** p$E$ is determined by the potential generated at an inert electrode that is immersed in the aqueous system. Such readings may be unstable due to the non-equilibrium conditions that frequently exist.

## ADDITIONAL RESOURCES

1. Butler, J.N. with Cogley, D.R., *Ionic equilibrium: solubility and pH calculations*, John Wiley and Sons, New York, 1998.

2. Martell, A.E. and R.D. Hancock, *Metal complexes in aqueous solutions*, Plenum Press, Inc., New York; 1996.

3. Brookins, D.G., *Eh–pH diagrams for geochemistry*, Springer-Verlag, New York; 1988.

## PROBLEMS

1. Consider the case of sulphurous acid ($H_2SO_3$), formed when sulphur dioxide gas dissolves in water.

   (a) Without calculation, but referring to the appropriate dissociation constants, carefully sketch a single variable species distribution diagram (versus pH) for this acid in water.

   (b) Calculate accurately the concentration of all the sulphite species at pH = 7.0.

2. Set up a spreadsheet calculation to examine the acid–base speciation of arsenate over the pH range 0 to 14. What is the % distribution of species in acid mine drainage whose pH is 4.9?

3. Use standard thermodynamic relations and assumptions to show that

$$n(pE^\circ) = \log K_{eq}$$

4. Consider the equilibrium of an aqueous solution containing solid manganese dioxide as indicated in the following equation.

$$MnO_4^- \text{ (aq)} + 4H_3O^+ \text{ (aq)} + 3e^- \rightleftharpoons MnO_2 \text{ (s)} + 6H_2O$$

   How are the solution's pH and pE affected by the individual addition of each of the following:

$$MnO_2, \quad CO_2 \text{ (g)}, \quad Fe^{2+} \text{ (aq)}, \quad \text{and } CaCO_3?$$

5. A potential measurement in a paddy field yields a reading of −278 mV versus the saturated calomel reference electrode. What is the pE value for this soil?

6. Iron toxicity has occasionally been observed in rice plants grown in lowlands (submerged soils). Use the simple pE / pH diagram for iron in Fig. 10.P.1 to explain how this can happen.

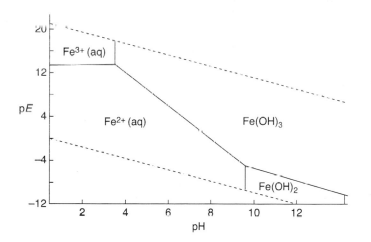

**Fig. 10.P.1** The pE / pH diagram for iron.

7. Iron (III) hydroxide can act as an oxidizing agent as indicated by the half-reaction

$$Fe(OH)_3 + 3H_3O^+ \text{ (aq)} + e^- \rightleftharpoons Fe^{2+} \text{ (aq)} + 6H_2O$$

   (a) Assuming pH = 7 and otherwise standard conditions, can the oxidation of $NH_4^+$ to $NO_3^-$ be effected by this reaction in the hydrosphere?

(b) Could the reaction bring about the oxidation of $HS^-$ to $SO_4^{2-}$ at a pH of 9 (again, under otherwise standard conditions)?

8. Consider the following aqueous uranium species

$$UO_2^{2+} \rightleftharpoons U^{4+} \rightleftharpoons UOH^{3+}$$

Calculate the $UOH^{3+}$ concentration under typical pE and pH conditions for acid mine drainage waters assuming $C_U = 1 \times 10^{-5}$ mol $L^{-1}$. $\Delta G_f^\circ$ values / kJ mol$^{-1}$ are:

| | |
|---|---|
| $UO_2^{2+}$ | $-989.5$ |
| $UOH^{3+}$ | $-810.0$ |
| $U^{4+}$ | $-579.3$ |

9. Figure 10.P.2 is a partially completed pE / pH diagram for arsenic species.
   (a) Calculate the equation for the line (shown) for the boundary $H_2AsO_4^-$ (aq) / $H_3AsO_3$ (aq).
   (b) Calculate equations for the lines (not shown) for the boundaries $H_3AsO_4$ / $H_2AsO_4^-$ and $H_3AsO_4$ / $H_3AsO_3$.

   Assume a temperature of 25°C. For boundaries use $C_{species} = 1 \times 10^{-4}$ mol $L^{-1}$.

   (c) Comment on the predicted As species in seawater and acid mine drainage. In both situations assume that the solution is well aerated. $\Delta G_f^\circ$ values / kJ mol$^{-1}$ are:

| | |
|---|---|
| $H_3AsO_4$ | $-769.3$ |
| $H_2AsO_4^-$ | $-748.8$ |
| $H_3AsO_3$ | $-640.0$ |
| $H_2AsO_3^-$ | $-587.7$ |
| $H_2O$ | $-237.2$ |

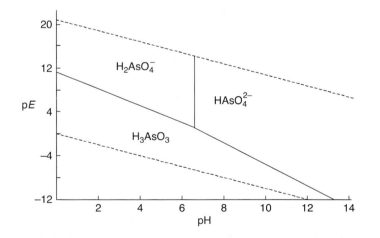

**Fig. 10.P.2** The pE / pH diagram for some arsenic species.

# Chapter 11
# Gases in water

The focus of this chapter is on how gases distribute themselves between air and water and how the properties of water change accordingly. Here we will discuss:

- Henry's law for 'simple gases' such as oxygen or nitrogen—gases that do not react with water
- the specific case of carbon dioxide, a gas that reacts with water and thus creates a more complex equilibrium system
- how to define alkalinity and acid neutralization capacity and their relation to environmental issues.

In the first section of the book we examined some of the ways in which the nature and concentrations of gases in the atmosphere control critical environmental phenomena—not only outdoor- and indoor-air quality, but also radiation at the Earth's surface and, through this, global climate. We also saw that gases, by dissolving in water droplets, can change the composition of rain and other forms of precipitation. This latter example of air chemistry affecting water chemistry is typical of the kinds of processes that connect together the compartments of the environment. It underlines the importance in environmental chemistry of keeping the broad picture in mind, even when we are focusing on some limited component.

This chapter will investigate other ways in which gases distribute themselves between the atmosphere and the hydrosphere. All gases (like oxygen or carbon dioxide) and other volatile compounds whose vapours may be present at low concentration in the atmosphere (like those from gasoline) distribute themselves between air and water. A quantitative description of the distribution depends on properties of the substance itself, such as its vapour pressure, solubility, and ability to react with water and other components in the hydrosphere. The distribution also depends on the properties of the air / water environment, such as atmospheric conditions, and the aqueous solution composition.

One frequently encountered situation is the need to describe the extent to which a gas that is present in a steady-state mixing ratio in the atmosphere dissolves in natural waters. In many such cases it can be assumed that there is an equilibrium between the two phases. A second and opposite type of situation is to describe the extent to which neutral molecules in water volatilize or move into the atmosphere. If the atmosphere is enclosed and limited in volume, an equilibrium calculation may suffice here as well. More commonly, volatilization takes place into the open atmosphere, for example, when an organic pesticide dissolved in a water body evaporates to some degree into the air on a warm and windy day. In these situations, equilibrium is never attained and the rate of volatilization is markedly influenced by air and water movement and temperature, along with the other factors noted above. In this chapter, we will be examining only the equilibrium type of situation.

## 11.1  Simple gases

### Henry's law describes the equilibrium relation between gases in air and water

In the present context, we define a simple gas as one like oxygen or nitrogen, which dissolve in water in their usual form as $O_2$ or $N_2$, respectively, and do not undergo any further chemical reaction with the water. At equilibrium, the concentration of a simple gas is described by a Henry's law relationship. Henry's law is based on Raoult's law and is valid when there are small (typically millimolar or less) concentrations in the water. In its fundamental form, it is written as

$$P_g = KX_1 \tag{11.1}$$

where $P_g$ is the partial pressure of the gas in the bulk atmosphere / Pa, $K$ is the Henry's law constant / Pa ($K$ is a function of the temperature, the gas, and the solvent; in the environmental context, the solvent is always water), and $X_1$ is the equilibrium mole fraction (a dimensionless number) of the solute in the bulk liquid (aqueous) phase.

For our purpose, a more useful (and sufficiently accurate at low concentrations in environmental samples) form uses a reciprocal version of eqn 11.1 and employs molar units for the aqueous solution concentration.

$$[G]_l = K_H P_g \tag{11.2}$$

where $[G]_l$ is the equilibrium concentration of solute in the bulk liquid (aqueous) phase / mol $L^{-1}$, $K_H$ is the Henry's law constant / mol $L^{-1}$ $Pa^{-1}$, and $P_g$ is, as in eqn 11.1, the partial pressure of the gas in the bulk atmosphere / Pa.

When referring to other data sources where Henry's law is applied to environmental calculations, it is important to be aware of the particular algebraic form of the equation employed as well as the units used to express concentration in the two phases (see Problem 11.6). These can always be determined by careful examination of the units for the Henry's law constant. Values of $K_H$, as defined in eqn 11.2, for various gases are given in Table 11.1.

**Table 11.1**  Henry's law constants for selected gases dissolved in water at 25°C.[a]

| Gas | $K_H$ / mol $L^{-1}$ $Pa^{-1}$ |
|---|---|
| $O_2$ | $1.3 \times 10^{-8}$ |
| $N_2$ | $6.4 \times 10^{-9}$ |
| $CH_4$ | $1.4 \times 10^{-8}$ |
| $CO_2$ | $3.3 \times 10^{-7}$ |
| $SO_2$ | $1.2 \times 10^{-5}$ |
| $H_2S$ | $9.9 \times 10^{-7}$ |
| $NH_3$ | $5.9 \times 10^{-4}$ |
| $Hg$ | $9.2 \times 10^{-7}$ |
| $Cl_2$ | $8.9 \times 10^{-7}$ |
| $C_6H_6$ (benzene) | $1.8 \times 10^{-6}$ |

[a] These are consensus values taken from an extensive listing: Sander, R. *Compilation of Henry's Law Constants for inorganic and organic species of potential importance in environmental chemistry*. http://www.mpch-mainz.mpg.de/~sander/res/henry.html accessed November 2009.

## The concentration of oxygen in natural waters

A very important environmental parameter is the concentration of oxygen in water that is in close contact with the surrounding atmosphere. We can calculate its equilibrium value at 25°C using eqn 11.2 and the value of $K_H$ from Table 11.1. In order to use the equation, however, the partial pressure of oxygen in the actual *moist* atmosphere must be calculated. To do this, first consult Table 8.1, which gives mixing ratios as a percentage for important gases in the atmosphere. The mixing ratio, 20.9% is equivalent to a fractional mixing ratio of 0.209 and, based on the definition of mixing ratio; this is the mole fraction of oxygen *in a dry atmosphere*. However, because the atmosphere is in contact with water, water vapour will be present. As a first estimate, we assume that the atmospheric mixing ratio of water can be obtained from the vapour-pressure curve shown in Fig. 8.1. If we subtract the partial pressure of water from the total pressure, the difference is the total pressure ($P_{dry}$) of the dry components of the atmosphere. Multiplying $P_{dry}$ by the mixing ratio of oxygen, we obtain the partial pressure of oxygen in the actual atmosphere.

---

**Example 11.1 The oxygen ($O_2$) concentration in water**

$$P_{O_2} = \frac{\overbrace{P_{dry}}}{\left(P^o - P_{H_2O}\right)} X_{O_2} \tag{11.3}$$

where $P^o$ = atmospheric pressure at 25°C. $P_{H_2O}$ at 25°C is $3.2 \times 10^3$ Pa.

$$P_{O_2} = (1.01 \times 10^5 - 3.2 \times 10^3)\text{Pa} \times 0.209$$
$$= 2.04 \times 10^4 \text{ Pa}$$

Using eqn 11.2,

$$[O_2]_{aq} = 1.3 \times 10^{-8} \text{ mol L}^{-1} \text{ Pa}^{-1} \times 2.04 \times 10^4 \text{ Pa}$$
$$= 2.7 \times 10^{-4} \text{ mol L}^{-1}$$
$$= 8.5 \text{ mg L}^{-1}$$

---

This concentration of 8.5 mg L$^{-1}$ or 8.5 ppm at 25°C is characteristic of well-oxygenated water such as surface water in a clean lake or in a fast-flowing river. These are the types of aqueous environments that appear on the upper parts of pE / pH diagrams. Low pE values occur when the oxygen has been consumed. In Chapter 15 we will see that this occurs most commonly in waters containing large amounts of degradable organic matter. Many natural and anthropogenic-affected environments can lead to this situation, for example, in a swamp or in a water body where untreated organic waste has been discharged. In such situations the concentration of oxygen will be less that 8.5 mg L$^{-1}$.

As in this example, in other cases where the mixing ratio of a gas is given, it is important to remember that the ratio applies to a dry atmosphere. For accurate calculations, a correction for partial pressure of water, such as we made here, can be done. Frequently, however, because the correction is small (about 3% in our example), it is neglected.

For most gases in water, the Henry's law constant increases with decreasing temperature and therefore the solubility of gases like oxygen also increase as the temperature goes down. This implies that oxygen volatilization from water is endothermic and dissolution is exothermic; that, of course, is the case. At 5°C oxygen solubility is approximately 12.4 mg L$^{-1}$, while at 30°C it is 7.5 mg L$^{-1}$. The solubility also depends on the ionic strength of the solution and is somewhat lower under high-salinity (sea water) conditions.

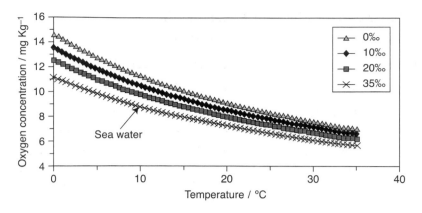

**Fig. 11.1** Variation of oxygen concentration in water as a function of temperature and salinity. Salinity / % NaCl △ 0 ◆ 1 ■ 2 × 3.5.

## Molecular oxygen in sea water

An important species that is dissolved in sea water but not listed in Table 9.1 is molecular oxygen. Typical concentrations range from 5.7 to just over 14.5 mg kg$^{-1}$ depending on both temperature and salinity (salinity was defined in Chapter 10). The lower value represents a condition of 35‰ salinity and 35°C, while the higher value represents a condition approaching 0‰ salinity and 0°C. Clearly, it will be estuarine environments, where salinity changes dramatically, that will have the greatest variations of oxygen concentrations. In the open ocean, where we may consider the salinity to be constant, at near 35‰, the temperature alone will determine the extent of oxygen solubility. The oxygen concentration in the open ocean approaches 11 mg kg$^{-1}$ at 0°C (Fig. 11.1). Because of the salinity of sea water, however, the Henry's law calculation presented above is not fully adequate to accurately determine the oxygen concentration. An equation developed by Benson and Krause[1] provides a way for calculating the solubility of oxygen in sea water.

$$\ln C_s = A + B/T + C/T^2 + D/T^3 + E/T^4 - S \times (F + G/T + H/T^2)$$

where the units of $C_s$ are $\mu$mol kg$^{-1}$, $T$ is temperature in Kelvin, $S$ is salinity in ‰, and the constants are: $A = -1.3529996 \times 10^2$; $B = 1.572288 \times 10^5$; $C = -6.637149 \times 10^7$; $D = 1.243678 \times 10^{10}$; $E = -8.621061 \times 10^{11}$; $F = 2.0573 \times 10^{-2}$; $G = -1.2142 \times 10^1$; and $H = 2.3631 \times 10^3$.

## Henry's law used for 'reverse' calculations

Henry's law calculations may also be used in a reverse fashion to determine the vapour pressure of a gas above water that contains a volatile chemical. Consider an enclosed but partially filled container with 100 mL of an aqueous solution containing 0.5 g of acetone.

---

**Example 11.2 Vapour pressure of a dissolved compound in a closed system**
The acetone concentration is

$$0.5\text{g} / (58.1 \text{ g mol}^{-1} \times 0.1 \text{ L}) = 0.0861 \text{ mol L}^{-1}$$

---

[1] Benson, B.B, and Krause, D., Jr., The concentration and isotopic fractionation of oxygen dissolved in freshwater and seawater in equilibrium with the atmosphere. *Limnol. Oceanogr.*, **29** (1984), 620–32.

The value of $K_H$ for acetone is $3.9 \times 10^{-3}$ mol L$^{-1}$ Pa$^{-1}$.

$$[Ac]_{aq} = K_H P_{ac} \tag{11.4}$$

The equilibrium vapour pressure of acetone above the solution is then

$$P_{ac} = [Ac]_{aq} / K_H = 0.0861 \text{ mol L}^{-1} / 3.9 \times 10^{-3} \text{ mol L}^{-1} \text{Pa}^{-1}$$

$$= 22 \text{ Pa}$$

---

As we noted above, this 'reverse' type of Henry's law calculation assumes an equilibrium situation and cannot be used where evaporation occurs into the open constantly replenished atmosphere. There are other features that may also limit the establishment of equilibrium. For some gases, such as oxygen, nitrogen, carbon dioxide, methane, and nitrous oxide, diffusion through the liquid surface film in contact with the surrounding atmosphere is slow and controls the rate of transfer from the atmosphere to water, while diffusion in the gas phase controls the reaction rate in the case of sulphur dioxide, sulphur trioxide, and hydrogen chloride. In all cases, the rate of establishment of equilibrium depends on the degree of contact between the two phases.

The following sample problem is a case where the reverse calculation has some validity because of the enclosed nature of the atmosphere.

---

### Example 11.3 TCE vapour pressure in a subsurface soil

TCE (trichloroethylene, CHCl $=$ CCl$_2$) is an important solvent widely used for degreasing parts in the electrical and electronic industries. Careless disposal onto land has resulted in many contaminated groundwater sources around the world.[2] For TCE, the Henry's law constant, log $K_H$, at 25°C is reported[3] to be 1.03 (units of $K_H$ are L atm mol$^{-1}$). For groundwater containing 450 ppm TCE, calculate the partial pressure (Pa) of TCE in adjacent air-filled pores in the soil.

First, convert the value of $K_H$ (with units of L atm mol$^{-1}$) to $K_H$ (with units of Pa L mol$^{-1}$).

$$\log K_H = 1.03, \text{ so then } K_H = 10.7 \text{ L atm mol}^{-1}$$

$$K_H = 1 / (101\,325 \text{ Pa atm}^{-1} \times 10.7 \text{ L atm mol}^{-1}) = 9.26 \times 10^{-7} \text{mol L}^{-1} \text{ Pa}^{-1}$$

Then, convert the aqueous concentration in ppm to mol L$^{-1}$.

The molar mass of TCE is 131.4 g mol$^{-1}$,
so, 450 ppm $= 450$ mg L$^{-1} = 3.4$ mmol L$^{-1} = 3.4 \times 10^{-3}$ mol L$^{-1}$
Now, use eqn 11.2 to calculate the partial pressure

$$[G]_l = K_H P_g \quad \text{or} \quad P_g = [G]_l / K_H, \text{ and}$$

$$P_{TCE} = 3.4 \times 10^{-3} \text{ mol L}^{-1} / 9.26 \times 10^{-7} \text{ mol L}^{-1} \text{Pa}^{-1} = 3700 \text{ Pa}$$

The equilibrium argument is legitimate in this case, because soil atmospheres at depth are very stable and air exchange with the open atmosphere above is extremely slow.

---

**Main point 11.1** The equilibrium relations between gaseous and water-soluble species of volatile compounds are described by Henry's law. For simple species, i.e. those that do not react with water, this law is sufficient to evaluate the equilibrium distribution. Many non-ionic organic compounds are in the simple species category.

---

[2] TCE is the major groundwater contaminant in sites around the United States. For a description of the Superfund Program for clean-up of these sites see http://www.epa.gov/superfund/, accessed November 2009.
[3] Schwarzenbach, R.P., P.M. Gschwend, and D.M. Imboden, *Environmental organic chemistry*, John Wiley and Sons, Inc., New York; 1993.

## 11.2 Gases that react with water

### Carbon dioxide in water

When gases like oxygen dissolve in water, their molecular form does not change. In that sense they are 'simple' and a straightforward Henry's law calculation can be used to calculate their equilibrium partitioning between air and water. Many gases of environmental importance, including sulphur dioxide and ammonia, however, react with water as they go into solution and this can increase their solubility to a far greater extent than would be estimated by the fundamental Henry's law calculation. The influence of these reactions must therefore be taken into account in calculating solubilities. Perhaps the most important environmental example of a reacting gas is carbon dioxide; when it dissolves it undergoes reactions so that four different species can be present (eqn 11.5). As a group, these are called 'carbonate species'. Calculating the total solubility of carbon dioxide and individual concentrations of all carbonate species requires modification of the simple Henry's-law calculation.

To do the calculation, it is necessary to look at the sequence of reactions, as shown in eqn 11.5.

$$\begin{array}{cccc}
\text{(a)} & \text{(b)} & \text{(c)} & \text{(d)} \\
K_H = 3.3 \times 10^{-7} & K_r = 2.0 \times 10^{-3} & K_{a1}' = 2.0 \times 10^{-4} & K_{a2} = 4.7 \times 10^{-11}
\end{array}$$

$$CO_2\,(g) \rightleftharpoons CO_2\,(aq) \rightleftharpoons H_2CO_3\,(aq) \rightleftharpoons HCO_3^-\,(aq) \rightleftharpoons CO_3^{2-}\,(aq)$$

$$K_{a1} = K_{a1}' \times K_r = 4.5 \times 10^{-7}$$

$$(11.5)$$

- Step (a) describes the equilibrium between atmospheric $CO_2$ and dissolved $CO_2$ described by Henry's law.
- Step (b) is the equilibrium between dissolved $CO_2$ and its aquated form, $H_2CO_3$. The extent of this equilibrium is described by $K_r = [H_2CO_3] / [CO_2\,(aq)]$ that has a value of $2.0 \times 10^{-3}$.
- Step (c) describes the loss of the first proton of $H_2CO_3$.

The small value of the equilibrium constant for step (b) indicates that most of the undissociated acid is actually present as $CO_2$ (aq). For this reason, when we describe the way in which carbon dioxide in water acts as an acid, we usually use $K_{a1}$, which combines steps (b) and (c), as expressed by $K_r$ and the constant $K_{a1}'$ for the loss of the first proton of $H_2CO_3$ and write the reaction as

$$CO_2\,(aq) + 2H_2O \rightleftharpoons HCO_3^-\,(aq) + H_3O^+\,(aq)$$

- Step (d) is for the loss of the second proton of carbonic acid. This completes the sequence of steps relating the gaseous and aqueous forms of carbon dioxide as was shown above.

### Calculating the solubility of carbon dioxide and related species

The calculation of carbon dioxide solubility then proceeds as follows. In the moist atmosphere where its mixing ratio is 387 ppmv, the partial pressure (25°C) of carbon dioxide is

$$P_{CO_2} = (1.01 \times 10^5 - 3.2 \times 10^3) \times 387 \times 10^{-6}\ \text{Pa} = 37.8\ \text{Pa}$$

Applying eqn 11.2, the concentration of $CO_2$ (aq) is then calculated to be

$$[G]_l = [CO_2\,(aq)] = 3.3 \times 10^{-7}\ \text{mol L}^{-1}\ \text{Pa}^{-1} \times 37.8\ \text{Pa} = 1.2 \times 10^{-5}\ \text{mol L}^{-1}$$

The other aqueous carbonate species are determined using the standard acid–base relations. In these and other calculations, the square brackets refer to aqueous molar concentrations, an approximation for activities as described in Chapter 10.

$$[H_2CO_3 (aq)] = K_r \times [CO_2 (aq)] = 2.4 \times 10^{-8} \text{ mol L}^{-1} \tag{11.6}$$

$$K_{a1} = \frac{[HCO_3^-][H_3O^+]}{[CO_2]} = 4.5 \times 10^{-7} \tag{11.7}$$

If, besides carbon dioxide, there are no additional sources of acid or carbonate species,

$$[HCO_3^-] - [H_3O^+] - ([CO_2 (aq)] \times K_{a1})^{1/2} = 2.3 \times 10^{-6} \text{ mol L}^{-1} \tag{11.8}$$

$$[CO_3^{2-}] = \frac{[CO_2 (aq)] \times K_{a1} \times K_{a2}}{[H_3O^+]^2} = 4.8 \times 10^{-11} \text{ mol L}^{-1} \tag{11.9}$$

Since $H_2CO_3$ (aq) is always present in a low concentration compared to $CO_2$ (aq)—0.2% of the latter value as defined by the $K_r$ value—its concentration is usually not calculated. The relative contributions of the $HCO_3^-$ and $CO_3^{2-}$ species depend on the nature of the solution, including its pH. In the present example, the total solubility of all carbonate species is $1.4 \times 10^{-5}$ mol L$^{-1}$ with $CO_2$ (aq) and $HCO_3^-$ (aq) being the quantitatively important forms.

Note that the above calculation indicates that the hydronium ion concentration of water in equilibrium with atmospheric carbon dioxide is $2.3 \times 10^{-6}$ mol L$^{-1}$ (eqn 11.8) corresponding to a pH of 5.64. This is the pH of 'clean' rain or of any other pure water sample in equilibrium with atmospheric carbon dioxide as has been discussed in Chapter 5.

## The influence of carbon dioxide on the solubility of calcium carbonate

In many environmental situations, water is exposed to atmospheric carbon dioxide and is also in contact with limestone ($CaCO_3$) or other carbonate rocks. If the system is at equilibrium, the concentrations of all important species in the water are calculated as follows.

$$K_H = \frac{[CO_2]}{P_{CO_2}} = 3.3 \times 10^{-7} \tag{11.10}$$

$$K_{a1} = \frac{[HCO_3^-][H_3O^+]}{[CO_2]} = 4.5 \times 10^{-7} \tag{11.11}$$

$$K_{a2} = \frac{[CO_3^{2-}][H_3O^+]}{[HCO_3^-]} = 4.7 \times 10^{-11} \tag{11.12}$$

$$K_{sp} = [Ca^{2+}][CO_3^{2-}] = 5 \times 10^{-9} \tag{11.13}$$

In any solution, the total ionic positive and negative charge must be equal. In the present case, this charge balance is described by eqn 11.14.

$$2[Ca^{2+}] + [H_3O^+] = [HCO_3^-] + 2[CO_3^{2-}] + [OH^-] \tag{11.14}$$

In the pH range 6–9

$$[H_3O^+] \ll [Ca^{2+}]$$

and, from Fig. 1.2, both

$$2[CO_3^{2-}] \text{ and } [OH^-] \ll [HCO_3^-]$$

Therefore, eqn 11.14 simplifies to

$$2[Ca^{2+}] = [HCO_3^-] \tag{11.15}$$

From eqns 11.10, 11.11, and 11.12,

$$[CO_3^{2-}] = \frac{K_{a1} \times K_{a2} \times K_H \times P_{CO_2}}{[H_3O^+]^2} \qquad (11.16)$$

From eqns 11.13 and 11.16,

$$[Ca^{2+}] = \frac{K_{sp}[H_3O^+]^2}{K_{a1} \times K_{a2} \times K_H \times P_{CO_2}} \qquad (11.17)$$

From eqns 11.10, 11.11, 11.15, and 11.17,

$$[HCO_3^-] = \frac{P_{CO_2} K_H K_{a1}}{[H_3O^+]} = \frac{2K_{sp}[H_3O^+]^2}{K_{a1} \times K_{a2} \times K_H \times P_{CO_2}} \qquad (11.18)$$

From eqn 11.18

$$[H_3O^+]^3 = \frac{K_H^2 K_{a1}^2 K_{a2} P_{CO_2}^2}{2K_{sp}} \qquad (11.19)$$

We showed above that the atmospheric partial pressure of carbon dioxide is 37.8 Pa and the equilibrium aqueous concentration is $1.2 \times 10^{-5}$ mol L$^{-1}$.

From eqn 11.19,

$$[H_3O^+] = \left( \frac{(3.3 \times 10^{-7})^2 \times (4.5 \times 10^{-7})^2 \times 4.7 \times 10^{-11} \times (37.8)^2}{2 \times 5 \times 10^{-9}} \right)^{1/3}$$

$$= 5.3 \times 10^{-9} \text{ mol L}^{-1}$$

$$pH = -\log[H_3O^+] = 8.28$$

From eqn 11.18,

$$[HCO_3^-] = \frac{3.3 \times 10^{-7} \times 4.5 \times 10^{-7} \times 37.8}{5.3 \times 10^{-9}} = 1.06 \times 10^{-3} \text{ mol L}^{-1}$$

From eqn 11.16,

$$[CO_3^{2-}] = \frac{4.5 \times 10^{-7} \times 4.7 \times 10^{-11} \times 3.3 \times 10^{-7} \times 37.8}{(5.3 \times 10^{-9})^2}$$

$$= 9.4 \times 10^{-6} \text{ mol L}^{-1}$$

From eqn 11.17,

$$[Ca^{2+}] = \frac{5 \times 10^{-9} \times (5.3 \times 10^{-9})^2}{4.5 \times 10^{-7} \times 4.7 \times 10^{-11} \times 3.3 \times 10^{-7} \times 37.8} \qquad (11.17)$$

$$= 5.3 \times 10^{-4} \text{ mol L}^{-1}$$

The calculations confirm that the two most important species contributing to mass balance are $Ca^{2+}$ and $HCO_3^-$. The pH of the solution is 8.28. Undissociated $CO_2$ and $CO_3^{2-}$ are present only in small concentrations. Again, check that this is consistent with the description of carbonate speciation that was shown in Fig. 1.2.

Where $P_{CO_2}$ levels differ from the atmospheric values, the solubility of calcium carbonate is correspondingly affected. This situation arises when biological respiration releases large amounts of carbon dioxide (as, for example, in soil containing a substantial population of microorganisms), or where photosynthesis takes place decreasing the $P_{CO_2}$ (as, for example, in a water body where algae are actively growing). The plot of $Ca^{2+}$ versus $P_{CO_2}$ given in Fig. 11.2 shows how $P_{CO_2}$ affects the solubility of calcium carbonate.

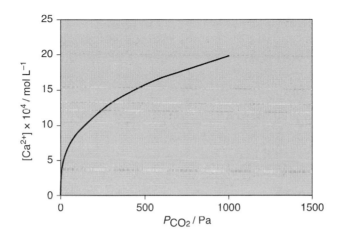

**Fig. 11.2** Aqueous solubility of calcium carbonate as a function of the partial pressure of carbon dioxide in the atmosphere.

## Lake Nyos—an environmental disaster

On 21 August 1986 at 7:30 pm, the community inhabiting a region close to Lake Nyos in the north-western part of Cameroon (Fig. 11.3) was startled by a series of strange rumbling sounds. Around the same time a white cloud mysteriously appeared above the surface of the lake and remained suspended there. Soon after, and without warning, a great plume of water spewed upward from the lake surface. Within minutes the entire population of the vicinity had passed into unconsciousness. Sometime later many people regained consciousness but they found that 1700 others, as well as most of the cattle population, had succumbed and died.

This mysterious and almost unprecedented tragedy has become the subject of intense scientific investigation. Many details related to what happened remain unclear, but it is known there was a massive release of over 240 000 t of carbon dioxide from the lake. So much gas was released that the water level in Lake Nyos dropped by one metre. Being about 1.5 times more dense than air, the carbon dioxide settled in the valley surrounding the lake and this caused the large numbers of people to suffocate from lack of oxygen.

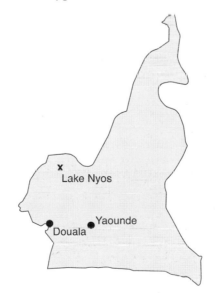

**Fig. 11.3** Location of Lake Nyos (**x**) in Cameroon.

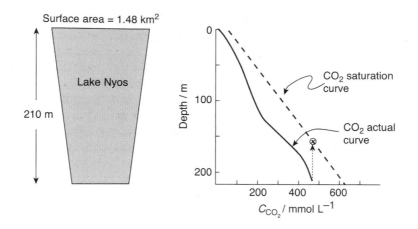

**Fig. 11.4** Cross-section of Lake Nyos shown as a truncated cone with horizontal scale compressed. Graph shows the carbon dioxide concentration profile (solid line) and saturation curve (dashed line).

Lake Nyos lies along a fault line and is of volcanic origin. It has a surface area of 1.48 km² and, with a vertical profile approximating that of a truncated cone, reaches a depth of up to 210 m (Fig. 11.4). Vents under the water allow continuous entry of carbon dioxide at a rate estimated to be 10 m³ min⁻¹. The lake can hold about 1.5 km³ of the gas in solution so that saturation would be reached in just over 20 years.

Opinion is not unanimous regarding the reasons for the carbon dioxide release. One widely accepted theory suggests that a cold rain that had fallen for several days prior to the disaster cooled the surface water increasing its density, causing it to sink and forcing deeper water upwards. The maximum concentration (mmol L⁻¹) of carbon dioxide in the water as a function of $h$, the depth in metres, has been calculated to be[4]

$$[CO_2(aq)] = 40 + 2.9h$$

This is shown by the dashed line in Fig. 11.4.

Therefore, at the original depth of 200 m, the concentration limit is 620 mmol L⁻¹ and the actual concentration, 475 mmol L⁻¹, is less than this saturation value. The depth at which this latter concentration equals the saturation value is 150 m. Therefore, any deep water raised to a level higher than 150 m deep would release carbon dioxide gas. It is believed that the massive overturn of the lake caused an expulsion of the gas along with ejection of water upward and on to the surrounding shore. Soluble iron in the iron (II) form was brought up from the deep reducing regions and oxidized, producing a large (200 m diameter) red spot in the centre of the lake. The red colour was presumably due to precipitated hydrous iron (III) oxide.

$$4Fe(HCO_3)_2(aq) + O_2 + 2H_2O \rightarrow 4Fe(OH)_3 + 8CO_2 \tag{11.20}$$

To prevent a recurrence of this tragic event, large polyethylene pipes have been installed in Lake Nyos so that deeper water continuously siphons to the surface. Gas bubbles out of this supersaturated water and there is therefore an on-going slow release of carbon dioxide.

[4] Nojiri, Y., Gas discharge at Lake Nyos. *Nature*, **346** (1990), 323.

## A freeware program for equilibrium calculations

Earlier in Chapter 10 we determined the species distribution of an aqueous acid at specific pH values and we saw how we can use a spreadsheet to calculate and plot the species distribution over a range of pH. Just above, we determined the solubility of calcium carbonate in water containing dissolved carbon dioxide in equilibrium with carbon dioxide in the atmosphere. The calculation was complex and if we wish to examine the influence of changing partial pressure of carbon dioxide on calcium carbonate solubility, we would have to carry out many such calculations. With *PHREEQC for windows* (see Appendix D.1 for more information) we are able to do this with relative ease, and generate the data necessary to create a diagram such as Fig. 11.2. Shown below is the specific file code for the program that will calculate the solubility of calcite ($CaCO_3$) as affected by the partial pressure of carbon dioxide. This file code is a modified version of the example file, *U_Gex2,* provided with the program and returns the data and the plot of $Ca^{2+}$ concentration (mmol $L^{-1}$) *vs.* $CO_2$ partial pressure (Pa) similar to that shown in Fig. 11.2.

### Example of file code

| | |
|---|---|
| 1 | # *Returns the data and graph of calcite solubility vs. partial pressure of carbon dioxide* |
| 2 | # |
| 3 | SOLUTION 1 |
| 4 | EQUILIBRIUM_PHASES 1 |
| 5 | Calcite |
| 6 | REACTION |
| 7 | CO2 1 |
| 8 | 0.01 in 101 steps |
| 9 | USER_GRAPH |
| 10 | –headings CO2 Ca |
| 11 | –chart_title |
| 12 | –axis_scale x_axis 0 1500 |
| 13 | –axis_scale y_axis 0 2 |
| 14 | –initial_solutions |
| 15 | –axis_titles P_CO2 / Pa Calcium / mM |
| 16 | –start |
| 17 | 10 *graph_x* 10^*si*('CO2(g)')*101325 |
| 18 | 20 *graph_y tot*('Ca')*1e3 |
| 19 | –end |
| 20 | END |
| 21 | □ |

| $CO_2$ / Pa | $Ca^{2+}$ / mmol $L^{-1}$ |
|---|---|
| 1 | 0.16 |
| 3 | 0.23 |
| 8 | 0.32 |
| 18 | 0.41 |
| 33 | 0.50 |
| 55 | 0.59 |
| 82 | 0.68 |
| 116 | 0.77 |
| 157 | 0.85 |
| 205 | 0.93 |
| 260 | 1.01 |
| 320 | 1.09 |
| 388 | 1.17 |
| 461 | 1.24 |
| 541 | 1.31 |
| 626 | 1.38 |
| 716 | 1.45 |
| 812 | 1.52 |
| 912 | 1.58 |
| 1018 | 1.65 |

Problem 19 at the end of this chapter suggests a PHREEQC calculation that allows you to compare the precipitation of calcium carbonate as its two polymorphic forms, calcite and aragonite.

---

**Main point 11.2** For species that take part in reactions in aqueous solution, it is necessary to invoke additional relations along with Henry's law in order to determine the total concentration and species distribution in the water. Gases with acid–base properties, like carbon dioxide, have solubilities that depend on pH.

## 11.3 **Alkalinity**

### Acid neutralizing properties of water

Arising out of knowledge of the chemistry of the carbonate species in water is the concept of alkalinity. *Alkalinity is a measure of the ability of a water body to neutralize acidity* and is very important in predicting the extent of acidification of lakes and rivers. A definition of alkalinity takes into account the important proton-accepting components of most natural waters and is given by eqn 11.21.

$$\text{alkalinity} = \underbrace{[OH^-] + [HCO_3^-] + 2[CO_3^{2-}]}_{\text{proton acceptors}} - \underbrace{[H_3O^+]}_{\text{proton donor}}$$

(11.21)

A related but broader concept is the acid neutralizing capacity (ANC), which takes into account the fact that a wide range of other proton-accepting species may be present in water of differing origins.

$$\text{ANC} = \underbrace{[OH^-] + [HCO_3^-] + 2[CO_3^{2-}] + [B(OH)_4^-] + [H_3SiO_4^-] + [HPO_4^{2-}] + [HS^-] + [NOM^-] + \cdots}_{\text{proton acceptors}} - \underbrace{[H_3O^+] - 3[Al^{3+}] - \cdots}_{\text{proton donors}}$$

(11.22)

In this definition, NOM refers to the natural organic matter in the water column.

The boron, silicon, phosphorus, and sulphur species that are able to contribute to acid-neutralizing capacity are ones whose p$K$ values are in the range of that of natural water.

The usual reactions are

$$B(OH)_4^- + H_3O^+ \rightarrow H_3BO_3 + 2H_2O \tag{11.23}$$

$$H_3SiO_4^- + H_3O^+ \rightarrow Si(OH)_4 + H_2O \tag{11.24}$$

$$HPO_4^{2-} + H_3O^+ \rightarrow H_2PO_4^- + H_2O \tag{11.25}$$

$$HS^- + H_3O^+ \rightarrow H_2S + H_2O \tag{11.26}$$

$$NOM^- + H_3O^+ \rightarrow NOMH + H_2O \tag{11.27}$$

The corresponding p$K_a$ values for $H_3BO_3$, $Si(OH)_4$, $H_2PO_4^-$, and $H_2S$ are 9.14, 9.66, 7.21, and 7.04, respectively. While the values are close enough to the pH of various natural waters, the contribution of these species to neutralization is, in almost all situations, minimal because their concentrations are generally too small to have a significant effect. In many water bodies, then, alkalinity is approximately equal to the ANC, which implies that the only proton-accepting species present in significant concentrations are the carbonate species and / or hydroxyl ions. Because a titration does not distinguish between species, measurements of alkalinity are actually measurements of ANC. Nevertheless, the term alkalinity is often used in the broader sense.

Typical values of alkalinity span a range from less than 50 to greater than 2000 μmol L$^{-1}$. Water bodies found in regions where the bedrock contains limestone are those that have the largest alkalinity.

Among the additional species included in an ANC calculation, the natural organic matter (NOM) concentration is highly variable and sometimes makes an important contribution. While the chemical structure of NOM depends on its source and history, a typical value for proton-accepting sites on NOM is 10 mmol g$^{-1}$. Therefore, a lake containing 10 μg mL$^{-1}$ NOM (and no other proton donors or acceptors) would have an ANC of approximately

$$10 \text{ μg mL}^{-1} \times 10 \text{ mmol g}^{-1} = 100 \text{ μmol L}^{-1}$$

due to NOM alone. A more detailed picture of the nature of NOM and its ability to act as a proton acceptor is given in Chapter 12.

## Alkalinity as a buffer in natural water

How do carbonate species act as a buffer in water systems? To answer this, consider two samples of water, each having pH = 7. The first one is pure H$_2$O and the second has sufficient carbonate species to give an alkalinity of 2000 μmol L$^{-1}$.

If 1.0 mmol L$^{-1}$ of H$_3$O$^+$ in the form of strong acid is added to the pure water, the pH drops to 3.0. If, on the other hand, the same concentration of H$_3$O$^+$ is added to the high-alkalinity water, the situation is more complex.

We first have to be aware of what species are present in the water before the acid is added. Look again at Fig. 1.2. It should now be clear how this figure has been developed. Moving from low to high pH, there are three species, CO$_2$, HCO$_3^-$, and CO$_3^{2-}$, respectively. On the pH axis, the intersection points of the curves occur at the p$K_{a1}$ (= 6.35) and p$K_{a2}$ (= 10.33) values for the carbonate system. The distribution diagram shows that, in the water at pH 7, there is a negligible amount of CO$_3^{2-}$, a small amount of dissolved CO$_2$, and most of the carbonate species are in the form of HCO$_3^-$. The alkalinity must therefore be due essentially to HCO$_3^-$ alone.

When 1.0 mmol L$^{-1}$ of H$_3$O$^+$ is added to 2000 μmol L$^{-1}$ (= 2.0 mmol L$^{-1}$) of HCO$_3^-$, a reaction to produce 1.0 mmol L$^{-1}$ of dissolved CO$_2$ occurs:

$$\text{HCO}_3^- \text{ (aq)} + \text{H}_3\text{O}^+ \text{ (aq)} \rightarrow \text{CO}_2 \text{ (aq)} + 2\text{H}_2\text{O} \tag{11.28}$$

1.0 mmol L$^{-1}$ of HCO$_3^-$ remains. The hydronium ion concentration in the water can then be calculated by rearranging eqn 11.7.

$$[\text{H}_3\text{O}^+] = \frac{K_{a1} \times [\text{CO}_2 \text{ (aq)}]}{[\text{HCO}_3^-]} = \frac{4.5 \times 10^{-1} \times 1 \times 10^{-3}}{1 \times 10^{-3}} = 4.5 \times 10^{-7} \text{ mol L}^{-1}$$

This corresponds to a pH of 6.35.

In the pure water, addition of this large input of acid resulted in a pH drop of 4 units (from pH 7 to 3) corresponding to a 10 000-fold increase in hydronium ion concentration. In contrast, for the water that had considerable alkalinity, the pH drop was 0.65 units meaning that the hydronium ion only increased only by a factor of 4.5. Clearly, alkalinity gives water its buffering ability.

## Alkalinity and pH

To illustrate the difference between pH and alkalinity, consider the two following hypothetical examples involving water and carbonate species alone.

---

### Example 11.4 **Alkalinity of two simple water samples**

Water sample 1 is water with pH = 9, but no carbonate or other dissolved proton donors or acceptors are present. The concentrations of the four species needed to calculate alkalinity are:

$$[\text{H}_3\text{O}^+] = 10^{-9} \text{ mol L}^{-1}$$

$$[\text{OH}^-] = 10^{-5} \text{ mol L}^{-1}$$

$$[\text{HCO}_3^-] = [\text{CO}_3^{2-}] = 0 \text{ mol L}^{-1}$$

Alkalinity can then be calculated using eqn 11.21

$$\text{alkalinity} = [OH^-] + [HCO_3^-] + 2[CO_3^{2-}] - [H_3O^+]$$

$$= 10^{-5} + 0 + 0 - 10^{-9}$$

$$= 10^{-5} \text{ mol L}^{-1}$$

$$= 10 \ \mu\text{mol L}^{-1}$$

Water sample 2 has a slightly lower pH of 8.3 but it contains dissolved $NaHCO_3$ at a concentration of 0.010 mol $L^{-1}$. By consulting Fig. 1.2, you can see that, at pH 8.3, $HCO_3^-$ is the only significant carbonate species present in water. The concentrations of all species contributing to alkalinity are therefore

$$[H_3O^+] = 10^{-8.3} \text{ mol L}^{-1}$$

$$[OH^-] = 10^{-5.7} \text{ mol L}^{-1}$$

$$[HCO_3] = 0.010 \text{ mol L}^{-1}$$

Rearranging and using eqn 11.12,

$$[CO_3^{2-}] = \frac{K_{a2} \times [HCO_3^-]}{[H_3O^+]} = \frac{4.7 \times 10^{-11} \times 0.010}{10^{-8.3}} = 9.4 \times 10^{-5} \text{ mol L}^{-1}$$

The alkalinity of water sample 2 is then

$$\text{alkalinity} = 10^{-5.7} + 0.010 + 2 \times 9.4 \times 10^{-5} - 10^{-8.3}$$

$$= 0.01 \text{ mol L}^{-1} = 10\,000 \ \mu\text{mol L}^{-1}$$

Clearly, then, while water sample 2 has a *slightly lower pH* than water sample 1, it has a *much higher alkalinity*.

---

These hypothetical cases help us to see a fundamental difference between the expression of acid–base properties in terms of pH and alkalinity. The pH can be considered to be an *intensity factor* that measures the concentration of alkali or acid immediately available for reaction. In contrast, the alkalinity is a *capacity factor* that is a measure of the ability of a water sample to sustain reaction with added acids or base. Note also that it is possible to have a negative value of alkalinity. Negative values imply that extensive acidification has already occurred so that no significant amount of proton acceptors remains in the water, but rather there is an excess of proton-donating species.

## Measurement of alkalinity and its environmental significance

Alkalinity is readily measured by titration of a water sample with acid. The endpoint for the carbonate to hydrogen carbonate plus hydroxyl to water part of the titration occurs near pH 8, and such an experiment gives a value for *carbonate alkalinity*. If the titration is continued to pH 4.5, the hydrogen carbonate is further protonated to form aqueous carbon dioxide and titration to this pH yields a value for *total alkalinity*. Actually, these experiments measure the concentration of all species that are titrated up to the particular endpoint pH values. While the method is usually considered as a means of determining alkalinity, it is really a method for determining ANC.

The use of a titration to measure alkalinity has meaning additional to that given by the analytical result. Acidification of a lake in its natural setting is itself analogous to a

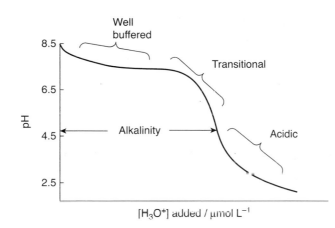

**Fig. 11.5** Lake status in terms of acid–base properties.

macroscale titration and lakes are sometimes termed well-buffered, transitional, or acidic (Fig. 11.5) depending on their position on the 'titration curve'. The alkalinity then is a measure of the extent of the well-buffered region. When a lake approaches the transitional category, only very little additional acidification is required to drive it past the endpoint with its precipitous drop in pH. Such lakes are considered to be highly sensitive to acid inputs from natural or anthropogenic sources. Acidic lakes are, of course, ones that have lost essentially all their alkalinity and are characterized by very small positive or even negative alkalinity values.

A sensitivity classification (Table 11.2) may be expressed in terms of alkalinity using units of mol L$^{-1}$ proton-accepting capacity. A high value of alkalinity means that a large amount of acid would have to be added to bring the lake into the transitional and acidic ranges. An alternative way of reporting alkalinity uses the stoichiometric neutralization reaction between carbonate and protons, and gives values in mg L$^{-1}$ CaCO$_3$ or mg L$^{-1}$ Ca. The relation between alkalinity expressed as proton-accepting capacity and as calcium carbonate (molar mass = 100 g mol$^{-1}$) is based on the following reasoning / calculation:

$$1 \text{ mg CaCO}_3 = 1000 \text{ μg which, on a molar basis, is}$$

$$1000 \text{ μg} / 100 \text{ g mol}^{-1} = 10 \text{ μmol CO}_3^{-2}$$

Since each carbonate is capable of neutralizing two hydronium ions, 10 μmol of CO$_3^{2-}$ is equivalent to 20 μmol proton-accepting capacity. Therefore, the alkalinity of a solution containing 20 μmol L$^{-1}$ of proton-accepting species could also be reported as 1 mg L$^{-1}$ CaCO$_3$.

While the latter two conventions seem to imply that the only significant proton-accepting species in the water body are in the carbonate family, this will not actually be the case in most situations. The HCO$_3$ species is usually more important unless the pH is very high.

**Table 11.2** Sensitivity classification of lakes expressed by various measures of alkalinity.

| Sensitivity | Proton acceptors / μmol L$^{-1}$ | CaCO$_3$ / mg L$^{-1}$ | Ca$^{2+}$ / mg L$^{-1}$ |
| --- | --- | --- | --- |
| High sensitivity | <200 | <10 | <4 |
| Moderate sensitivity | 200–400 | 10–20 | 4–8 |
| Low sensitivity | >400 | >20 | >8 |

The following example shows a simple prediction for a highly sensitive isolated lake. This lake is found in a region of granitic bedrock where there is a complete absence of carbonate-containing minerals. Many lakes with the properties noted below have been identified in the Canadian Shield region as well as at other locations around the world.

| | |
|---|---|
| Alkalinity | 25 $\mu$mol L$^{-1}$ proton-accepting capacity, 1.25 mg CaCO$_3$ L$^{-1}$ |
| Average depth | 20 m |
| Area | 37 ha (not needed for calculation) |
| Rainfall | 1000 mm mean annual amount; average pH = 4.0 |

Consider a square column representing a section of the lake with surface area of 1 dm$^2$ as shown in Fig. 11.6.

$$V_{\text{column}} = 20 \text{ m} \times 10 \text{ dm m}^{-1} \times 1 \text{ dm}^2 = 200 \text{ dm}^3 = 200 \text{ L}$$

This volume contains 200 L $\times$ 25 $\mu$mol L$^{-1}$ = 5000 $\mu$mol of proton neutralizer. The volume of rainfall falling on the horizontal surface of this section of the lake during 1 year would equal the annual depth of rainfall times the surface area.

$$V_{\text{rainfall}} = 1000 \text{ mm} \times 1 \text{ dm}^2 = 10 \text{ L}$$

This volume contains 10 L $\times$ 10$^{-4}$ mol L$^{-1}$ H$_3$O$^+$ = 1000 $\mu$mol of protons.

Therefore, the calculation predicts that the lake would become acidic in about 5 years. In doing this, there are many assumptions including that the lake is self-contained with neither inlet nor outlet, that there is no interaction with sediment or bedrock, and that there are no losses of acidity through seepage to groundwater. In real situations, drainage could import additional acidity. Neutralization by slow reactions with sediment or bedrock could retard acidification.

While this calculation applies to an actual lake—and there are many others like this, especially in northern Canada, Scandinavia, and Russia—many lakes are much less sensitive. The wide range of alkalinity values for water bodies around the world is illustrated by the values that are given in Table 11.3.

The highly anomalous value for Lake Magadi requires some comment. This lake and several others like it are situated in the lowest part of the Rift Valley in East Africa. Lake Magadi is

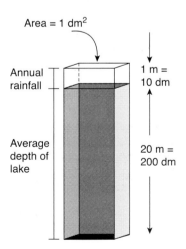

**Fig. 11.6** A square column representing a section of a lake.

**Table 11.3** Alkalinity values of selected water bodies around the world.

| Water body | Alkalinity / mg L$^{-1}$ CaCO$_3$ |
|---|---|
| Lake Baikal (Russia)—open lake | 104 |
| Lake Ontario (Canada–USA)—open lake | 110 |
| Rhine River (Germany–Holland border) | 225 |
| Lake Victoria (Uganda)—near Jinja | 52 |
| Lake Magadi (Kenya)—surface water | 280 000 |

supplied with groundwater that originates from rainfall in the nearby highlands. The groundwater is enriched with sodium, carbonate, and other ions extracted from the bedrock and soil, and these species are further concentrated due to evaporation in the hot arid climate of the region. A solid compound that is essentially sodium carbonate called 'trona' precipitates out and the 'lake' is mostly solid interspersed with pools of brine and interstitial water. The water has a pH greater than 10 and extremely high alkalinity.

### Adirondack lakes and acid precipitation

There is promising evidence[5] that acidification of lakes due to acidic rainfall can be reversed by controlling the acid inputs. Acid-sensitive lakes in the Adirondack Mountains of the eastern United States are beginning to show signs of recovery from acidification 20 or 30 years after regulations regarding emissions of sulphur dioxide were put in place. By the year 2000, at least 60% of the lakes covering an area from New York to Virginia have developed significantly improved acid-neutralizing capacity, with an average increase over an 8-year period of 1.60 μeq L$^{-1}$ y$^{-1}$. The increasing ANC correlates well with a declining rate of sulphuric acid deposition into the lakes, clear evidence that this is associated with the reduced emissions of sulphur dioxide from power and steel manufacturing plants upwind of the lake area.

A somewhat surprising finding is that nitrate levels are also declining in the lakes, while emissions of NO$_x$ and deposition of nitric acid in the area have continued at a constant rate. One theory is that the large amount of carbon dioxide produced by fossil fuel combustion is increasing growth of plants within the watershed. The more rapid photosynthesis is accompanied by enhanced rates of nitrate uptake, limiting the amount of this acid nutrient that runs off or leaches out of soil and accumulates in the lakes. It is also observed that dissolved organic carbon levels are increasing in the lakes, perhaps due to enhanced growth (and decay) of biomass in the water bodies. As we will show in the next chapter, this organic matter has the ability to form complexes with soluble metal ions in the water. Aluminium is found in acidic lakes, sometimes at levels that are toxic to fish. When present in organically bound forms, however, the toxicity is reduced.

---

**Main point 11.3** The alkalinity of a water body is a measure of the water's ability to neutralize inputs of acid. Alkalinity is defined in terms of the dissociation products of water and of aqueous carbon dioxide. The acid-neutralizing capacity of water takes into account additional soluble species, including natural organic matter and some inorganic compounds, that are also able to neutralize acid.

---

[5] Anonymous, Adirondack lakes recovering from acid rain, *Environ. Sci. Technol.*, **37** (2003), 202A.

The discussion here has centred on aqueous alkalinity in its role as a suite of components that neutralize acidity in water. Related to this, but with a different emphasis is that alkalinity also acts as a natural sink for sequestering carbon dioxide from the atmosphere.[6] There is some possibility that this ability could be enhanced by specific modifications of ocean chemistry. The simple view of this complex issue is that carbonate alkalinity reacts with dissolved carbon dioxide (which is, of course, in equilibrium with atmospheric carbon dioxide) by

$$CO_2 \text{ (aq)} + CO_3^{2-} \text{ (aq)} + H_2O \longrightarrow 2HCO_3^- \qquad (11.29)$$

Therefore, if it were possible to increase local concentrations of carbonate in sea water, the water's carbon dioxide uptake ability would also increase. To do this would require very careful technological investigations around many issues such as selecting the form of carbonate (solid or dissolved aqueous) that would be employed and the way it would be added and mixed. And most importantly, it would be essential to seriously consider how attempts to bring about major changes in ocean chemistry would affect the entire sensitive ecology of the sites where such modifications would be made? By no means can this kind of major technological fix be considered as a simple solution to the problems of excessive anthropogenic releases of atmospheric carbon dioxide.

---

## ADDITIONAL RESOURCES

1. Drever, J.I., *The geochemistry of natural waters: surface and groundwater environments*, 3rd edn, Prentice Hall, Englewood Cliffs, New Jersey; 1997.

---

## PROBLEMS

1. Lake Titicaca is situated at an altitude of 3810 m in the Bolivian Andes. Determine the atmospheric pressure and calculate the solubility of oxygen in the lake at a temperature of 5°C. (The Henry's law constant at 5°C is $1.9 \times 10^{-8}$ mol L$^{-1}$ Pa$^{-1}$)

2. The solubility of oxygen in water at 25°C is approximately 8.5 mg L$^{-1}$. At $P°$ and the same temperature, what is the volume of gas occupied by 8.5 mg of oxygen?

3. At 30°C, the solubility of oxygen in water is 7.5 mg L$^{-1}$. Consider a water body at that temperature containing 7.0 mg L$^{-1}$ oxygen. By photosynthesis, 1.5 mg of carbon (as carbon dioxide) is converted to organic biomass ($\{CH_2O\}$) during a single hot day. Is the amount of oxygen produced at the same time sufficient to exceed its aqueous solubility?

4. Thermal power plants, whether powered by fossil fuel or nuclear energy, discharge large quantities of cooling water to a lake or river. Discuss the meaning of *thermal pollution* in the context of this chapter.

5. The estimated atmospheric carbon-dioxide concentration in the Northern Hemisphere in 1950 was 310 ppmv. It may be predicted with some certainty that the concentration in 2010 will be 390 ppmv. Calculate the pH of *pure* rain that would be in equilibrium with the carbon dioxide in each of the 2 years cited, and comment on the contribution that carbon dioxide makes towards precipitation acidity.

6. One of the alternative expressions for Henry's law when applied to oxygen is

$$P_{O_2} = K'_H X_{O_2}$$

---

[6] Lackner, Klaus S., Carbonate chemistry for sequestering carbon dioxide, *Ann. Rev. Energy Environ.*, **27** (2002), 193–232.

where $P_{O_2}$ is the pressure of oxygen in the gas phase and $X_{O_2}$ is its mole fraction in solution. Calculate the value of the Henry's law constant, $K'_H$, at 25°C in units of MPa.

7. If sulphur dioxide ($SO_2$) is bubbled through water, the following reactions take place

$$SO_2 + H_2O \rightleftharpoons H_2SO_3$$
$$H_2SO_3 + H_2O \rightleftharpoons HSO_3^- + H_3O^+ \quad K_a = 1.23 \times 10^{-2}$$

If the $SO_2$ is in a gas stream at a concentration of 10.9 ppbv, while the total gas pressure is 1 atm and the pH of the resulting solution is 4.89, what must the Henry's law constant be for $SO_2$ in water at this temperature?

8. Hydrogen sulphide gas has toxicity comparable to hydrogen cyanide, but because of its intense odour its presence is observed at concentrations well below toxic levels. A mixing ratio of 100 ppmv can be lethal to humans over an extended time. What are the equilibrium concentrations of $H_2S$ and $HS^-$ in water in an atmosphere containing that concentration of the gas? What is the solubility of $H_2S$ when the gas is bubbled through water? Assume 25°C in both calculations

9. The solubility of oxygen in sea water at 15°C and 35‰ is 7.9 mg kg$^{-1}$. Calculate the Henry's law constant, $K_H$, under these conditions.

10. A septic tank leaks into the soil. The organic matter decomposes to produce carbon dioxide. Gas flow is restricted in the soil so that there is no air exchange with the atmosphere above. The carbon dioxide develops a pressure of 350 Pa. Calculate the pH of the soil water, assuming no buffering ability.

11. A groundwater sample obtained near a septic tank has a pH of 6.90 and $HCO_3^-$ concentration of 8.25 mmol L$^{-1}$. Calculate the partial pressure of carbon dioxide in the associated soil atmosphere.

12. Methane and carbon dioxide are produced under anaerobic conditions in wetlands by fermentation of organic matter approximated by the following equation

$$2\{CH_2O\} \rightarrow CH_4 + CO_2$$

Calculate the total pressure ($P°$ plus that due to water) at a depth of 5 m. Gas bubbles are evolved at that pressure and remain in contact with water at the sediment surface long enough so that equilibrium is attained. Calculate the methane concentration (mol L$^{-1}$) in the interstitial water at 25°C.

13. Calculate the alkalinity of water that is in equilibrium with atmospheric carbon dioxide and contains no additional species except those resulting from that equilibrium.

14. A sample of water from Lake Huron (one of the Great Lakes in North America) has a pH of 7.34 and an alkalinity of 1.21 mmol L$^{-1}$ as calcium carbonate. Assume equilibrium with the calcium carbonate in the sediment and calculate the concentration of calcium ion (in mg L$^{-1}$) in the water (25°C). What assumptions are made in carrying out the calculation?

15. In a soil atmosphere, the mixing ratio of carbon dioxide is often much higher than that in the normal atmosphere due to respiration (release of carbon dioxide) by microorganisms. If the mixing ratio of carbon dioxide in the air-occupying soil pores were 5000 ppmv, calculate the pH of the associated soil solution (25°C), assuming no other sources of proton donors or acceptors.

16. At a depth of 200 m in Lake Nyos, what will be the pressure of the gas, carbon dioxide, as it is released into the water? Use Henry's law to calculate the solubility of carbon dioxide under these conditions?

17. In Mexico City, the concentration of sulphur dioxide sometimes reaches 200 μg m$^{-3}$. Calculate the pH of an aqueous aerosol in equilibrium with this gas at 25°C. As a second case, assume that all of the sulphur dioxide is oxidized in the aerosol by ozone and hydrogen peroxide to produce sulphuric acid. Calculate the pH after such reactions have occurred.

18. Surface water in an open area of the Indian Ocean adjacent to Kenya has an alkalinity of 2320 $\mu$mol L$^{-1}$ and a total carbonate concentration of $2.03 \times 10^{-3}$ mol L$^{-1}$. Calculate the concentrations of $HCO_3^-$ and $CO_3^{2-}$, and also the pH of this sea water.

19. In line 5 of the code supplied for the PHREEQC calculation, replace calcite with aragonite. How does the graph differ using this polymorph of $CaCO_3$? Note: the database has $K_{sp}$ of calcite = $3.31 \times 10^{-9}$, and that of aragonite = $4.61 \times 10^{-9}$.

# Chapter 12
## Organic matter in water

The focus of this chapter is on the occurrence and nature of dissolved organic matter found in various aquatic systems and its influence on environmental processes. We will accomplish this by:

- investigating the origins and occurrence of organic matter in water
- discussing environmental functions that are related to organic matter
- introducing humic material—its properties and its interactions.

Organic matter (OM), in dissolved or particulate forms, is found in every water body—oceans and fresh water of all types. Clear evidence of its presence is the characteristic yellow-brown colour of a bog or swamp and, to a lesser extent, of some lakes and rivers. Even 'clean' water, such as that from a deep lake in a remote part of the world or from the open ocean, contains at least a small fraction of OM—typical concentrations being in the range of 1 to 3 mg L$^{-1}$.

Most analytical methods for measuring organic matter in water actually determine the carbon content. As we saw in the previous chapter, carbon, the essential element of organic compounds, is also present in the aqueous environment as a component of inorganic species in the carbonate family. In measuring the carbon content then, it is necessary to distinguish between organic and inorganic forms. Nevertheless, through reactions such as photosynthesis, respiration, and oxidative degradation, the various species can be interconverted. As you might expect, many reactions that bring about synthesis or degradation of organic matter are biological reactions.

Carbon is present in the atmosphere as carbon dioxide, in water as various carbonate species like $HCO_3^-$ and $CO_3^{2-}$, and as carbonate minerals in sedimentary rocks. Quantitatively, these species are the major carbon-containing components of air, water, and the terrestrial environment respectively. The amounts of organic carbon compounds that are present in the three compartments are usually smaller, but organic compounds play a role far more important than their small concentrations would suggest. Estimates of the total amounts of carbon-containing compounds in the atmosphere, in water, and on land are given in Table 12.1. Appendix A.3 breaks down the average plant biomass (in kg C per m$^2$) in terms of various types of ecosystems, and also documents the productivity of each of these systems.

**Fermi question**
Estimate the area of swampy wetland that would be required to supply sufficient biomass to meet the modest yearly energy (for heating and cooking) needs of one person. Information in Appendix A.3 will be helpful in making this estimate.

**Table 12.1** Carbon pools in the global environment[a].

| Carbon reservoir | Mass of carbon / Pg[b] |
|---|---|
| Atmosphere | 720 |
| Terrestrial environment | |
|     Plants | 830 |
|     Soil surface detritus | 60 |
|     Soil organic matter | 1400 |
|     Peat | 500 |
|     Fossil fuels | 5000 |
| Aquatic environment (oceans) | |
|     Living organisms | 3 |
|     Dissolved organic matter | 1000 |
|     Dissolved inorganic matter | 37 000 |
| Sedimentary carbonate material | 20 000 000 |

[a] Data from Bolin, B., Requirements for a satisfactory model of the global carbon cycle and current status of modelling efforts, in *The changing carbon cycle: a global analysis* (ed. J.R. Trabalka and D.E. Reichle), Springer-Verlag Publishers, New York; 1986.
[b] Pg = petagram = $10^{15}$ g.

While organic forms of carbon represent only a small proportion of the global carbon reservoir, they are instrumental in many reactions and influence environmental chemistry far beyond their mass contribution. The processes by which living organic matter is formed, its many interactions, and then the ways in which it degrades are particularly significant. Further discussion of relations between the biological components in the global carbon cycle is given in Chapter 15 and elsewhere throughout the book. The present discussion relates to the chemistry of non-living organic species in the water column.

Various ways of classifying OM in water help us to understand the relations between the various chemical types of OM and the natural or anthropogenic origins of particular compounds. Where the OM is derived from natural sources, it is often referred to as natural organic matter (NOM). The OM is further divisible into two broad categories—dissolved organic matter (DOM) and particulate organic matter (POM).

---

**A classification of aqueous organic matter**

- OC—organic carbon, the carbon component of organic matter
- OM—organic matter, approximate relation is OM $\sim 1.7 \times$ OC
- TOC(M)—total organic carbon (matter); readily measured by a carbon analyser
- NOC(M)—natural organic carbon (matter); in most cases synonymous with TOC(M)
- DOC(M)—dissolved organic carbon (matter)
- POC(M)—particulate organic carbon (matter); operationally distinguished from DOC(M) by filtration through a 0.45-μm nominal pore size filter; TOC = POC + DOC
- BOC(M)—biodegradable organic carbon (matter)

---

## 12.1 **Origins of organic matter in water**

The OM that is of natural origin is derived primarily from plant and / or microbial residues. On land, plants grow, sometimes shed leaves, and die, leaving roots within the upper soil layers and 'litter' on the soil surface. Great numbers of microorganisms also flourish within the soil. When they die their biomass adds to the soil organic content. Quantitatively smaller, but of considerable biological importance, is the contribution from substances secreted by living macro- and micro-organisms. Such molecules, when released in the region directly adjacent to the microbe or plant root (the rhizosphere), play important roles in facilitating or preventing uptake of chemicals by the organisms.

In their original or chemically modified form, the residues of organic matter produced on land are available to be transferred from the soil into the hydrosphere. Transport usually occurs due to rainfall that runs off or percolates through the soil column carrying soluble and particulate OM to streams, lakes, and oceans or into groundwater. The organic matter that originates from terrestrial sources makes up an important fraction of the total OM especially in small continental water bodies.

Organic matter is also produced in situ within a water body. Wetlands, both natural and constructed, are a prime example. There, the luxuriant growth of vegetation produces a thick mat of aerial material and roots that, upon death, are deposited in the water. The top layer of the swamp sediment consists almost entirely of organic material in various stages of decay and degraded material goes into the water column in particulate or dissolved forms. Other water bodies, rivers, lakes, and oceans, support the growth of aquatic plants and animals to a smaller degree and their organic remains also become part of the total aquatic system. As they do in soil, microorganisms inhabit the hydrosphere in great numbers, depending on environmental conditions. Some, like algae, are photosynthetic organisms that require sunlight and oxygen, and therefore grow in the near-surface layers of water. Bacteria and other types of microorganisms are especially abundant in the sediments. All microorganisms require a supply of nutrients in order to grow and multiply. During their survival period the various plants and microorganisms release some of their products to the water column. Microscopic animals also release soluble organic matter from their bodies. On a global scale it is estimated that around 10% of microbial activity in water goes to the production of DOM. Appendix A.3 includes productivity information related to the formation of organic carbon compounds in water bodies.

After dying, the organic residues from larger plants and animals, as well as the microbial biomass, become chemically modified by a variety of decomposition and new synthesis processes. Many of the reactions to produce altered organic species are facilitated by the presence of living microorganisms in the soil or water. Some of the organic species generated through degradation and synthesis move into the water column in soluble or particulate forms.

Besides the natural sources, there are human inputs that contribute to the organic matter in water. These include large volumes of poorly defined wastes, such as domestic sewage or pulp-mill effluent, that are sometimes discharged directly or after treatment into rivers, lakes, and oceans. Besides the bulk effluents, anthropogenic sources also supply specific organic compounds—agricultural chemicals, medicinals, and products or by-products of industrial processes. The range of these is as broad as the range of organic chemistry itself.

The distinction between natural and human sources is not always simple. Falling clearly into the natural category are the organic products that result from leaf decay, while release of nitrilotriacetic acid (NTA) into a waste stream from a detergent-manufacturing facility is an example of an indisputably anthropogenic event. Other cases are less clear-cut. Trichloromethane (chloroform) is a chlorinated hydrocarbon that we might assume is produced only via industrial processes. However, it has been shown that millions of tonnes of this compound are also formed through natural reactions each year. In fact over 1500 organochlorines have been identified as natural products found in living organisms. While many of these are low molar mass compounds such as mono- and trichloromethane and 2,4,6-trichlorophenol, there are also complex, relatively high molar mass chlorinated organics that occur naturally in the environment. There is therefore a substantial list of chemicals—not just organochlorines—that have natural as well as anthropogenic origins.

When we consider the chemistry of organic matter that is found in the hydrosphere, you will see that it is frequently convenient to subdivide the species into two categories based on molecular size.

- Discrete small molecules, like monosaccharides or low molar mass organic acids, whose chemical structure and properties are amenable to individual specific study. Individual polluting species like chemical pesticides fall into this category as well.

- Macromolecules, which are treated as classes in terms of their general structural properties and reactivity. Characterization of macromolecules is often based on an operational definition—that is, a definition arising out of a particular analytical protocol—and not based on fundamental structural properties.

Many of the natural forms of organic matter fall in the macromolecule category, and we will focus on these kinds of compounds in the present chapter

> **Main point 12.1** Organic matter from diverse sources is present to varying degrees in dissolved and particulate forms in all natural waters. The OM originates from both natural and anthropogenic sources. Specific organic species can be identified, usually in low concentration, in samples, but much of the organic matter is present as poorly defined broad classes of material.

## 12.2 Environmental issues related to aqueous organic matter

### Toxicity of specific organic compounds

Organic matter in water is of environmental importance for several reasons. For one thing, particular compounds may be toxic in varying degrees to living organisms, including humans. Polyaromatic hydrocarbons, polychlorinated biphenyls, and dioxins are all well known, and have received considerable publicity and study because of their contribution to real and alleged environmental problems. Residues of pesticides and their metabolic products can also be carried into water. Due to their widespread application in agriculture and forestry, even in urban settings, it is important to understand how they move and react in the environment. Some aspects of the environmental chemistry of these synthetic organic compounds will be discussed in Chapter 20. As a general name, we refer to these as biocides or xenobiotic compounds.

### Reaction with other aquatic species

There are also less direct ways in which organic matter influences environmental processes and these also merit study. For example, inorganic tin undergoes alkylation in aquatic environments to form compounds such as monomethyl tin ($CH_3Sn^{3+}$) and dimethyl tin (($CH_3)_2Sn^{2+}$). The alkylation process is a biological one in that it takes place in the fish gut or via microorganisms in the water column. The organotin product species are more toxic to aquatic biota than are the original inorganic tin compounds; the enhanced toxicity is usually attributed to their ability to move across cell membranes. Toxicity becomes greater as the number of organic groups increases in the series $R_nSn^{(4-n)+}$ for $n = 1$ to 3. Toxicity is also inversely related to the length of

R and is at a maximum where R is a methyl or ethyl group. There are many additional examples of organometallic compounds and metal–organic ligand complexes in environmental situations. Invariably, the environmental chemical and toxicological properties of the compounds are different from those of the free metal. In some cases toxicity is enhanced; in other cases it is reduced by reaction with organic matter in the hydrosphere.

### Consumption of oxygen

A third environmental feature of aqueous organic matter, in particular the bulk residues of plants and animals or some industrial discharges, is that the non-living organic material can be oxidized by oxygen and other oxidizing agents in water. Therefore, when released into a water body, the bulk OM degrades, consuming oxygen and leaving the system in an oxygen-deprived, anaerobic state. This condition generates a low p$E$ environment and therefore can change the chemistry of the entire system. The loss of oxygen also creates stress on many aquatic organisms including fish. Such conditions are frequently observed downstream from points where high OM wastewater is discharged.

Each of these environmental aspects related to aqueous organic matter will come up in later sections of the book. For now we will look at chemical properties of the broadly defined naturally occurring macromolecules present in soluble, colloidal, and sedimentary forms in the hydrosphere. Closely related macromolecules are also important in the terrestrial environment.

> **Main point 12.2** Environmental consequences of several types are associated with organic matter in water. In some cases, aqueous organic matter is toxic. Frequently, it takes part in reactions with other aquatic species, thus affecting the behaviour of those species. When present in high concentration, OM can create anaerobic conditions in the water.

## 12.3 Humic material

Humic material (HM) is a form of environmental organic matter of plant or microbial origin. The humic material is not made up of discrete, well-defined molecules but is a class of substances that are produced and reside in soil and water, forming a major component of both the terrestrial (soil organic matter) and aquatic (natural organic matter) carbon pools (Table 12.1). In the hydrosphere, HM typically makes up about 50% of the dissolved organic matter (DOM) in surface water,[1] as well as much of the organic sediment. Because individual molecules cannot be identified, humic material (also called humate or humus) is subdivided in an operational sense into three classes or categories.

- Fulvic acid (FA) is the fraction of humic matter that is soluble in aqueous solutions that span all pH values.
- Humic acid (HA) is insoluble under acid conditions (pH 2) but soluble at elevated pH.
- Humin (Hu) is insoluble in water at all pH values.

### Formation of humic material—degradative pathway

Humic substances are formed via a complex sequence of only partly understood reactions. Several hypotheses have been proposed to account for their synthesis in nature. According to one set of hypotheses, plant biopolymers are modified through degradation to form the central

---

[1] As an approximation, the remaining 50% of DOM consists of low molar mass acids such as oxalic, citric, formic, and acetic acids (~25%), neutral compounds much of which is carbohydrate material (~15%), and other species (~10%).

core of humic substances. These theories propose that labile macromolecules such as carbohydrates and proteins are degraded and lost during microbial attack, while refractory compounds or biopolymers—for example, lignin, paraffinic macromolecules, melanins, and cutin—are selectively modified and transformed to produce a high molar mass material that is a precursor of humin. Further oxidation of these materials adds to its oxygen content in the form of typical functional groups like carboxylic acids and, as this process continues, the molecules become small enough and hydrophilic enough to be soluble in alkali. Eventually the molecules become even smaller and sufficiently oxygen-rich to dissolve in both acid and base. The degradative pathway then follows this sequence:

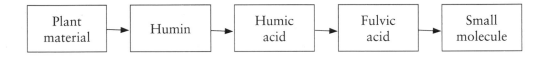

The extensive degradation produces structures that retain some original features but also have considerable structural dissimilarity from the parent material.

## Formation of humic material—synthetic pathway

As an alternative, hypotheses based on ideas of condensation polymerization suggest that plant biopolymers are initially degraded to small molecules after which these molecules are repolymerized to form humic substances. It has been proposed that polyphenols synthesized by fungi and other microorganisms, together with those liberated from the oxidative degradation of lignin, undergo oxidative polymerization. A consequence of this scheme is that fulvic acid would be a precursor of humic acid and then of humin (the reverse of the degradative theory) as shown by:

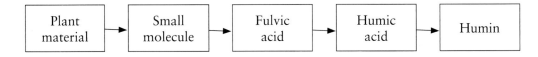

The hypothesis could explain the observed considerable similarity of humic materials formed from a diversity of precursor macromolecules in differing environments.

The two proposed pathways have overlapping features and both routes could contribute to the actual formation of humic material depending on the environment. For instance, the condensation polymerization scheme might predominate in wet sediments and in the aquatic environment in general, whereas the conditions in soils that are subject to harsh continental climates could favour an oxidative degradation process.

While the detailed features of the pathway are not entirely clear, in either case the formation of humic material does involve degradation of fresh organic matter. When oxidative decomposition processes go to completion the endproducts are carbon dioxide and water and we describe the reaction in simplified form as

$$\{CH_2O\} + O_2 \rightarrow CO_2 + H_2O \tag{12.1}$$

assuming that the bulk of the OM input has elemental composition similar to that of carbohydrate material. Note that this overall reaction is the same one that describes combustion of biomass. Other organic components of plant material also produce carbon dioxide and water as principal products of oxidative decomposition. You can see why large-scale deforestation has a dual effect on increasing the carbon dioxide concentration in the atmosphere—first by removing the growing trees and their ability to sequester carbon dioxide from the air and second by leaving behind organic debris, some of which decomposes completely to release gaseous carbon dioxide.

When plant residues are buried, oxygen is much less available and the final product of decomposition contains carbon in a reduced chemical state. Deep burial of large masses of organic matter is the first step in the formation of the different fossil fuels. Over very long periods of time, most of the oxygen in the original biological material is lost and a suite of hydrocarbons as well as carbon itself is the final product.

The humic material that is produced by the humification processes can be considered to be an intermediate in the overall sequence of degradation and / or resynthesis reactions. But, unlike many intermediates in chemical reactions carried out in the laboratory, HM is a stable intermediate, especially under conditions of limited oxygen availability. Carbon-14 dating of HM from a variety of sources has indicated a range of age from about 20 years for stream humate, up to 500 or 1000 years for soil humus and much longer for that associated with buried peat or coal deposits.

## Composition and structure

### Elemental analysis

Analysis of a wide variety of humic substances shows that it usually contains carbon, hydrogen, oxygen, and nitrogen within ranges (%) as follows:

| C | O | H | N | Inorganic elements (ash) |
|---|---|---|---|---|
| 45–60 | 25–45 | 4–7 | 2–5 | 0.5–5 |

Because the carbon content is frequently closer to 60%, an approximate factor of 1.7 is used as an estimate to convert mass values from organic carbon (OC) to organic matter (OM). This factor is also employed for forms of organic matter other than humic material in both water and soils.

---

**Example 12.1 Conversion of DOC to DOM concentrations**

Calculate the dissolved organic matter (DOM) concentration in a lake that is determined to have 2.3 mg L$^{-1}$ of dissolved organic carbon (DOC).

The DOM content would be estimated as

$$1.7 \times 2.3 \text{ mg L}^{-1} = 3.9 \text{ mg L}^{-1}$$

---

As noted above, it is organic carbon that is usually measured analytically. A brief discussion of these and other parameters used as measures of organic matter in water is given in Chapter 15.

While the elemental concentration ranges indicated above cover most analyses of the three types of humic substances, the carbon content of material from a given source usually increases in the series

$$\text{FA} < \text{HA} < \text{Hu}$$

The oxygen content of the same set of materials follows the reverse trend. Table 12.2 shows the range of chemical features for fulvic acid and humic acid.

The environment in which the material is formed also has some influence on the composition, with carbon content being greater in soil humus than in humate from lakes and oceans. The reverse is true for oxygen and nitrogen. In sedimentary material, there is some evidence that the relative proportion of Hu compared to HA and FA increases with depth. Depth is directly related to age of the sediment and so a greater age of the buried material is reflected in a larger carbon content. Such trends are common but have not been observed in all situations.

**Table 12.2** Chemical features of fulvic and humic acid[1].

|  | Fulvic acid | Humic acid |
|---|---|---|
| Carbon / % by weight | 40.7–50.6 | 53.6–58.7 |
| Hydrogen / % by weight | 3.8–7.0 | 3.2–6.2 |
| Nitrogen / % by weight | 0.9–3.3 | 0.8–5.5 |
| Oxygen / % by weight | 39.7–49.8 | 32.8–38.3 |
| Sulphur / % by weight | 0.1–3.6 | 0.1–1.5 |
| Molar mass / daltons | 500–2000 | 2000–5000 |
| Atomic ratio H / C | ~1.3 | ~0.8 |
| Atomic ratio O / C | ~0.8 | ~0.5 |

[1] Data from various sources, reported in Eby, G.N., *Principles of environmental geochemistry*. Brooks / Cole-Thomson Learning, Pacific Grove, CA, 2004.

The presence of oxygen and nitrogen in HM is an indication that certain functional groups are present in the humate molecules. We will now consider the nature of this functionality. The most important oxygen-containing functional groups along with typical ranges of content (mmol functional group per g of humic material) are carboxyl (2–6), phenolic-OH (1–4), alcoholic-OH (1–4), carbonyl, both ketones and quinones (2–6), and methoxyl (0.2–1). The functional groups are the sites of many of the important reactions with which HM is involved.

- It is these groups that, individually or together, enable specific reaction of humic substances with inorganic elements and with other organic molecules in soil–water systems.
- The functional groups, particularly carboxyl and phenolic groups, are also major contributors to the cation exchange properties of soils and sediments. In addition to being present in the above groups, both oxygen and nitrogen appear as bridging units and in ring structures.

### Spectroscopy as an aid for structure determination

Quantitative structural information is obtained using classical chemical procedures. Spectroscopic methods, including infra-red (IR), carbon-13 nuclear magnetic resonance ($^{13}C$ NMR) and $^{13}C$ MAS (magic-angle spinning) NMR spectroscopy, can be employed to independently indicate the presence of structural features in the material. In addition to confirming the presence of various functional groups, these spectroscopic methods indicate that there is both aliphatic and aromatic character to HM. Figure 12.1 shows IR spectra obtained on fulvic and humic acid from a Kenyan soil, and Table 12.3 lists information that can be used to interpret this and other such spectra.

Infra-red spectra of many other samples of humic material from around the world have been reported and, in their principal features, are similar to the ones shown here. The common features in the IR bands for various samples of both FA and HA suggest that all types of humic materials share a great deal of similarity in their structural features. The observed bands are consistent with the presence of the expected chemical units, including phenolic and aliphatic −OH groups and carbonyl and carboxyl groups, and there is evidence of aromaticity and double bonds in aliphatic units. Quantitative assessment of the 'concentration' of functional groups is not possible from the IR spectrum.

Carbon-13 NMR spectra for the same Kenyan humate samples are shown in Fig. 12.2. The broad and overlapping peaks that appear on the NMR spectra shown are characteristic of all spectra of humic material. This is largely due to the presence of paramagnetic species such as

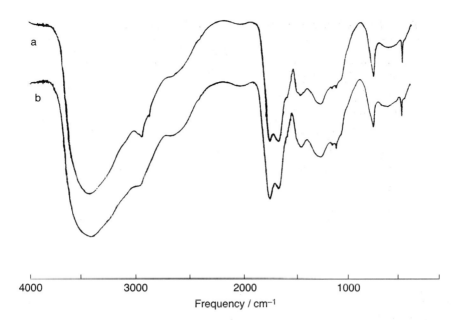

**Fig. 12.1** IR spectra of (a) humic acid and (b) fulvic acid from soil obtained near Lake Nakuru in Kenya. (Moturi, M.C.Z., Studies on humic substances extracted from neutral and alkaline soils and sediments from Kenya, MSc Thesis, Queen's University, Kingston, Ontario; 1991.)

**Table 12.3** Some of the prominent bands observed in IR spectra of humic and fulvic acids.

| Absorption band / cm$^{-1}$ | Assignment |
|---|---|
| 3400 | Hydrogen-bonded O−H stretching vibrations from phenolic and aliphatic OH groups |
| 2920 | Asymmetric stretching vibrations of aliphatic C−H bands in −CH$_3$ and −CH$_2$ units |
| 2860 | Symmetric C−H stretch of aliphatic bonds in −CH$_3$ and −CH$_2$ units |
| 1720 | C=O stretching vibrations from −COOH and probably ketonic carbonyls |
| 1630 | May be due to C=C stretching in aromatic rings, asymmetric stretching of COO$^-$, hydrogen-bonded C=O, or C=C stretching alkenes conjugated with carbonyl groups or other double bonds |
| 1540 | Aromatic C=C stretching vibrations or N−H deformations |
| 1420 | O−H bending vibrations of alcohols, carboxylic acids, and phenols, or C−H deformations of −CH$_2$ and −CH$_3$ aliphatic groups |
| 1240 | C−O stretching and O−H deformation of −COOH |
| 1050 | May be due to O−H deformations and C−O stretching in polysaccharides, phenolic and alcoholic groups, or Si−O of silicate impurities |
| 940 | Aromatic C−H out-of-plane bending vibrations |
| 800 | Aromatic C−H out-of-plane bending vibrations |

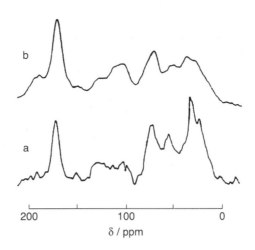

**Fig. 12.2** Carbon-13 NMR spectra of (a) humic acid and (b) fulvic acid from soil obtained near Lake Nakuru in Kenya. (Moturi, M.C.Z., Studies on humic substances extracted from neutral and alkaline soils and sediments from Kenya, MSc Thesis, Queen's University, Kingston, Ontario; 1991.)

$Fe^{3+}$ and stable organic radicals from residual soil components in the samples. Nevertheless, useful information is still obtainable and chemical shift assignments have been made as in Table 12.4.

Although the size of peaks depends on the number of carbon atoms in a particular structural unit, it is not advisable to make quantitative interpretations from $^{13}C$ NMR spectra because peak size also depends on the microenvironment of the atoms and on variations in instrumental conditions. In the spectra shown, there is clear evidence of the presence of aliphatic groups (30 ppm), alkoxyl groups (55 ppm), carbohydrate-type functionality (75 ppm), aromatic components (the broad band from about 100–140 ppm), and carboxylic groups (175 ppm). NMR spectra of other humic material from varied sources around the world show similar features.

### A hypothetical structure

Using all these types of information, it becomes possible to develop an idea of the molecular structure of humic substances. Figure 12.3 shows a portion of a *hypothetical structure of a generic humate molecule*. All of the features—the aliphatic and aromatic character, the varied functionality, the polymeric nature—are represented here.

It is, of course, essential to realize that Fig. 12.3 does not depict a specific humate molecule and, in fact, there is no such thing as a single specific humate species in any of the three classes.

**Table 12.4** Chemical shift assignments of $^{13}C$ NMR spectra of fulvic and humic acids[a].

| Chemical shift / ppm | Assignment |
| --- | --- |
| 0–50 | Unsubstituted saturated aliphatic C |
| 10–20 | Terminal methyl groups |
| 15–50 | Methylene groups in alkyl chains |
| 25–50 | Methine groups in alkyl chains |
| 29–33 | Methylene C, $\alpha$, $\beta$, $\delta$, and $\varepsilon$ from terminal methyl groups |
| 35–50 | Methylene C of branched alkyl chains |

**Table 12.4** (*Continued*)

| Chemical shift / ppm | Assignment |
|---|---|
| 41–42 | α-carbons in aliphatic acids |
| 45–46 | $R_2NH_3$ |
| 50–95 | Aliphatic carbon singly bonded to one O or N atom |
| 51–61 | Aliphatic esters and ethers, methoxy, ethoxy |
| 57–65 | Carbons in $CH_2OH$ groups, $C_6$ in polysaccharides |
| 65–85 | Carbon in CH(OH) groups, ring C of polysaccharides, ether bonded aliphatic C |
| 90–110 | Carbon singly bonded to O atoms, anomeric C in polysaccharides, acetal or ketal |
| 110–160 | Aromatic and unsaturated C |
| 110–120 | Unsubstituted aromatic C, aryl H |
| 118–122 | Aromatic C ortho to O-substituted aromatic C |
| 120–140 | Unsubstituted and alkyl-substituted aromatic C |
| 140–160 | Aromatic C substituted by O or N, aromatic ether, phenol, aromatic amines |
| 160–230 | Carbonyl, carboxyl, amide, ester C |
| 160–190 | Carboxyl C |
| 190–230 | Carbonyl C |

[a] Most of the data is taken from Hayes, M.H.B., P. MacCarthy, R.L. Malcolm, and R.S. Swift (eds.), *Humic substances II: In search of structure*, J. Wiley and Sons, Chichester; 1989.

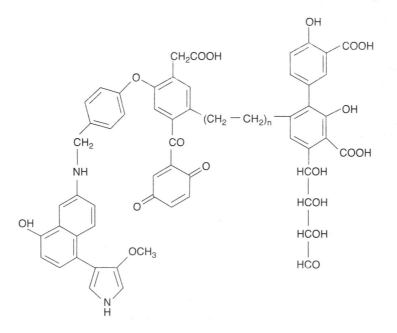

**Fig. 12.3** Structure of a generic molecule of a humic substance.

proportions of the different structural components vary from situation to situation and
s the molar mass. The molecule in the figure has a molar mass of 1056 daltons, and this
be typical of some fulvic acid materials. If, however, we isolate a sample of 'pure' fulvic
cid, it will be made up of molecules with a range of molar mass values as will similar samples
of humic acid or humin. The molar mass of polydisperse humic material is difficult to measure,
but average values ranging from a few hundred daltons for fulvic acid to hundreds of thousands
of daltons for humin have been estimated.

---

**Main point 12.3** Humic substances, natural material of both terrestrial and aquatic origin, comprise about half of the organic matter present in many waters. Humic substances are made up
of three classes of macromolecules having a range of molar mass and structural characteristics.
Common features of all humic material are aliphatic and aromatic components and functional
groups that impart acid–base properties and complexing ability to the molecules.

---

### Literature link  Characterization of freshwater dissolved organic matter

"Dissolved organic matter (DOM) is a complex, heterogeneous mixture found ubiquitously in nature.
It comprises a major mobile fraction of organic carbon on Earth and is an intimate link between the
terrestrial and aquatic environment" is the opening statement in a paper *Major structural components
in freshwater dissolved organic matter*, by B. Lam, A. Baer, M. Alaee, B. Lefebvre, A. Moser, A. Williams
and A. J. Simpson (Environ. Sci. Technol. **41**, 8240–8247 (2007). The purpose of the research
described here was to examine the structure of specific samples of DOM and compare these with
well-characterized samples found elsewhere. This information was then used to hypothesize regarding the source of the organic material that served as precursor for the DOM present in the water. The
sampling site for the study was Lake Ontario, one of the Great Lakes in eastern North America. As a
group, these five lakes represent the largest freshwater system on the planet and play a linking role in
cycling carbon from terrestrial sources in that part of the world to the marine environment.

Sample preparation included isolation of the OM onto an anion exchange resin (why use anion
exchange for this purpose?), elution from the resin with 0.1 M sodium hydroxide, and finally dialysis to
remove salts before freeze drying. Because 100 dalton cutoff tubing was used for dialysis, the retained
DOM would be substantially humic material (HM) and other medium and large macromolecules that
may be precursors or degradation products of the HM.

Characterization of the recovered material was done using a variety of sophisticated 2- and
3-dimensional proton and carbon-13 NMR techniques. This enabled measuring long-range proton–carbon correlations that provided support for making structural assignments of the DOM. Conventional solid
state proton NMR experiments showed that major structural features of the macromolecules included
aliphatic, aromatic, carbohydrate and carboxyl-rich alicyclic components. Modified 2-dimensional $^{13}C$
NMR confirmed the presence of these features and added further rich structural information. For example, the methoxy group, which is often a very intense signal in soil organic matter was not observed in
this lake-derived material, pointing either to a non-soil origin or that a transformation had occurred in
the aquatic environment.

The DOM also bore considerable similarity to previously studied lake and marine materials including
a Pacific Ocean sample that showed the same three major components—aliphatics, carbohydrates and
carboxyl-rich alicyclics.

Further multidimensional NMR experiments assisted in developing a more detailed picture of the
origins of this DOM. In particular, heteronuclear multiple bond correlation (HMBC) data provided
information on long-range (up to 3 bonds) $^{1}H-^{13}C$ connectivity that assisted in evaluating the
organization of individual H-C units. This and other data showed that the samples had significant features characteristic of terpenoids—compounds that are found in abundance in nature. The

carboxyl-rich alicyclic structures show two key features: a *cyclic terpenoid* backbone and a high degree of carboxylation. While materials derived from cyclic terpenoids may be the major contributor to Great Lakes DOM, another fraction, clearly derived from *linear terpenoids*, was also clearly observed.

Because both terrestrial and aquatic vegetation are sources of these structurally similar precursor compounds, the question remains open as to the original sources of the DOM that is found in the Lake Ontario water.

## Forms of humic materials

Humic materials as a group are found in the aqueous and terrestrial environments in a variety of forms and associations.

- Free HM consists of soluble or insoluble forms of the material itself.
- Complexed HM is chemically bound to metals, other inorganic species such as phosphate, or organic molecules. The complexed HM can also be either in solution or in particulate form.
- Surface-bonded HM is chemically bonded to other solids such as clay minerals or iron and aluminium oxides. In this way the surface of the inorganic material is altered so that its chemical properties are determined largely by the organic coating. The HM can then react in a manner similar to that of the pure particulate HM itself.

## Aqueous humic material as a proton acceptor

As their names indicate, the free forms of humic materials are acids. The acidic character is associated largely with the carboxylate and phenolic groups. The former have $pK_a$ values in the range between 2.5 and 5 depending on the proximity of electronegative atoms with respect to the carboxyl groups, while the phenolic hydrogens have $pK_a$ values around 9 or 10. The carboxyls, therefore, are sufficiently strong acids that they remain substantially deprotonated when dissolved in water at a low concentration. If the water body is poorly buffered because it lacks other proton acceptors such as hydrogen carbonate, dissolved humic material will acidify the water so that it has a typical pH of 5.5 to 6.5. Since the aqueous pH remains substantially higher than the $pK_a$ of the carboxylic acid groups, the humic material itself is negatively charged and, in some cases, may be the major source of anionic charge in the dissolved phase. Consider the following example.

---

### Example 12.2 Humic material contribution to charge balance in water

A sample of water from the Canadian Shield containing a dissolved humic material at a total concentration of 8.0 mg $L^{-1}$ has measurable concentrations ($C_X$) of the following ions:

| | | | |
|---|---|---|---|
| $H^+$ | pH = 5.88 | $Cl^-$ | 0.138 mg $L^{-1}$ |
| $NH_4^+$ | 3.6 µg $L^{-1}$ (as N) | $NO_3^-$ | 7.0 µg $L^{-1}$ (as N) |
| $Na^+$ | 75.9 µg $L^{-1}$ | $HCO_3^-$ | 14.4 µg $L^{-1}$ (as C) |
| $K^+$ | 50.8 µg $L^{-1}$ | $SO_4^{2-}$ | 59.4 µg $L^{-1}$ (as S) |
| $Mg^{2+}$ | 0.124 mg $L^{-1}$ | | |
| $Ca^{2+}$ | 0.569 mg $L^{-1}$ | | |

Any water sample must have a balanced concentration of positive and negatively charged species giving a net ionic charge of zero. (We can express charge in mol $L^{-1}$, just as we would for any physical entity.) If we convert the mass concentrations above into molar concentrations and then carry out a charge-balance calculation we find that there is an apparent large excess of positive (cationic) charge. The charge due to each cation and anion is calculated in the table below.

| | Concentration / $\mu mol\ L^{-1}$ | Charge / $\mu mol\ L^{-1}$ |
|---|---|---|
| $H^+$ | 1.3 | 1.3 |
| $NH_4^+$ | 0.2 | 0.2 |
| $Na^+$ | 3.3 | 3.3 |
| $K^+$ | 1.3 | 1.3 |
| $Mg^{2+}$ | 5.1 | 10.2 |
| $Ca^{2+}$ | 14.2 | 28.4 |
| Total positive charge | | 44.7 |
| $Cl^-$ | 3.9 | 3.9 |
| $NO_3^-$ | 0.5 | 0.5 |
| $HCO_3^-$ | 1.2 | 1.2 |
| $SO_4^{2-}$ | 1.2 | 3.8 |
| Total negative charge | | 9.4 |

Considering only the species indicated, therefore, there is a major charge imbalance as a net positive charge of $44.7 - 9.4 = 35.3\ \mu mol\ L^{-1}$ is indicated. Since electroneutrality must obtain in the solution, a balancing negative charge of $35.3\ \mu mol\ L^{-1}$ must exist. To account for this we must include in the calculation the contribution due to dissolved humic material that was present at a concentration of $8.0\ mg\ L^{-1}$.

If this material has a carboxyl group concentration of $C_{coo-}\ \mu mol\ g^{-1}$, the associated negative charge is

$$8.0\ mg\ L^{-1} \times C_{COO-}\ mmol\ g^{-1} = 8\ C_{COO-}\ \mu mol\ L^{-1}.$$

At pH = 5.88, about 2 pH units above the usual carboxyl group $pK_a$ value, we would expect that close to 100% of the carboxyl groups would be deprotonated. Therefore, assuming that the required negative charge comes exclusively from the organic matter,

$$8\ C_{COO-} = 35.3\ \mu mol\ L^{-1}$$

*and*

$$C_{COO-} = \frac{35.3\ \mu mol\ L^{-1}}{8\ mg\ L^{-1}}$$
$$\approx 4.4\ mmol\ g^{-1}$$

---

The calculated concentration of carboxyl groups in the dissolved organic matter is therefore $4.4\ mmol\ g^{-1}$, which is well within the usual range for humic material of $2–6\ mmol\ g^{-1}$. It should be emphasized that this type of calculation is only approximate as there is a possibility of error associated with each of the analytical measurements that, in sum, leads to considerable uncertainty in the final result. We have also assumed that there are no additional significant contributions from other charged species in the water.

When we considered the concepts of alkalinity and acid-neutralization capacity in the previous chapter, it was noted that the difference between the two was, in some cases, due in large part to organic matter. The previous example illustrates this, as the slightly acidic lake had a very small hydrogen carbonate concentration and a more important proton acceptor was the dissolved humic material. The actual alkalinity titration measurement might account for an indefinite proportion of the dissolved humic matter—indefinite because the $pK_a$ values of HM cover a range from 2.5 to 5 and so only part of these would be protonated during a titration to an endpoint at pH 4.5, as is specified for the determination of alkalinity. Nevertheless, humic material is the most important buffering constituent in some slightly acidic lakes with large DOM concentrations.

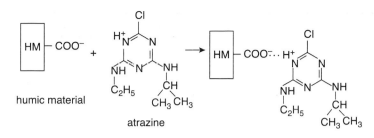

**Fig. 12.4** Atrazine bonded to HM by electrostatic forces.

## Humic material as a complexing agent for metal ions

Besides playing a role as a proton acceptor and in contributing to charge balance in aqueous systems, the dissolved humic substances also react with metals in solution through the formation of ionic or covalent bonds. The strength of the metal–organic interaction is measured in terms of a stability constant, $K_f$, of the form described in Chapter 10. The actual value of $K_f$ depends on the nature and number of binding sites on the HM, on the properties of the metal, and on other environmental factors such as pH and the presence of other competing ligands. We will say more about this in the next chapter.

## Reactions between humic material and small organic molecules

Humic substances are also capable of reacting with many specific anthropogenically derived organic compounds that are found in water and in soil. Organic pesticides are a prime example. Interaction occurs via several mechanisms; van der Waals forces are always a source of weak attractions. Stronger interactions occur due to processes such as the following.

- Reaction between positively charged species and negatively charged humate sites (Fig. 12.4). The substrate in the example in Fig. 12.4 is atrazine, a widely used herbicide in the triazine class. Under typical environmental conditions (pH <8) the atrazine molecule is positively charged due to protonation of one of the ring nitrogens and this allows for reaction with the carboxyl groups of the humic material.

- Hydrogen bonding (Fig. 12.5). A variety of hydrogen-bonding reactions are possible, involving oxygen- and nitrogen-containing functional groups of both the humic material and the organic pesticide molecule. In the example shown, the insecticide carbaryl is hydrogen-bonded to HM using the two electronegative atoms, nitrogen and oxygen, in the carbamate structure.

- Salt linkage or ligand exchange (Fig. 12.6). In this example, assuming a near-neutral pH of 6 to 8, the chlorophenoxy herbicide 2,4-dichlorophenoxyacetic acid (2,4-D), would be present in solution with its carboxyl group deprotonated—as is also the case for the carboxyls of the

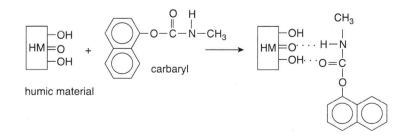

**Fig. 12.5** Binding of the pesticide carbaryl to HM through hydrogen bonding.

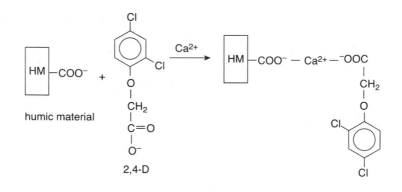

**Fig. 12.6** Binding of the pesticide 2,4-dichlorophenoxyacetic acid to HM through a metal-ion bridge.

humic material. Transition metals or even common cations such as $Ca^{2+}$ found in aqueous systems, including water within the soil matrix, then are able to form a metal bridge or salt linkage so that the herbicide is bound to the soluble or particulate humic material.

- Hydrophobic interactions (Fig. 12.7). For essentially non-polar molecules like the well-known (and controversial) insecticide dichlorodiphenyltrichloroethane (DDT), the molecules find compatible settings within mostly hydrocarbon regions of humate molecules. The association involves a net energy loss since the small hydrophobic molecule no longer disrupts the ordered structure of the water solvent after it moves from the aqueous phase to the humic substance. This type of process is often referred to as a hydrophobic interaction.

The combination of all these factors contributes to organic solute binding by humic substances. An even more detailed description of the interactions requires taking into account other environmental factors such as solution pH and ionic strength.

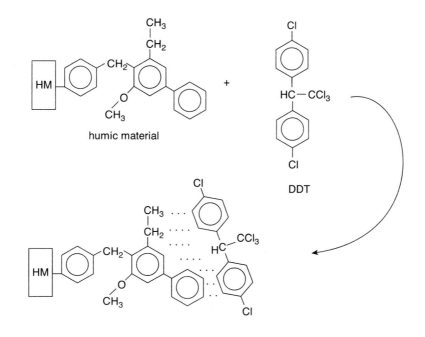

**Fig. 12.7** Binding of the insecticide DDT to HM through hydrophobic interactions. An essentially hydrophobic portion of the hypothetical humate molecule is specifically shown in this figure.

The behaviour of most small neutral organic molecules that are present in water is strongly influenced by the presence of humic material. As a consequence of the solute-binding processes, otherwise insoluble organic substances may be brought into solution through association with soluble material. The opposite process also occurs. If and when the humate becomes insoluble—as could occur when riverine sources encounter sea water in estuaries—the aggregated material may bring organic solutes down with it. In this case, and also by direct association with solid humic substances, such solutes are 'immobilized'. Therefore, depending on circumstances, humate either enhances solubility or removes from solution organic compounds like pesticides. Furthermore, once the association is established, reactivity of the solute is altered. Examples of chemical reactions, such as the hydrolysis of atrazine, have been cited where degradation is enhanced by association with HM. There are also cases where degradation is inhibited by the same association—for example the base hydrolysis of DDT.

## Humic material associated with soil or sediment

Humic material is therefore found in water in free form or complexed with metal ions or other organic species. The third form in which HM occurs in the environment is in association with other solids in soil / water or sediment / water systems. The solid complexes are frequently studied using various types of advanced microscopy. A particular association is that with clay minerals and it is well known that humic substances form coatings on such mineral surfaces.

The surface bonding occurs through specific covalent interactions between metals ($Al^{3+}$, $Fe^{3+}$) on the clay surface or metals ($Ca^{2+}$, $Al^{3+}$) in the adjacent solution and the humate molecules, as well as via weaker hydrogen bonds (Fig. 12.8). Because both the clay surface and the humic material have net negative charges, simple electrostatic interactions cannot play a role.

The fact that a portion of the humic material may be associated with the clay mineral fraction in a water / solid system adds a further dimension to the complications involved in describing the chemistry of the humate. As a result of their interaction, the surface properties of both the clay minerals and the humate are altered, thus affecting subsequent interactions with other metals and organics, and this in turn influences the reactivity of the retained species.

**Main point 12.4** Humic material affects the charge balance and alkalinity of water, and it interacts with metals, small neutral organic molecules, and sediment / soil solids. In doing so it can alter the mobility and reactivity of these substances.

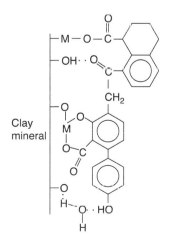

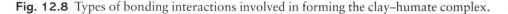

**Fig. 12.8** Types of bonding interactions involved in forming the clay–humate complex.

## ADDITIONAL RESOURCES

1. Ghabbour, E. and G. Davies (eds.), *Humic substances: Nature's most versatile materials*, Taylor & Francis, New York; 2004.

2. Schwarzenbach, R.P., P.M. Gschwend, and D.M. Imboden, *Environmental organic chemistry*, 2nd edn, Wiley and Sons, Hoboken, New Jersey; 2003.

3. Aiken, G.R. (ed.), *Humic substances in soils, sediments and water: geochemistry, isolation and characteristics*, J. Wiley and Sons, New York; 1985.

4. Olk, D., The International Humic Substances Society, *A partial bibliography of recent humic literature*, 2006, http://ihss.gatech.edu/ihss2/news.html, accessed October 2009.

## PROBLEMS

1. Use diagrams of a type similar to those in Figs 12.4–12.7 to show the kinds of interactions you might expect between humic material and phenanthrene, trichloroethylene (TCE, CHCl = CHCl$_2$), copper (II), and copper (II) complexed with oxalate.

2. Explain how molecular size and oxygen concentration differences could explain the relative solubilities of humic acid and fulvic acid.

3. The $^{13}$C NMR spectra of six humic acid samples from marine and estuarine sediments are shown in Fig. 12.P.1 (From Hatcher, P.G. and Orem, W.H., Structural interrelationships among humic substances in marine and estuarine sediments, In *Organic marine geochemistry* (ed. M.L. Sohn), American Chemical Society, Washington, DC; 1986). The samples are from various locations—the Potomac River, on the coast of the United States; New York Bight, 16 km off the New York coast, on the continental shelf; Walvis Bay, from the deep continental shelf on the Namibian coast; Mangrove Lake, from a marine lake in Bermuda.

   We indicated that accurate quantitative analysis of solid samples by NMR is problematic, but it is appropriate to develop semi-quantitative ideas based on the spectra: Which HA samples appear to be lowest in aromatic carbons? Where marine algal material is the principal source of HA, the structure is predominantly aliphatic—which samples may have the highest aromatic content? Why? The Mangrove Lake surface sediment HA appears to be enriched in carbohydrate material. Why might the concentration decline with depth? To what might the small peak at 50 ppm be due?

4. Water from a lake in an area with limestone bedrock has a chemical composition as follows:

| | |
|---|---|
| Calcium | 95 mg L$^{-1}$ |
| Magnesium | 13 mg L$^{-1}$ |
| Sodium | 17 mg L$^{-1}$ |
| Potassium | 4 mg L$^{-1}$ |
| Hydrogen carbonate (as HCO$_3^-$) | 338 mg L$^{-1}$ |
| Sulphate (as SO$_4^{2-}$) | 7 mg L$^{-1}$ |
| Chloride | 12 mg L$^{-1}$ |
| Fluoride | 0.2 mg L$^{-1}$ |
| Nitrate (as NO$_3^-$) | 3 mg L$^{-1}$ |
| pH | 6.8 |

If the organic matter (HM) content of the water is 12 mg L$^{-1}$, show that charge balance is adequately maintained.

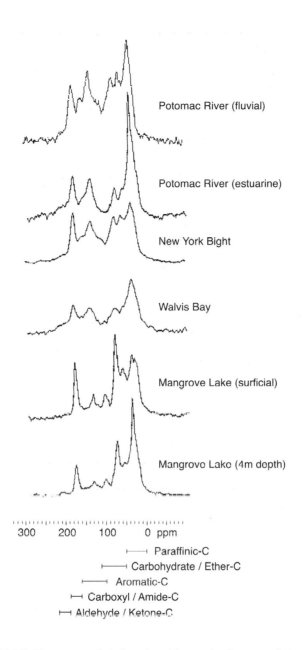

Paraffinic-C

Carbohydrate / Ether-C

Aromatic-C

Carboxyl / Amide-C

Aldehyde / Ketone-C

**Fig. 12.P.1** Carbon-13 NMR spectra of six humic acid samples from marine and estuarine sediments.

5. The chemical nature of the HM in very deep (up to > 1000 m) cores of sediments from the Black Sea have been studied in detail in order to understand the processes that occur over very long periods of time. Average percentage values of carbon, hydrogen, and oxygen concentrations in two depth ranges are:

|         | C    | H   | O    |
|---------|------|-----|------|
| Shallow | 56.3 | 5.2 | 31.9 |
| Deep    | 60.9 | 5.2 | 27.9 |

What do these results suggest about the long-term processes occurring within the organic sediments? (These data are summarized from work by A.Y., Huc, B.M. Durand, and J. Monin as reported in Rashid, M.A., *Geochemistry of marine humic compounds*, Springer, New York; 1985.)

6. The equivalent weight of HM is sometimes defined as the molar mass per mol of carboxylic plus phenolic groups. For a particular sample of fulvic acid, there are 4.2 and 2.2 mmol of carboxylic and phenolic groups, respectively, per gram of the FA. What is the equivalent weight?

7. The dissolved organic matter isolated from a natural water sample has the following composition:

   C   44.2% by weight

   H   6.1% by weight

   O   46.7% by weight

   N   1.7% by weight

   Molar mass 1900 daltons.

   Is the DOM likely to be fulvic acid or humic acid? Explain your reasoning.

8. Consider both the degradative and synthesis reaction series for the formation of various types of humic material. Can both theories be used to explain the relative amounts of carbon and oxygen in the three classes of HM as described in the text?

9. Look up the structure of the following pesticides and show how they could interact with generic humic material.

   | | | | |
   |---|---|---|---|
   | Atrazine | Chlorpyrifos | Chlorothalonil | Malathion |
   | Fenitrothion | Chlordane | Aldicarb | Trifluralin |

# Chapter 13
# Metals and semi-metals in the hydrosphere

The focus of this chapter is on the nature, occurrence, and behaviour of dissolved metals and semi-metals found in various aquatic systems, and their influence within the environment. Here we will discuss:

- the types of metals and semi-metals that occur in water
- how various classification schemes are used to explain complexation behaviour of metals, including an environmental classification
- how chemical speciation influences the bioavailability of metals and semi-metals
- the elements calcium, copper, and mercury as examples for understanding the environmental chemical behaviour of different classes of metals.

Of all the environmental issues involving water, contamination by toxic metals is one of the most visible. This problem is often referred to as 'heavy-metal pollution', a term we do not recommend as will be explained in due course. Not only is the issue visible, it is of critical importance to ecosystems and people throughout the world. One extremely important example is the widespread contamination of groundwater in Bangladesh and eastern India by the semi-metal (or metalloid) arsenic. In this region of South Asia, many millions of people are affected, and it has been described as 'the largest mass poisoning in the world, perhaps in history'. Arsenic contamination of surface and groundwater also occurs to a more limited extent in many other countries with the element sometimes originating from natural geological sources and sometimes associated with human activities. Contamination of water by other metals and semi-metals occurs at locales throughout the world, and later in this chapter we will examine the chemistry of three important metals, calcium, copper, and mercury.

Arsenic, element 33, is a semi-metal appearing below phosphorus in the periodic table. On the Earth's surface, its crustal abundance averages out at approximately $1.5~\mu g~g^{-1}$, mostly in association with sulphide minerals. In water, its chemistry is akin to that of phosphorus, with 3 and 5 oxidation states being most common. The background concentration in both oceans and fresh water is about $1.5~\mu g~L^{-1}$ with the arsenic typically in the form of $AsO_3^{3-}$ (arsenite) and $AsO_4^{3-}$ (arsenate) species. See Chapter 10, Problem 9 for a partial pE / pH diagram for arsenic species.

### The arsenic problem in South Asia

During the 1970s, residents of eastern India and Bangladesh were strongly encouraged to tap the groundwater supply of the Ganges–Brahmaputra–Meghna Delta region as an alternative to using surface water. The rationale was that groundwater is much less subject to bacterial contamination and its consumption would limit the incidence of water-borne diseases among those who used it for domestic purposes. The same groundwater was also widely used as a source of irrigation. What was not known was that many of the new wells supplied water whose arsenic content was substantially above that allowed in drinking-water standards. Beginning in the early 1980s, patients began to show up at clinics throughout the eastern subcontinent with skin lesions that eventually developed into more serious symptoms and in many cases led to death. Estimates are that the health of 20 to 30 million people is at risk in this region as a consequence of arsenic contamination in their drinking water.

The origin of the arsenic in water pumped from the tubewells is still uncertain. In rocks and sediments of the Delta, arsenic is found in small amounts as arsenides ($As^{3-}$, such as $Zn_3As_2$ or AlAs) in association with a variety of sulphide-containing minerals. Most of these minerals are highly insoluble, at least when considering their simple dissolution. Three hypotheses (which include apparently contradictory components) have been put forward to explain the means by which arsenic moves from the insoluble forms to soluble species in the water.

1. It is suggested that lowering of the water table through excessive groundwater withdrawal brings the sedimentary minerals into contact with air, thus enabling their oxidation. Under these conditions arsenopyrite will oxidize, releasing iron (II) and (III), sulphate, arsenite, and arsenate into solution.

2. Reduction of hydrated $Fe_2O_3$ minerals to which arsenic is adsorbed has been postulated to occur where there are organic-rich sediments that create reducing conditions.

3. Release of arsenite and arsenate that is sorbed to other minerals could occur due to displacement from sorption sites by phosphate. Elevated levels of phosphate, derived from fertilizers, are present to some degree in the groundwater of this area.

There is evidence to justify each of these hypotheses, and it is still not clear which one or ones may be the most important causes for arsenic release. Identifying causes is complicated by the apparently random distribution of contaminated wells among and within villages in the area.

The presence of arsenic contamination of groundwater is obviously a problem of the utmost importance in that region of the world. No clear remedy has, however, been developed, in part due to the uncertainties in understanding the cause. Various treatment methods involving the use of coagulants, adsorbents, and ion exchangers have been tried with limited practical success. Such methods require additional devices or materials and in all cases the removal of arsenic from the water produces a toxic residue or sludge that must be disposed of. In most cases, the immediate solution has been to identify the unsafe wells by painting them red and safe ones green.

This is but one example of a serious problem associated with element contamination. Around the world many examples involving other metals and semi-metals have been observed and described.

## 13.1 Metals in the aqueous environment

Metals and semi-metals make up about 75% of all the elements on the periodic table, and they are found as ions and complex compounds throughout the hydrosphere. The concentration of metal species in various types of water covers a wide range. In the oceans, the concentration of sodium ion is approximately 0.48 mol L$^{-1}$ (Table 9.1, 10 760 mg kg$^{-1}$) and magnesium ion is present in sufficiently large amounts that extraction from marine water is a viable source of the element. Other sea-water metal concentrations range down to ultratrace levels (Appendix B.1).

Metals in fresh water are usually present in small concentrations. In water bodies located in regions with carbonate bedrock, calcium ion concentrations are in the mmol $L^{-1}$ range, but most other elements are found in much lower concentrations. There are special situations, however, where unusually high levels of metals are found. The saline lakes in the Rift Valley of East Africa are an example, in that they contain extremely large amounts of alkali and alkaline earth elements. The concentration of sodium carbonate in Lake Magadi in Kenya (previously described in Chapter 11, in terms of alkalinity) greatly exceeds the solubility limit and much of the 'lake' is actually solid, with only scattered pools of saturated brine.

There are other special situations as well. Pit lakes—abandoned open pit mines that penetrate below the water table—often have elevated levels of metal ions derived from the ore minerals. For example, some pit lakes from the Robinson Mining District in Nevada, USA have copper and zinc concentrations as high as 0.6 and 0.8 mmol $L^{-1}$, respectively. In waterlogged soils, the redox-sensitive elements, iron and manganese, are reduced to soluble $+2$ species and their concentrations may then also approach the mmol $L^{-1}$ range in the interstitial water.

The environmental chemistry of metals is of interest and importance for several reasons:

- Besides being dissolved in water (to varying degrees) metals are present in the solid components of the Earth and there are ongoing interactions between the soluble forms and the solid phase materials. The extent to which metals remain in solution is one factor that determines their mobility in the environment

- Some metals such as potassium and calcium are important nutrients required in substantial amounts by plants, animals, and microorganisms

- Other metals such as copper and zinc are also nutrients, but the amounts of these that are required by organisms are very small

- These micronutrients, if present in excessive amounts, can be toxic so there is a range of concentrations, sometimes narrow, that is suitable for supporting life processes

- Still other metals such as cadmium and mercury are not essential nutrients for most organisms and even very small concentrations can be toxic to many living organisms.

In general, a description of the pathways and cycles through which metals in water interact with the associated soils, sediments, and biota is referred to as the subject of *metal biogeochemistry*.

The discussion in this chapter will centre on developing general ideas that allow us to understand the range of behaviour exhibited by various metals in different environmental situations.

Metals exist in the aqueous environment in a variety of forms—as aquo complexes that are fully protonated or partially deprotonated, as complexes with inorganic ligands such as chloride and carbonate, and as complexes with naturally occurring organic molecules including discrete molecules that have low molar mass like citrate as well as macromolecules such as fulvic acid.[1] In a given situation, the particular *species distribution* of a metal depends on the properties of the metal itself as well as on the availability and nature of potential ligands. Other bulk aqueous solution properties including the ionic strength, pH, and redox status also play a role in defining species distribution. It is helpful to assemble some of the vast collection of knowledge about individual cases into some general principles. To do this we will make use of basic concepts of inorganic chemistry.

---

[1] No distinction is made here or in subsequent discussions between true complexes and ion pairs. The usual definitions are as follows. In a complex, there is a direct covalent bond between the metal and the ligand. In an ion pair, both species are surrounded by a hydration sphere, but they maintain combined existence as a single unit held together by electrostatic forces.

## 13.2 **Classification of metals**

The term 'heavy metals' is frequently encountered in the environmental literature and is usually used to designate metals that are toxic because they cause adverse biological reactions. Originally, the expression had a scientifically legitimate origin as it was coined to refer to metals such as lead and mercury. Lead has an atomic mass of 207.2 g mol$^{-1}$ and a specific gravity of 11.34; the corresponding values for mercury are 200.59 g mol$^{-1}$ and 13.55. They are, therefore, heavy metals in every sense of the name. The terminology, however, is sometimes applied indiscriminately and it is not unusual to find a list of heavy metals that includes elements like aluminium (atomic mass 26.98 g mol$^{-1}$ and specific gravity 2.70) as well as semi-metals like arsenic. Aside from the semantic problem, there is no chemical basis for deciding which metals should be included in this category.

### Traditional classifications of metals

A number of categories or classifications of metals have been proposed for general use, not just for environmental purposes. Among the useful classifications is one that divides metals into type A, type B, and transition metals as first described by Ahrland *et al.* (1958).[2] In this system, type A metal ions are those with inert gas (d$^0$) electronic configurations. Such ions are characterized by spherical symmetry and low polarizability. The grouping is almost synonymous with the hard-sphere metal category of Pearson (1963).[3] Included in this grouping are the environmentally important cations Na$^+$, K$^+$, Mg$^{2+}$, Ca$^{2+}$, and Al$^{3+}$.

#### Type A metal ions

For type A metal ions, an electrostatic model to a large extent explains the stability of metal–ligand complexes—that is, the stability of these complexes is positively correlated with the charge squared to radius ($Z^2/r$) ratio for both the metal ion and the ligand species. Therefore, highly charged small ions form stronger complexes, as shown by the relative stability of most complexes with the alkaline-earth metals

$$Mg^{2+} > Ca^{2+} > Sr^{2+} > Ba^{2+}$$

Some of the properties of type A cations are:

- a preference for O- or F-containing ligands over sulphur and higher halides—for example, $Al(H_2O)_5(OH)^{2+}$ and $Al(H_2O)_4F_2^+$ are important soluble aluminium species. Weak complexes with oxyanions such as $SO_4^{2-}$, $NO_3^-$, and oxygen-containing functional groups (e.g. $-COOH$, $\diagdown C=O$) of organic molecules are also found;
- the metal ion may form insoluble OH$^-$, CO$_3^{2-}$, or PO$_4^{3-}$ compounds—for example, CaCO$_3$ and AlPO$_4$ are important solid forms of these elements;
- complexes with OH$^-$ are more stable than those with HS$^-$ or S$^{2-}$; metal sulphide soluble complexes or precipitates are not important;
- complexes with F$^-$, Cl$^-$, Br$^-$, and I$^-$ tend to be weak; the limited stability is in the order F$^-$ > Cl$^-$ > Br$^-$ > I$^-$;
- complexes with H$_2$O are more stable than those with NH$_3$ or CN$^-$.

#### Type B metal ions

Type B metal ions are ones with $nd^{10}$ and $nd^{10}(n+1)s^2$ type electron configurations. Such cations have readily distorted electronic distributions; in other words, they exhibit high polarizability.

[2] Ahrland, S., J. Chatt, and N.R. Davies, The relative affinities of ligand atoms for acceptor molecules and ions, *Quart. Rev. Chem. Soc.*, **12** (1958), 265–76.
[3] Pearson, R.J., Hard and soft acids and bases, *J. Am. Chem. Soc.*, **85** (1963), 3533–9.

This class is closely equivalent to the soft-sphere metal ion category of the Pearson system. Included in the group are $Ag^+$ ($Kr(4d^{10})$), $Zn^{2+}$($Ar(3d^{10})$), and $Pb^{2+}$($Xe(4f^{14}\ 5d^{10}\ 6s^2)$).

For type B ions, covalent bonding plays a role in complex formation and therefore an electrostatic model alone is unable to explain stability relations. On the other hand, a major factor affecting complex stability is the ability of the metal to accept electrons from the ligand. Therefore, high electronegativity (en) (in relative terms) of the metal ion and low electronegativity of the ligand donor atom would be consistent with high stability of a complex with type B metal ions.

For example, in the IIB (or 12) group of elements, the Pauling electronegativities (values in brackets) are in the order

$$Zn(1.6) < Cd(1.7) < Hg(1.9)$$

and complex stability is generally in the same order.

The trend of electronegativity of ligand donor atoms, and stability of their complexes (as measured by the overall stability constant $\beta_f$) with type B metal ions is

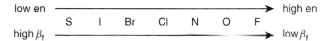

Therefore, type B cations exhibit the following complexation properties:

- in general, these metal ions form more stable complexes than do the type A cations;
- complex stability with the halides is in the order $I^- > Br^- > Cl^- > F^-$. This is the reverse of the situation observed with type A metal ions;
- complexes with ligands containing nitrogen are more stable than those with oxygen—for example, $NH_3$ is favoured over $H_2O$, and $CN^-$ is favoured over $OH^-$;
- complexes with sulphide or organosulphides are common and stable. Such compounds are frequently insoluble;
- complexes with carbon, such as organometallic complexes, are observed. The mercury compounds, $CH_3Hg^+$ and $(CH_3)_2\ Hg$, are well known examples.

### Transition metal ions

Transition metal cations are those with $nd^x$ ($0 < x < 10$) electronic configurations. Pearson has described a borderline class that is similar to, but not identical with that of the transition metal ions. Defined either way, this class exhibits properties that are intermediate between those of the type A and B classes. Borderline metal ions are able to form complexes with all types of donor ligands, with the relative importance depending on a number of factors. Second-row transition elements usually show more type B character than do those in the first row, and type B character also tends to increase somewhat as one moves from left to right on the periodic table.

Electrostatic factors play a role in determining the stability of borderline metal ions and this is reflected in a general trend (Fig. 13.1, straight line) showing increasing stability for high-spin octahedral complexes with increasing atomic number of the $+2$ ions of the first transition-metal series.

The increase in stability can be attributed to a stronger electrostatic effect because of increasing $Z^2/r$ (the (charge)$^2$/ radius) ratio associated with decreasing atomic radius across the series. The other important factor that leads to deviations from the straight line is crystal field stabilization energy. Placement of electrons in a $t_{2g}$ d orbital stabilizes the ion (as for $Sc^{2+}, Ti^{2+}, V^{2+}$), while electrons in $e_g$ d orbitals (as in $Cr^{2+}, Mn^{2+}$) reduce stability. This argument is a simplification; for example, $Cu^{2+}$ complexes tend to have greater stability than do those of $Ni^{2+}$ due to a different structure of the complexes.

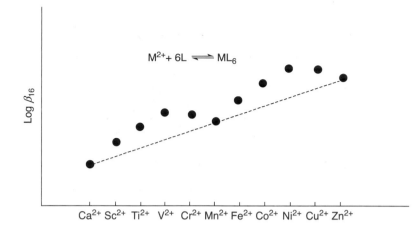

**Fig. 13.1** Trend in overall stability constants $\beta_{f6}$ for high-spin octahedral complexes involving +2 ions of transition (borderline) metals. As one moves from calcium to zinc across the periodic table, there is a gradual decrease in radius associated with the *lanthanide contraction*. Therefore, for +2 ions the value of $Z^2/r$ increases with increasing atomic number in this row of elements.

It should be noted that the trend presented here assumes a consistent +2 oxidation state. For some of these metals, this oxidation state is rarely found in the hydrosphere. Vanadium (II), for example, is virtually unknown in the environment, the common forms being vanadium (V) as the highly deprotonated $VO_2^+$ cation or the $HVO_4^{2-}$ anion and, under reducing conditions, vanadium (IV) as the $VO^{2+}$ species.

### An environmental classification of metals

A more recent attempt at classifying environmentally important metals has been made by Nieboer and Richardson (1980).[4] You will see that their system builds on the earlier concepts of Ahrland and Pearson but takes into account bonding due to both covalent and ionic interactions. To do this, a *covalent index* is plotted versus an *ionic index* (Fig. 13.2).

The covalent index $X_m^2 r$ ($X_m$ = metal-ion electronegativity, $r$ = ionic radius of metal) is a reflection of the ability of the metal to accept electrons from a donor ligand, and becomes the chemical parameter used to differentiate between type A, borderline, and type B metal ions. Values of the index are smallest for type A and largest for type B ions. The ionic index $Z^2/r$ is the same as that used to describe stability of complexes with type A ions. It measures the possibility of ionic-bond formation and thus more highly charged species tend to be found on the right-hand side of the diagram. These are also the species that tend to act as Brønsted acids as noted earlier. Overall, considering a natural sample containing a variety of ligands with different donor atoms (for example, humic material), the tendency to exist in complexed form might be expected to follow an angled trend as shown in Fig. 13.2.

Some of the functional groups found in naturally occurring ligands and involved in forming metal complexes are listed later in this chapter (Fig. 13.3).

### Observations about environmental properties of metals

It is interesting that elements such as potassium and calcium that serve as macronutrients for microorganisms, plants, and animals are found in the type A class. When dissolved in water and in their interaction with complexing ligands, they are usually found to be associated with

---

[4] Nieboer, E. and D.H.S. Richardson, The replacement of the nondescript term 'heavy metals' by a biologically and chemically significant classification of metal ions, *Environ. Pollut. (Series B)*, **1** (1980), 3.

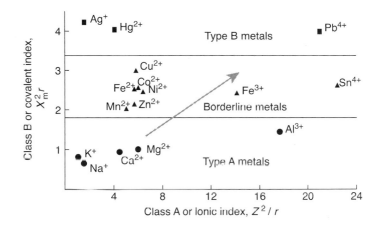

**Fig. 13.2** Classification of some metal ions of environmental importance (redrawn from Nieboer, E. and D.H.S. Richardson, The replacement of the non-descript term 'heavy metals' by a biologically and chemically significant classification of metal ions, *Environ. Pollut. (Series B)*, **1** (1980), 3). Subdivisions based on the traditional, Ahrland et al. and Pearson classifications (see footnotes 2 and 3, this chapter) are indicated by the horizontal lines. The stability of complexes increases with increasing ionic and / or covalent index, as shown by the arrow.

oxygen electron donors. Most biological micronutrients, including manganese, copper, and zinc, are found in the borderline group. In contrast to the type A metal ions, the borderline ions form stable complexes with a variety of electron donor atoms including oxygen, nitrogen, and sulphur. The type B metal ions include several that are known to be toxic to organisms. In general (there are exceptions), toxicity increases in the order type B > borderline > type A metal ions. While type B ions have a strong affinity for sulphur donor atoms, they also form more stable complexes with oxygen-donating compounds than do borderline and type A metal ions. The ability to form methylated derivatives that are stable in aqueous solutions is another feature that is characteristic of metals in this class. Methyl derivatives of type A metals decompose in water and the same is true of most borderline metals.

> **Main point 13.1** The tendency of a metal ion to form complexes with other species present in water depends in part on the nature of the metal itself. Various classification schemes have been proposed to explain and predict the favoured types of ligands and degree of complex formation. The schemes are usually based on the ability of metal ions to form ionic and covalent bonds with various types of ligands.

## 13.3 Types of complexes with metals—metal speciation in the hydrosphere

### Aquo complexes

This is the simplest form in which a metal ion can exist in water. Where no other ligand is available to form complexes with a metal in aqueous solution, it exists as an aquo complex, as we showed in Chapters 9 and 10. As was noted in the classification discussion, type A metal ions favour coordination with oxygen-containing ligands and so aquo complexes are particular common with these elements.

In the case of metal-aquo complexes, however, depending on circumstances coordinated water molecules may lose a proton. The degree to which deprotonation occurs is, to a large extent, a property of the metal ion under consideration. The pH of the solution is also important

**Table 13.1** Values of $Z^2/r$ and $pK_{a1}$ for aquo complexes of selected metal ions[a].

| Metal ion | $Z^2 r^{-1}/nm^{-1}$ | $pK_{a1}$ | Metal ion | $Z^2 r^{-1}/nm^{-1}$ | $pK_{a1}$ |
|-----------|---------------------|-----------|-----------|---------------------|-----------|
| $Na^+$ | 8.6 | 14.48 | $Ni^{2+}$ | 48 | 9.40 |
| $K^+$ | 6.6 | >14 | $Cu^{2+}$ | 46 | 7.53 |
| $Be^{2+}$ | 68 | 6.50 | $Zn^{2+}$ | 46 | 9.60 |
| $Mg^{2+}$ | 47 | 11.42 | $Cd^{2+}$ | 37 | 11.70 |
| $Mn^{2+}$ | 48 | 10.70 | $Hg^{2+}$ | 34 | 3.70 |
| $Fe^{2+}$ | 43 | 10.1 | $Al^{3+}$ | 133 | 5.14 |
| $Co^{2+}$ | 45.2 | 9.6 | $Fe^{3+}$ | 115 | 2.19 |

[a] In each case, the radius, $r$, used is for the 6-coordinate high spin metal aquo complex. $pK_a$ values are taken from Yatsimirksii, K.B. and V.P. Vasil'ev, *Instability constants of complex compounds*, Pergamon, Elmsford, New York; 1960.

in determining whether protons are lost. This type of reaction is an acid–base reaction and the general equation is

$$M(H_2O)_a^{b+} + H_2O \rightleftharpoons M(H_2O)_{a-1}(OH)^{(b-1)+} + H_3O^+ \qquad (13.1)$$

An alternative simplified description of the reaction omits the waters of hydration in the formulae.

$$M^{b+} + 2H_2O \rightleftharpoons MOH^{(b-1)+} + H_3O^+ \qquad (13.2)$$

Further deprotonation steps may also occur. Since these reactions ultimately involve a separation of two positive charges, they are favoured in the case of more highly charged ions and smaller metal ions, usually expressed in combination as the $Z^2/r$ ratio, where $Z$ is the numerical value of the charge and $r$ is the ionic radius in nm. The inverse relation between $pK_{a1}$ and $Z^2/r$ holds up well for the type A elements, but other factors are more important in the case of transition-metal ions (especially heavy ones). Table 13.1 lists the ratio and the $pK_{a1}$ values for selected metal aquo complexes.

Using its usual definition, the $pK_{a1}$ is the pH at which the aquo complex is present with half in the fully protonated form and half having lost a single proton. In solutions having pH below that of $pK_{a1}$, the metal ion is coordinated largely by water molecules, while a hydroxy group replaces a water molecule when the pH is greater than the $pK_{a1}$. From these data, therefore, it is evident that the waters of hydration surrounding metal ions with a single positive charge exist exclusively in protonated form throughout the entire pH range. Of the +2 ions, deprotonation occurs more readily for smaller species (due to the larger value of $Z^2/r$). In aqueous solutions with pH > 5.7, $Be(OH)^+$ would be more important than $Be(H_2O)^{2+}$, while the $MgOH^+$ and $CaOH^+$ species occur only at very high pH. Deprotonated species begin to assume significance in environmentally common situations for +3 ions including $Fe^{3+}$ and $Al^{3+}$. One consequence of this is that such ions can lose more than one proton, eventually becoming neutral and insoluble in water. A sample calculation illustrates this point.

Iron (III) exists in pure water as an aquo complex, $Fe(H_2O)_6^{3+}$ (for simplicity, we write its structure as $Fe^{3+}$ without the coordinated waters). Being a +3 ion, however, there is considerable tendency for deprotonation to occur. In the first step, $Fe(H_2O)_5(OH)^{2+}$ or simplified form $Fe(OH)^{2+}$ is the product and in the second step, $Fe(H_2O)_4(OH)_2^+$, i.e. $Fe(OH)_2^+$ is formed. As well as the two deprotonated products, a bridged dimer can also be formed from two of the second deprotonated iron species. This *polynuclear species* of iron has the structure shown, but can be written in simplified form as $Fe_2(OH)_2^{4+}$.

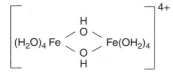

Aluminium and some of the other metals are also known to form polynuclear species.

Taking into account the four species indicated in the equations below, we will calculate their equilibrium concentrations in pure water at pH = 7.00. The relevant equilibria and equilibrium constants are as follows.

$$Fe(OH)_3 \rightleftharpoons Fe^{3+} + 3OH^- \qquad K_{sp} = 1.6 \times 10^{-39} \qquad (13.3)$$

$$Fe^{3+} + 2H_2O \rightleftharpoons FeOH^{2+} + H_3O^+ \qquad K_{a1} = 6.3 \times 10^{-3} \qquad (13.4)$$

$$FeOH^{2+} + 2H_2O \rightleftharpoons Fe(OH)_2^+ + H_3O^+ \qquad K_{a2} = 3.2 \times 10^{-4} \qquad (13.5)$$

$$2Fe^{3+} + 4H_2O \rightleftharpoons Fe_2(OH)_2^{4+} + 2H_3O^+ \qquad K_{ad} = 1.3 \times 10^{-3} \qquad (13.6)$$

Note again that the waters of hydration associated with the iron are not shown.

---

**Example 13.1  Concentrations of iron species in pure water**
Using reaction 13.3,

$$[Fe^{3+}][OH^-]^3 = K_{sp}$$

$$[Fe^{3+}][10^{-7.00}]^3 = 1.6 \times 10^{-39}$$

$$[Fe^{3+}] = 1.6 \times 10^{-18} \text{ mol L}^{-1}$$

Using reaction 13.4,

$$\frac{[FeOH^{2+}][H_3O^+]}{[Fe^{3+}]} = 6.3 \times 10^{-3} \text{ mol L}^{-1}$$

Now substitute in the value of $[Fe^{3+}]$ calculated above

$$[FeOH^{2+}] = \frac{6.3 \times 10^{-3} \times 1.6 \times 10^{-18}}{1.0 \times 10^{-7}}$$

$$= 1.0 \times 10^{-13} \text{ mol L}^{-1}$$

Similarly, calculations are done to determine the equilibrium concentrations of the two additional species.

$$[Fe(OH)_2^+] = 3.2 \times 10^{-10} \text{ mol L}^{-1}$$

$$[Fe_2(OH)_2^{4+}] = 3.3 \times 10^{-25} \text{ mol L}^{-1}$$

Therefore, the total concentration of the various aquo species of $Fe^{3+}$ in water at pH = 7.00 is $3.2 \times 10^{-10}$ mol L$^{-1}$ (= $1.8 \times 10^{-8}$ g L$^{-1}$ = 0.018 ppb), and the most significant species is $Fe(OH)_2^+$.

---

Clearly, iron (III) is extremely insoluble in pure water. The solubility is greatly enhanced when ligands are present to form a stable complex. In the hydrosphere, these ligands are frequently derived from the natural organic matter (NOM) present in the water. However, in general, simple iron (II) species are more soluble than the corresponding iron (III) species, so that reducing, low pE conditions lead to increased concentrations of the element in the hydrosphere. Refer back to Fig. 10.P.1 which shows that the most important forms of iron in the hydrosphere are insoluble $Fe(OH)_3$ in oxygenated water and soluble $Fe^{2+}$ in water that has been depleted of oxygen.

Soluble species are not the only forms in which metals exist in the hydrosphere. Depending on the element and other environmental conditions, a significant portion may be present in association with suspended material.

---

**Main point 13.2** The simplest dissolved forms of any metal in the hydrosphere are the aquo complexes. To varying degrees, deprotonation of these complexes occurs, especially in the case of metal species with a large charge squared to radius ratio. Species such as aluminium (III) therefore act as Brønsted acids in aqueous media.

---

### Other complexes with inorganic ions and molecules

In most natural waters that have not been altered by significant human interventions, there are few inorganic species that can act as ligands and form complexes with dissolved metals. Carbonate and bicarbonate, depending on pH, are exceptions and are found in surface water everywhere. Chloride and sulphate are also common ions in natural systems, especially in seawater where their concentrations are 0.029 and 0.56 mol $L^{-1}$, respectively. Saline groundwater is increasingly being found around the world. In parts of southern and western Australia, unusually large concentrations are present—for example in the gold-mining region of western Australia, concentrations of naturally occurring chloride about three times greater than that in seawater have been observed.

Table 13.2 gives the principal inorganic species of some environmentally important metal ions. In compiling the table, it was assumed that the metals are present within the typical 'normal' concentration range in water and that the water contains carbonate, sulphate, and chloride at levels approximately equal to that found in average river water. Additional species associated with the marine environment are also noted.

Several of the general features of the classifications described above are underlined by the data in the table.

- All the metals form aquo complexes in the aqueous media. Metals such as sodium and potassium have a small value of $Z^2/r$ and the coordinated water molecules remain protonated in all situations. For metals with a larger ratio (e.g. Al(III)), deprotonation occurs more readily and, with metals in very high oxidation states (eg. Cr(VI) and Mo(VI)), oxyanions are the principal species.

- Those metals that are subject to redox reactions are present as different species in oxidizing (high pE) compared to reducing (low pE) environments.

- In sea water, the high concentration of chloride and, to a lesser extent, sulphate favours formation of complexes with these ligands in the case of certain metals.

- The tendency for formation of complexes with ligands other than water is generally in the order: type B metal ions (e.g. Ag(I), Hg(II)) > borderline metal ions (e.g. Mn(II), Zn(II)) > type A metal ions (e.g. Ca(II), Al(III)).

### Complexes with humic material

Metals in water are also frequently found in association with organic matter. As illustrated in Fig. 12.3 humic material (HM), whether it is dissolved in water or present as part of the solid phase in soils and sediments, has functional groups that are capable of acting as ligands in forming complexes with metals. Some metals such as the alkaline-earth elements react by forming relatively weak ionic bonds at the negative sites on the deprotonated humic molecules. On the other hand, other elements such as Cu(II), Pb(II), and the trivalent metals Al(III) and Fe(III) have large stability constants. Complexation in these cases involves covalent bonding, and bidentate chelates are probably important. Figure 13.3 shows some of the functional groups associated with humate molecules.

**Table 13.2** Principal inorganic aqueous species of environmentally important metal ions[a] (only inorganic species are considered and coordinated water molecules are not included in the formulae).

| | pH = 4 Oxidizing environment | pH = 4 Reducing environment | pH = 7[b] Oxidizing environment | pH = 7[b] Reducing environment | pH = 10 Oxidizing environment | pH = 10 Reducing environment |
|---|---|---|---|---|---|---|
| Sodium | $Na^+$ | $Na^+$ | $Na^+$ | $Na^+$ | $Na^+$ | $Na^+$ |
| Potassium | $K^+$ | $K^+$ | $K^+$ | $K^+$ | $K^+$ | $K^+$ |
| Magnesium | $Mg^{2+}$ | $Mg^{2+}$ | $Mg^{2+}$, $MgSO_4^0(sw)$ | $Mg^{2+}$, $MgSO_4^0(sw)$ | $Mg^{2+}$ | $Mg^{2+}$ |
| Calcium | $Ca^{2+}$ | $Ca^{2+}$ | $Ca^{2+}$, $CaSO_4^0(sw)$ | $Ca^{2+}$, $CaSO_4^0(sw)$ | $Ca^{2+}$ | $Ca^{2+}$ |
| Aluminium | $Al^{3+}$, $AlOH^{2+}$ | $Al^{3+}$, $AlOH^{2+}$ | $Al(OH)_2^+$, $Al(OH)_3^0$, $Al(OH)_4^-(sw)$ | $Al(OH)_2^+$, $Al(OH)_3$, $Al(OH)_4^-(sw)$ | $Al(OH)_4^-$ | $Al(OH)_4^-$ |
| Vanadium | $H_2VO_4^-$, $VO_2^+$ | $VO^{2+}$ | $H_2VO_4^-$, $HVO_4^{2-}$, $V_{10}O_{28}^{6-}$ | $VO^{2+}$ | $VO_4^{3-}$ | |
| Chromium | $HCrO_4^-$ | $CrOH^{2+}$ | $HCrO_4^-$, $CrO_4^{2-}$ | $CrOH^{2+}$, $Cr(OH)_2^+$ | $CrO_4^{2-}$ | $Cr(OH)_4^-$ |
| Manganese | $Mn^{2+}$ | $Mn^{2+}$ | $MnO_2^-$, $MnCl^+(sw)$ | $Mn^{2+}$, $MnCl^+(sw)$, $MnSO_4(sw)$ | $MnC_2^0$ | $MnCO_3^0$ |
| Iron | $FeOH^{2+}$, $Fe(OH)_2^+$ | $Fe^{2+}$ | $Fe(OH)_3^0$ | $Fe^{2+}$, $FeCO_3^0$ | $Fe(OH)_4^-$ | $FeOH^+$, $Fe(OH)_2^0$ |
| Cobalt | $Co^{2+}$ | $Co^{2+}$ | $Co^{2+}$, $CoCO_3$ | $CoCO_3$ | $Co_3O_4$ | $CoCO_3$ |
| Nickel | $Ni^{2+}$, $NiSO_4^0$ | $Ni^{2+}$ | $Ni^{2+}$, $NiHCO_3^+$, $NiCl^+(sw)$ | $Ni^{2+}$, $NiHCO_3^+$, $NiCl^+$ sw | $NiOH^+$, $Ni(OH)_2^0$, $NiCO_3$ | $NiOH^+$, $Ni(OH)_2^0$, $NiCO_3$ |
| Copper | $Cu^{2+}$ | $Cu^{2+}$ | $Cu^{2+}$, $CuOH^+$, $CuHCO_3^+$, $CuCl^+(sw)$ | $Cu^{2+}$, $CuOH^+$, $CuHCO_3^+$, $CuCl^+(sw)$ | $Cu(OH)_2^0$, $Cu(CO_3)_2^{2-}$ | $Cu(OH)_2^0$, $Cu(CO_3)_2^{2-}$ |
| Zinc | $Zn^{2+}$ | $Zn^{2+}$ | $Zn^{2+}$, $Zn(OH)_2^0$, $ZnCl^+(sw)$ | $Zn^{2+}$, $Zn(OH)_2^0$, $ZnCl^+(sw)$ | $Zn(OH)_2^0$ | $Zn(OH)_2^0$ |
| Molybdenum | $HMoO_4^-$ | | $HMoO_4^-$ | | $HMoO_4^-$, $MoO_4^{2-}$ | |
| Lead | $Pb^{2+}$, $PbSO_4^0$ | $Pb^{2+}$ | $Pb^{2+}$, $PbOH^+$, $PbHCO_3^+$, $PbCl^+(sw)$, $PbSO_4^0(sw)$ | $Pb^{2+}$, $PbOH^+$, $PbHCO_3^+$, $PbCl^+(sw)$, $PbSO_4^0(sw)$ | $Pb(OH)_2$, $PbCO_3$, $Pb(CO_3)_2^{2-}$ | $Pb(OH)_2$, $PbCO_3$, $Pb(CO_3)_2^{2-}$ |
| Mercury | $HgOH^+$, $Hg(OH)_2^0$, $HgCl_2^0$ | $Hg^0$ | $Hg(OH)_2^0 \geq HgCl_2^0$, $HgCl_4^{2-}(sw)$, $HgCl_3^-(sw)$ | $Hg^0$ | $Hg(OH)_2^0$ | $Hg^0$ |

[a] Sulphate, chloride, and carbonate in concentrations that approximate those found in average river water are assumed to be present in the water. The metal concentrations are also assumed to be in the range of those found in 'normal' water. Some neutral species such as $Fe(OH)_3$ and $MnO_2$ are highly insoluble and will be present as colloids even when the metal concentration is very small.

[b] The notation (sw) indicates that this additional species is present in sea water. Note that sea water pH is approximately 8.

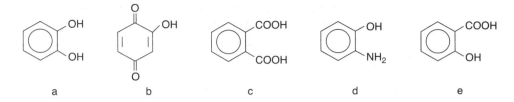

**Fig. 13.3** Examples of humic material functional groups available for complexation reactions. Structures c (phthalate) and e (salicylate) are thought to be particularly important players in chelate-formation processes. Small amounts of other elements like sulphur and phosphorus are also found in humic materials and can also contribute to their complexing ability.

A possible reaction between lead and a portion of an HM molecule is given in reaction 13.7. The chelate formed with the salicylate functional group is a stable six-membered system

$$Pb^{2+} + HM - \bigcirc - COOH \longrightarrow HM - \bigcirc - C=O + 2H^+ \qquad (13.7)$$

The waters of hydration have been omitted for clarity.

The stability constant, $K_f$, for this reaction has been determined to be approximately $10^6$. In a natural situation, the extent of complexation depends on a number of factors.

- The nature of the metal ion. Because bonding may involve both ionic and covalent forces, various properties of the metal affect stability as we have seen earlier. Most important among these are the ionic index—so that trivalent ions like $Al^{3+}$ tend to be strongly bonded and monovalent ions like $Na^+$ and $K^+$ have only a weak association—and the ability to form covalent linkages as indicated by the covalent index. The alkaline earth metal ions $Ca^{2+}$ and $Mg^{2+}$ react to a much smaller extent than divalent transition metal ions such as $Cu^{2+}$ or $Pb^{2+}$ due to their inability to bond covalently.

- Ambient solution pH. A low solution pH means competition from hydronium ions for sites on the functional groups that react with metals and therefore a reduced tendency for complex formation. For example, aluminium in the pore water of acidic soils is present in the form of the aquo complex, depending on pH in a partially deprotonated form. On the other hand, when the water has near neutral pH, any soluble species are likely to be in complexed (usually with soluble HM) form. Conformational changes occurring at low pH may further inhibit reaction with metals.

- Ionic strength. The ability of humic material to react with transition metals is inversely related to solution ionic strength. There are two reasons for this: competition for ligand sites from the cations (especially alkaline earth cations) which contribute to the increased ionic strength; and availability of anions (like $Cl^-$, $SO_4^{2-}$, and $HCO_3^-$) to react with metals, thus inhibiting metal–humate reactions.

- Availability of functional groups. This depends on both the concentration and nature of the humic material. An estimate of the *maximum* complexation capacity may be made by assuming that metals react in a 1:1 ratio with humic functional groups.

Given that humic materials contain a heterogeneous collection of functional groups that vary from source to source, it becomes difficult to estimate the extent of complex formation between

**Table 13.3** Conditional stability constants (pH 5.0) for soluble fulvic acids and selected metal ions[a].

|        | $Mg^{2+}$ | $Ca^{2+}$ | $Mn^{2+}$ | $Co^{2+}$ | $Ni^{2+}$ | $Cu^{2+}$ | $Zn^{2+}$ | $Pb^{2+}$ |
|--------|-----------|-----------|-----------|-----------|-----------|-----------|-----------|-----------|
| $K_f'$ | $1.4 \times 10^2$ | $1.2 \times 10^3$ | $5.0 \times 10^3$ | $1.4 \times 10^4$ | $1.6 \times 10^4$ | $1.0 \times 10^4$ | $4.0 \times 10^3$ | $1.1 \times 10^4$ |

[a] Schnitzer, M and S.U. Khan, *Soil organic matter*, Elsevier Scientific Publishing Company, Amsterdam; 1978.

particular metal ions and humic acid in general. Nevertheless, stability constants for such reactions have been determined and are useful in a semi-quantitative sense for estimating the degree to which complexes might be expected to form in a given situation. Because protonated functional groups have a wide range of $pK_a$ values, it is best to report the constants as conditional stability constants ($K_f'$) defined at a particular pH. A further refinement would be to establish the ionic strength associated with the defined values. Table 13.3 gives values for $K_f'$ for soluble fulvic acid (FA) and several metal ions at pH 5.

---

**Example 13.2 Concentration of binding sites (complexation capacity) in fulvic acid**
Consider water of pH 5, containing 85 μg $L^{-1}$ (1.45 μmol $L^{-1}$) of nickel and 8 mg $L^{-1}$ of soluble fulvic acid. In order to calculate the concentration of complexed nickel, we must know or estimate the concentration of functional groups capable of binding with nickel. As we saw in Chapter 12, a reasonable estimate for this is around 5 mmol $g^{-1}$ of the fulvic acid. This then means that the concentration of potential binding sites is

$$8 \text{ mg L}^{-1} \times 5 \text{ mmol g}^{-1} = 40 \text{ μmol L}^{-1}.$$

The 40 μmol $L^{-1}$ corresponds to an ability to form soluble complexes with about 2.3 mg $Ni^{2+}$ $L^{-1}$, far more than is available in the water.

---

It is essential to realize that determining the concentration of binding sites allows us to estimate only the *potential* for reacting with a metal. The calculated value is called the complexation capacity and is characteristic of a water body at a given time. Typical values for complexation capacity for particular types of water bodies are:

| | |
|---|---|
| rivers | 1–2 μmol $L^{-1}$ |
| lakes | 2–5 μmol $L^{-1}$ |
| ponds | 5–15 μmol $L^{-1}$ |
| swamps | >15 μmol $L^{-1}$ |

The complexation capacity determined in the above example is clearly characteristic of water containing a large concentration of soluble organic matter.

The *extent* to which reaction actually occurs depends on the other factors as noted previously. Using the conditional stability constant in Table 13.3, we can calculate the percentage of nickel that would be complexed with FA in the present situation. Without showing the charges and forms of the species involved, the general reaction of nickel ($Ni^{2+}$) with FA can be represented as

$$Ni + FA \rightarrow NiFA \tag{13.8}$$

For the reaction, the conditional stability constant is

$$K_f' = \frac{[NiFA]}{[Ni_u][FA_u]} = 1.6 \times 10^4 \tag{13.9}$$

In this formulation, $[Ni_u]$ refers to the total concentration of all soluble nickel species that are not complexed with fulvic acid and $[FA_u]$ is similarly the concentration all of the fulvic acid not complexed with nickel. The latter concentration is expressed in terms of functional groups available for complexation and we are assuming that no other metals are present to compete for these sites.

---

### Example 13.3 Free and complexed Ni in water containing fulvic acid

Calculate the concentration of free and complexed nickel in the water described in Example 13.2. Since the ligand is available in large excess, we can make the approximation that the concentration of binding sites on fulvic acid (called $C_{FA}$) is equal to $[FA_u]$, $4.0 \times 10^{-5}$ mol L$^{-1}$, as calculated above. The total concentration of nickel is 1.45 μmol L$^{-1}$. Let the value of uncomplexed nickel, $[Ni_u] = u$.

$$\underset{\underset{Ni_u}{\uparrow}}{\frac{(1.45 \times 10^{-6} - u)}{u \times 4.0 \times 10^{-5}}} \overset{NiFA}{\underset{FA_u}{}} = 1.6 \times 10^4$$

Therefore

$$u = [Ni_u] = 8.8 \times 10^{-7} \text{ mol L}^{-1}$$

and

$$[NiFA] = 5.7 \times 10^{-7} \text{ mol L}^{-1}$$

---

About 40% of the nickel in this water is complexed by the fulvic acid. In a real environmental situation, the calculation would be complicated by the fact that other metals present would compete with nickel for binding sites on the soluble organic matter. We must also restate the caveat that these calculations assume equilibrium conditions.

## Complexes with ligands of anthropogenic origin

In addition to the naturally occurring ligands—sulphate, chloride, anions of organic acids, humic material—that are found in water in various environments, there are also individual compounds resulting from industrial, general urban, and agricultural activities that find their way into surface and groundwater. Some of these are also able to complex with metals. When an insoluble compound is formed, the effect is to remove the metal from the aqueous phase. If the compound is soluble, the mobility of the metal in the environment is increased. There are a number of situations where enhanced solubility is of major concern.

Complexing agents that are released as effluents into water bodies include the following.

- Ammonia—resulting from the decay of nitrogen-containing organic wastes.
- Sulphide, sulphite, and sulphate—discharged from pulp and paper mills, which commonly use sulphur-based compounds as a component in pulp bleaching processes.
- Phosphate—present in human waste and a constituent of some detergents; therefore present in municipal wastewater. Also released mostly in particulate form from agricultural run-off where phosphate fertilizers have been used. The phosphate cycle is discussed in the next chapter.
- Cyanide—used in various industrial processes including extraction of gold from ore minerals.
- EDTA (ethylenediaminetetraacetic acid)—used for industrial cleaning, in the photographic, textile, and paper industries, and in detergents in some countries such as Germany.
- NTA (nitrilotriacetic acid)—a detergent builder employed as a phosphate substitute in some countries such as Canada and several countries in northern Europe.

### Nitrilotriacetic acid in water

The behaviour of NTA in water provides an interesting case study of the environmental effect of introducing a human-produced complexing agent into natural water. Where its use is permitted, NTA is present in detergents at a concentration of about 15% by mass in order to act as an efficient binder for calcium and other ions associated with water hardness. Sequestering the metals improves the action of the surfactants in the detergent. As we shall see, it is the same property of being able to form strong complexes with metals that makes NTA a valuable component in detergents that also makes it a potential problem when present in natural water bodies.

The structural formula of NTA (molar mass = 191.8 daltons) is shown here.

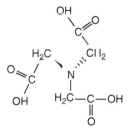

As can be seen from the formula, NTA is a triprotic carboxylic acid. The acid dissociation constants are $pK_{a1} = 1.66$, $pK_{a2} = 2.95$, and $pK_{a3} = 10.28$.

Complexation with metal ions is a simple one-step reaction in which the NTA chelate is coordinated with the metal in a tetrahedral configuration involving the three carboxyl groups and the nitrogen atom. The complex with copper is shown in Fig. 13.4. Table 13.4 gives the stability constants, as $\log K_f$, for the 1:1 complexes of a number of metals with NTA.

There is concern that when NTA is present in an urban wastewater stream, it will be released into water bodies where it could enhance the solubility of metals, especially the borderline and type B metal ions whose presence at elevated concentrations in drinking water is undesirable.

When wastewater is subjected to biological treatment processes, such as the activated sludge process (Chapter 16), it has been shown that a considerable amount of NTA is degraded by microbial digestion to produce carbon dioxide, water, inorganic nitrogen, and cellular mass. Breakdown is slower under low-temperature conditions. Where the wastewater contains relatively high concentrations of metals such as copper, nickel, and lead, these metals form chemically stable complexes with NTA. In contrast to the free ligand, the metal complexes are quite resistant to microbial degradation.

Concentrations of NTA in water just below waste streams have been found to reach several hundred $\mu g \, L^{-1}$, especially in cold climates where biodegradation is incomplete. In Example 13.4 we will consider a situation where dissolved copper (II) is present in a waste stream that contains a small residual concentration of NTA. We will determine the species in which these two entities exist in the water, neglecting the presence of other possible complexing agents or metal ions.

In order to carry out the calculation, we make use of the formation reaction for the complex, using $H_3T$ as an abbreviation for the triprotic NTA.

$$Cu^{2+} (aq) + T^{3-} (aq) \rightarrow CuT^- (aq) \quad \log K_f = 12.96 \tag{13.10}$$

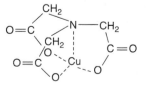

**Fig. 13.4** The tetrahedral complex formed between nitrilotriacetic acid (NTA) and copper.

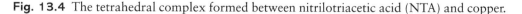

**Table 13.4** Log $K_f$ values for the 1:1 complexes between selected metals and NTA[a].

| Metal ion | log $K_f$ | Metal ion | log $K_f$ |
|-----------|-----------|-----------|-----------|
| $Mg^{2+}$ | 5.43 | $Mn^{2+}$ | 7.44 |
| $Ca^{2+}$ | 6.45 | $Cu^{2+}$ | 12.96 |
| $Fe^{2+}$ | 8.83 | $Zn^{2+}$ | 10.66 |
| $Fe^{3+}$ | 16.26 | $Pb^{2+}$ | 11.4 |

[a] Aderegg, G., Critical survey of stability constants of NTA complexes, *Pure Appl. Chem.*, 54 (1982), 2693.

While this is the reaction corresponding to the definition of formation constant, in the water itself only a fraction of the copper and NTA are in the simple forms shown. We can calculate those two fractions using methods described earlier.

It is important to recall that copper, like any other metal ion, exists in the aqueous solution as a hydrated species, and deprotonation of the coordinated waters can take place. For copper in water, the first (and only significant) deprotonation step has a $pK_{a1}$ value of 7.53 (Table 13.1). For the 4-coordinated aquo complex, the reaction can be written as

$$Cu(H_2O)_4^{2+} \text{ (aq)} + H_2O \rightleftharpoons Cu(H_2O)_3(OH)^+ \text{ (aq)} + H_3O^+ \text{ (aq)} \quad (13.11)$$

The formula, $Cu(H_2O)_4^{2+}$, can be simplified as $Cu^{2+}$ and $Cu(H_2O)_3(OH)^+$ as $CuOH^+$.

The fraction of species present as the fully protonated $Cu^{2+}$ is calculated using the method described in Section 10.1. In the present case, the calculation is much simpler since the copper aquo complex can be thought of as a monoprotic acid. We can write its dissociation equilibrium using the most simplified formula:

$$K_{a1} = \frac{[CuOH^+][H_3O^+]}{[Cu^{2+}]} \quad (13.12)$$

---

### Example 13.4 The extent of copper–NTA complexation in a waste stream

Consider the following situation. The total NTA and copper concentrations in wastewater effluent (pH = 7.5) are 100 μg L$^{-1}$ (= $5.2 \times 10^{-7}$ mol L$^{-1}$) and 2.0 mg L$^{-1}$ (= $3.1 \times 10^{-5}$ mol L$^{-1}$), respectively.

$$\alpha_{Cu^{2+}} = \frac{[Cu^{2+}]}{[Cu^{2+}] + [CuOH^+]} = \frac{[Cu^{2+}]}{[Cu^{2+}] + \dfrac{K_{a1}[Cu^{2+}]}{[H_3O^+]}}$$

$$= \frac{[H_3O^+]}{[H_3O^+] + K_{a1}} = \frac{10^{-7.5}}{10^{-7.5} + 10^{-7.53}} = 0.52 \quad (13.13)$$

The NTA exists in four possible forms with varying degrees of deprotonation. Again, refer back to the calculation of phosphate speciation in Chapter 10. The algebra here follows the same pattern and the fraction of the $T^{3-}$ form is calculated as follows.

$$\alpha_{T^{3-}} = \frac{T^{3-}}{H_3T + H_2T^- + HT^{2-} + T^{3-}} = \frac{K_1K_2K_3}{[H_3O^+]^3 + K_1[H_3O^+]^2 + K_1K_2[H_3O^+] + K_1K_2K_3} \quad (13.14)$$

$$= 1.66 \times 10^{-3}$$

The $K_f$ expression (for eqn 13.10) can then be written in the following form

$$K_f = \frac{[CuT^-]}{[Cu^{2+}][T^{3-}]} = \frac{[CuT^-]}{[Cu_u]\alpha_{Cu^{2+}}[NTA_u]\alpha_{T^{3-}}} \quad (13.15)$$

where $[Cu_u]$ and $[NTA_u]$ refer to the molar concentrations of the uncomplexed forms of copper and NTA, respectively.

The conditional stability constant applies to specific solution conditions and is defined as

$$K_f' = \frac{[CuT^-]}{[Cu_u][NTA_u]} = K_f \alpha_{Cu^{2+}} \, \alpha_{T^{3-}} \tag{13.16}$$

In the present case, for water at pH 7.5,

$$K_f = 10^{12.96} = 9.1 \times 10^{12}$$

$$K_f' = 9.1 \times 10^{12} \times 0.52 \times 1.66 \times 10^{-3} = 7.9 \times 10^9$$

Since, in the present example, copper is in a 60-fold excess compared to NTA, essentially all of the copper is in the uncomplexed form and $[Cu_u] = C_{Cu}$. Therefore, the ratio of copper complexed with NTA to uncomplexed NTA is

$$\frac{[CuT^-]}{[NTA_u]} = K_f' \times C_{Cu} = 7.9 \times 10^9 \times 3.1 \times 10^{-5} = 2.4 \times 10^5$$

The large ratio of NTA complexed with copper ($CuT^-$) compared with uncomplexed NTA ($NTA_u$) shows that essentially all of the NTA is in the metal-bound form. Such forms are stable with respect to further biodegradation, keeping the metal in solution, and this is one of the concerns regarding the release of this water-softening agent into the environment.

---

**Main point 13.3** Natural and anthropogenic ligands present in water play an important role in forming complexes with metals in the hydrosphere. The tendency to form such complexes depends on the nature and concentration of the various ligands that are present in a particular sample. Soluble humic material is found in water environments everywhere and is probably the most important naturally occurring ligand involved in complex formation.

## Metal species and bioavailability

Much of the interest in the speciation of metals in water is related to issues of bioavailability and toxicity. A large number of studies involving different metals in locations throughout the globe have shown clearly that it is not enough to measure the total concentration of metal in a water sample in order to assess the risk of creating a toxic effect. Several models have been developed in an attempt to describe how metal uptake into organisms occurs, including the role played by speciation in determining the extent of uptake. The biotic ligand model (BLM) is one such theory that has gained wide, but not universal acceptance. To understand the basis of the theory, think of a single cell or a collection of cells making up a more complex organism. In order to create a toxic effect, a contaminant metal first has to associate itself, through some kind of binding process, with the cell wall. In other words, the cell wall acts as a ligand that is able to complex with the metal holding it there and thus removing it from solution. Subsequent steps include transfer of the metal into the cell where it can react with enzymes and other molecules involved in metabolism. The fundamental feature supporting the BLM is that there is a critical concentration of contaminant at the organism surface that leads to toxicity.

If molecules on the cell wall exhibit a complexing function, it may be necessary that they compete with complexing species that are already bound to the metal keeping it in solution. For this reason, the BLM predicts that metal-ion availability for uptake will depend not on the total concentration of metal to which the organism is exposed, but rather on the concentration of free, uncomplexed metal—that is, the metal that is most readily able to accept binding to a cell-wall ligand.

The BLM has been shown to provide good predictability in relating free metal-ion concentration to metal uptake by fish and invertebrates and there is ongoing research to modify and extend it to other species and special environments. There are, however, a number of environmental situations where correlations between aqueous free metal concentration and uptake are clearly not met.[5] These include non-equilibrium conditions, the presence of certain types of organic ligands or of thiols or thiosulphate in the water, complexation with assimilable organic compounds such as citrate, extremely hard water, and metal binding to colloids. In spite of these and other limitations, the biotic ligand model has proved to be a vast improvement over models based on total metal concentrations.

## 13.4 Three metals—their behaviour in the hydrosphere

The literature on the environmental properties and behaviour of metals is extensive. There is considerable information about baseline levels in various types of water and many examples exhibiting contamination have been described in detail. In this section, we will focus on how principles of inorganic chemistry outlined above are expressed in three important cases, including a type A, a borderline, and a type B metal.

### Calcium

Calcium, at 3.6%, is the fifth most abundant element in the Earth's crust (Appendix B.1). Important mineral forms of this element include limestone ($CaCO_3$) and dolomite ($CaMg(CO_3)_2$) as well as aluminosilicate minerals such as the feldspar anorthite ($CaAl_2Si_2O_8$) and the clay mineral montmorillonite (Chapter 14). Weathering of limestone and, to a lesser extent, other minerals leads to some dissolution in water. The solubility of the particular mineral and other environmental factors—particularly the concentration of carbonate species in water—determine the final concentration of calcium in a specific water body.

Aside from the metallic form, calcium has a single common oxidation state, +2, and so it is not directly influenced by the redox status of the water. It is a type A metal and its behaviour in the hydrosphere is characteristic of that class. It is therefore usually associated with an oxygen-donor ligand environment, which means that, in addition to the aquo species, it is able to form complexes with species such as phosphate, carbonate, and sulphate when they are present. Except in soil solutions, phosphate is generally found in such low concentrations that its reaction with calcium is insignificant. Where carbonate (usually as hydrogen carbonate) and sulphate are in the mmol $L^{-1}$ concentration range, their complexes with calcium are of some importance.

As shown by the stability constant (Table 13.3), calcium interacts only to a limited extent with dissolved humic material. Bonding presumably involves the oxygen-containing functional groups. Depending on the concentration of the dissolved HM, in water whose pH is near neutral, a small but significant portion of the calcium may be in the complexed form. Under somewhat acid conditions, hydrogen ions compete effectively for sites on the humate and the calcium reverts to the simple ionic form. Calcium also forms weak complexes with anions of some organic acids—oxalate and citrate being two examples. For example, in a solution containing 2.1 mg $L^{-1}$ $Ca^{2+}$ and 2.3 mg $L^{-1}$ oxalate at pH 6.5, the fraction of calcium in soluble complexed forms is about 0.1%. The stability constants, $K_{f1}$ and $K_{f2}$, used in making this calculation were $4.6 \times 10^1$ and $1.1 \times 10^1$. Other metals in water will, of course, compete with calcium for sites on the complexing ligand.

Most suspended and sedimentary mineral material contains calcium at the per cent level. Dissolution of calcium from these solids is enhanced under acidic conditions. Therefore, internal

---

[5] Chapman, P.M., F. Wang, C.R. Janssen, R.R. Goulet, and C.N. Kamunde, Conducting ecological risk assessments of inorganic metals and metalloids: current status, *Hum. Ecol. Risk Assess.*, **9** (2003), 641–97.

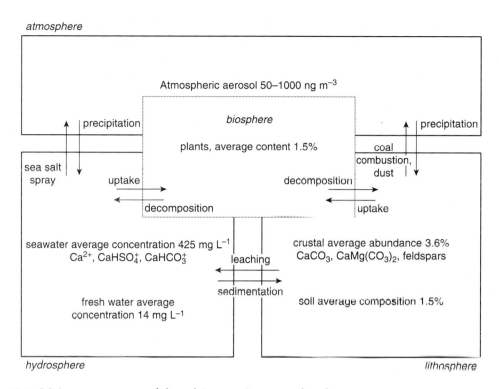

**Fig. 13.5** Major components of the calcium environmental cycle.

or external processes that produce acidity—acid rain, nitrification, and so on—increase the concentration of calcium in the associated water. This is especially true where the calcium is present as carbonate minerals. In soils, calcium is usually the principal ion that occupies exchange sites on the organic and mineral material (see Chapter 18). The exchangeable calcium is also readily displaced by acidity in water that is passing through the soil.

Figure 13.5 summarizes major components of the environmental cycle for calcium.

## Copper

Copper is an economically important element but it is found in only trace quantities (global average abundance is 50 $\mu$g g$^{-1}$) in the Earth's crust. (It is perhaps surprising that many elements, well-known due to their economic importance, are actually less abundant in the Earth's crust than are some less well-known elements. Compare the abundance values in Appendix B.1, for example, for copper and cerium.) For both plants and animals copper is required as an essential trace nutrient, but excessive amounts are toxic. Mineral forms include the free metal, a number of silicate and oxide species, and mixed copper / iron sulphide minerals such as chalcopyrite ($CuFeS_2$).

In both fresh and sea water, the principal oxidation state of copper is the +2 state.

Monovalent copper (I) species are also known but they dissociate to form copper (0) and copper (II) in most instances. An exception is in sea water where, under reducing conditions, copper (I) chloro species are stable.

Copper is a borderline metal and therefore has a good ability to form complexes with a variety of ligands. Ligands with nitrogen donor atoms are especially favoured. In natural fresh water that is in equilibrium with the atmosphere, the aquo complex of copper is the principal species at low pH. In the neutral pH range, partially deprotonated forms become important as does a

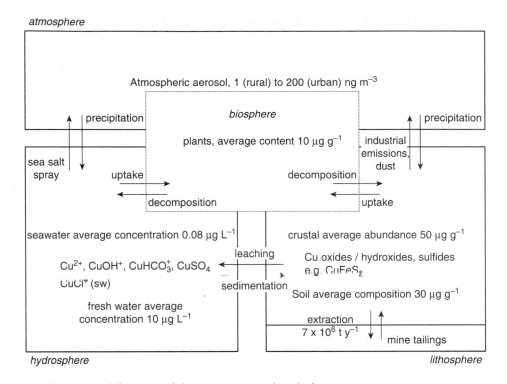

**Fig.13.6** The principal features of the environmental cycle for copper.

complex with hydrogen carbonate. At still higher pH values, further deprotonation takes place and a complex involving two carbonate ions is formed (Table 13.2).

As with many borderline and type B metals, the aqueous chemistry of copper is dominated by its reaction with organic matter. In Saginaw Bay, Lake Huron, in North America, for example, it has been shown that more than 98% of the copper is present in the form of stable complexes with OM.[6] In many cases, these complexes are between the metal and dissolved humic material. It is likely that the small fraction of nitrogen in this organic matter plays an important role in binding with copper and creating a stable species. When the humic material is present in soluble form in water, it serves to increase the solubility of copper. Conversely, particulate humic material—either suspended or precipitated—serves to remove soluble copper from the water column.

Significant features of the global cycle of copper are summarized in Fig. 13.6

## Mercury

The environmental chemistry of mercury has been studied extensively, principally because of several incidents of poisoning resulting in chronic illness and death. Perhaps the best known case occurred in Minamata, a Japanese fishing village, during the middle of the twentieth century. Hundreds of people died from ingesting fish that contained upwards of 100 $\mu$g g$^{-1}$ of mercury. The source of the mercury was from waste $Hg^{2+}$ catalyst from polyvinyl chloride manufacture that was discharged into the ocean bay.

In the Earth's crust, the concentration of mercury is three orders of magnitude smaller than that of copper, the global average value being 50 ng g$^{-1}$. Where it is found in nature, it is in elemental form or combined with sulphur; the most important mineral is cinnabar, a form of

[6] Bazzi, A., J.T. Lehman, and J.O. Nriagu, Chemical species of dissolved copper in Saginaw Bay, Lake Huron, with square wave anodic stripping voltammetry, *J. Great Lakes Res.*, **28** (2002), 466–78.

mercury (II) sulphide ($\alpha$-HgS). In the hydrosphere and associated sediments it can exist as 0, +1, and +2 species depending on redox and other environmental conditions. The mercury (II) forms comprise the most important aqueous species when aerobic conditions obtain.

Mercury aquo complexes tend to deprotonate readily. Therefore, even in moderately acidic conditions at pH 4, the mono and dihydroxy complexes are the two dominant forms, if no other complexing agents are present. Mercury (II) is a type B metal ion. One consequence of this is its strong affinity for sulphur as evidenced in the most common mineral form of the element. The tendency to form complexes with other chelating ligands is also strong. This is clear in Fig. 1.3 where we saw that, even when the chloride ion concentration in well water was as low as 9.5 $\mu$g mL$^{-1}$, the principal mercury (II) species is $HgCl_2$ (aq). Mercury (II) also has a strong affinity for organic ligands in water. In the pore water of forest soils that are rich in organic matter, essentially all of the dissolved mercury exists as complexes with the soluble organic matter. The stability constants for the 1:1 complex involving mercury and humate have been estimated to be in the range of $10^{18}$ to $10^{20}$. Binding may involve the nitrogen and sulphur groups that are present in small amounts in the humate, as well as the more common oxygen-containing functional groups.

Mercury (II) also has the ability to form a bond with carbon in the form of methyl mercury species. Methylation occurs under anaerobic conditions via a microbially mediated pathway that involves transfer of a methyl group from methylcobalamin, a derivative of vitamin $B_{12}$, to the mercury atom. This reaction requires reducing conditions as the methylcobalamin is formed through the action of methane-generating bacteria. This kind of situation will be described in Chapter 15. In the anaerobic environment, a competitive reaction occurs in which mercury (II) is reduced to the metallic state or, in the presence of sulphate-containing sediments, reacts with sulphide (formed during reduction) to form a very insoluble mercury (II) compound.

An alternative pathway occurs under aerobic conditions inside bacterial cells where the mercury binds with an enzyme, after which a methyl group is transferred to the mercury.

Both pathways are favoured by a moderately low pH and result in the production of monomethyl mercury (II) ($CH_3Hg^+$). This ion readily forms complexes with a variety of ligands. For example, compounds such as $CH_3HgCl$ and $(CH_3Hg)_2S$ are formed and are highly stable with respect to breaking the Hg–C bond. To some extent, however, chemical, photochemical, and biotic degradation does occur.

Further methylation also takes place to produce dimethyl mercury ($(CH_3)_2Hg$). This neutral compound is highly insoluble and it is volatile. Once in the atmosphere, it eventually photolyses to regenerate ionic mercury (II) and is rained out.

Environmental methylation is not unique to mercury. Methylated derivatives of tin (IV) and lead (IV) have been observed in natural, unpolluted situations; the mechanism involves oxidative methylation of the lead (II) or tin (II) ions by the methyl carbonium ion. This route is in contrast to that for mercury (II) where the transfer occurs via a methyl carbanion from methylcobalamin.

---

### Example 13.5 **Mercury in the human body**

The half-life of methyl mercury in the human body is estimated to be close to 100 days. Suppose a person weighing 65 kg eats 0.5 kg of fish containing 0.5 $\mu$g g$^{-1}$ mercury three times a week over an extended time. What is the steady-state average concentration of mercury in this person's body? (Note that, in many jurisdictions, 0.5 $\mu$g g$^{-1}$ is defined as the maximum allowed concentration in fish.)

$$\text{residence time} = \frac{\text{total mass of mercury}}{\text{rate of intake}} \quad \text{(see Example 1.1)}$$

$$\text{rate of intake} = 500 \text{ g (fish)} \times 0.5 \text{ }\mu\text{g (mercury) g}^{-1} \text{ (fish)} \times 10^{-3} \text{ mg }\mu\text{g}^{-1} \times 3 / 7 \text{ day}^{-1}$$

$$= 1.1 \times 10^{-1} \text{ mg (mercury) day}^{-1}$$

$$100 \text{ days} = \frac{\text{total mass of mercury}}{1.1 \times 10^{-1} \text{ mg day}^{-1}}$$

$$\text{Total mass of mercury (in the steady state)} = 100 \text{ days} \times 1.1 \times 10^{-1} \text{ mg day}^{-1}$$
$$= 11 \text{ mg}$$

The concentration of mercury in the body is then

$$11 \text{ mg (mercury)} / 65 \text{ kg (body)} = 0.17 \text{ mg kg}^{-1}, \text{ or } 0.17 \text{ ppm}$$

This is the *average concentration* of mercury. Note, importantly, that accumulated mercury is unevenly distributed in various organs in the body. It is mercury that accumulates in the brain that leads to the neurological dysfunctions associated with diseases such as have been observed at Minamata.

Figure 13.7 shows the principal features of the global cycle of mercury. This element is unique amongst metals in that its environmental cycle includes an important contribution from gas-phase species.

### Mercury in the Amazon Basin

Since the late 1970s when major deposits of gold were discovered in the Brazilian Amazon region, there has been a 'gold rush' bringing in persons hoping to make a fortune on finding and recovering gold from the soils and river sands. The operations involve over a million people in an area of about $170\,000$ km$^2$ and several hundred tonnes of gold are obtained annually.

The technology used is simple. A common recovery process begins with dredging of the river sand. After sieving, the material is passed through carpeted riffles to retain the heavier particles. In another method, soils or sediment are excavated and any consolidated material is crushed before being subjected to centrifugal separation to produce a concentrate.

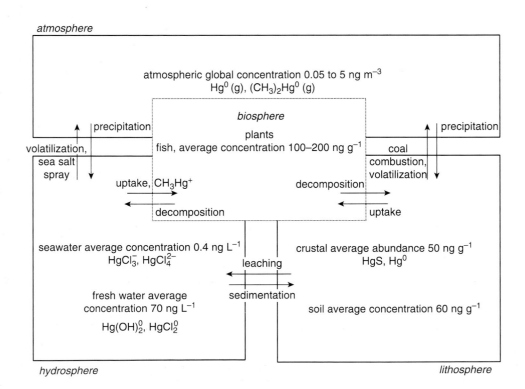

**Fig. 13.7** Major components of the environmental cycle of mercury.

The separated concentrates are placed in amalgamation drums where they are mixed with mercury, which is able to dissolve gold. The solution with mercury as solvent is called an amalgam; it is very dense and is separated from the concentrate matrix by panning. The mercury is then recovered from the amalgam by 'roasting' it in partially enclosed retorts or even in the open air leading to massive atmospheric emissions of the element. The 'pure' gold that is produced and supplied to the market may still contain up to 5% (by mass) mercury and, therefore, reburning at other sites is frequently required. Without proper recovery, this adds to the atmospheric release in the small towns where gold dealers locate. It has been estimated that each year more than 100 tonnes of mercury are lost to the atmosphere in the recovery and purification operations.

While global background atmospheric concentrations of mercury are typically 0.05 to 5 ng m$^{-3}$, levels in the Brazilian mining areas have been found to range from 20 to 500 ng m$^{-3}$. In towns near dealers' shops, atmospheric concentrations reach as high as 3 μg m$^{-3}$ and inside the shops, levels of up to 300 μg m$^{-3}$ have been measured. The World Health Organization limit for public exposure has been set at 1 μg m$^{-3}$.

The atmospheric mercury is oxidized to mercury (II) by photochemical reactions involving ozone and water vapour. In its ionic form, mercury (II) is rained out on to land or water, where it takes part in other reactions. In other areas of the Amazon Basin remote from mining, where atmospheric deposition is the only source, sediment concentrations have been positively correlated with organic matter content, reflecting the element's tendency to form complexes with ligands from the humate material. Conditions in the Amazon Basin—high temperature, high biological activity, slightly acidic conditions, a plentiful supply of organic matter—favour methylation processes. Methylated compounds are therefore also present.

Additional mercury, perhaps half the amount that is released to the atmosphere, is returned to the river as part of the sediments. The element has low mobility and therefore sediment concentrations near sources of mining are high. Small streams feeding into the Madeira River (a major tributary of the Amazon and a centre of the gold deposits) have shown sedimentary levels up to 20 μg g$^{-1}$. In the short term, the mercury is quite unreactive and remains in the elemental form, sometimes visible in exposed sediments during the dry season. Near dredged areas in the river itself, concentrations are an order of magnitude lower, and only a few km downstream the sediment concentration declines by a further factor of ten. Lateral transport is mainly in the form of suspended material and most movement downstream occurs in 'white water' during the rainy season when the suspended solids load is at a maximum. Most of the mercury is associated with suspended matter and the total concentration in the murky water has been measured to be as large as 13 μg l$^{-1}$; the dissolved concentration is small. In surface sediments, oxidizing conditions enable slow oxidation of mercury to form soluble mercury (II) chloro species. In these forms, some is accumulated by the plants growing in the nutrient-rich waters. Where sediments contain abundant organic matter methylation can occur, and this form of the element is taken up by carnivorous fish. Elevated levels of mercury are observed in both plants and fish in rivers near the mining regions. Some of these are important components of the diet of people living in the area.

> **Main point 13.4** Calcium (type A metal), copper (transition metal), and mercury (type B metal), are illustrative of the kinds of aqueous associations and behaviour of all metals within the three classes.

## 13.5 Metals associated with suspended matter in water

Thus far in the chapter, we have considered metal ions only when they are present in dissolved forms. In groundwater, filtered through layers of permeable soil material, water is usually clear, indicating that any metals present are, in fact, dissolved. In other compartments of the hydrosphere, however, there are situations where a considerable fraction of the metal exists as suspended particulate material. You can picture a fast-flowing river carrying an abundant sediment load that gives it a distinct milky appearance—for example the 'White Amazon' mentioned above or the 'White Nile' in north-eastern Africa. A portion of any metal in the river is in the form of dissolved species, but additional metal is also found in the suspended matter. What is dissolved and what is suspended is a complicated matter, which will be discussed in Chapter 14. Often the distinction is based on a laboratory procedure where filtration through a membrane with nominal pore size of 0.45 μm is used for the separation.

Suspended material consists of typical soil minerals and organic matter, especially those in the fine (clay-sized) fraction. The elements present in such inorganic and organic structures usually include relatively high concentrations of alkali and alkaline-earth metals, aluminium, and iron, along with smaller amounts of other metals depending on the particular materials involved. The sediment therefore carries a considerable fraction of major structural elements in suspended form compared with that dissolved in the water. On the other hand, some elements of environmental interest, particularly those usually found in trace amounts, may not be present in significant amounts in the original mineral structures of the suspended sediment but instead become associated with the sediments as adsorbed species on the surface of the fine particles.

The ratio of suspended to dissolved cations in rivers is quite variable, with values typically ranging from less than 0.1 to much greater than 1. As an example, the mean particulate and soluble concentrations of four metals in two rivers in south-eastern Quebec Province in Canada are given in Table 13.5.

Metals in suspended sedimentary material of various origins are often considered to be less available for uptake by organisms than are the dissolved forms of the same element. While this is broadly true, the actual picture of availability is in many cases more complicated. For one thing, surface-adsorbed elements are more available than structural elements. For example, when a sediment-bearing river discharges water into an estuary, a significant fraction of the adsorbed metal cations are displaced by the very high concentration of sodium ions via an ion-exchange reaction and become part of the solution phase.

$$M^{n+}solid^- + nNa^+ (aq) \longrightarrow Na^+solid^- + M^{n+} (aq) \tag{13.17}$$

Likewise, elements associated with iron or manganese (hydr)oxide particles can be 'remobilized' if the particles themselves go into solution as a result of reductive dissolution such as was discussed in Chapter 10. On the other hand, suspended organic material can decompose into

**Table 13.5** Concentrations and ratios of particulate (p) and soluble (s) cadmium, copper, lead, and zinc in the Yamaska and St Francois rivers[a].

| | Concentration / μg L$^{-1}$ | | | | | |
|---|---|---|---|---|---|---|
| | Yamaska River | | | St Francois River | | |
| | Particulate (p) | Soluble (s) | Ratio p / s | Particulate (p) | Soluble (s) | Ratio p / s |
| Cadmium | 0.4 | <0.1 | >4 | 0.14 | <0.2 | 0.7 |
| Copper | 0.9 | 1.0 | 0.9 | 4.1 | 4.6 | 0.9 |
| Lead | 1.5 | 1.5 | 1 | 3.8 | <1 | >4 |
| Zinc | 4.2 | 3.2 | 1.3 | 12.2 | 6.7 | 1.8 |

[a] Campbell, P.G.C., A. Tessier, and M. Bisson, Anthropogenic influences on the speciation and fluvial transport of trace metals, in *Management and control of heavy metals in the environment*, Proceedings of an International Conference, London, UK, September 1979, C.E.P. Consultants, Edinburgh; 1979.

smaller, soluble molecules under oxidizing conditions, and this too can lead to release of soluble forms of complexed metals.

Some organisms can also take up colloid-bound metals directly from suspended material. Most obvious examples are filter-feeders like clams and mussels in water bodies and earthworms in soil. It is interesting that excretions from these organisms may include more soluble forms of rejected elements, thus contributing to the alteration of the biogeochemical cycles in a small but locally significant way.

> **Main point 13.5** The forms in which metals are found in water depend on the nature of the metal itself and on all the other components within the aqueous environment in which the metal is located. These include components that are of both natural and anthropogenic origin and substances that are dissolved as well as suspended material. The resulting speciation of the element is a major factor in determining its availability for biological uptake.

## ADDITIONAL RESOURCES

1. Landner, L. and R. Reuther, *Metals in society and in the environment,* Kluwer Academic Publishers, Dordrecht, the Netherlands, 2004.

2. Klee, R.J. and T.E Graedel, Elemental cycles: a status report on human or natural dominance, *Ann. Rev. of Environ. Res.,* **29** (2004), 69–107.

3. Wright, D.A. and P. Welbourn, *Environmental toxicology,* Cambridge Environmental Chemistry Series no. 11, Cambridge University Press, Cambridge; 2001.

## PROBLEMS

1. Plot the distribution (versus pH) of the zinc aquo complex and the first four deprotonated species of this complex. The $pK_a$ values are $pK_{a1} = 9.2$, $pK_{a2} = 7.9$, $pK_{a3} = 11.3$, $pK_{a4} = 12.3$. Note the unusual feature that $pK_{a2}$ is smaller than $pK_{a1}$. Comment on how this affects the distribution.

2. The stepwise formation constants for the complexes $PbOH^+$ (aq) and $Pb(OH)_2$ (aq) from $Pb^{2+}$ (aq) are $2.0 \times 10^6$ and $4.0 \times 10^4$, respectively. The reactions can be written in simple form as

$$Pb^{2+} \text{ (aq)} + OH^- \text{ (aq)} \rightleftharpoons PbOH^+ \text{ (aq)}$$

and

$$PbOH^+ \text{ (aq)} + OH^- \text{ (aq)} \rightleftharpoons Pb(OH)_2 \text{ (aq)}$$

Calculate the $pK_{a1}$ and $pK_{a2}$ values for deprotonation of the aquo complex of lead (II), and determine the fractional concentration of the two most important species at pH 7.0.

3. Without calculation, sketch a distribution diagram, versus pH, for uncomplexed NTA species in water, covering the entire pH range from 0 to 14. What will be the most important species in water with near neutral pH?

4. A freshwater sample contains 160 ppb $Cu^{2+}$ and 4.3 ppm fulvic acid which has 4.0 mmol $g^{-1}$ sites available for complexation. Calculate the equilibrium concentrations of free and complexed copper and of free fulvic acid. Use the conditional stability constant in Table 13.3.

5. Examine Table 13.4, which reports data related to the stability of complexes between various metals and NTA. Comment on the relative values in the table in terms of the environmental classification of metals. Predict the approximate $pK_f$ values for nickel (II) and mercury (II) based on that classification.

6.  In lake water containing 0.9 mmol L$^{-1}$ calcium and 12 µg L$^{-1}$ fulvic acid, determine the fraction of the fulvic acid that is bound to calcium, assuming that this is the only metal present in a significant concentration. The pH of the water is 5.0.

7.  Note the electronic structure of cadmium (II) and indicate the group (type A, borderline, or type B) in which it should be placed. Predict the principal inorganic forms of the element in fresh and sea water under both oxidizing and reducing conditions.

8.  Artificial and natural wetlands are frequently used as catchment basins for urban run-off and storm water. The sediments in these basins have some ability to extract and retain soluble metals from the water as it flows through the pond. For lead, cadmium, and zinc, use information in this chapter and elsewhere to predict the affinity of each of these metals for the carbonate, the iron oxide, and the organic matter fractions of sedimentary material suspended and settled in the pond.

# Chapter 14
# Environmental chemistry of colloids and surfaces

The focus of this chapter is on the chemistry of colloids and surfaces and how the interface between the solid and aqueous environments plays a crucial role in the dispersion of species in aquatic systems. We will accomplish this by:

- reviewing the size and surface properties of solid particles
- introducing two adsorption isotherm relations, Langmuir and Freundlich
- discussing aqueous phosphate chemistry and its environmental consequences
- introducing the distribution constants $K_d$, $K_{OM}$, $K_{OW}$, their environmental relevance, and the terms relating to bioconcentration
- investigating the types of colloidal material in the natural environment, with emphasis on clay minerals.

On several occasions earlier in the book, we have indicated that it is not easy to assign discussions about particular processes to unique compartments of the environment. The relation between water and soil is a good example that underlines this difficulty. Soil processes determine metal solubility in the associated pore water and, through leaching this affects the composition of lakes and rivers. The atmosphere can also be involved. There is an intimate connection between atmospheric carbon dioxide levels and concentrations of carbonate species in water; soil minerals and organic matter also take part in the complex carbonate interrelations.

In this chapter, connections between the hydrosphere and the terrestrial environment are especially clear and obvious. The subject matter is concerned with the behaviour of the solid phase, in particular the finely divided solid material, when it is in contact with water. The principles discussed are important in terms of suspended solids and sediments in water bodies (Fig. 14.1(a)). They are also important for understanding the relation between soil and its pore water (Fig. 14.1(b)).

Consider the following situation. The concentration of phosphorus in a body of water—say, a pond or small lake must be measured. A sample, or more likely many samples, are obtained using an appropriate sampling protocol. In the laboratory, analysis for phosphorus is then to be carried out. Several well-established standard methods for phosphorus analysis are available and these can be applied in any laboratory that has basic analytical equipment.

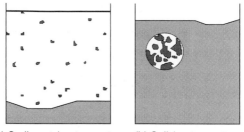

(a) Sediment / water system    (b) Soil / water system

**Fig. 14.1** Environmental colloids interaction with water: (a) in sediment / water systems where particles are suspended in the water column and / or deposited as settled sedimentary material; (b) in soil / water systems made up of a soil matrix that is permeated with water in some or all of the pore space.

Therefore, while measuring phosphorus concentrations may seem to be a simple process, there are, in fact, enormous complications. Of these, one of the most important is the issue concerning which forms of phosphorus should be considered to be *in* the water. On the one hand, there are species, most commonly forms of inorganic orthophosphate as well as organic phosphorus compounds that unquestionably are dissolved constituents of the aqueous system. But in addition, there is inorganic and organic sedimentary material temporarily suspended in the water at the time of sampling and this solid-phase material too contains phosphorus in a variety of forms. Between the obviously soluble and clearly insoluble components is a whole range of material having different densities and particle sizes. The question regarding analysis for phosphorus concentration in a water sample then requires decisions about which size fractions should be considered 'phosphorus *in* the water'.

This is the type of question we will deal with in the present chapter. In general, the subject matter is chemistry connected with the interface between dissolved and solid phases in environmental systems.

## 14.1 Sizes of environmental solid particles

Figure 14.2 illustrates size-range distributions of some component particles of a natural aqueous system. An arbitrary but reasonable classification of soluble, colloidal, and precipitated forms based on particle size diameters is also given. In this classification, the diameter of colloidal-sized particles spans the range from 10 nm to 10 μm.

With regard to the phosphorus analysis problem, there are two connected questions related to the sizes described in the figure. One is the *practical experimental issue* of how to separate the desired size fractions. Recent developments in centrifugal and filtration technologies have provided a variety of options in this area, but for many routine analytical purposes there is a consensus that filtration through a 0.45-μm controlled pore size filter provides an appropriate division between 'soluble' and 'insoluble' fractions (Fig. 14.2 shows that 0.45 μm is near the centre of the colloidal region). For more detailed studies, other divisions could be more appropriate.

The other question raises the issue of the *environmental significance* of this experimentally defined size separation. To refer again to the phosphorus example, we answer the question by first considering the purpose of determining the concentration of phosphorus in a water body. Probably the most likely reason is that that element is one of the major nutrients supporting growth of algae and aquatic plants and, where this growth is excessive, eutrophication occurs. Amongst the essential nutrients, phosphorus is most frequently the limiting factor and it therefore

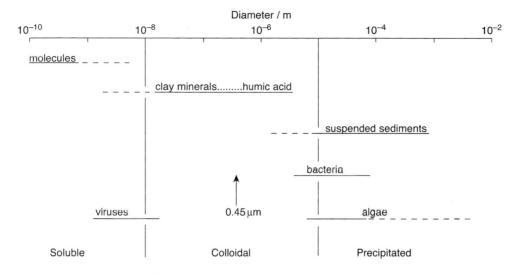

**Fig. 14.2** Size classification of materials in the hydrosphere.

controls the rate of eutrophication. Of the phosphorus species, soluble orthophosphate ($PO_4^{3-}$, and its protonated forms) is the primary nutrient source of the element. But, over a longer time span, other phosphorus-containing species too may release soluble orthophosphate to growing organisms. Included in this category could be easily decomposable non-living biomass and colloidal-sized inorganic matter that supports adsorbed phosphorus on its surface. Much less likely to be available for plant or algal growth would be phosphorus that is present as a structural component of primary minerals like the very insoluble mineral apatite ($Ca_5(PO_4)_3(F,Cl,OH)$). At least some of the components containing *biologically available* phosphorus appear in the fractions of soluble and colloidal-sized material that passes through the 0.45-μm filter. Most of the structural phosphorus is usually found in the larger-sized suspended residual portion. Therefore, the experimental filtration through a 0.45-μm device provides an approximate but far from perfect separation of biologically accessible forms from resistant forms.

The example of phosphorus is a good illustration of the need to know more about the forms of this or other elements that are associated with solid-phase materials suspended in the hydrosphere. Because much of the suspended material is in the colloidal size range, we will focus on the chemistry of naturally occurring colloids.

## 14.2 Surface properties of colloidal materials

### Specific surface area

An environmentally important characteristic of very small particles is their enormous specific surface area, which is usually expressed using units of square metres of surface per gram of particles ($m^2 g^{-1}$). Figure 14.3 illustrates in a simple way the origin of the large surface area of small-sized materials. The cube with 1 cm dimensions has a total surface area of 6 cm² but, if it is subdivided into one trillion ($10^{12}$) cubes having 1 μm dimensions, the total surface area of the same mass of material expands to 6 m², larger by a factor of 10 000. Although the calculation is based on a simplified and artificial situation, it uses a particle size that is within the colloidal range and gives values that approach the low end of specific surface areas of some naturally

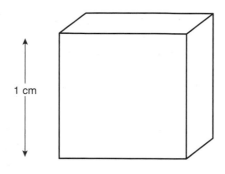

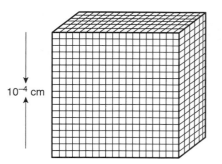

Assume cube with dimension = 1 cm and density = 1 g mL$^{-1}$

Surface area = 6 x 1 cm$^2$ = 6 cm$^2$

Specific surface area = 6 cm$^2$ g$^{-1}$

$\qquad$ = 6 x 10$^{-4}$ m$^2$ g$^{-1}$

Assume subdivision of cube into smaller cubes with dimension 1μm = 10$^{-4}$ cm and density = 1 g mL$^{-1}$

Total surface area = 6 x (10$^{-4}$)$^2$ x (10$^4$)$^3$

$\qquad\qquad\qquad\qquad$ $A$ $\qquad$ $n$

$\qquad\qquad$ = 6 x 10$^4$ cm$^2$

Specific surface area = 6 m$^2$ g$^{-1}$

**Fig. 14.3** Specific surface area of colloidal-size cubes. Typical measured values for natural materials are: kaolinite, 5–20 m$^2$ g$^{-1}$; montmorillonite, 700–800 m$^2$ g$^{-1}$; and fulvic and humic acids, 700–10 000 m$^2$ g$^{-1}$.

occurring colloids. Real colloids would, of course, have highly irregular geometries and this adds to the surface area.

---

**Main point 14.1** In the hydrosphere, the colloidal size range comprises material that falls between soluble and precipitated fractions. It includes a variety of minerals and organic matter. A fundamental property of colloids is that they have a very large specific surface area.

---

## Surface charge

As a consequence of their extremely large surface areas, colloids exhibit unique properties that depend on the particular chemical nature of the colloid. A common and environmentally important feature is that the colloid surface is capable of adsorbing[1] molecules or ions from the surrounding solution. The adsorption then temporarily (if reversible) or permanently (if irreversible) serves to remove such species from solution. In a lake or other water body, adsorption brings about a decrease in the concentration of the soluble species, while, in water percolating through the soil, adsorption on to the solid phase attenuates the movement of particular solutes.

There are several ways in which adsorption can occur. One is due to electrostatic attraction to a charged surface. Many common environmental colloids, for example, the clay minerals, have

---

[1] The term *adsorption* implies a surface reaction, while *absorption* suggests that the retained material is taken into the interior of the particle. Because these processes often overlap, especially in the case of porous solids like much natural organic matter particles in water, the term *sorption* is often used to account for the uncertainty as to what processes are involved.

a negative surface charge that is relatively constant in magnitude. This property enables the clay surface to attract positively charged species (cations) like $Ca^{2+}$ and retain them in an electrostatically adsorbed form. In the case of other solids, the charge is not fixed, but depends on the properties of the surrounding solution. Metal oxides, such as iron and aluminium oxides, are an important class of environmental solids found in soils and sediments. They are a good example of variable-charge materials. Whether the surface is protonated and therefore positive or deprotonated and negative depends on the pH of the associated water. This is illustrated in Fig. 14.4.

Humic material also has variably charged surfaces in that deprotonation of carboxyl groups results in a negative charge (a common situation), while protonation of amino groups (less likely) generates a positive charge. The position of the protonation / deprotonation equilibrium is a unique property of each material—not only of its chemical composition, but also of the way in which it has been formed.

To measure this property, the pH value at which surface positive and negative charges associated with protonation / deprotonation equilibria, are just balanced, the pH of a series of samples is measured after the material has been equilibrated in the presence of different added amounts of acid and base (Fig. 14.5). The set of measurements is repeated in solutions containing various concentrations of non-reactive (indifferent) electrolyte. Ideally, the plot gives a point where all the curves intersect and the pH value at this point is called the zero point of charge (zpc), point of zero charge (pzc), or $pH_0$. We will use the latter term. Some $pH_0$ values of common environmental colloids are given in Table 14.1.

For variable-charged substances such as those listed in the table, where the pH of the ambient solution is less than the colloid $pH_0$, the surface becomes protonated and the colloid therefore attains a net positive surface charge. In this state it is able to electrostatically attract and adsorb negative species (anions). Conversely, where the surrounding solution pH is greater than the colloid $pH_0$, the colloid surface is negative and has an electrostatic affinity for positive species (cations). Another way of saying this is that the lower the $pH_0$ value, the more likely the colloid is to possess a negative surface charge. In many environmental circumstances the ambient pH is not far from 7, and most of the materials listed in the table have a negatively charged surface. In some common situations however, the hydrated iron and aluminium oxides develop a positive

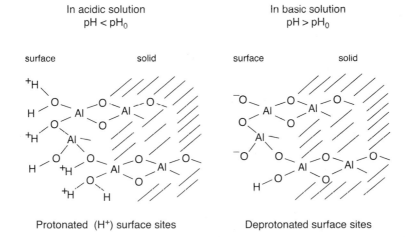

Hydrated—$Al_2O_3$

**Fig. 14.4** Surface charge of hydrated alumina in aqueous environments of varying acidity.

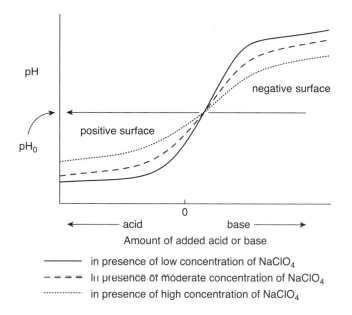

**Fig. 14.5** Experimental measurement of $pH_0$.

surface. In either case, the charged surface interacts with ions in the surrounding solution in an important and interesting way.

## The electrical double layer

The charge properties of a colloid surface are often described in terms of an *electrical double layer*. Double-layer theory has been developed with varying degrees of sophistication and a simple version is illustrated in Fig. 14.6.

As described above, a colloid has charge associated with the surface species, such as protonated or deprotonated functional groups, or other charged atoms in contact with the solution. The charge serves to attract oppositely charged *counterions* from the surrounding solution and these form a 'layer' adjacent to the colloid surface. This leads to ion-exchange properties that will be described below. One should not think of this layer as a clearly defined set of ions that are exclusively of the opposite charge to that of the surface. Instead, the picture is of a

**Table 14.1** $pH_0$ values of various natural colloids.

| Colloid | $pH_0$ | |
|---|---|---|
| $SiO_2$ | 2.0 | |
| $MnO_2$ | 2–4.5 ⎫ | |
| $Fe_2O_3$ hydrated | 6.5–9 | |
| Goethite | 6.5–9 ⎬ | Actual value depends on mode of formation and age of precipitate |
| Haematite | 8.5 | |
| $Al_2O_3$ hydrated | 5–9 ⎭ | |
| Humic material | 4–5 | |
| Bacteria | 2–3 | |

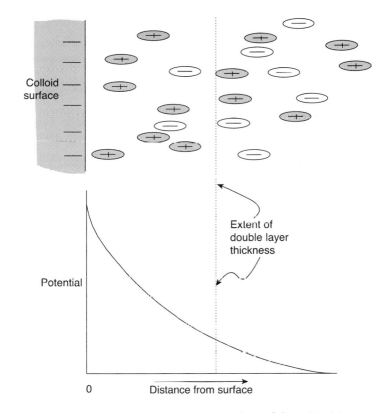

**Fig. 14.6** The electrical double layer. In this case, the surface of the colloid has a net negative charge and this attracts positive counterions to the region adjacent to the surface. The thickness of the layer is defined as the distance from the surface where the charge is reduced to $1/e = 0.37$ times its original value.

pre-ponderance of the oppositely charged counterions near the surface with the proportions gradually reverting to the normal situation where positive and negative species are balanced as one moves out into the bulk solution. As this happens, the potential, which is a maximum at the surface, decreases to zero. The 'thickness' of the counterion layer is defined as the distance at which the potential has decreased to $1/e$ (0.37) of its value at the surface.

A colloidal system, such as a lake or river containing suspended clay particles, is stable because the small charged particles repel each other. As they move about in the liquid medium due to thermal motion, they are unable to come sufficiently close to overcome the repulsive forces of the surface charge and aggregate into larger, settleable units. The result is a long-lasting suspension of the colloid.

There are several mechanisms by which colloids are destabilized, causing them to settle out of the solution. One mode of destabilization results when there is a high concentration of electrolyte in the water. This provides a source of counterions that accumulate around the solid surface, effectively reducing the thickness of the double layer. The availability of many counterions ensures that they are present in large concentration so that the surface potential falls more rapidly to zero (Fig. 14.7). A well-known and important situation where these processes apply occurs when a river discharges into an ocean estuary. Sedimentation is frequently observed in the estuary due to destabilization of the river colloids as they encounter the high-salt sea water.

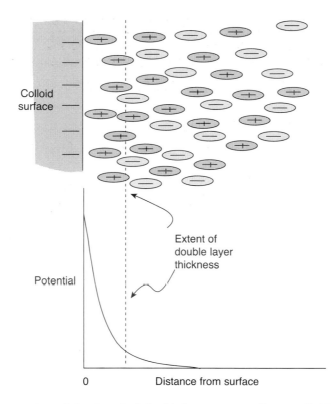

**Fig. 14.7** The compression of the electrical double layer surrounding a colloid particle by a high salt concentration of ions enables neutralization of the surface charge over a small distance. Compare with Fig. 14.6.

## Surface adsorption—electrostatic retention

Because many environmental colloids have a negatively charged surface, electrostatic attraction for cations is an important phenomenon and this is usually described in terms of an ion-exchange process. In changing environments, these counterions can be replaced by others. For example, potassium ions in solution might exchange with surface sodium ions.

$$\text{colloid}^-{:}Na^+ + K^+ \text{ (aq)} \rightleftharpoons \text{colloid}^-{:}K^+ + Na^+ \text{ (aq)} \tag{14.1}$$

The equilibrium position of the reaction depends on the nature of the colloid, and also depends on the nature (principally, charge density) and concentration of the dissolved species adjacent to the colloidal particle. In soils and sediments, the relative importance of cations occupying exchange sites is frequently $Ca^{2+} > Mg^{2+} > K^+ > Na^+$. In low-pH environments these adsorbed ions may be partially displaced by $Al^{3+}$ and / or $H_3O^+$ because these two species become important solution constituents under highly acidic conditions.

A measure of the number of negative exchange sites (which is the same as the amount of positive charge that can be accommodated) on a given amount of material is termed the cation exchange capacity (CEC), the units of which are centimoles of positive charge on the surface per kilogram of solid (cmol $(+)$ kg$^{-1}$). We will say more about this very important soil and sediment property in Chapter 18. Positively charged surfaces are less common in environmental solids, but when they occur, as in the case of some iron and aluminium hydrous oxides, the anion exchange capacity is also sometimes measured.

> **Main point 14.2** Small colloidal particles have a very large specific surface area. Some environmental colloids carry a fixed or variable charge on the surface. In these cases, the charged surface attracts oppositely charged ions from the adjacent solution. Such colloids therefore have properties of an ion-exchange material.

## Surface adsorption—specific binding

A second mechanism for retention of species on colloid surfaces is associated with *specific binding*. The adsorption phenomena that are based on electrostatics can be treated as physical processes where charge density on both the colloid and solution species determines the extent of adsorption—meaning that small highly charged species are more likely to be retained. In contrast to this, specific binding involves the forming of covalent chemical bonds between the solution species and the surface atoms of the colloid. Covalent bonding is not related to surface charge but requires a suitable combination of electron donor and receptor atoms on the solution and surface species. The adsorption of fatty acids by a hydrated iron oxide surface is an example of adsorption involving covalent bond formation (reaction 14.2).

$$\text{Fe—OH} + \text{HOOC—R} \longrightarrow \text{Fe—OH} \cdots \underset{\text{Fe—O}}{\overset{\text{O}}{\underset{}{\parallel}}} \text{C—R} + \text{H}_2\text{O} \qquad (14.2)$$

For this type of chemical reaction, the equilibrium may lie far to the right and, while adsorption is favoured under particular pH conditions, it is not simply dependent on whether there is a positive or negative charge on the surface of the solid. Such chemical adsorption processes are, to a significant extent, irreversible and are frequently called *specific adsorption*. Here the adsorption depends on a degree of chemical, as opposed to electrostatic, affinity between the colloid and the adsorbed species.

Metal ions are also capable of forming specific bonds with oxide surfaces as illustrated in the example with zinc (reaction 14.3)

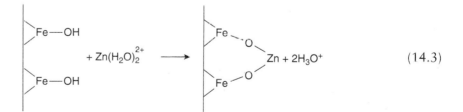

$$\text{Fe—OH} + \text{Zn(H}_2\text{O)}_2^{2+} \longrightarrow \underset{\text{Fe}}{\overset{\text{Fe}}{}} \underset{\text{O}}{\overset{\text{O}}{}} \text{Zn} + 2\text{H}_3\text{O}^+ \qquad (14.3)$$

When specific adsorption occurs, the nature of the surface is altered and so the extent of protonation or deprotonation as measured by the $pH_0$ also changes. Covalent binding of a cation to the surface makes the surface more positive, shifting the colloid $pH_0$ to a higher value, while binding of an anionic species produces a downward shift in $pH_0$.

As noted above, the quantity used to measure the number of sites available for *electrostatic retention* of ions is the exchange capacity, most commonly the cation exchange capacity (CEC). To quantitatively describe *specific adsorption*, a variety of mathematical descriptions have been developed for particular cases. We will examine two of the relations that are fundamental and are frequently used.

> **Main point 14.3** Specific chemical reactions whereby soluble species form covalent bonds with the atoms on the surface of a colloid are another mechanism of retention.

## 14.3 Quantitative descriptions of adsorption I

### The Langmuir relation

The Langmuir relation assumes that the solid (adsorbent) surface has a specific number of sites each of which is capable of reacting with and binding to a solution molecule, the adsorbate. All of the sites are considered to be equivalent and, when all are occupied, no further adsorption can occur—in other words, adsorption is limited to monolayer coverage (Fig. 14.8). Another way of saying this is that the quantity adsorbed ($C_s$) reaches the maximum quantity adsorbable ($C_{sm}$) when all sites are occupied (see eqn 14.4).

A mathematical expression of the Langmuir relation is

$$\frac{C_s}{C_{aq}} = \frac{bC_{sm}}{1 + C_{aq}b} \tag{14.4}$$

where $C_s$ = quantity adsorbed by the suspended solid, soil, or sediment / mol g$^{-1}$, $C_{aq}$ = equilibrium aqueous solution concentration / mol L$^{-1}$, $b$ = binding constant / L mol$^{-1}$, which depends on the physical and chemical nature of the solid material, and $C_{sm}$ = maximum quantity adsorbable / mol g$^{-1}$, which depends on the nature of the solid as well as the concentration of surface sites. (Note that any unit of mass or quantity may be chosen for use in the numerator of the $C_s$ and $C_{aq}$ terms as long as they are the same for both the solid and the solution.) A plot of the above relation is given in Fig. 14.9.

The fraction of active sites occupied by adsorbate under specific conditions is given the symbol $\theta$.

$$\theta = \frac{C_s}{C_{sm}} = \frac{bC_{aq}}{1 + bC_{aq}} \tag{14.5}$$

Further algebraic manipulations provide another useful form (eqn 14.9) of the relation

$$C_s = \frac{bC_{aq}C_{sm}}{1 + bC_{aq}} \tag{14.6}$$

$$\frac{1}{C_s} = \frac{1 + bC_{aq}}{bC_{aq}\,C_{sm}} \tag{14.7}$$

$$\frac{1}{C_s} = \frac{1}{bC_{aq}\,C_{sm}} + \frac{bC_{aq}}{bC_{aq}\,C_{sm}} \tag{14.8}$$

$$\frac{1}{C_s} = \frac{1}{C_{sm}} + \frac{1}{bC_{aq}\,C_{sm}} \tag{14.9}$$

When plotted, eqn 14.9 produces a linear plot that may be used to obtain values of $C_{sm}$ and $b$ (Fig. 14.10).

The phosphorus in the water example (shown below) is treated in a way that assumes Langmuir behaviour for the adsorption process. Phosphate adsorption by sedimentary material has often been described in terms of the Langmuir relation. In the laboratory, two portions of a single sedimentary material are equilibrated with different concentrations of orthophosphate in water and the equilibrium concentrations in the solution and on the solid are determined. The equilibrium values are shown in Table 14.2.

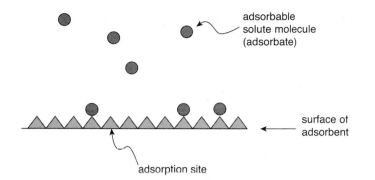

**Fig. 14.8** The Langmuir adsorption process. The diagram shows that there are a limited number of sites on the surface, where adsorption can occur. Once these are completely occupied with adsorbate, no further sorption can occur.

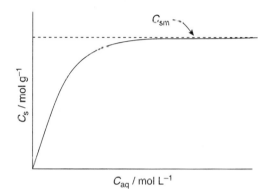

**Fig. 14.9** A graphical representation of the Langmuir relation showing that adsorption continues to a maximum concentration ($C_{sm}$) on the solid surface. $C_s$ and $C_{aq}$ are the equilibrium concentrations of adsorbate on the solid and in solution, respectively.

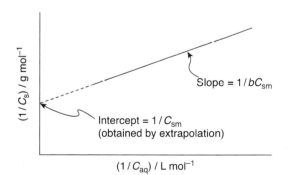

**Fig. 14.10** Linearized form of the Langmuir equation. The values of $C_{sm}$ and $b$ may be determined from the values of the slope and intercept.

**Table 14.2** Equilibrium values for phosphorus in water and sediment.

| | Phosphorus concentration | |
|---|---|---|
| | In water, $C_{aq}$ / mol L$^{-1}$ (ng mL$^{-1}$) | In sediment, $C_s$ / mol g$^{-1}$ (µg g$^{-1}$) |
| Sample 1 | $4.0 \times 10^{-7}$ (12) | $2.0 \times 10^{-7}$ (6.2) |
| Sample 2 | $1.3 \times 10^{-7}$ (4) | $1.0 \times 10^{-7}$ (3.1) |

---

**Example 14.1  Maximum phosphate adsorption as determined using the Langmuir relation**
Using data from Table 14.2 and eqn 14.9 with mol based units for subsample 1, the first equation is

$$\frac{1}{2 \times 10^{-7}} = \frac{1}{C_{sm}} + \frac{1}{4 \times 10^{-7}\, bC_{sm}}$$

$$C_s \qquad C_{aq}$$

$$= \frac{4 \times 10^{-7}\, b + 1}{4 \times 10^{-7}\, bC_{sm}}$$

$$2bC_{sm} = 4 \times 10^{-7}\, b + 1$$

$$5 \times 10^{6}\, C_{sm}\, b = b + 2.5 \times 10^{6} \qquad \text{(i)}$$

and, for subsample 2,

$$\frac{1}{1 \times 10^{-7}} = \frac{1}{C_{sm}} + \frac{1}{1.3 \times 10^{-7}\, bC_{sm}}$$

$$= \frac{1.3 \times 10^{-7}\, b + 1}{1.3 \times 10^{-7}\, bC_{sm}}$$

$$1.3bC_{sm} = 1.3 \times 10^{-7}\, b + 1$$

$$1 \times 10^{7}\, C_{sm}\, b = b + 7.7 \times 10^{6} \qquad \text{(ii)}$$

Multiplying eqn (i) by 2 and subtracting eqn (ii) gives

$$0 = b - 2.7 \times 10^{6}$$

$$b = 2.7 \times 10^{6}\,\text{L mol}^{-1}$$

By substituting this value into eqn 14.8 we obtain a value for $C_{sm}$.

$$C_{sm} = 38 \times 10^{-7}\,\text{mol g}^{-1}\ (12.8\ \text{µg g}^{-1})$$

---

This is a measure of the maximum concentration of phosphate that is expected to be adsorbed on the surface of the sedimentary material. In a sense, it is a measure of the capacity of the sediment to remove phosphate from solution. While the Langmuir relation allows for predictions of this sort, it gives no indication of the processes and mechanisms of the removal process.

In the general case Table 14.2, shown previously, can be written in the form of Table 14.3. The values of $b$ and $C_{sm}$ can then be calculated from the equations

$$b = \frac{(C_s C'_{aq} - C'_1 C_{aq})}{C_{aq} C'_{aq}\, (C'_s - C_s)} \qquad (14.10)$$

$$C_{sm} = \left( \frac{C_s C'_{aq}}{bC_{aq}} + C_s \right) \qquad (14.11)$$

**Table 14.3** Values for concentration of dissolved species in water ($C_{aq}$) and in sediment ($C_s$)—the general case.

|  | Concentration | |
| --- | --- | --- |
|  | Water | Sediment |
| Sample 1 | $C_{aq}$ | $C_s$ |
| Sample 2 | $C'_{aq}$ | $C'_s$ |

**Main point 14.4** The Langmuir relationship describes adsorption in terms of solution species occupying a fixed number of equivalent sites on the surface of a solid colloid.

## Phosphorus environmental chemistry and eutrophication

We began the chapter by noting that phosphorus exists in many forms in water. This is an appropriate place for us to look at its environmental chemistry in more detail. The global phosphorus cycle is described in Fig. 14.11. There are no common gaseous forms of the element and it is found in the atmosphere only in association with dust particles. In water, phosphorus, derived from a variety of inorganic and organic sources, hydrolyses to release orthophosphate species that have various degrees of protonation depending on the aqueous pH. Refer again to Fig. 10.1 (reproduced in Fig. 14.12) to see examples of how speciation changes with pH. Terrestrial

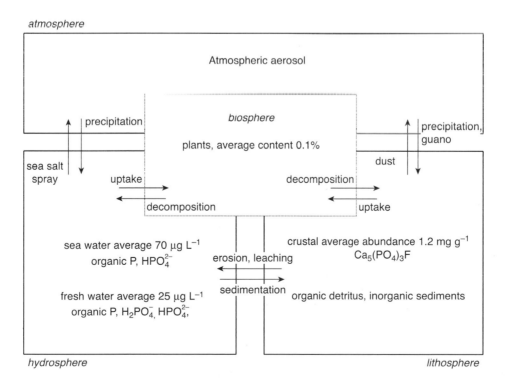

**Fig. 14.11** The phosphorus cycle.

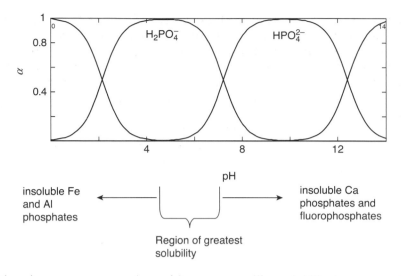

**Fig. 14.12** Phosphorus aqueous species and factors controlling solubility in water and the associated suspended material, sediment, or soil.

phosphorus is composed of a number of specific minerals including apatite, $Ca_5(PO_4)_3(F, Cl, OH)$, and vivianite, $Fe_3(PO_4)_2 \cdot 8H_2O$. As mineral species and as a minor constituent of organic matter, it is also present in all soils, where it cycles through plants as they grow and decay. Being an essential nutrient for plants, additional amounts of phosphate are frequently supplied as fertilizers in agriculture.

In water, phosphorus solubility is controlled by the availability of iron and aluminium under acid conditions and of calcium under alkaline conditions; each of these metals forms insoluble phosphates (Fig. 14.12). Where the pH is on the slightly acid side, phosphorus has its maximum solubility, and the pre-dominant aqueous species under these conditions is $H_2PO_4^-$.

The environmental significance of phosphorus arises out of its role as an important nutrient for both plants and microorganisms. In considering substances that pollute the hydrosphere, one category is the nutrient elements—especially those required in relatively large amounts by micro- and macroorganisms, namely, nitrogen, phosphorus, and potassium. These nutrients become pollutants when excessive concentrations of them in water stimulate the growth of unwanted aquatic organisms, including plants and algae. While the organisms are growing, photosynthesis occurs and they serve as oxygen-generating sources (reaction 14.12, left to right). However, when the plants or algae die and decay, their decomposition leads to consumption of oxygen, causing carbon dioxide to be released and creating an anaerobic and mildly acidic environment (reaction 14.12, right to left).

$$CO_2 + H_2O \rightleftharpoons \{CH_2O\} + O_2 \tag{14.12}$$

When nutrient-rich conditions lead to excessive biological productivity, the water body is referred to as being *eutrophic*. Visible evidence of eutrophication is the murky green colour of an algal bloom or the thick mat of plants growing up to the surface in shallow water. In contrast, an *oligotrophic* water body is one where the growth of aquatic organisms, especially algae, occurs only to a limited extent. When a formerly oligotrophic lake rapidly (typically over a period of several years) becomes eutrophic, the situation is usually considered to be undesirable as it inhibits the development and growth of higher forms of life, notably desirable fish species.

We should not assume that intense algal growth is undesirable *per se*. In fact, algae are a food source for fish and, when present in moderate amounts, can act as support for a thriving population. In order to ensure an adequate supply of algae, inland fish producers in China, India, and elsewhere actually add nutrients (in the form of inorganic or organic fertilizers) to the water to stimulate microbiological activity. In these instances the decomposition aspect of the algal life cycle takes place in the digestive system of the fish and therefore does not contribute to eutrophication.

Then, too, we should remember that eutrophication is a natural process that occurs over geological time as deep, clear, well-aerated water bodies gradually are supplied with nutrients and decomposable organic material and fill with sediment from erosion of the surrounding land surfaces. Over this time, the plant population changes from being totally aquatic to wetland types to purely terrestrial. It is the human-induced premature and often rapid *cultural* eutrophication that we frequently consider to be undesirable.

Of the nutrient elements that contribute to eutrophication, most commonly the limitation to growth is the small amount of phosphorus, and low-phosphorus water bodies, when supplied with excessive amounts of the element, tend to become eutrophic. There are some less common situations where it is nitrogen or other elements that limit growth.

The sources of phosphorus that finds its way into lakes and rivers may be divided into two categories—point sources, which are well-defined discharges from factories or outlets of municipal sewer systems, and diffuse sources, which include run-off from rural and urban landscapes.

Except in specific circumstances, the major type of point source of phosphorus discharge is municipal wastewater. Within the municipal discharges, phosphorus has two principal origins. The first is from commercial soaps and detergents. These have traditionally contained condensed polyphosphates that serve to sequester calcium ion, which is in high concentration in 'hard water'. The phosphates form a complex with calcium, thus preventing it from forming a gummy precipitate with the surface-active ingredients in the detergent.

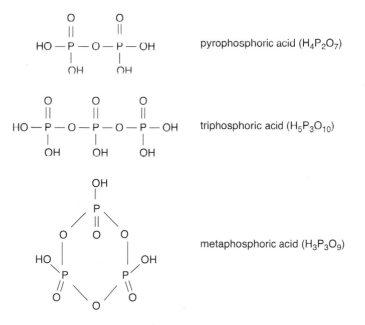

Condensed polyphosphates include pyrophosphoric acid, triphosphoric acid, and metaphosphoric acid with the triphosphoric acid in a trisodium form being most commonly used in detergent formulations. Typically, detergents contain from 5 to 50% $P_2O_5$ by weight. The polyphosphates hydrolyse rapidly in aqueous solutions to release orthophosphate species (eqn 14.13).

$$HO-\overset{\overset{\displaystyle O}{\|}}{\underset{\underset{\displaystyle O^-}{|}}{P}}-O-\overset{\overset{\displaystyle O}{\|}}{\underset{\underset{\displaystyle O^-}{|}}{P}}-O-\overset{\overset{\displaystyle O}{\|}}{\underset{\underset{\displaystyle O^-}{|}}{P}}-OH + 2H_2O \longrightarrow 3HO-\overset{\overset{\displaystyle O}{\|}}{\underset{\underset{\displaystyle O^-}{|}}{P}}-OH \qquad (14.13)$$

The actual orthophosphate product of the reaction adjusts to be in the acid–base form consistent with the pH of the receiving water (Fig. 14.12). Recent legislation in many jurisdictions has limited the use of phosphates in detergents but its replacement by alternatives is not without environmental consequences. Nitrilotriacetic acid (NTA), whose chemistry was described in Chapter 13, is one of the alternative sequestering agents.

A second source of phosphorus in municipal wastewater is human wastes. Human faeces is made up of about 25% organic matter (the remainder is mostly water). An 'average person' generates approximately 100 g dry weight of faeces each day and it contains about 0.03% nitrogen (largely of bacterial origin) and 0.005% phosphorus. The corresponding figures for urine are production of 1200 g per day containing 0.5% organic carbon, 1.0% nitrogen, and 0.03% phosphorus. Given these figures, a small urban area of 100 000 persons would produce 36.5 kg of phosphorus daily. If we assume that water use per capita is 0.20 m$^3$ (a typical figure for some high income countries, which also have a comprehensive sewage system), the concentration of phosphorus derived from human wastes in sewage influent would be approximately 2 mg L$^{-1}$. This would contribute significantly to the approximate 10 mg L$^{-1}$ P average, typical of wastewater in many cities. However, the total specific value fluctuates widely depending on season, time of day, and weather, among others. If municipal wastewater remains untreated, all the phosphorus would be discharged to the receiving water and this can be a major contributor to eutrophication. Where treatment is done, it ranges from use of settling ponds or lagoons where the natural biological processes cause uptake and removal of limited amounts of pollutants, to primary, secondary, or tertiary treatment plants using a variety of physical, chemical, and / or biological processes. These processes are described in Chapter 16. In all cases there is some discharge of phosphorus that inevitably finds its way to streams, rivers, ponds, and lakes. To some degree, the extent of phosphorus release can be controlled by upgrading the treatment process.

Much more difficult to control are the many diffuse sources of phosphorus. These include all forms of urban, agricultural, and forest run-off and leachates. The element is present in low concentrations as soluble organic and inorganic forms and also associated with clay-sized material that is carried into the hydrosphere by erosion.

## Restoration of a eutrophic water body—the Bay of Quinte

There are many water bodies, large and small, throughout the world where phosphorus-controlled eutrophication is evident. One example is the Bay of Quinte (BOQ; Fig. 14.13) at the north-eastern end of Lake Ontario in Canada.

During the 1960s and 1970s, the eutrophic status of this portion of the lake was evidenced by dense algal blooms that blocked light from reaching the rooted aquatic plants. This resulted in reduced growth of plants that, in turn, contributed to shifts and instability in the fish communities. Drinking water was also affected by taste and odour problems.

The bay is surrounded by several small cities as well as a considerable extent of agricultural and recreational land. Over half of the phosphorus is from point sources, principally from five wastewater treatment plants located around the bay. Diffuse sources including run-off from agricultural and forested land and from urban areas makes up the rest of the annual loading of the element. Since 1978 considerable effort has gone into controlling release from both types of source. Provincial regulations strictly limited the amount of phosphorus allowed in detergents and other cleaning agents. Improvements were made in wastewater treatment plants in accordance with concentration targets of 0.5 μg mL$^{-1}$ (as P) set for phosphorus in the effluent.

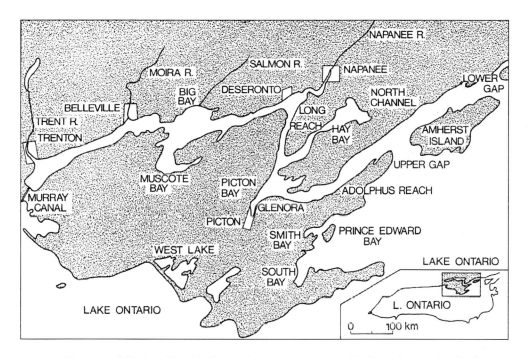

**Fig. 14.13** The Bay of Quinte (BOQ), shown here as the narrow Z-shaped water body that extends eastward from Trenton, is situated on the north side of Lake Ontario. While it is part of the lake, a large peninsula isolates it from the main water body and there is only limited water exchange with the open lake. The BOQ is a long (~120 km) serpentine stretch of water, bordered by several towns and small cities as well as agricultural and recreational properties. In different ways, all of these contribute to phosphorus loading in the bay.

Encouragement and incentives were provided to improve farm practices in a way that would reduce phosphorus release through run-off and leaching. These included the adoption of conservation tillage in crop production, care in manure storage, improvements in household sewage systems, and minimizing livestock access to pasture adjacent to the shoreline.

As a consequence of these measures, during three decades the phosphorus concentration in the BOQ declined to less than half of its previous value and this has led to a marked improvement in water quality in the BOQ as shown in Table 14.4. Phytoplankton volume and *chlorophyll a* mass concentrations are both indicators of microbial growth, which in turn depends on the availability of nutrient phosphorus. The vertical light extinction index, $\varepsilon_{PAR}$, is a measure of the absorbance of photosynthetically available radiation (PAR) by turbidity in the water per vertical metre depth; more light absorbed signifies greater turbidity.

These data may be compared with a general classification of the trophic status of lakes as shown in Table 14.5. The classification includes lakes that support little biological activity (*ultraoligotrophic*) to ones where the productivity of organisms is excessively high (*hypertrophic*). Based on the phosphorus data, one can see that, in the years since control measures were instituted, the BOQ has moved from the eutrophic to the mesotrophic category. Major improvements (reductions) in microbial biomass production are also indicated, although classification according to this criterion indicates somewhat different categories from those defined by phosphorus levels.

The BOQ trends therefore make it clear that, in the years after the implementation of controls (1977), there has been significant improvement in water quality—a conclusion supported by all four types of measurements.

**Table 14.4** Water-quality data related to eutrophication, for the Bay of Quinte.[a]

| | Mean values during | | | | | |
|---|---|---|---|---|---|---|
| | 1971–77 | 1978–83 | 1984–89 | 1990–94 | 1995–99 | 2000–04 |
| Total P / $\mu g\ L^{-1}$ | 78 | 49 | 43 | 37 | 35 | 33 |
| Phytoplankton biomass / $g\ m^{-3}$ | 13 | 6.9 | 9.1 | 6.9 | 6.2 | 3.7 |
| Chlorophyll *a* / $\mu g\ L^{-1}$ | 36 | 21 | 27 | 23 | 13 | 13 |
| Vertical light extinction, $_{PAR}$/ $m^{-1}$ | 1.88 | 1.43 | 1.57 | 1.40 | 1.16 | 0.97 |

[a]Marten Koops, Great Lakes Laboratory for Fisheries and Aquatic Sciences, Fisheries and Oceans Canada, 2009.

It is now recognized that further improvements may have to take into account sedimentary release of accumulated phosphorus. In the sediment, the element is found as a component of some primary minerals, occluded within the amorphous hydrous oxides of iron and aluminium, adsorbed on the surface of minerals such as calcite and quartz, and associated with the organic component of the sediment. In the BOQ about 50–75% of the sedimentary phosphorus is associated with calcium carbonate and 20–40% is in the organic fraction. Some of the latter category, especially that found in living biomass, appears to be active in terms of refluxing or releasing phosphorus to more available, that is, more soluble, forms. Endogenously produced phosphatase enzymes allow certain bacteria to mineralize organic phosphorus, releasing soluble orthophosphate into the water column. It is estimated that the extent of reflux is equivalent to the release of approximately 36 kg P $d^{-1}$, in early summer and 72 kg P $d^{-1}$ in late summer when decomposition of accumulated biomass is at a maximum. These amounts are several times larger than the 19 kg P $d^{-1}$ that are now estimated to be released from wastewater treatment plants serving the urban areas along the bay.

If sedimentary reflux maintains elevated phosphorus levels in the bay, the introduction of even tighter regulations on phosphorus discharge might have little immediate effect on water quality. Consideration has been given to other strategies that would affect sedimentary release; these include enhancing the flow of water through the bay in order to 'flush out' the system, dredging portions of highly contaminated sediment, or containing the surface phosphorus by covering the

**Table 14.5** Trophic status of lakes,[a] including corresponding Bay of Quinte values before (1972–77) and after (2000–04) phosphorus controls were introduced.

| | Total P / $\mu g\ L^{-1}$ | Chlorophyll *a* / $\mu g\ L^{-1}$ |
|---|---|---|
| Ultraoligotrophic | <4 | <1.0 |
| Oligotrophic | 4–10 | 1.0–2.5 |
| Mesotrophic | 10–35 | 2.5–8 |
| *Bay of Quinte 2000–04* | *33* | *13* |
| Eutrophic | 35–100 | 8–25 |
| *Bay of Quinte 1972–77* | *78* | *36* |
| Hypertrophic | >100 | >25 |

[a] Rast, W. and M. Holland, Eutrophication of lakes and reservoirs: a framework for making management decisions. *Ambio*, **17** (1988), 2–12.

sediment with uncontaminated material. However, each of these possibilities has major practical limitations and is unlikely to be used on a large scale.

In other lakes around the world also, sedimentary reflux has been shown to be a principal means by which eutrophic conditions persist when phosphorus input levels are reduced.

> **Fermi question**
> Estimate the percentage reduction in phosphorus concentration in untreated (raw) wastewater that would be expected in an urban area by legislating that all detergents should be phosphate free.

> **Main point 14.5** Phosphorus is an important nutrient for plants and microbes. When present in excess it contributes to eutrophication of water bodies. Much of its chemistry and the required control technologies are based on the strong interaction between phosphorus and various environmental colloids.

## 14.4 Quantitative descriptions of adsorption II

### The Freundlich relation

A second relation used to describe adsorption on environmental colloids is the empirical Freundlich equation.

$$C_s = K_F C_{aq}^{n_F} \tag{14.14}$$

where $C_s$ – quantity adsorbed per unit mass / mol g$^{-1}$, $C_{aq}$ = equilibrium solution concentration / mol L$^{-1}$, and $K_F$ / L g$^{-1}$ and $n_F$ (a dimensionless number) are empirical Freundlich constants ($n_F$ is usually < 1). (Again, any unit of mass or quantity may be chosen for use in the numerator of $C_s$ and $C_{aq}$ terms as long as the same unit is chosen for both the solid and the solution.)

To linearize the equation, logarithms are taken

$$\log C_s = \log K_F + n_F \log C_{aq} \tag{14.15}$$

and plots of the type shown in Fig. 14.14 are then made.

The Freundlich relation differs from that of Langmuir in that it does not consider all sites on the adsorbent surface to be equal but rather adsorption becomes progressively more difficult

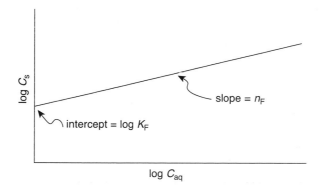

**Fig. 14.14** A linearized Freundlich isotherm plot. Values of the constants, $K_F$ and $n_F$ are determined from the intercept and the slope of the plot.

as more and more adsorbate accumulates. Furthermore, it is assumed that, once the surface is covered, additional adsorbed species can still be accommodated. In other words, multilayer adsorption is predicted by this relation.

The equation does not imply any particular mechanism of adsorption. It is purely empirical, and has been found to be most satisfactory for small molecules present in a very low concentration range.

Values of $K_F$ and $n_F$ have been determined for some particular cases. For a forest soil, it has been reported[2] that $K_F = 0.0324$ L g$^{-1}$ and $n_F = 0.82$ for cadmium. From this we may calculate the equilibrium concentration of adsorbed cadmium in the presence of a 4.0 μg L$^{-1}$ solution concentration.

---

**Example 14.2 Cadmium adsorption described by a Freundlich relation**

$$C_s = K_F C_{aq}^{n_F} \tag{14.14}$$

$$C_s = 0.0324 \text{ L g}^{-1} \times (4.0 \text{ μg L}^{-1})^{0.82}$$

$$C_s = 0.10 \text{ μg g}^{-1}$$

This is the concentration of adsorbed cadmium in equilibrium with a 4.0 μg L$^{-1}$ solution concentration.

---

A slightly different problem is to calculate the concentration of cadmium that would be adsorbed by 10 g of soil under equilibrium conditions *starting with* 1 L of a 4.0 μg L$^{-1}$ solution of the metal ion.

---

**Example 14.3 Cadmium adsorption from water containing a limited amount of the element**
At equilibrium

$$C_{aq} = (4.0 - \text{adsorbed quantity}) \text{ μg L}^{-1}$$

$$C_s = 0.0324 \text{ L g}^{-1} ((4.0 - \text{adsorbed quantity}) \text{ μg L}^{-1})^{0.82}$$

Solving the last equation directly is difficult, but it may be done simply enough using an iterative procedure. A starting point is to use the adsorbed concentration calculated above, i.e. $C_s = 0.10$ μg g$^{-1}$. For this, the first estimate of total mass adsorbed on 10 g of soil would be 1.0 μg and this is subtracted from the 4.0 μg present in the original 1 L of solution.

$$C_{s1} = 0.0324 (4.0 - 1.0)^{0.82}$$

$$C_{s1} = 0.080 \text{ μg g}^{-1}$$

Therefore, the total mass of Cd in 10 g of soil = 0.80 μg. Repeating the substitution,

$$C_{s2} = 0.0324 (4.0 - 0.8)^{0.82}$$

$$C_{s2} = 0.084 \text{ μg g}^{-1}$$

Therefore, the total mass of Cd in 10 g of soil = 0.84 μg. Again

$$C_{s3} = 0.083 \text{ μg g}^{-1}$$

and once more

$$C_{s4} = 0.083 \text{ μg g}^{-1}$$

---

[2] Sidle R.C. and L.T. Kardos, Adsorption of copper, zinc and cadmium by a forest soil. *J. Environ. Qual.*, **6** (1977), 313–7.

We now have the correct value. The concentration of cadmium adsorbed on the soil is $0.083\ \mu g\,g^{-1}$ and that remaining in the equilibrium solution is $3.2\ ng\,mL^{-1}$.

---

> **Main point 14.6** The Freundlich relationship describes a situation where multilayer adsorption is possible and is usually applicable where the solution contains low concentrations of small molecules.

## 14.5 Quantitative descriptions of adsorption III

### A simplified Freundlich relation—the distribution coefficient $K_d$

Like other substances, small organic molecules in water distribute themselves between the aqueous phase and the solid, particulate materials suspended as colloids or in the sediments. A Freundlich relation often provides a satisfactory description of the equilibrium distribution, especially when, as is frequently the case, the organic solute is present in very small concentrations. Under these conditions $n_F$ in the expression $C_s = K_F C_{aq}^{n_F}$ can be approximated to have a value of 1. This implies that all sites responsible for retention of the solute molecule are similar and that the initial substance retained on the surface neither enhances nor inhibits subsequent retention. The modified equation is then written as

$$C_s = K_d C_{aq} \tag{14.16}$$

Based on this relation, $K_d$, given by eqn 14.16, is the simplified distribution coefficient describing the partitioning of the solute between the solid phase and water.

$$K_d = \frac{C_s}{C_{aq}} \tag{14.17}$$

This basic relationship has been expanded upon and used in many environmental studies.

The value of $K_d$ depends on the organic solute itself, on the chemical and physical nature of the solid phase, and also on other environmental properties such as temperature and solution ionic strength. A large $K_d$ signifies a strong interaction between soluble species and the solid.

To calculate the distribution of a solute between solution and suspended or settled solid in an environmental situation, one needs to make use of the value of $K_d$. However, the infinite range of all potential properties and their combinations makes it impossible to tabulate values of $K_d$ so that they can be applied in a meaningful way to different situations. Nevertheless, two related 'reportable' parameters have been developed to describe the partitioning of small neutral organic species in particulate / aqueous systems, and these can be used to calculate the $K_d$ in particular situations. The two parameters are the organic matter / water partition coefficient ($K_{OM}$) and the octanol / water partition coefficient ($K_{OW}$). Environmental scientists and engineers make use of $K_{OM}$ and $K_{OW}$ data to describe and predict movement of organic contaminants such as pesticides in soil / water and sediment / water systems.

We will discuss the chemical basis and some applications of the $K_{OM}$ and $K_{OW}$ parameters and then show how they can be used to estimate the value of $K_d$. It is essential to first recognize the importance of sediment and soil organic matter in retaining these organic contaminants.

### Sorption of organic species by environmental solids

Suspended and settled sediments and soils are comprised of both mineral and organic components. As we showed earlier, there are various physical and chemical mechanisms by which these environmental solids can adsorb and retain species from the surrounding solution. In the case of small organic solutes, all of the mechanisms can apply to varying degrees. Consider the mineral and organic components of the solids.

- Mineral components. The surfaces taking part in sorption include those of inorganic species such as the hydrous oxides and clay minerals, which have hydroxyl groups extending into the aqueous solution from the mineral face. A combination of van der Waals, induced-dipole:dipole, dipole:dipole, and hydrogen-bonding forces (these are listed in increasing order of strength) are responsible for the binding of species to the surface. Where specific binding occurs, retention of small organic molecules by mineral surfaces has been recognized as being important but, in most cases, the sorption is very weak. This is because there is a strong attraction between the mineral and water, and retention of a solute must simultaneously involve displacement of water molecules from the surface.

- Organic components. The more active surfaces with respect to neutral organic solutes are those of the natural organic macromolecules present in suspended or sedimentary solids. We have seen in Chapter 12 that the organic matter includes humic material derived from plant or microbial sources. This material has polar properties because of its oxygen-containing functionality. However, also important in the present context, humic matter has major hydrocarbon regions located within the molecule where non-polar solutes encounter little competition from surface-bound water molecules. Thus, small hydrophobic solutes can effectively 'dissolve' within the interior non-aqueous medium. In this sense the retentive forces go beyond being surface forces and the process is more akin to absorption than to adsorption. It is for this reason that the generic term *sorbed* or *sorption* is especially appropriate here.

It is worth reminding ourselves too that humic material (HM) itself may be bound to mineral phases where it forms a coating. Because the mineral surface is blocked from directly reacting with other solutes, this accentuates the importance of the solid organic component as a sorbent (Fig. 12.8). The HM is retained on the surface by hydrogen bonds between surface groups, such as SiOH and electronegative atoms in the otherwise hydrophobic organic molecule. Other weaker van der Waals forces make a smaller contribution to the energy of binding. The result is that while the colloidal particle may have an inorganic core it develops an organic, humate-like surface.

As a consequence, for colloids suspended in water or present in sediments and soils, an organic matter content as low as 1% can dominate the sorptive properties with respect to small neutral organic species so that the contribution from the mineral fraction is negligible.

The $K_{OM}$ and $K_{OW}$ parameters are therefore defined under the assumption that the organic component of the solid is the dominating component responsible for sorption of small organic solutes.

## The $K_{OM}$ coefficient

Thinking of both organic and mineral adsorbants together, in the general case, the equilibrium concentration of substrate in the soil or sedimentary material, $C_S$, is given by the relation

$$C_S = f_{OM} \times C_{OM} + f_{MM} \times C_{MM} \tag{14.18}$$

where $C_{OM}$ and $C_{MM}$ are the concentrations of the solute in the organic matter and mineral matter, respectively, and $f_{OM}$ and $f_{MM}$ are the fractions of these two components in the whole soil or sediment. As we have been emphasizing, in many situations involving neutral organics, the amount that is associated with the mineral material, $C_{MM}$, is so small that the relation reduces to

$$C_S = f_{OM} \times C_{OM} \tag{14.19}$$

Combining equations 14.17 and 14.19, the distribution coefficient, $K_d$ can then be expressed as

$$K_d = \frac{f_{OM} \times C_{OM}}{C_{aq}} \tag{14.20}$$

Now we can define a $K_{OM}$ partition coefficient to describe the distribution of organic compounds between water and the solid organic matter within the associated suspended material, sediment or soil.

$$K_{OM} = \frac{C_{OM}}{C_{aq}} \tag{14.21}$$

The relation between $K_d$ and $K_{OM}$ is then given by 14.22,

$$K_d = f_{OM} \times K_{OM} \tag{14.22}$$

And, from eqn 14.17,

$$C_S = f_{OM} \times K_{OM} \times C_{aq} \tag{14.23}$$

It is important to understand why $K_{OM}$ values are so useful. While every particular combination of soil (or sediment) and organic solute has its own $K_d$ value, it has been found that $K_{OM}$ values for particular chemicals are relatively constant ($\pm 0.3$ log $K_{OM}$ units) over a considerable variety of sediments and soils. This confirms that there is a large degree of similarity in the nature of solid organic matter—substantially humic material—present in many environmental situations. It is useful, then, to tabulate and report $K_{OM}$ values as they can be applied in many situations around the world. An example will illustrate some useful features of these concepts.

---

### Example 14.4 Relation between $K_d$ and $K_{OM}$

The compound $p$-dichlorobenzene has been found to have $K_{OM}$ ~630. For a soil containing 1.6% organic matter, the distribution coefficient can be calculated as follows.

$$K_d = f_{OM} \times K_{OM}$$
$$K_d = (1.6/100) \times 630 = 0.016 \times 630$$
$$K_d = 10$$

---

In a simple way, this means that, if 100 L of water containing 5 ppm dichlorobenzene is spilled on a 0.2 m³ volume of the dry soil, it would distribute itself at equilibrium according to the following calculation.

---

### Example 14.5 Partitioning of dichlorobenzene between soil and associated water

$$\text{Mass of dichlorobenzene} = 5 \text{ mg L}^{-1} \times 100 \text{ L}$$
$$= 500 \text{ mg}$$
$$\text{Mass of soil} = 0.2 \text{ m}^3 \times 1200 \text{ kg m}^{-3}$$
$$= 240 \text{ kg}$$

(A typical soil bulk density is 1200 kg m⁻³, see Chapter 18.)

If $x$ is the mass in mg of dichlorobenzene in the soil after equilibration, then

$$K_d = \frac{C_S}{C_{aq}} = \frac{x/240}{(500 - x)/100} = 10$$
$$x = 480 \text{ mg}$$

Therefore, 480 mg are sorbed to the soil matrix and the concentration in the soil once equilibrium is achieved is

$$480 \text{ mg} \div 240 \text{ kg} = 2.0 \text{ ppm}$$

In the water, 20 mg remain, giving a concentration of

$$20 \text{ mg} \div 100 \text{ L} = 0.20 \text{ ppm}$$

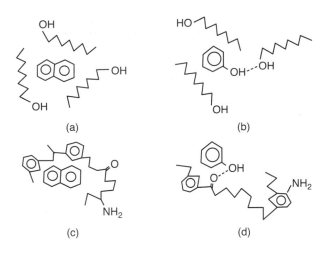

**Fig. 14.15** Representation of the ability of *n*-octanol to solubilize both (a) non-polar (naphthalene) and (b) polar (phenol) molecules. (c) and (d) Similar representation showing how a portion of a humate molecule can interact with the same species. Hydrophobic portions of the octanol and HM tend to associate with hydrophobic organic solutes, while hydrogen bonding and other interactions favour reaction with polar groups that may be present on these solutes.

Repeating the calculation on a different soil with 2.5% organic matter content, the value of $K_d$ would be 16 and the corresponding concentrations are 2.0 ppm in the soil 0.13 ppm in the water.

Again, we must emphasize two things—the use of $K_{OM}$ values[3] in calculations such as these implies that the organic component of the solid phase is the only significant contributor to uptake of the solute. This is often a good assumption and, when true, tabulated $K_{OM}$ values for different chemicals can be used in a variety of sediment- or soil-water situations. There is, however, an important limitation in the $K_{OM}$ concept. The limitation is that $K_{OM}$ values are not easily determined by direct experimental measurement.

## The octanol–water partition coefficient $K_{OW}$

In order to determine $K_{OM}$ values indirectly then, one additional equilibrium constant, the octanol–water partition coefficient, $K_{OW}$, has been defined. This is a constant that can be measured in the laboratory, and its value is connected with the $K_{OM}$ value through a variety of empirical relationships.

Octanol (actually *n*-octanol, $CH_3(CH_2)_7OH$) is an amphiphilic solvent—by this we mean that it has both hydrophilic and hydrophobic character. As such, water has a substantial solubility in the solvent (molar ratio 0.25) while the solubility of *n*-octanol in water is much less (molar ratio is $8 \times 10^{-5}$). The amphiphilic character gives *n*-octanol a solvating ability not unlike that of humic material and other naturally occurring organic colloids (Fig. 14.15) in that it has some ability to associate with both polar and non-polar compounds.

[3] There is another distribution parameter, the $K_{OC}$ value, which is often used as an alternative for the $K_{OM}$ and shares the same usefulness and limitations. Like the $K_{OM}$, the $K_{OC}$ describes the distribution of solute in the aqueous phase and that associated with the organic matter solid, but the amount of solid is expressed as *mass of its carbon component*, not the total mass. Assuming that organic carbon makes up 60% of organic matter (Section 12.3), OM = $1.7 \times$ OC and $K_{OC} = 1.7 \times K_{OM}$. Throughout this book, we will use only $K_{OM}$ values, but where $K_{OC}$ appears in other literature, the conversion can readily be made.

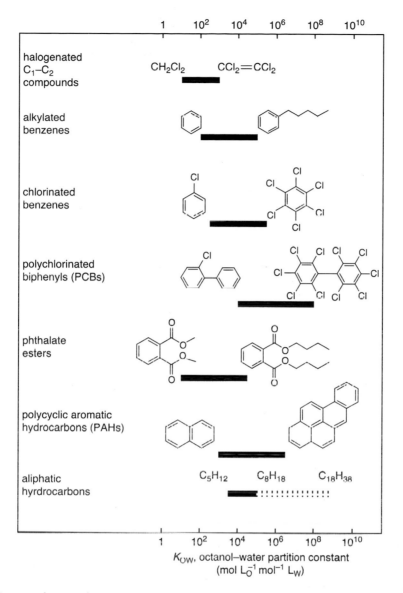

**Fig. 14.16** Range of octanol–water partition coefficients for classes of organic compounds. (Redrawn from Schwarzenbach, R.P., P.M. Gschwend, and D.M. Imboden, *Environmental organic chemistry*, John Wiley and Sons, Inc., New York; 1993.)

Taking into account these properties of octanol, the equilibrium concentrations of an organic solute can be measured in a binary mixture of octanol and water. The $K_{OW}$ is then defined as

$$K_{OW} = \frac{C_O}{C_{aq}} \tag{14.24}$$

where $C_O$ is the equilibrium molar solubility of the solute in octanol and $C_{aq}$ the corresponding solubility in water. The octanol / water partition coefficient, $K_{OW}$, provides a relatively convenient means of predicting the partitioning behaviour of a hydrophobic organic substance between being dissolved in water and being sorbed by solid organic matter associated with the water. Figure 14.16 shows ranges of $K_{OW}$ values for various classes of (usually synthetic) organic compounds. The range covers more than 10 orders of magnitude and several features are qualitatively predictable.

Small $K_{OW}$ values—favouring the aqueous phase—are characteristic of low molar mass species and species that contain electronegative atoms like oxygen in their structure. Hydrocarbons or compounds with large carbon:oxygen ratios, especially high molar mass ones, have large $K_{OW}$ values and tend to be associated with the suspended or precipitated organic matter in the environment.

In summary, large values of $K_{OW}$ imply a tendency for an organic solute to be associated with the organic solid phase while small values suggest that the solute will remain in solution. Octanol / water partition coefficients may be estimated by calculation as well as being experimentally measurable in laboratory experiments (see Problem 14.10). Tabulated values for many compounds are available.[4] Using $K_{OW}$ in a relative sense enables us to make qualitative predictions and comparisons about the environmental behaviour of different compounds.

Because of octanol's ability to act as a good surrogate for the solid organic material of water / particulate systems, we could therefore expect there to be a relation between $K_{OW}$ and $K_{OM}$. Logarithmic relations have been established experimentally[5] and take the form

$$\log K_{OM} = a \log K_{OW} + b \tag{14.25}$$

It is perhaps surprising that, for a wide range of chemical classes, a single version of eqn 14.25 shows a fairly good correlation between the two parameters. Karickhoff used the equation with coefficients as shown in eqn 14.26

$$\log K_{OM} = 0.82 \log K_{OW} + 0.14 \tag{14.26}$$

for a variety of aromatic hydrocarbons, chlorinated hydrocarbons, chloro-*S*-triazines, and phenylureas. Overall the correlation coefficient, $r^2$, was 0.93. Even better correlations can be established when an individual equation is used for each particular class of compounds.

## Bioconcentration factors[6]

Bioaccumulation, is a widely recognized phenomenon whereby organisms, everything from microorganisms to plants to polar bears and whales, accumulate particular contaminant chemicals in their tissues. Uptake occurs via a variety of mechanisms depending on the organism and the chemical. In the aqueous environment filter feeders like molluscs extract nutrients and along with the nutrients they also extract other chemical entities that are not metabolized and accumulate in the soft flesh or hard shells. Microorganisms, aquatic plants, fish, and other animals living

---

**Terms used to describe uptake of chemicals by organisms**

- **Bioaccumulation.** The processes by which an organism takes up and retains a contaminant through multiple exposure routes; it depends on the rate of intake versus the rate of elimination (through urine or faeces) or breaking down of the chemical via metabolic processes

- **Biomagnification.** A process or series of processes that results in a chemical becoming increasingly concentrated at successively higher trophic levels of a food chain or food web

- **Bioconcentration.** The increase in concentration of a chemical in an organism compared to that in the ambient medium in which it lives—for example, a higher concentration of contaminant in fish compared with that in the water

- **Bioconcentration factor (BCF).** Used to quantify the accumulation of chemicals in organisms; it is the ratio of the contaminant in the organism to that in the ambient medium

---

[4] Tewari, Y.B., M.M. Miller, S.P. Wasik, and D.E. Martire, Aqueous solubility and octanol / water partition coefficient of organic compounds at 25.0°C, *J. Chem. Eng. Data*, **27** (1982), 451–4.

[5] Karickhoff, S.W., Semi-empirical estimation of sorption of hydrophobic pollutants on natural sediments and soils, *Chemosphere*, **10** (1981), 833–46.

[6] Porteous, A., *Dictionary of environmental science and technology*, John Wiley and Sons, Chichester; 1992.

in water are also capable of accumulating soluble metals and organic chemicals. One concern is that the assimilated chemicals can be further accumulated and concentrated by other species that use the original organism as a food source. Eventually, at higher levels in the food chain, the amount of accumulated chemical may be large enough to be toxic.

Many chlorinated compounds are chemically and biologically stable and are relatively lipophilic. For these reasons, they are susceptible to accumulation in the fatty tissue of organisms. In the Firth of Clyde off western Scotland, for example, 1989 measurements of various chlorinated compounds, including polychlorinated biphenyls (PCBs) and the insecticides dieldrin and DDT, showed that concentrations in the soft tissue of mussels were several orders of magnitude higher than concentrations in the ocean water. For DDT, the aqueous concentration was as low as 1 ng $L^{-1}$ while that in the mussels reached 300 $\mu$g $kg^{-1}$. The ratio of these two numbers, using a common mass basis, is $3 \times 10^{-4}$ g (DDT) $kg^{-1}$ (mussel) / $1 \times 10^{-9}$ g (DDT) $kg^{-1}$ (water) = 300 000 and is referred to as the bioconcentration factor (BCF) for uptake of DDT by mussels from water.

There is usually a direct relationship between the octanol / water coefficient and the value of the BCF. As should be evident, the relation is based on the idea that octanol has an ability to dissolve lipophilic compounds in a way that can be related to a similar ability of some types of biological tissue.

An example is illustrated by a study of various organochlorine contaminants in water, fish, and seal in Lake Baikal in eastern Siberia (Fig. 14.17).[7] Persistent compounds such as DDT and polychlorinated biphenyls are somewhat volatile and can travel over long distances within the atmosphere. As a result, even regions where such compounds have never been used can exhibit significant levels in water and, because they are hydrophobic, the compounds bioaccumulate to higher levels in the fatty tissue of animals living in the water. This is evident in the Lake Baikal study. The BCF values used in the figure are ratios of the concentration (g $kg^{-1}$) of various compounds in Baikal seal blubber to their concentration (g $L^{-1}$) in the ambient water. You can see that there is a reasonably good correlation for a wide range of organochlorines between tabulated log $K_{OW}$ values and log BCF. Such correlations serve a predictive function in that biouptake may be estimated for other halogenated compounds for which $K_{OW}$ values are

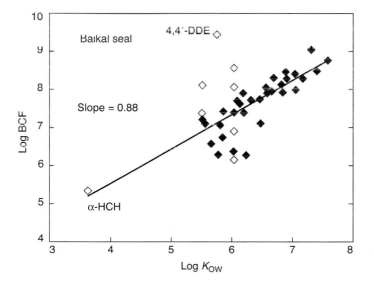

**Fig. 14.17** Correlation between octanol / water coefficient and bioconcentration factor in the Baikal seal.

[7] Kuklick, J.R., T.F. Bidleman, L.L. McConnell, M.D. Walla, and G.P. Ivanov, Organochlorines in the water and biota of Lake Baikal, Siberia, *Environ. Sci. Technol.*, **28** (1994), 31–7.

available. However, there are limitations to this kind of exercise. You can see from the figure that the BCF of some compounds is not well described by the correlation.[8] For 4,4'-DDE, a BCF of about $10^7$ is predicted, while the experimental value is closer to $10^9$.

As a general approximation, bioaccumulation of small organic molecules becomes important and of concern when the BCF value is 10 000 or above.

---

**Main point 14.7** The significance of the various partition coefficients and the relationship between them can be summarized as follows:

$K_{OW}$ describes the distribution of small organic molecules between octanol and water. Values can be determined by experiment or calculation; tabulated values are available and can be mathematically related to $K_{OM}$ and BCF.

$K_{OM}$ describes the distribution of small organic molecules between the organic fraction of solids and water in different sediment- or soil-water systems. For a specific situation, the fundamental $K_d$ can be calculated from the $K_{OM}$.

$K_d$ coefficients also are used to calculate equilibrium concentrations of small organic molecules in the solid and aqueous components of sediment- or soil-water systems. Unlike the $K_{OM}$, the $K_d$ value is unique for each situation.

The biocentration factor is a measure of the ratio of concentration of a particular species in a living organism to that in the ambient medium.

---

## 14.6 Colloidal material in the natural environment

There is a wide variety of colloidal components found in different environments. Some colloids have wide-ranging distribution throughout the world. Others are formed under specific conditions and their presence is clear evidence of those conditions.

### Colloids formed under specific environmental conditions

#### High p*E* (oxidizing) environment

In water, iron and manganese are soluble under reducing conditions as iron (II) and manganese (II) species. Under an oxidizing (high p*E*) regime, however, these elements are oxidized to form colloidal $Fe_2O_3 \cdot xH_2O$ and $MnO_2 \cdot xH_2O$, which can remain suspended or gradually settle to form a sedimentary deposit. In the oceans, this phenomenon is associated with hydrothermal vents that release effluent with a high content of reduced iron and manganese. When these species encounter an oxygen-rich environment, the insoluble minerals are formed as colloids that eventually precipitate and are incorporated into the sediment. In lakes and rivers, smaller amounts of reduced iron are derived from various sources such as effluent from some mining or mineral processing operations. In an oxidizing aqueous environment, the colloidal hydrous oxides are again formed.

We will see in the next chapter, that many types of microorganisms are distributed throughout all water bodies as well as within soil pore water. Under well-aerated conditions (high pE) and where there is a plentiful supply of nutrients they may be especially abundant. Many of these organisms have sizes within the colloidal range and remain suspended in the water. Upon their death, they eventually settle and become a part of the sediment.

#### Low p*E* (reducing) environment

Under reducing conditions, especially in sea water, a process occurs that is essentially the reverse of the one described above. Such environments include deep water sediments where the only

---

[8] The significance and limitations of associative relationships such as this one are discussed in Smith, J. and P. Smith, *Environmental modelling*, Oxford University Press, Oxford, UK; 2007.

source of oxygen in the water is immediately adjacent to the sediment surface. If the sediment contains reducing materials such as organic matter, the oxygen is used up and anaerobic conditions result. The insoluble hydrous oxides of iron (III) and manganese (IV) are then reduced to more soluble iron (II) and manganese (II) species. At the same time sulphate, which is abundant in sea water and present in smaller amounts in fresh water, is reduced to sulphide. As a consequence of these simultaneous processes, colloidal iron and manganese sulphides are precipitated from solution and are therefore found in anaerobic sediments. The reduction reactions are of microbiological origin and will be discussed in the next chapter.

### Low pH (acidic) environment

Where waters containing dissolved organic matter encounter an acidic environment, acid-insoluble humic acid is precipitated from solution as a colloidal deposit. Sediments deposited in other situations contain organic matter of highly variable amount and composition. In lakes and coastal regions of oceans, sediment may contain from 1% to more than 20% organic matter, much of which is carried in as particulate material from terrestrial sources. Smaller amounts are found buried in deep ocean sediments. Organic compounds are also produced in the water. The growth of both microorganisms and larger organisms depends on nutrients and energy from dissolved constituents and from sunlight. After their life cycle is complete, the residual organic matter forms part of the particulate fraction, remaining suspended or sinking to the lake or ocean floor. While passing through the water column and while in the sediments, decomposition and resynthesis take place so that other compounds are formed including the aqueous equivalent of terrestrial humic material. The origin of humic substances found in sediments can often be determined, since (for example) terrestrial material contains a higher proportion of aromatic subunits, is more oxidized, and therefore has a greater concentration of acidic functional groups than does humic material produced in situ in the aqueous (particularly marine) environment.

### High pH (alkaline) environment

Calcium carbonate is a frequently observed colloid formed in an alkaline environment—even one that has only a temporary existence.

The following example illustrates a mechanism by which such material is formed.

---

### Example 14.6 **Calcium carbonate solubility in water**
Consider a shallow eutrophic water body that has pH = 7.20, $[HCO_3^-] = 1.06 \times 10^{-3}$ mol L$^{-1}$, and $[Ca^{2+}] = 1.5 \times 10^{-3}$ mol L$^{-1}$. These properties are characteristic of lakes located in a region where the bedrock and sediments contain limestone.

Under these conditions

$$[CO_3^{2-}] = \frac{K_{a2} \times [HCO_3^-]}{[H^+]} = \frac{4.7 \times 10^{-11} \times 1.06 \times 10^{-3}}{6.3 \times 10^{-8}} = 7.9 \times 10^{-7} \text{ mol L}^{-1}$$

The reaction quotient,

$$Q_{sp} = [Ca^{2+}][CO_3^{2-}] = 1.5 \times 10^{-3} \times 7.9 \times 10^{-7} = 1.2 \times 10^{-9}$$

which is less than $5 \times 10^{-9} = K_{sp}$. Because the reaction quotient is smaller than the solubility product, this indicates that $CaCO_3$ is sufficiently soluble so that no precipitate forms.

---

Production of biomass in eutrophic lakes (expressed in terms of carbon) on days with intense sunlight is frequently in the range 1000–2000 mg C m$^{-2}$ d$^{-1}$. Consider that in the present case the production rate is 1700 mg C m$^{-2}$ d$^{-1}$. We will assume that this production is mostly in the top 50 cm of the water column, where the sunlight is most intense.

At pH 7.20, most of the carbonate is in the hydrogen carbonate form and algal photosynthesis can be described by the following reaction.

$$HCO_3^- + H_2O \xrightarrow{h\upsilon} \{CH_2O\} + OH^- + O_2 \qquad (14.27)$$

---

### Example 14.7 Effect of photosynthesis on calcium carbonate solubility

During the period of photosynthesis in the volume ($1\ m^2 \times 0.5\ m$) defined above, the amount of consumed hydrogen carbonate (and hydroxide ion produced) is

$$1700\ mg \times 1 \times 10^{-3}\ g\ mg^{-1}/12\ g\ mol^{-1} = 0.142\ mol$$

The molar concentration changes of these species are therefore

$$0.142\ mol/500\ L = 2.8 \times 10^{-4}\ mol\ L^{-1}$$

To determine how this affects the water chemistry, we should calculate the concentrations of other species before and after the production of biomass.

$$[H_3O^+] = 6.3 \times 10^{-8}\ mol\ L^{-1}$$

$$[HCO_3^-] = 1.06 \times 10^{-3}\ mol\ L^{-1}$$

$$[CO_2] = \frac{[H_3O^+][HCO_3^-]}{K_{a1}} = \frac{6.3 \times 10^{-8} \times 1.06 \times 10^{-3}}{4.5 \times 10^{-7}}$$

$$= 1.5 \times 10^{-4}\ mol\ L^{-1}$$

and, from Example 14.6,

$$[CO_3^{2-}] = 7.9 \times 10^{-7}\ mol\ L^{-1}$$

$2.8 \times 10^{-4}\ mol\ L^{-1}$ of has been used up, but the $2.8 \times 10^{-4}\ mol\ L^{-1}$ of hydroxide that is produced is sufficient to react with all of the aqueous carbon dioxide, producing an equivalent quantity of hydrogen carbonate. The additional $1.3 \times 10^{-4}\ mol\ L^{-1}$ of hydroxide converts the same amount of hydrogen carbonate to carbonate. The net result of these reactions is

$$[CO_3^{2-}] = 1.3 \times 10^{-4}\ mol\ L^{-1}$$

$$[HCO_3^-] = 1.06 \times 10^{-3} - 2.8 \times 10^{-4} + 1.5 \times 10^{-4} - 1.3 \times 10^{-4}$$

$$= 8.0 \times 10^{-4}\ mol\ L^{-1}$$

$$[H_3O^+] = \frac{[HCO_3^-]K_{a2}}{[CO_3^{2-}]} = \frac{8 \times 10^{-4} \times 4.7 \times 10^{-11}}{1.3 \times 10^{-4}}$$

$$= 2.9 \times 10^{-10}\ mol\ L^{-1}$$

This corresponds to a pH of 9.54

Under the new conditions,

$$Q_{sp} = 1.3 \times 10^{-4} \times 1.5 \times 10^{-3}$$

$$= 2.0 \times 10^{-7}$$

i.e. $Q_{sp} > 5 \times 10^{-9} = K_{sp}$. Because the solubility product of calcium carbonate has now been exceeded, insoluble calcium carbonate is precipitated out of the water.

---

It is through this biogeochemical process that large amounts of colloidal calcite are produced in lakes and, ultimately, these become massive deposits of sedimentary limestone.

This phenomenon is associated with biomass synthesis in water during the daytime. Besides the alterations in carbonate chemistry, the pH rises substantially. At night, there is no photosynthesis and the pH declines, but dissolution of calcium carbonate under these conditions is slow, so that most of it remains in suspended or precipitate form.

Calcium carbonate, silica, and smaller amounts of other minerals also arise from deposition of the remains (detritus) of some types of plankton and, to a smaller extent, the bones of fish and the shells of crustaceans.

> **Main point 14.8** Colloidal material found in water is derived from terrestrial sources or is generated in situ. A variety of colloid compositions is observed depending on the environment in which it has been formed.

## Clay minerals

The above natural colloid types—hydrated iron and manganese oxides, sulphide minerals, organic material, and carbonate minerals—have formed and continue to form in particular environments but there is an additional category of colloids that is distributed widely almost everywhere throughout the globe. This category comprises the clay minerals, which are a suite of aluminosilicate minerals with a layered lattice structure. All are products produced by physicochemical weathering of primary minerals. They are usually of terrestrial origin and have been carried to the hydrosphere as eroded material in run-off and by winds or glaciers.

The clay minerals (also called phyllosilicates) have as a common structural feature $SiO_4$ tetrahedra linked together in a planar structure by three of the oxygen atoms. The 'sheet' thus formed is then joined via the additional oxygen to octahedral units of aluminium surrounded by six oxygens or hydroxyl groups. The tetrahedral and octahedral layers together form what is referred to as a 1:1 layer clay mineral structure. This is characteristic of several particular clay minerals including kaolinite (Fig. 14.18).

### 1:1 clay minerals

Kaolinite and other clay minerals that are products of weathering of primary minerals, are frequently encountered in water, sediment, and soil and they are present as very small particles

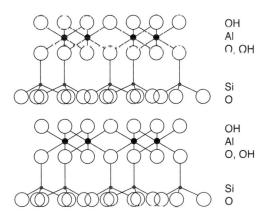

**Fig. 14.18** Crystal structure of kaolinite. Each sheet is made up of silicon–oxygen tetrahedra attached to one side of an aluminium–oxygen octahedral sheet to form a 1:1 layer. The octahedra are arranged in pairs and the sheet is referred to as a dioctahedral type. The distance between two layers in kaolinite is about 0.7 nm.

within the colloidal size fraction. Kaolinite is derived from orthoclase feldspar in the following reaction—a very slow process that only occurs over a geological time-span.

$$\underbrace{2KAlSi_3O_8 \text{ (s)}}_{\text{orthoclase}} + 2H_3O^+ \text{ (aq)} + 7H_2O \rightarrow \underbrace{Al_2Si_2O_5(OH)_4}_{\text{kaolinite}} \text{(s)} + 4H_4SiO_4 \text{ (aq)} + 2K^+ \text{ (aq)} \tag{14.28}$$

The $H_4SiO_4$ (aq) and $K^+$ (aq) are removed in solution during the weathering process.

The electrostatic adsorption capability for kaolinite arises from its negative charge, which is a consequence of two factors.

- On the edges of units, broken bonds leave oxygen atoms with excess negative charge, the magnitude of which is a function of the number of such exposed atoms, which in turn depends on the particle size of the clay.

- The other, usually less important source is from dissociation of –OH groups that are located in the octahedral layer. This is a pH-dependent phenomenon.

The negative charges generated in these two ways are balanced by cations held by electro-static forces at the surface of the mineral in the adjacent ambient solution, and the total cation-exchange capacity (CEC) for kaolinite is in the range 3–15 cmol (+) $kg^{-1}$.

### 2:1 and other clay minerals

A second clay mineral is montmorillonite (also called smectite) which is one of the class of 2:1 phyllosilicates (Fig. 14.19). The class is so named because the layers consist of an aluminium octahedral sheet sandwiched between two silica tetrahedral sheets. A negative charge is developed on montmorillonite for the two reasons noted above, but there is a third overriding factor as well.

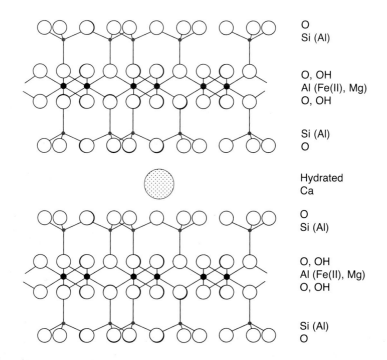

**Fig. 14.19** Crystal structure of montmorillonite. The 2:1 sheet consists of a dioctahedral sheet sandwiched between two tetrahedral sheets. There is a considerable degree of substitution of +2 ions for $Al^{3+}$ in the octahedreal sheet, giving a substantial net negative charge to the clay. The large distance between individual layers (~1.4 nm) allows hydrated cations such as calcium to readily exchange.

- A large amount of negative charge is due to isomorphous substitution of aluminium for silicon atoms in the tetrahedral layers, which occurs to a limited extent, and of iron (II) or magnesium for aluminium in the octahedral layer, which occurs to a considerable extent. The net result of these substitutions of 3+ ions for 4+ ions and of 2+ ions for 3+ ions is a reduction in positive charge—in other words, excess negative charge. Because much of this negative charge is located in the interior of the 2:1 layer, the cations that balance it are only loosely held at the surface and therefore are readily exchangeable.

The CEC of montmorillonite is very large, ranging between 80 and 150 cmol (+) kg$^{-1}$ (Fig. 14.19). A summary of structures and properties of these and other clay minerals is given in Fig. 14.20.

The CECs of various environmental colloids including clay minerals are given in Table 14.6. The bracketed numbers are 'averages' that may be used in making rough estimates of cation-exchange capacities for particular samples. Note that the CEC values in the table apply to solid material in the colloidal size fraction—larger-sized materials make a negligible contribution to the exchange properties.

---

### Example 14.8 Calculation of a total CEC value for a sediment

Consider sedimentary material containing 8% organic matter and 41% clay minerals. Of the latter, 70% is kaolinite and 30% is chlorite. The CEC may be estimated to be

$$0.08 \times 200 + 0.41 \times 0.70 \times 8 + 0.41 \times 0.30 \times 25 = 21 \text{ cmol (+) kg}^{-1}$$

organic matter    kaolinite    chlorite

---

This approximate calculation assumes that each sediment component contributes independently to the cation exchange properties, whereas it is known that interactions between the components may alter these properties.

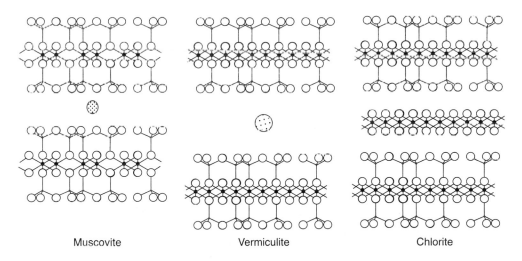

Muscovite                    Vermiculite                    Chlorite

**Fig. 14.20** Crystal structure of three clay minerals. *Muscovite* is similar to montmorillonite, but there is more substitution of Al$^{3+}$ for Si$^{4+}$ in the tetrahedral sheets and much less substitution in the octahedral sheet. As a result, interlayer cations such as potassium are tightly held, the interlayer spacing is smaller (~1.0 nm), and the clay does not readily swell on wetting. *Vermiculite* is also similar to montmorillonite. The octahedral sheets are both di- and trioctahedral types. The interlayer spacing is quite large (~1.3 nm) and hydrated cations can readily exchange. *Chlorite* is a 2:1:1 clay mineral with a trioctahedral layer positioned between two typical 2:1 layers. Spacing between adjacent 2:1 layers is ~1.4 nm.

**Table 14.6** Cation-exchange capacities (CEC) of various environmental materials found in the colloidal size fraction.

| Material | CEC range (average) / cmol (+) kg$^{-1}$ |
| --- | --- |
| Kaolinite | 3–15 (8) |
| Halloysite | 4–10 (8) |
| Montmorillonite | 80–150 (100) |
| Chlorite | 10–40 (25) |
| Vermiculite | 100–150 (125) |
| Hydrous iron and aluminium oxides | ~4 |
| Feldspar | 1–2 (2) |
| Quartz | 1–2 (2) |
| Organic matter | 150–500 (200) |

[a] Most values from Birkeland, P.W., *Pedology, weathering, and geomorphological research*, Oxford University Press, New York; 1974.

---

**Main point 14.9** The clay minerals are a class of colloidal silicate minerals found suspended in water and in soils and sediments. An important property of clay minerals is their ability to act as cation exchangers and this is quantified by measurements of their cation-exchange capacity.

---

## ADDITIONAL RESOURCES

1. Schwarzenbach, R.P., P.M. Gschwend, and D.M. Imboden, *Environmental organic chemistry*, 2nd edn, Wiley and Sons, Hoboken, New Jersey; 2003.

2. Chiou, C.T., *Partition and adsorption of organic contaminants in environmental systems*, John Wiley and Sons, Chichester; 2002.

3. Macdonald, R., D. Mackay, and B. Hickie, Contaminant amplification in the environment, *Environ. Sci. Technol.*, **36** (2002), 457A.

4. MacKay, D., W-Y. Shiu, and K-C. Ma, *Illustrated handbook of physical-chemical properties and environmental fate for organic chemicals*, CRC Press LLC; 1997.

5. Stumm, W., *Chemistry of the solid–water interface*, John Wiley and Sons, Inc, New York; 1992.

6. A very comprehensive searchable compendium of log $K_{OW}$ values is available from the Canadian National Committee for CODATA (CNC?CODATA) at http://logkow.cisti.nrc.ca/logkow/, accessed October, 2009.

---

## PROBLEMS

1. Based on what you know about the structure of humic material, explain why it is more likely to have a net negative charge rather than net positive charge.

2. Show that a cation-exchange capacity (CEC) value in units of cmol (+) kg$^{-1}$ is numerically the same as one using units of meq 100 g$^{-1}$. An explanation is provided in Chapter 18.

3. A sediment sample containing 47% clay and 3.2% organic matter was found to have a cation-exchange capacity of 10.5 cmol (+) kg$^{-1}$. Assuming that the organic matter has a CEC of 210 cmol (+) kg$^{-1}$, show that the clay could be kaolinite but not montmorillonite.

4. A particular lake sediment sample is made up of 23% organic matter and 77% mineral fraction, the latter containing: 42% clay minerals; 24% silt; 11% sand. The clay minerals are 90% kaolinite and 10% halloysite. Calculate an approximate cation-exchange capacity (cmol (+) kg$^{-1}$).

5. Among the forms of phosphorus found in sediments are:

   (a) organic phosphorus;

   (b) apatite (($Ca_5(PO_4)_3(F,Cl,OH)$) is one representation for this mineral);

   (c) adsorbed on hydrated iron (III) oxides;

   (d) associated with clay minerals;

   How would you expect the solubility of each of these forms to be affected by the ambient p$E$ of the water?

6. (a) Use Fig. 10.P.1 to suggest how phosphorus release could occur from buried iron-rich sediments.

   (b) In laboratory experiments, it has been shown that phosphorus is released from sediment under dark conditions,[9] but is accumulated in the sediment when it is exposed to light. Suggest an explanation.

7. The behaviour of lead in soil / water systems has been described in terms of Langmuir adsorption. From laboratory experiments using 'pure' solids, the following values for Langmuir parameters have been estimated.

   $$\text{For clay} \qquad b = 3200 \text{ L mol}^{-1}$$
   $$C_{sm} = 4 \times 10^{-5} \text{ mol g}^{-1}$$
   $$\text{For organic matter} \qquad b = 10\,000 \text{ L mol}^{-1}$$
   $$C_{sm} = 4 \times 10^{-4} \text{ mol g}^{-1}$$

   (a) Use these values to calculate the concentration (in mg L$^{-1}$ or ppb) of water-soluble lead in equilibrium with clay containing 5 ppm adsorbed lead, and with organic matter also containing 5 ppm adsorbed lead. Both soil concentrations are those found at equilibrium.

   (b) Use the information from this question to predict the relative mobility of lead (released from gasoline combustion) deposited on clay-rich versus organic-rich roadside soils.

8. Under what conditions do the Langmuir and Freundlich equations reduce to having the same form?

9. The $K_{OW}$ function is frequently used to describe the movement of pesticides in the surface soil pore water, but it is used less frequently to describe the behaviour of an organic contaminant in groundwater. Explain.

10. To determine the value of $K_{OW}$, one method is the shake-flask procedure whereby the molecule of interest is placed in a flask containing octanol and water and shaken to establish equilibrium and then the concentration is measured in both phases. This direct method has practical problems associated with it. As an alternative, the $K_{OW}$ can be determined by relating it to retention times in reverse phase liquid chromatography. Explain the basis of this method.

11. The correlations between $K_{OW}$ and $K_{OC}$ imply that organic carbon in all soils has similar properties. Other relations that have been suggested include correlations between $K_{OC}$ and the ratio (N + O) / C of soil organic matter and between $K_{OC}$ and percentage aromaticity in the organic fraction of the soil. Suggest situations where these correlations might be appropriate.

---

[9] Moore, P.A. Jr., K.R. Reddy, and D.A. Graetz, Phosphorus geochemistry in the sediment–water column of a hypereutrophic lake, *J. Environ. Qual.*, **20** (1991), 869–75.

12. The solubility in water ($S$) and log $K_{OW}$ values for several pesticides are as follows:

| | $S$ / mg L$^{-1}$ | log $K_{OW}$ |
|---|---|---|
| DDT | 0.0028 | 6.0 |
| Aldrin | 0.08 | 5.8 |
| Parathion | 19 | 3.7 |
| Atrazine | 38 | 2.6 |
| Carbaryl | 73 | 2.4 |

Using your understanding of the basis of the $K_{OW}$ coefficient, comment on the relationship between the trends shown here.

13. In Atlantic Canada, like elsewhere around the world, aquaculture, specifically salmon farming, has established itself as a viable industry. However, there are concerns that the high fat content of this fish leaves it susceptible to significant **bioaccumulation** or **biomagnification** of persistent organic pollutants from nutritional feed stocks used in the industry. Calculate the concentration of PCBs that would **bioconcentrate** into the fish if the total concentration of PCBs = 0.05 pg L$^{-1}$ in the ocean water (density 1.015 kg L$^{-1}$) where the fish are raised. Use the value of log $K_{OW}$ = 7.11 for the PCBs. (note the use of the different terms above; refer to the definitions given in Section 14.5)

14. Is the log $K_{OW}$ value used in the previous question reasonable? Refer to the webpage 'LOGKOW© (ICUS / CODATA) a databank of evaluated octanol water partition coefficients' found at http://logkow.cisti.nrc.ca/logkow/index.jsp, accessed November, 2009, for specific log $K_{OW}$ values for various PCB congeners (and many other compounds).

15. The average concentration of pentachlorophenol (PCP, formula $C_6HCl_5O$) in rainbow trout, raised in a Northern Ontario, Canada, fish hatchery was found to be 2.7 ng L$^{-1}$. Assume the only process involved is bioconcentration and calculate the PCP concentration in the fish habitat. Refer to the webpage given in the previous question.

# Chapter 15
## Microbiological processes

The focus of this chapter is on the role microorganisms play in facilitating environmental processes. Here we will discuss:

- the classification systems and properties of microorganisms
- their role in organic matter degradation processes in terms of their energy relationships ($p\mathcal{E}°$ (w) and $\Delta G°$ (w))
- the important environmental cycles of carbon, nitrogen and sulphur.

Up to now, it is only in passing that we have made a distinction between *abiotic* and *biotic* environmental processes. Abiotic reactions are ones that occur through purely physical and / or chemical means. Such reactions would take place even in a completely sterile environment, if one existed. Biotic reactions, on the other hand, have a biological component. Photosynthesis, by which inorganic carbonate species are converted to forms of organic carbon in green plants or other organisms, is a typical example of a biotic environmental reaction. In many other cases, the biological component involves *microorganisms*; biota classified as microorganisms are often very small, with dimensions frequently in the micrometre size range, so they are frequently invisible. They are found in great numbers almost everywhere—in water, in soil, even in the air. In fact, it has been pointed out that the total mass of microorganisms in water and soil on the Earth is many times greater than that of all the larger animals put together.

Microorganisms play an essential role in facilitating many chemical reactions that occur in the natural environment. A case in point is a process through which ammonium ion is converted to nitrate in water or soil.

$$\text{NH}_4^+ \text{ (aq)} + 2\text{O}_2 + \text{H}_2\text{O} \rightarrow \text{NO}_3^- \text{ (aq)} + 2\text{H}_3\text{O}^+ \text{ (aq)} \tag{15.1}$$

This important oxidation reaction is called nitrification. Although nitrification can be described in chemical terms using a straightforward equation, it is not a single-step reaction and it does not occur to a significant extent as an abiotic, purely chemical process. The reaction has a $\Delta G°$ value of –266.5 kJ, indicating that it is a highly favoured process in terms of thermodynamics. Nonetheless, as is well known, an aqueous solution of ammonium chloride or some other source of ammonium ions is kinetically stable in the presence of air and can be kept without oxidizing appreciably for years. This is not true in natural water or in soil where oxidation of ammonium occurs at a measurable rate due to the mediation of microorganisms. The term mediation is used to signify that certain classes or specific species of microorganisms are the location of metabolic processes that result in the net chemical transformation shown in the equation. Often, particular enzymes within the microorganism are involved in catalysing the reaction.

In the case of nitrification, it is *Nitrosomonas* sp. and *Nitrobacter* sp. bacteria that make use of the ammonium ion as a substrate for their own metabolism, enabling them to survive and grow. In an aerobic environment in the presence of these bacteria, ammonium is readily oxidized to nitrate, usually in a matter of days.

For this reaction and in many other situations, *environmental chemistry* is actually *environmental microbiological chemistry*. The situations include degradation, synthesis, and other chemical transformations. The chemical species involved as reactants include a wide variety of naturally occurring carbon compounds as well as species containing nutrient elements like nitrogen, phosphorus, and sulphur, and even metals like iron or manganese.

In the present part of our survey of environmental chemistry we will introduce some of the fundamental terminology of microbiology and consider overall aspects of a number of important environmental reactions in which microorganisms play a major role. We will not be examining details of the metabolic transformations that occur *within* the organisms themselves. While the focus here is particularly on aqueous microorganisms, much of what is said applies in the terrestrial environment as well.

# 15.1 Classification of microorganisms

In water, free-floating organisms are termed plankton; these are further subdivided into plant and animal components—phytoplankton and zooplankton, respectively. Microorganisms that exist in or on the sediment at the bottom of a water body form part of the benthos and are called benthic organisms. The soil environment too is home to communities of microorganisms.

There are several systems that can be applied in the classification of aquatic and soil microorganisms. Each system is useful in being able to describe particular features relevant to studies in environmental chemistry. We begin by looking at definitions within some of the systems.

## Classification based on phylum of microbe

### Bacteria

Bacteria are probably the most abundant type of microorganism found in both aqueous and terrestrial environments. Bacteria exist in a variety of sizes with diameters ranging from 0.2 to 50 μm, but most have dimensions less than 50 μm. Being of colloidal size, they have a large surface:volume ratio. Exposed to the surrounding solution, on their surface are deprotonated acidic groups and thus bacteria carry a negative charge under common environmental conditions. There are many species of bacteria, taking forms ranging from spherical through ellipsoidal to rod-like. They are found in a variety of environments—ones that are exposed to the atmosphere and oxygen-rich (aerobic) as well as those that are isolated from the atmosphere and therefore devoid of oxygen (anaerobic). As nutrients, bacteria require a variety of chemical species including the common ones such as nitrogen, phosphorus, and potassium. As carbon sources required for their own growth, bacteria most commonly modify and incorporate preformed molecules from other organic materials.

The population (nature and number) of bacteria in a given environment depends on many factors including temperature, pH, electrolyte concentration, nutrient supply, and, in the case of soils, moisture content. In sediments and soils, bacterial count is positively correlated with organic matter content and can range from several million to several hundred million individual bacteria per gram of solid. However, because of their very small size, on a mass basis even the larger number corresponds to less than 0.1% of the total mass of the solid material. In the water column itself bacterial numbers range from $5 \times 10^4$ to $5 \times 10^7$ mL$^{-1}$ depending on the location, season, position in the water column, and the trophic status of the water.

### Fungi

Fungi are a ubiquitous and diverse category of microorganism, commonly thought to be associated almost exclusively with the terrestrial environment. This is because they tend to favour an aerobic, usually somewhat dry, situation. However, fungi also inhabit most oxygenated parts of pools, lakes, rivers, and the oceans. They are found in a wide range of forms and sizes including multicellular species (for example, mushrooms) readily visible without magnification. Fungi always require pre-synthesized compounds as carbon sources; in using these molecules to grow, they break down the original substance and incorporate the transformed molecules into their own structure. Fungi therefore play a key role in the degradation of litter in soil—an essential link in the chain of reactions resulting in formation of humic material. Microorganisms that live on decaying organic matter are called saprophytes. In contrast to bacteria, most of which usually prefer neutral to slightly alkaline environments, fungi are better adapted to acidic conditions.

In contrast to bacteria and other microorganisms, fungi can be very large in size. One extreme example of this is a fungus recently found in the Malheur National Forest in Oregon, USA. Living underground at a depth of about a meter, this single organism, the 'honey mushroom' or *Armillaria ostoyae*, extends over an area of about 10 km$^2$. Mushroom-like structures above ground are evidence of the massive organism below.

This is an exceptional example. However, the number concentration of fungi in soil is usually several orders of magnitude smaller than that of bacteria but, being much larger in size, on a mass basis they may be the most important type of microorganism in the terrestrial environment. Many fungal species have a filamentous structure, so they too have a very large ratio of surface area:volume in spite of their large overall dimensions.

## Actinomycetes

Actinomycetes are a class of unicellular organisms. At one stage of their development they take the form of fine, branching filaments similar to the fungi. In a later stage, they develop into a population of bacteria-like species. They are found in both aquatic and terrestrial situations; in the latter case they prefer aerobic, relatively dry conditions. Actinomycetes are more abundant in warm climates and are particularly numerous in tropical grasslands where they may make up about 30% of the total microbial population. In cooler, wetter areas, they contribute much less to the overall numbers. They are sensitive to acidity but are able to tolerate higher salt concentrations than many bacteria. Like fungi, actinomycetes are frequently involved in degradation processes in both water and soil, but they also are producers of natural antibiotics that control the populations of coexisting microorganisms.

## Algae

Algae are chlorophyll-containing organisms that range in size from microscopic individuals to large plant-like structures. They are abundant in the upper epilimnion of eutrophic water bodies as well as in the moist surface layer of soil. While fungi and actinomycetes are essentially degraders of pre-existing organic structures, the algae act as producers of new organic matter. They therefore play an important role in converting inorganic carbonate into organic forms. In this way they are an essential link in the global carbon cycle. Another important role is to act as agents of non-symbiotic nitrogen fixation.

As they convert inorganic carbon into organic compounds, other inorganic nutrients and even non-nutrient species are assimilated in order to build up the algal molecular structure. Because algae accumulate non-essential elements, they are sometimes used as integrating environmental monitors. By comparing the elemental content of algae growing in clean and polluted water, unusually high levels of metals like lead, cadmium, or mercury can be detected. Where excessive values are observed, it indicates that high average concentrations of the contaminant in the ambient water were present throughout the growing season.

Algae can cope with great temperature extremes and, besides being present in lakes and rivers in all parts of the Earth; they have been found in polar ice as well as in hot springs.

### Protozoa

Protozoa are the simplest form of animal life; they are commonly 5–50 μm in size and are found both in water and in the ground. In soil, they live in thin films of water on the surface of particles. They also play an important role in the microbial processes that go on in wastewater treatment facilities. Besides having a requirement for an adequate supply of moisture, they prefer a habitat that is warm (18 to 30°C), well oxygenated, and with pH between 6 and 8, although they can exist throughout the range 3 to 10. Being consumers (sometimes referred to as grazers or predators) of bacteria and other microorganisms, protozoa regulate the total microbial population.

## Classification based on ecological characteristics

### Autochthonous microorganisms

A mixed population of microorganisms is present in any environment. Some species are active, while others are dormant. As we have indicated earlier, the numbers of different species at any time depend on many environmental conditions. In a stable situation such as a mature forest, the microbial population is also stable, in balance with the surroundings, and with numbers determined by the available food supply and other properties of the local environment. The typical microorganisms that are indigenous in a particular area like this are called the autochthonous microorganisms.

### Zymogenous microorganisms

If the environmental situation is changed, such as by an influx of a fresh nutrient supply, particular types of microorganisms that can take advantage of the new situation may proliferate, at least until the situation returns to its original stable structure. The highly active but fluctuating population is called a *zymogenous* or *allochthonous population*.

An example of the latter situation is one that occurs after a heavy rainfall when a small nutrient-poor water body is inundated with a quantity of nitrate dissolved in run-off from adjacent fields or urban areas. In the new situation, algae that can make use of the nitrate are able to flourish. The growth of algal numbers is, however, temporary and, when the nitrate has dissipated, the baseline conditions resume, with more stable chemistry and the autochthonous microbial population.

## Classification by carbon source

As a microbial population grows, the organisms incorporate carbon into the structure of their bodies. The structural material of microbes consists of organic molecules—carbohydrates, proteins, lipids, etc., of which carbon makes up about half of the molecular mass. Depending on the microorganism, the carbon required to produce these compounds can be obtained from either inorganic or organic sources. On this basis, they can be further classified according to the source of carbon that is required in order for them to function and build up their own species.

### Autotrophs

Autotrophs are microorganisms that are capable of growing in a completely inorganic medium using carbonate species as their sole source of carbon. Earlier, we referred to autotrophs as producers or synthesizers. The carbonate may be in the form of atmospheric carbon dioxide or aqueous carbon dioxide, hydrogen carbonate or carbonate. Green plants (of course, these are *macroorganisms*), most algae, and some bacteria are all classified as autotrophs. As subdivisions within this category, there are *photoautotrophs*—algae are a well-known example—that use solar radiation as an energy source for their synthetic processes. On the other hand, *chemoautotrophs* derive energy from chemical oxidation reactions. Most bacteria are chemoautotrophs.

An example is *Nitrosomonas*, the bacterium that facilitates the first step of nitrification by deriving energy through oxidizing ammonium to nitrite.

$$2NH_4^+ (aq) + 3O_2 + 2H_2O \rightarrow 2NO_2^- (aq) + 4H_3O^+ (aq) \tag{15.2}$$

## Heterotrophs

Heterotrophs are microorganisms that make use of pre-synthesized organic compounds as a carbon source. These organisms are degraders or consumers of other living matter—plant and animal residues in water and soil and many organic waste materials. The heterotroph category includes fungi, actinomycetes, protozoa, and most bacteria. Specific microbes are also important as degraders of synthetic organic molecules such as pesticides. For example, the degradation of the carbamate pesticide isopropyl-*N*-phenylcarbamate, IPC, is facilitated by bacterial species such as *Arthrobacter*.

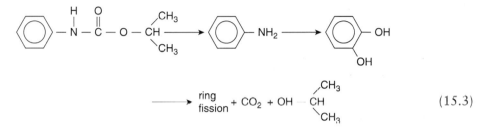

$$\tag{15.3}$$

When degradation of organic molecules proceeds, often through multiple steps, to the final inorganic products such as carbon dioxide, ammonia, and sulphate, the process is referred to as ultimate degradation or mineralization.

## Classification by source of electron acceptor

This method of classification considers the way in which microorganisms take part in and facilitate redox reactions.

### Aerobes

Aerobes are a large category of microorganisms that use molecular oxygen, either in gaseous form or dissolved in water, as the electron acceptor for their oxidation reactions. In the process, the oxygen is reduced and hydronium ions are consumed.

$$O_2 + 4H_3O^+ (aq) + 4e^- \rightarrow 6H_2O \tag{15.4}$$

In order for aerobic species to be active, the environment must be in contact with a plentiful supply of oxygen-containing air. This will be the case in surface waters, especially those that are in constant motion such as a turbulent river or a wave-covered lake. A porous surface soil will also harbour a large population of aerobic microbes.

### Anaerobes

Air transfer is limited in deep water and compacted soil well below the surface so that oxygen becomes depleted. In these situations, the strictly aerobic organisms are no longer able to function. However, oxidation reactions can continue through the activity of *anaerobic* microorganisms. Anaerobes have the capability of effecting oxidation without molecular oxygen. Instead, they use other electron-poor species, for example, sulphate, to accept electrons from the reduced substrate. The electron acceptor species is converted into a more reduced form.

$$SO_4^{2-} (aq) + 9H_3O^+ (aq) + 8e^- \rightarrow HS^- (aq) + 13H_2O \tag{15.5}$$

Bacteria make up the largest number of anaerobes.

Within the anaerobic category are *obligate anaerobes*, for which oxygen is toxic so they can only function in the absence of oxygen. The toxicity of oxygen is due to the absence of cytochromes and catalase, allowing for the accumulation of toxic hydrogen peroxide. There are also *facultative anaerobes*, which are organisms that are able to make use of either oxygen or other electron acceptors for oxidation.

The degradation of organic matter in the sediment of a lake, in a forest soil, in a landfill for solid waste and in a wastewater treatment plant all take place via various sequences of oxidation steps. The oxidation of ammonia, shown above, is another example. Depending on the specific environment, the microorganisms that enable these reactions to occur may be either aerobes or anaerobes.

This is an appropriate place to point out a distinction between the terms anaerobic and anoxic. The former refers to a situation that is not exposed to air, including oxygen. On the other hand, anoxic conditions are ones where there is an absence of air and also of other oxygen-containing agents such as nitrate and sulphate. Equation 15.5 represents a reaction that occurs in an anaerobic environment, but it is not an anoxic process.

## Classification by temperature preference

All microorganisms follow an activity versus temperature relationship of the form shown in Fig. 15.1. As illustrated in this plot, microbial growth rate increases with increasing temperature according to the Arrhenius relation, $\ln k = -E_a / RT +$ constant, where $k$ is the reaction rate constant. However, above a particular maximum temperature, enzymes within the microorganism become denatured and growth declines precipitously. The position of the maximum along the $x$-axis depends on the species of microorganism and they may be defined according to the value of $T_{\text{optimum}}$.

- *psychrophiles.* $T_{\text{optimum}} < 20°C$ (uncommon but found in polar environments);
- *mesophiles.* $T_{\text{optimum}}$ 20–45°C (most common);
- *thermophiles.* $T_{\text{optimum}} > 45°C$ (found in tropical environments and associated with situations like compost production).

## Classification by morphology

For purposes of identification and characterization, the shape, form, and colour of microorganisms are important characteristics noted by microbiologists. Some of the many shapes observed, and the nomenclature associated with these are:

- rod shape: bacilli;
- spherical shape: cocci;
- spiral shape: spirilla.

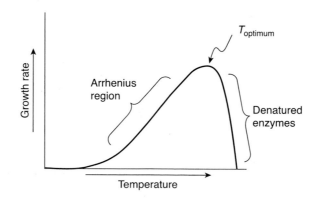

**Fig. 15.1** Growth rate versus temperature for microorganisms.

Many other morphological categories exist and some species change shape or exhibit a variety of forms. While useful for identification of species, this system of classification has no direct connection with the environmental roles played by the particular microorganisms.

---

**Main point 15.1** Microorganisms are small species of flora and fauna that populate all compartments of the environment. They are categorized in various ways including by phylum, environmental preference, carbon source, electron acceptor type, temperature requirements, or morphology.

---

## 15.2 **Microbiological processes—the carbon cycle**

Many of the natural environmental cycles operating within the Earth's environment have key links that are controlled in large part by microorganisms. When human interventions occur, the microbial population adjusts to the new environment, and the chemical reactions within that part of the cycle are affected by the changes in the situation. This point in our study of environmental chemistry is a good place to summarize broad chemical features in some of these important microbiological / chemical cycles. We will include major abiotic as well as biotic processes, but discussion will centre on the main biological processes that occur in water or on land. In the discussions, only the general types of reactions are considered; there are always a vast number of specific detailed reactions that are not included.

The carbon cycle (Fig. 15.2) is one of the great cycles that determines the forms of life and their interactions within the global environment. Important organic and inorganic carbon-containing species exist in water, on land, and in the atmosphere. Every organic compound and

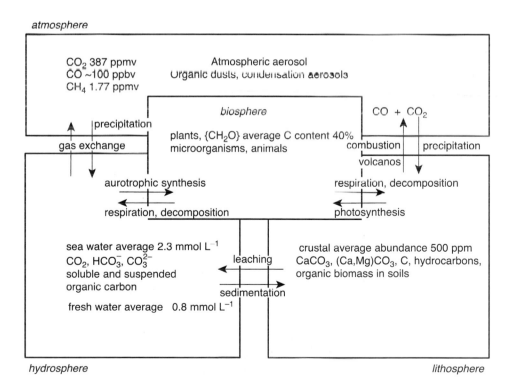

**Fig. 15.2** The carbon cycle—a summary of principal processes in the environment.

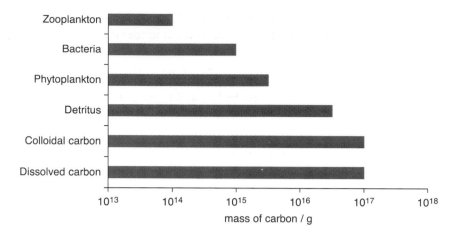

**Fig. 15.3** The forms of organic carbon present in the oceans.

all the inorganic carbonate species fit into this general picture in some way—through synthesis, transformation, and / or decomposition.

## Carbon species in water and on land

In the terrestrial environment, there are three major reservoirs of carbon (refer back to Table 12.1 and also to Appendix A.3). On a mass basis, by far the largest of these (20 000 Eg) is in carbonate rocks, limestone ($CaCO_3$) and dolomite (($Ca,Mg)CO_3$), and in the associated dissolved carbonate species. The second major carbon-containing deposit (5 Eg) is made up of material buried in the Earth in the form of fossil fuels of all three phases—solid coal, liquid petroleum, shale oil and tar sands, and natural gas. The third large terrestrial reservoir of carbon (1.7 Eg) is organic material in various stages of chemical modification, on or in the soil. Clearly, the fossil fuels and soil organic matter have a biological origin.

In the aquatic environment, biotic processes again are important in determining the forms of carbon that are present. Figure 15.3, on a logarithmic scale, shows the forms of organic carbon in the oceans. Ultimately, even inorganic carbonate species have a biological origin.

## Formation of carbonate minerals

In the previous chapter we showed, by calculation, how the daily photosynthetic cycle can be an agent through which organic carbon molecules are formed in a water body. While photosynthesis is taking place, an increase in pH occurs and calcium carbonate can be precipitated from solution. In the oceans, these kinds of processes operate on an immense scale. Carbonate species are present in all natural water arising from atmospheric carbon dioxide, remineralized organic matter, and dissolved carbonate minerals. The oceans are an especially large reservoir of carbon compounds (Fig. 15.2, Table 12.1) and play a major role in establishing and maintaining the global carbon balance. In the surface of oceans, through energy supplied by solar illumination, photoautotrophic microorganisms synthesize particulate organic matter using inorganic carbonate species from both water and the atmosphere as their carbon source. Simultaneously, additional carbonate is microbially incorporated to form the hard structures of some oceanic plankton. One net result of carbon assimilation is that surface waters develop a slightly higher pH and lower alkalinity than water below the thermocline.

In thermodynamic terms, the surface of the oceans is thus supersaturated with respect to the precipitation of calcium carbonate. The reaction that best describes features controlling solubility of this compound is

$$CaCO_3 + CO_2 (aq) + H_2O \rightleftharpoons Ca^{2+} (aq) + 2HCO_3^- (aq) \qquad (15.6)$$

The size of the equilibrium constant, a modified solubility product constant, $K'_{sp} (1)$, for reaction 15.6 is calculated as shown in Example 15.1. The conditions (1) are those for surface sea water.

---

**Example 15.1  The modified solubility product constant ($K'_{sp} (1)$) for calcium carbonate in sea water**

$$K'_{sp} (1) = \frac{K_{a1} \times K_{sp}}{K_{a2}} = \frac{1.0 \times 10^{-6} \times 5.9 \times 10^{-7}}{7.8 \times 10^{-10}} = 7.6 \times 10^{-4}$$

---

The values for the constants ($K_{a1}$, $K_{a2}$, and $K_{sp}$) are values for reactions in surface sea water.[1] Because they depend on the ionic composition of the water, the acid dissociation constants are different from the 'fresh water' values given in Appendix B.4 of this book.

---

**Example 15.2  Reaction quotient $Q'_{sp} (1)$ for calcium carbonate in sea water**

For a situation near the ocean surface, typical values of pH, alkalinity, carbon dioxide, and calcium ion concentration are 8.2, $2.1 \times 10^{-3}$ mol L$^{-1}$, $2.5 \times 10^{-5}$ mol L$^{-1}$, and $1.1 \times 10^{-2}$ mol L$^{-1}$, respectively. Calculate the reaction quotient, $Q'_{sp} (1)$, related to these conditions.

From reaction 15.6 and referring to Fig. 1.2,

$$Q'_{sp} (1) = \frac{[HCO_3^-]^2[Ca^{2+}]}{[CO_2]} = \frac{(2.1 \times 10^{-3})^2 \times 1.1 \times 10^{-2}}{2.5 \times 10^{-5}} = 2.0 \times 10^{-3}$$

---

The ratio of reaction quotient to solubility product constant is a measure of the degree of saturation and a value of the ratio greater than 1 indicates a condition of supersaturation. In this case the ratio is

$$\frac{Q'_{sp} (1)}{K'_{sp} (1)} = \frac{2.0 \times 10^{-3}}{7.6 \times 10^{-4}} = 2.6$$

indicating that the surface water sample is highly supersaturated with respect to calcium carbonate, and a colloidal precipitate is formed.

As microorganisms die and sink into deeper parts of the oceans, along with their carbonate shells, the solar flux declines to negligible values. The net reaction is biological (mostly anaerobic) decomposition of organic matter, thus converting the organic carbon forms back into soluble mineral species. Carbon dioxide is released and its hydrolysis causes a drop in pH of about half a unit through the thermocline. A recalculation of the solubility product constant for eqn 15.6 requires using a $K_{sp}$ for calcium carbonate that applies to pressures found at a depth of 2500 m (conditions (2)). The value of the $K_{sp}$ under these conditions is $1.3 \times 10^{-6}$, and the modified solubility product, $K'_{sp} (2)$, as calculated in Example 15.1 for reaction 15.6 is

$$K'_{sp} (2) = \frac{K_{a1} \times K_{sp}}{K_{a2}} = \frac{1.0 \times 10^{-6} \times 1.3 \times 10^{-6}}{7.8 \times 10^{-10}} = 1.7 \times 10^{-3}$$

---

[1] Stumm, W. and J.J. Morgan, *Aquatic chemistry: chemical equilibria and rates in natural waters*, 3rd edn., John Wiley and Sons, Inc., New York; 1996.

Note that we have used the original acid dissociation constant values. A more refined calculation would require that these also be modified to account for the conditions at depth.

The reaction quotient is calculated in the new situation using concentrations of the relevant species in that part of the ocean (as previously shown in Example 15.2).

$$Q'_{sp}(2) = \frac{[HCO_3^-]^2[Ca^{2+}]}{[CO_2]} = \frac{(2.4 \times 10^{-3})^2 \times 1.1 \times 10^{-2}}{4.5 \times 10^{-5}} = 1.4 \times 10^{-3}$$

In the deep-water situation, the ratio of the reaction quotient to the solubility product, $Q'_{sp}(2) / K'_{sp}(2)$, has become 0.82. Being less than 1 it indicates an undersaturated situation. On the basis of these changed thermodynamic conditions, one would predict spontaneous dissolution of the calcium carbonate microbe shells. The fact that calcium carbonate remains in ocean sediments indicates that the dissolution reactions are kinetically hindered. Little dissolution occurs in the time before the sediments are buried, and massive calcium carbonate deposits accumulate in this way.

## Biomass degradation

Another key process in the carbon cycle that is controlled almost exclusively by microbial activity is decomposition of dead biomass. In this context, the biomass being considered refers to plant, microbial, and, to a lesser extent, animal material—as, for example leaf, twig, and branch litter in a forest, root and straw residues in a cultivated field, or aquatic plants and microorganisms in a lake or swamp. This organic material is in large part made up of carbohydrate—in particular, cellulose and cellulose-related compounds. Decomposition is most commonly a process of oxidation and the half-reaction for complete oxidation is expressed in simple form as

$$\{CH_2O\} + 5H_2O \rightarrow CO_2\,(g) + 4H_3O^+\,(aq) + 4e^- \tag{15.7}$$

To explain conditions under which this reaction, the oxidation of organic matter, occurs in water, we need to introduce a new redox term, $pE°(w)$.

## $pE°$ (w) values—redox reactions at pH = 7

When the oxidative degradation takes place in water whose pH is near 7, we describe the redox potentials in terms of $pE°$ (w). The term $pE°$ was defined and described in Chapter 10. The superscript ° indicates a $pE$ under standard conditions, and this includes the condition that the activity of the hydronium ion is 1, equivalent to a pH in the system of 0. Clearly, in real environmental circumstances, a pH of 0 is highly unusual and it is more likely that the value be near neutrality. We therefore define $pE°$ (w) as a measure of $pE°$ at pH = 7 rather than pH = 0. In other words, we keep the definition of standard conditions (activity of all species in the reaction = 1) for everything except the hydronium ion that we set at $10^{-7}$.

---

### Example 15.3 The relation between $pE°$ and $pE°$ (w)
Consider the four-electron process described by the reaction

$$O_2 + 4H_3O^+\,(aq) + 4e^- \rightarrow 6H_2O \quad pE° = 20.80 \tag{15.8}$$

$$pE = pE° - \frac{1}{4}\log\frac{1}{P_{O_2}\,(a_{H_3O^+})^4}$$

The usual standard conditions include $P_{O_2} = 101.3$ kPa = 1 atm. At the boundary,

$$pE = pE° - \frac{1}{4}\log\frac{1}{(a_{H_3O^+})^4}$$

If we define $pE = pE°$ (w) when the pH is 7, then

$$pE = pE° \text{ (w)} = 20.80 - \frac{1}{4} = \log \frac{1}{(10^{-7})^4}$$

$$pE° \text{ (w)} = 20.80 - 7 = 13.80$$

For a general reaction

$$\text{Ox} + n_1\text{H}_3\text{O}^+ + n_2\text{e}^- \rightarrow \text{Red} \tag{15.9}$$

$$pE = pE° - \frac{1}{n_2} \log \frac{a_{\text{Red}}}{a_{\text{Ox}} \times (a_{\text{H}_3\text{O}^+})^{n_1}}$$

Applying standard conditions for all but $\text{H}_3\text{O}^+$,

$$pE = pE° - \frac{1}{n_2} \log \frac{1}{(a_{\text{H}_3\text{O}^+})^{n_1}}$$

and

$$pE = pE° \text{ (w)} = pE° - 7 \,(n_1/n_2) \tag{15.10}$$

where $n_1$ = the number of hydronium ions and $n_2$ = the number of electrons in the half-reaction considered.

---

Refer back to the degradation of biomass as defined by equation 15.7. This is an oxidation reaction but values of $pE°$ always refer to reduction reactions. In the present case the $pE°$ of $-1.20$ (Appendix B.5) refers to reaction 15.7 when it is read from right to left. For the reduction of $CO_2$, both $n_1$ and $n_2$ are 4; therefore, using eqn 15.10, $pE = pE° \text{ (w)} = -1.20 - 7 = -8.20$. For the oxidation of $\{CH_2O\}$ as in reaction 15.7, the $pE$ value at neutral pH is then $+8.20$. In order for oxidation to occur, the half-reaction must be coupled to an appropriate reduction half-reaction—that is, an electron-accepting process. By appropriate, we mean that a naturally occurring oxidizing agent must be available in sufficiently high concentration, and the $pE°$ for the overall process must be positive. We will consider a number of these oxidizing agents (electron acceptors) in order of decreasing $pE° \text{ (w)}$ values.

## Oxygen as the primary oxidizing agent for biomass degradation

Molecular oxygen, $O_2$, is the most important oxidizing agent in most aerobic environments. In the process of carrying out oxidation, it is reduced as shown by the half-reaction 15.11. The values of $pE° \text{ (w)}$ for this and subsequent reactions are given in Appendix B.5.

$$O_2 + 4H_3O^+ + 4e^- \rightarrow 6H_2O \quad pE° \text{ (w)} = +13.80 \tag{15.11}$$

The sum of reactions 15.7 and 15.11 gives the overall process that describes the oxidation of biomass by oxygen.

$$\{CH_2O\} + O_2 \rightarrow CO_2 + H_2O \quad pE° \text{ (w)}_1 = +22.00 \tag{15.12}$$

and, under the standard conditions described here, $pE° \text{ (w)}_1$ is the sum of the $pE° \text{ (w)}$ for the oxygen (reduction) half-reaction and the value of $pE \text{ (w)}$ for the carbohydrate (oxidation) half-reaction.

---

**Example 15.4 $pE° \text{ (w)}_1$ and $\Delta G° \text{ (w)}_1$ for oxidation of organic matter by oxygen**
Considering reaction 15.12,

$$pE° \text{ (w)}_1 = pE° \text{ (w)}_{O_2} \text{ (reduction)} + pE \text{ (w)}_{CH_2O} \text{ (oxidation)}$$

$$= 13.80 + 8.20$$

$$= +22.00$$

The Gibbs free energy for the oxidation at 298 K is a four-electron process and

$$\Delta G^\circ (w)_1 = -2.303nRTpE^\circ (w)_1 \approx -500 \text{ kJ}$$

The reaction as written allows for the oxidation of 1 mol of {CH$_2$O}

$$\Delta G^\circ (w, 1 \text{ mol})_1 \approx -500 \text{ kJ}$$

---

Reaction 15.12 is therefore highly favoured thermodynamically, and this confirms that oxidation of organic matter by oxygen should occur spontaneously under aerobic conditions. While abiotic biomass oxidation by this reaction is very slow, degradation does proceed at a reasonable rate when microorganisms that are capable of facilitating the redox process are present. Appropriate microorganisms include a wide range of heterotrophic species of bacteria, fungi, protozoa, and actinomycetes that usually are abundant in well-oxygenated water, sediment, and soil.

We have used cellulose, represented as {CH$_2$O}, as the model compound but biomass is a complex material made up of a wide range of components. Some are more readily degraded than cellulose, while others are more resistant to oxidation. To some degree, however, decomposition occurs by reactions similar to 15.12, with oxygen being the prime electron acceptor for microbial oxidative degradation.

### Nitrate as oxidizing agent

Where molecular oxygen is absent, in the anaerobic situation it is still possible for organic matter to be degraded via reactions involving other oxygen-containing species. Nitrate can serve as an electron acceptor, being reduced to nitrite, ammonium ion, nitrous oxide, or dinitrogen gas depending on environmental circumstances. The reduction reaction producing dinitrogen gas is important as a link in the global nitrogen cycle and the equation for it is

$$2NO_3^- \text{ (aq)} + 12H_3O^+ \text{ (aq)} + 10e^- \rightarrow N_2 + 18H_2O \quad pE^\circ (w) = 12.65 \quad (15.13)$$

When coupled with oxidation of carbohydrate (15.7), the overall reaction is

$$4NO_3^- \text{ (aq)} + 5\{CH_2O\} + 4H_3O^+ \text{ (aq)} \rightarrow 2N_2 + 5CO_2 + 11H_2O \quad (15.14)$$

The overall pE$^\circ$ (w) is calculated as shown before in Example 15.4.

$$pE^\circ (w)_2 = pE^\circ (w)_{NO_3^-} \text{ (reduction)} + pE (w)_{CH_2O} \text{ (oxidation)}$$

$$= 12.65 + 8.20$$

$$= +20.85$$

The Gibbs free energy for the reaction as written (a 20-electron process) is

$$\Delta G^\circ (w)_2 = -2.303 \, nRTpE^\circ (w)_2 = -2380 \text{ kJ}$$

Equation 15.14 accounts for the oxidative degradation of 5 mol of {CH$_2$O}. For the reaction written in terms of 1 mol of {CH$_2$O},

$$\Delta G^\circ (w, 1 \text{ mol})_2 = -2380 / 5 = -480 \text{ kJ}$$

The negative value −480 kJ indicates that nitrate is also capable of oxidizing organic matter. In the absence of oxygen, reaction 15.14 therefore becomes an important means by which biomass is oxidized as long as a source of nitrate is present in the soil or water. This is a good example of an anaerobic but not anoxic situation.

In Section 15.3, we will show that this reaction—called denitrification—is an important link in the nitrogen cycle. To proceed at a significant rate, it requires a large supply of organic matter and the presence of denitrifying organisms. These conditions apply in stagnant waters supplied with an abundance of decaying biomass. Zones of the ocean where upwelling of the deep waters occurs also produce abundant organic-rich sediments. Parts of the Atlantic and Pacific Oceans adjacent to the west coasts of Africa and South America are such regions.

## Sulphate as oxidizing agent

Where oxygen and nitrate are both absent (or have been consumed) and sulphate is present, the latter species may serve as an electron acceptor for the oxidation of organic matter.

$$SO_4^{2-} (aq) + 9H_3O^+ (aq) + 8e^- \rightarrow HS^- (aq) + 13H_2O \quad pE^\circ (w) = -3.75 \quad (15.15)$$

The overall reaction is obtained by combining 15.7 and 15.15.

$$2\{CH_2O\} + H_3O^+ (aq) + SO_4^{2-} (aq) \rightarrow HS^- (aq) + 2CO_2 + 3H_2O \quad (15.16)$$

The $pE^\circ$ (w) and Gibbs free energy are calculated as before

$$pE^\circ (w)_3 = pE^\circ (w)_{SO_4^{2-}} (reduction) + pE(w)_{CH_2O} (oxidation)$$

$$= -3.75 + 8.20$$

$$= +4.45$$

$$\Delta G^\circ (w, 1 \text{ mol})_3 = -102 \text{ kJ}$$

Again, the $\Delta G^\circ$ calculated here is for one mole of $\{CH_2O\}$. The negative value indicates that oxidation by sulphate is also thermodynamically favourable. This reaction is important, for example, in oxygen-depleted organic-rich marine sediments where there is an abundance of sulphate available in the sea water. Bacteria of the species *Desulphovibrio* are important mediators for this process.

## No oxidants present—self-oxidation of biomass

In an environment where no oxygen, nitrate, or sulphate are present, i.e. an anoxic situation, organic matter may still undergo oxidation by means of anaerobic reactions leading to a variety of products. The ultimate products of anaerobic decomposition include methane and occur via internal redox processes, or redox disproportion reactions. The overall reduction half-reaction is

$$CO_2 + 8H_3O^+ + 8e^- \rightarrow CH_4 + 10H_2O \quad pE^\circ (w) = -4.13 \quad (15.17)$$

Combining eqn 15.17 with the half-reaction for the oxidation of carbohydrate (15.7) gives the following overall redox process.

$$2\{CH_2O\} \rightarrow CH_4 + CO_2 \quad (15.18)$$

The $pE^\circ$ (w) and Gibbs free energy are

$$pE^\circ (w)_4 = pE^\circ (w)_{CO_2} (reduction) + pE (w)_{CH_2O} (oxidation)$$

$$= -4.13 + 8.20$$

$$= +4.07$$

$$\Delta G^\circ (w, 1 \text{ mol})_4 = -93 \text{ kJ}$$

The smaller negative free-energy value indicates that reaction 15.18 is thermodynamically the least favoured of the four that we have described. Nevertheless, it is a spontaneous process and is yet another reaction that effects the decomposition of organic matter. Therefore, decomposition of biomass can and does occur in environmental situations even in the absence of any other oxidizing agents. Species of actinomycetes are the pre-dominant anaerobes responsible for anaerobic biodegradation. The organisms required for the production of methane are called methanogens. This gas, sometimes called 'marsh gas', is frequently observed in swamps or other wetlands. We have seen that methane release from natural and artificial wetlands contributes almost half to the atmospheric input of this important greenhouse gas.

Anaerobic decomposition of this type is sometimes referred to as a 'fermentation reaction'. Fermentation to produce other products can also occur. In some of these reactions carboxylic acids and hydrogen gas are produced as, for example, according to eqn 15.19.

$$3\{CH_2O\} + 4H_2O \rightarrow CH_3COO^- (aq) + HCO_3^- (aq) + 2H_3O^+ (aq) + 2H_2 (g) \qquad (15.19)$$

Reactions like 15.19 represent a less complete oxidation of the original organic matter than when carbon dioxide and methane are formed.

## The sequence of biomass oxidation

In summary, there are at least four mechanisms by which biomass is oxidized. For any particular situation, the mechanism that operates depends on the availability of oxidants, on the value of $\Delta G$ (the above four reactions were given in order of decreasing negative value of $\Delta G^\circ$ (w, 1 mol)), and on the presence of appropriate microorganisms to facilitate the actual reaction. Assuming that the last factor is not limiting, a reaction sequence such as that in Fig. 15.4 would be expected if each oxidant were to react one after the other. The bracketed concentrations are actual data taken from a Niagara River water sample. In the sequence, oxygen is the primary oxidant followed by nitrate and then sulphate. During the time when oxygen is the oxidant, nitrogen in the biomass—often at a concentration of around 1%—is released from the degraded biomass as ammonium ion and it is then oxidized to produce nitrate, additional to that originally present in the water. Similarly, there may be a small increase in sulphate concentrations during the initial two stages of organic-matter oxidation.

The reactions we have discussed occur in soils as well as in water. With respect to soils, there is considerable interest in the redox processes that occur after they have been flooded. Flooded soils are a seasonal feature in many parts of the world and are also associated with the cultivation of rice—a practice that makes use of some 170 million ha of agricultural land in tropical countries worldwide.

In newly submerged soils, and also in the case of ocean and fresh-water sediments, oxygen is the primary oxidant. The rate at which it is used up depends on surface aeration and water

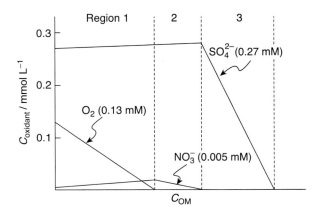

**Fig. 15.4** Reaction sequence for oxidation of organic matter in a lake or river. The bracketed values are the original concentrations of each oxidant. Data are taken from samples from the Niagara River. The $x$-axis is not to scale.

movement, and on production and use of oxygen by plants and microorganisms. After oxygen is depleted, there is a succession of reactions similar to that described above. Table 15.1 reports data describing a microbiological and chemical sequence that occurs after a soil is submerged. The observed data are consistent with the sequence.

When the dry soil is submerged with oxygen-saturated water, it has an initial pE in the aerobic range. The microbial population is small, but aerobic species multiply rapidly in the newly moist environment. The initially high oxygen concentrations decline rapidly as oxygen is consumed by the processes of decomposing biomass that produce carbon dioxide. In a matter of a few days, the oxygen is depleted. Nitrate and sulphate then become the active electron acceptors until they too are depleted. The pE drops dramatically and the activity of aerobes declines while anaerobic microorganisms become dominant. Eventually, measurable quantities of methane and some hydrogen are produced—evidence that fermentation reactions have become the means of biomass decay.

The flooded-soil case is an example where a substantial area of soil becomes reduced. Reducing conditions are also usually observed in organic-rich sediment, especially deeper in the sediment where there is no direct contact with the water above. On a much smaller scale too, it is possible to have reducing conditions in an otherwise aerobic situation. Where soil aggregates are several mm or larger in dimensions, the gas diffusion rate may be insufficient to maintain a significant presence of oxygen in the interior regions of the particle. In this microenvironment, the local pE value then actually reflects reducing conditions.

The reactions that we have been discussing are the dominant ones controlling the redox status in most water or water / soil environments and this helps to explain why pE values tend to be found in one of two general regions. The sets of high- and low-pE environments shown in Fig. 15.5 are evidence of this phenomenon, one that is sometimes referred to as *redox buffering*. Where oxygen is present, the pH is controlled by the oxygen / water half-reactions (region 'a' on Fig. 15.5) but, when it is used up, the pE drops substantially to the region 'b' where it is determined by the next available redox couple. In many cases this is the sulphate / hydrogen sulphide couple as, in the unpolluted hydrosphere, nitrate is present in relatively small concentrations. After sulphate is exhausted, a further small decline in pE occurs while the system moves into the fermentation region.

**Table 15.1** Chemical and microbiological changes occurring over a 23-day period after a soil is submerged.[a]

| Time / days | pE | Microbial population / $10^6$ g | | $C_{gas}$ / mL 100 $g^{-1}$ of soil | | | |
| | | Aerobes | Anaerobes | $O_2$ | $CH_4$ | $H_2$ | $CO_2$ |
|---|---|---|---|---|---|---|---|
| 0 | 7.6 | 34 | 22 | 3.2 | ND | ND | 83 |
| 1 | 3.7 | 220 | | 0.3 | ND | ND | 10 |
| 2 | −0.84 | 110 | 33 | ND | ND | 0.2 | 172 |
| 4.5 | −3.9 | 55 | 50 | ND | 0.3 | ND | |
| 6 | | | | ND | 2.2 | 3.6 | 280 |
| 8 | −4.2 | 53 | 170 | ND | 14.7 | 2.1 | |
| 10 | | | | ND | 21.4 | ND | 226 |
| 13 | −4.2 | 62 | 130 | ND | | | |
| 23 | | | | | 60.3 | 3.2 | |

[a] ND, not detected. From Takai, Y., T. Koyama, and T. Kamura, *Soil and Plant Food*, 2 (1956), 63–6 as reported in Alexander, M., *Introduction to soil microbiology*, John Wiley and Sons, Inc., New York; 1961.

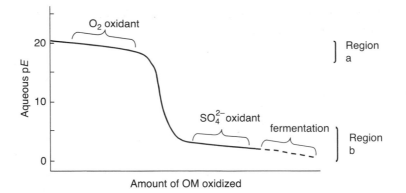

**Fig. 15.5** Redox buffering regions. The p*E* value is maintained at a relatively constant high value when oxygen is present; at a lower value by the other oxidant species where there is no oxygen.

## Biomass in lakes

Figure 15.6 shows 'idealized' oxygen and temperature profiles of water (not sediment) in two lakes—one oligotrophic and one eutrophic—that are subject to the seasonal variations

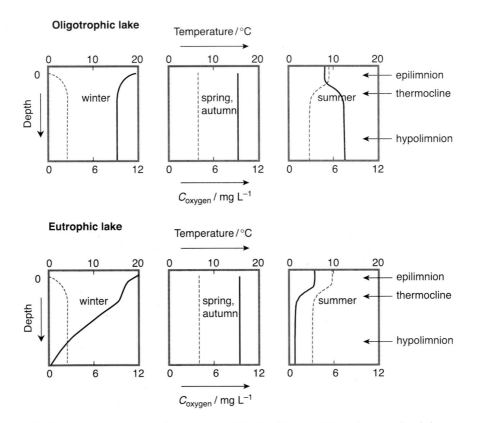

**Fig. 15.6** Idealized temperature and oxygen profiles in oligotrophic and eutrophic lakes, as seasons change in the temperate regions of the world. (Redrawn from Wetzel, R.G., Limnology, Saunders, Philadelphia; 1975.) _____ temperature, _____ $C_{oxygen}$

characteristic of temperate-climate regions. The three profiles for each lake correspond to the situations in winter, spring / autumn, and summer.

In this idealized situation, the temperature profiles in the two lakes are shown to be identical and follow the seasonal pattern explained in Chapter 9. Note the well-mixed, uniform situation in spring and autumn, and the presence of a thermocline in summer. In the oligotrophic lake, the oxygen concentration tracks, in an inverse sense, the temperature profile. This is because of the greater solubility of oxygen in cold compared to warm water.

The picture in a eutrophic lake is influenced by the same principles, but an additional factor is the presence of oxidizable biomass in the deeper regions of the water body. In winter, this leads to a stable situation with a steady decline in oxygen concentration from lake surface to sediment surface. The depletion is due to the low-level activities of degradative heterotrophic microorganisms that consume oxygen as they degrade biomass in the sediment. During this season, oxygen is replenished only by diffusion from surface water that is in contact with the atmosphere. That contact with air above may be made negligible by the ice cover. In spring and autumn, the rapid mixing due to overturn provides a relatively uniform oxygen-concentration profile as the element is distributed by both diffusion and convection. This is a period of considerable microbiological activity throughout the entire profile of the lake. The summer picture is quite different. The major decline below the thermocline is due to a stable, warm environment where oxygen has been depleted by microbial degradation yet not replenished due to the inability of water in the hypolimnion to mix with the oxygen-rich water of the epilimnion.

Degradation of non-living biomass is not the only microbial process controlling the dissolved oxygen content of water bodies. Living microorganisms, especially algae, also affect oxygen levels. In Chapter 11, we calculated the concentration of molecular oxygen at 25°C in a well-aerated water body and found it to be close to 8 mg $L^{-1}$. The growth of algae during daylight hours and their respiration at night creates a daily cycle with concentration of oxygen fluctuating around this value (Fig. 15.7). The cycle is readily explained in terms of the reversible chemical reactions corresponding to these two processes.

$$CO_2 + H_2O \xrightleftharpoons[\text{respiration}]{\text{photosynthesis}} \{CH_2O\} + O_2 \qquad (15.20)$$

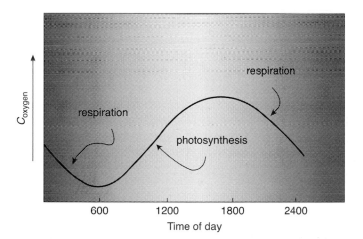

**Fig. 15.7** Daily variations in the concentration of oxygen in water containing algae. Actual values of concentration depend on the algal population and temperature of the water, among other factors.

As the figure shows, the algal growth cycle creates enhanced aqueous oxygen concentrations in daytime and diminished concentrations at night. Further consequences are changes in pH and possible precipitation of calcium carbonate as described earlier.

## Parameters to measure organic matter content in water

### Biological oxygen demand (BOD)

Various compounds found in water are substrates that can be degraded and contribute to oxygen depletion. Up to now, we have considered only the organic components of non-living biomass—dead algae and plants, and organic wastes from municipal or industrial sources—as being the principal reducing agents (electron donors) in water. However, inorganic species including ammonium, sulphide, and reduced forms of iron, chromium, manganese, and other metals may also use up oxygen.

Any material in a water body that can react with dissolved oxygen, contributes to what is termed the biological (or biochemical) oxygen demand (BOD). High concentrations of these oxidizable materials lead to anaerobic conditions—stagnant water that does not support higher life forms such as fish. Therefore, BOD is one very important and widely reported measure of natural water quality. Also, for industrial effluents, the BOD is measured in order to assess the extent to which these waste discharges could promote anoxic conditions when released into a water body. Analysis for BOD is done by adding appropriate heterotrophic microorganisms to a measured amount of diluted effluent or water sample, saturating it with air, incubating for 5 days, and then determining the oxygen that remains in the water. The concentration of oxygen used up (mg $L^{-1}$) during the experiment is the BOD of the diluted sample. On this basis, water quality may be described in terms of its BOD values as:

- very clean:       $<1$ mg $L^{-1}$ $O_2$
- fairly clean:     $1$–$3$ mg $L^{-1}$ $O_2$
- doubtful purity:  $3$–$5$ mg $L^{-1}$ $O_2$
- contaminated:     $>5$ mg $L^{-1}$ $O_2$

The term here, of course, refers only to the content of *degradable* organic matter or other chemical species. Many effluents have BOD values much greater than the maximum solubility (8 mg $L^{-1}$) of oxygen in water; for example, municipal sewage effluent having a BOD of 50 mg $L^{-1}$ means that this material, after being diluted 1 part effluent with 9 parts clean water, uses up 5 mg $L^{-1}$ of $O_2$ during the test.

### Chemical oxygen demand (COD)

Besides BOD, there are other parameters that are sometimes used to evaluate the organic matter content in a water body or effluent. The chemical oxygen demand (COD) measures the concentration of substances that can be oxidized by a chemical oxidant, acidified potassium dichromate, when it is refluxed with a sample at the boiling temperature. In the case of COD, the major contributor to oxygen demand is usually diverse types of soluble and particulate organic matter, and a simple representation of the oxidation is

$$3\{CH_2O\} + 16H_3O^+ \,(aq) + 2Cr_2O_7^{2-} \,(aq) \rightarrow 4Cr^{3+} \,(aq) + 3CO_2 + 27H_2O \quad (15.21)$$

---

**Example 15.5 COD determined from analytical data**

Suppose a 100.0 mL sample of effluent from a pulp and paper mill is taken for measurement of the COD. The sample is digested in an excess of acidified dichromate solution and, by back titration, it is found that $4.64 \times 10^{-4}$ mol of dichromate have been consumed in the chemical oxidation. Calculate the COD.

According to reaction 15.21, this is equivalent to

$$3/2 \times 4.64 \times 10^{-4} = 6.96 \times 10^{-4} \text{ mol of } \{CH_2O\}$$

Therefore, the concentration of $\{CH_2O\}$ in the original solution is

$$6.96 \times 10^{-4} \text{ mol} \times \frac{1000 \text{ mL L}^{-1}}{100 \text{ mL}} \times 1000 \text{ mmol mol}^{-1} = 6.96 \text{ mmol L}^{-1}$$

To convert this value into a measure of COD, we conceptually make use of eqn 15.12,

$$\{CH_2O\} + O_2 \rightarrow CO_2 + H_2O$$

which indicates the molar equivalence of $\{CH_2O\}$ and $O_2$.

Therefore, the potential for consumption of oxygen by this effluent is also 6.96 mmol L$^{-1}$, and the COD is

$$6.96 \text{ mmol L}^{-1} \times 32 \text{ g mol}^{-1} = 220 \text{ mg O}_2 \text{ L}^{-1}$$

### Total organic carbon

A third measure of organic matter concentration in water is the total organic carbon (TOC). Usually TOC is determined using instruments that combust the sample producing carbon dioxide. The amount of carbon dioxide produced is measured using IR spectroscopy. Further subdivisions of the aqueous TOC include dissolved organic carbon (DOC) and particulate organic carbon (POC) as described in the opening section of Chapter 12.

Continuing with our example, the COD test had measured that 6.96 mmol L$^{-1}$ $\{CH_2O\}$ were present in the effluent. The carbon concentration of the sample is then the TOC and equals

$$6.96 \text{ mmol L}^{-1} \times 12 \text{ g mol}^{-1} = 84 \text{ mg L}^{-1} \text{ or } 84 \text{ ppm}$$

We must remember again the assumptions made in these calculations.

- Organic matter is represented by the simple formula $\{CH_2O\}$.
- In comparing COD and TOC, we have neglected oxidation of inorganic components of water such as ammonia.
- All of the organic matter is oxidized in the chemical digestion.

Because these assumptions are never totally true, the equivalencies we have calculated will at best be approximate.

Since resistant compounds are oxidized under the harsh conditions of this experiment to a greater extent than in a BOD experiment, usually COD is greater than BOD often by a factor of around two or more.

## Biotic control of carbon dioxide in the atmosphere

Throughout the above discussion, it should be evident that biotic processes are a major player in determining the levels of carbon dioxide in the atmosphere. Carbon dioxide is released by respiration, decomposition, and combustion of biomass; the first two of these release processes are clearly biotic in nature. Acting as a counterbalance, carbon dioxide is incorporated by both plants and autotrophic microorganisms during synthesis of organic matter through the process of photosynthesis. In Chapter 8, we considered the possibility of growing trees as a means of sequestering atmospheric carbon dioxide and thus reducing its atmospheric concentration. Another related strategy that has been proposed for study is to take deliberate steps that will enhance the growth of photosynthetic microorganisms in the oceans.

About 20% of the Earth's oceans are in regions that receive plentiful solar radiation and have an ample supply of all the major nutrients, yet there is only limited growth of phytoplankton, much less than might have been expected. Studies within these 'high nitrate, low chlorophyll' (HNLC) areas point to iron as being the micronutrient that limits the rate of growth of photosynthetic species. The possibility, then, is that fertilization of the ocean water by iron will increase the rate of growth with consequent transfer of significant amounts of carbon dioxide from the atmosphere into the oceans where its ultimate fate should be deposition as carbonate sediments.

An experiment designed to examine this iron hypothesis has been carried out in the Pacific Ocean off the west coast of South America and 500 km south of the Galapagos Islands.[2] The surface water within a 64-km² area was fertilized with iron to a concentration of approximately 4 nmol L⁻¹ and growth of various microorganisms was followed over a 10-day period. As expected, there were increases in production of autotrophic organisms, and also in heterotrophic species that graze on the primary producers. A total net increase in accumulated biomass of about 60 µg C L⁻¹ was estimated within 15 m of the surface of the iron-enriched ocean.

> **Fermi question**
> What proportion of the annual global carbon dioxide release from combustion could, in principle, be sequestered by fertilizing 1% of the Earth's oceans in the manner described here?

While enhanced biological growth was clearly demonstrated in the experiment, the full implications of the strategy are far from clear. Once again, this is an example of a 'technological fix' to an environmental problem and those who are studying the concept recommend caution before considering implementation beyond an experimental scale.

> **Main point 15.2** Microorganisms play key roles in parts of the global carbon cycle including synthesis and degradation of many biomolecules. These processes are the central factors that determine the oxygen content of different aquatic environments. Because degradable organic matter in water consumes oxygen, it is an important parameter that determines water quality. Biological oxygen demand (BOD), chemical oxygen demand (COD), and total organic carbon (TOC) are three measures commonly used to measure water quality in this way.

## 15.3 **Microbiological processes—the nitrogen cycle**

A second great cycle that affects life in all its varieties on Earth is the nitrogen cycle. We have looked at the atmospheric forms of nitrogen and have seen that various nitrogen oxides play a central role in near-Earth gas-phase chemistry. They are important constituents of precipitation; they affect global climate and influence the stratospheric reactions that determine the ozone synthesis–decomposition balance.

In the hydrosphere and on land, nitrogen chemistry comes into play in many environmental situations. In several forms it is an essential nutrient for both aquatic and terrestrial plants and microorganisms and, as a constituent of proteins, it is essential to animals. But excessive concentrations of inorganic species can contribute to eutrophication of water bodies and are toxic to some organisms including humans. Here, we will look at some of the interactions between species in the hydrosphere and lithosphere. Again, emphasis will be on chemical processes that are mediated by microorganisms.

A $pE$ / pH diagram for the four aqueous inorganic nitrogen species is shown in Fig. 15.8. Under aerobic conditions, nitrate is the stable nitrogen species in both water and soil, but an

---

[2] Martin, J.H. and forty-three others, Testing the iron hypothesis in ecosystems of the equatorial Pacific Ocean, *Nature*, **371** (1994), 123–9.

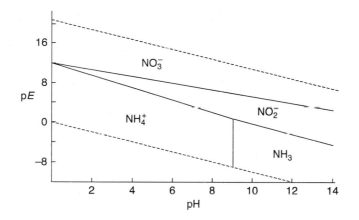

**Fig. 15.8** The pE / pH diagram for nitrogen species.

anaerobic low pE state leads to reduction from nitrate through nitrite to ammonia in its protonated and unprotonated forms depending on pH. Recall that nitrate was positioned as the second oxidant after oxygen in the series of agents that bring about oxidation of biomass in water. Because of the redox buffering effect described earlier, intermediate pE values are uncommon in water and nitrite is usually a transient species measured only in small concentrations.

The reactions that interrelate species in the atmospheric / aqueous / terrestrial nitrogen cycle are shown in Fig. 15.9. A number of atmospheric nitrogen compounds whose environmental behaviour had been discussed in Chapters 4 and 8 are included in this diagram.

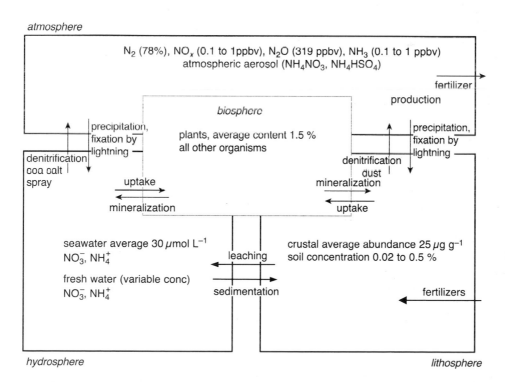

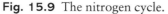

**Fig. 15.9** The nitrogen cycle.

We will consider briefly chemical aspects of the principal nitrogen transformations within and between environmental compartments in the global cycle.

## Nitrogen fixation

Nitrogen fixation reactions are those through which the gaseous dinitrogen molecule is converted into one of the 'fixed' forms of nitrogen associated with the aqueous or terrestrial environments—particularly nitrate or ammonium ions. The conversion requires breaking the strong $N \equiv N$ bond (the bond energy for the $N \equiv N$ triple bond is 945 kJ mol$^{-1}$); therefore a large energy input is required.

There are limited appropriate sources of this high level of energy in the atmosphere including lightning discharges, cosmic rays, and meteor trails. An estimated $5 \times 10^6$ t y$^{-1}$ of nitrogen is fixed via these natural phenomena.

Industrial production of ammonia using the well-known Haber process accounts for another $6 \times 10^7$ t y$^{-1}$ of nitrogen fixation.

$$N_2 + 3H_2 \rightarrow 2NH_3 \qquad (15.22)$$

In the Haber process, air is the nitrogen source and methane in natural gas is the source of hydrogen. The reaction is carried out at about $3 \times 10^4$ kPa pressure and 500°C in the presence of a nickel catalyst.

The hydrogen is usually obtained from natural gas by a two-step 'reforming process'

$$CH_4 + H_2O \xrightarrow{\text{heat}} 3H_2 + CO \qquad (15.23)$$

$$CO + H_2O \xrightarrow{\text{heat}} H_2 + CO_2 \qquad (15.24)$$

A large portion of the ammonia is applied directly to soil as a fertilizer and the remainder is used as the starting material for the manufacture of urea ($(NH_2)_2CO$, also a fertilizer) and other industrial products. Note that production of this commodity chemical requires natural gas both as a starting material and as the energy source for the synthesis processes. Ammonia manufacture is therefore a major player in the world energy economy.

Biological nitrogen fixation that is mediated by terrestrial microorganisms can involve bacteria like *Rhizobium* sp. that live in a symbiotic relation with nodules on the roots of particular species of plants such as legumes and alders. These organisms are capable of catalysing the conversion of atmospheric nitrogen into organic forms usable by plants. Likewise, there are at least 15 species of non-symbiotic, free-living microorganisms such as *Spirillum lipoferum* that are found in particular environments and are able to fix as much as 100 kg ha$^{-1}$ y$^{-1}$ of nitrogen. In total, it is estimated that $1.5 \times 10^8$ t y$^{-1}$ nitrogen is biologically fixed in soil.

Biological fixation can also occur through the agency of marine microorganisms including a variety of species of blue-green algae, *Azotobacter*, and *Clostridium*; the global amount has been estimated to be $1 \times 10^7$ t y$^{-1}$.

In total, about $2.3 \times 10^8$ t y$^{-1}$ of nitrogen is transferred from the atmosphere to the other compartments of the environment through fixation reactions; about 25% of this is a consequence of human activity.

## Denitrification

The ultimate balancing factor leading to the return of nitrogen from land and water into the atmosphere is denitrification. The denitrification reactions occur most commonly in stagnant fresh water and in deep, organically rich waters of the sea. They also occur to a significant extent under anaerobic conditions in the soil. There are several types of denitrification reactions. One of these is the reduction of nitrate to form nitrogen gas, a reaction we described earlier when showing how nitrate is able to act as an electron acceptor for the oxidation of organic matter. The process involves several steps with nitrite ion and nitric oxide being intermediates.

A number of facultative heterotrophic bacteria including species of *Pseudomonas* and *Achromobacter* mediate these processes. Because the organisms are heterotrophic, a plentiful supply of easily decomposable organic matter favours denitrification.

$$4NO_3^- \text{ (aq)} + 5\{CH_2O\} + 4H_3O^+ \text{ (aq)} \rightarrow 2N_2 + 5CO_2 + 11H_2O \qquad (15.14)$$

Where small amounts of oxygen are present, a second gaseous product of denitrification is nitrous oxide (reaction 15.25)

$$2NO_3^- \text{ (aq)} + 2\{CH_2O\} \text{ (aq)} + 2H_3O^+ \text{ (aq)} \rightarrow N_2O + 2CO_2 + 5H_2O \qquad (15.25)$$

The increases in atmospheric nitrous oxide mixing ratios that have been observed in the last decades are due, at least in part, to enhanced rates of denitrification influenced by the greater rates of application of nitrogenous fertilizers worldwide.

A third type of denitrification leads to release of ammonium ion, which is then available to be used for the synthesis of cell protein in the microorganisms themselves. When the product is incorporated into the cell structure, the reaction is referred to as *assimilatory denitrification*. However, under alkaline conditions, the ammonia remains in a deprotonated form through which it can be released to the atmosphere. Like the reactions to produce nitrogen and nitrous oxide gases, this is an example of *dissimilatory denitrification*. In the absence of dissolved oxygen, as in waterlogged soils, *Bacterium denitrificans* is one species that mediates the reaction to produce ammonium / ammonia.

$$NO_3^- \text{ (aq)} + 2\{CH_2O\} + 2H_3O^+ \text{ (aq)} \rightarrow NH_4^+ \text{ (aq)} + 2CO_2 + 3H_2O \qquad (15.26)$$

## Combustion

Fossil fuels and combustible biomass contain nitrogen in varying amounts depending on the nature of the fuel. When these materials are burned, nitrogen is returned to the atmosphere mostly in the form of $NO_x$ compounds. Additional $NO_x$ is generated even in the absence of nitrogen in the fuel, during combustion from combination of nitrogen and oxygen in the air. The extent of $NO_x$ production by this reaction is related positively to combustion temperature. These reactions have been discussed in more detail in Chapters 2 and 4.

## Ammonification

Organic matter contains nitrogen in amounts varying from less than one to several per cent depending on the organism and the type of tissue. Most of the nitrogen is in the form of amino acids in protein. When organic matter decomposes to smaller molecules in water and soil, the nitrogen is first released in a reduced form as ammonium ions or ammonia, depending on the ambient pH. The conversion of nitrogen from organic to inorganic forms is a type of mineralization called *ammonification*. Carbon–nitrogen bonds are relatively reactive and ammonification is therefore a rapid reaction.

$$2CH_3NHCOOH + 3O_2 + 2H_3O^+ \text{ (aq)} \rightarrow 2NH_4^+ \text{ (aq)} + 4CO_2 + 4H_2O \qquad (15.27)$$

## Nitrification

Nitrification is the microbiological reaction sequence we noted at the beginning of the chapter. Ammonium ion, present in water or in soil as a result of ammonification or added in the form of an ammonium-containing fertilizer (ammonium sulphate, ammonium nitrate, urea, and ammonia itself), is subject to oxidation in an aerobic environment. The optimum environmental pH for nitrification is between 6.5 and 8, and the reaction rate decreases significantly when the pH falls below 6. The reaction takes place in two steps, collectively called nitrification. (Note that there are several types of denitrification, only one (reaction 15.26) of which is essentially the reverse of nitrification.) Both steps (reactions 15.2 and 15.28) of the nitrification process are mediated by autotrophic bacteria.

$$2NH_4^+ \text{ (aq)} + 3O_2 + 2H_2O \xrightarrow{\textit{Nitrosomonas}} 2NO_2^- \text{ (aq)} + 4H_3O^+ \text{ (aq)} \qquad (15.2)$$

$$2NO_2^- \text{ (aq)} + O_2 \xrightarrow{\textit{Nitrobacter}} 2NO_3^- \text{ (aq)} \qquad (15.28)$$

The overall reaction (which we have already seen) is

$$NH_4^+ \text{ (aq)} + 2O_2 + H_2O \rightarrow NO_3^- \text{ (aq)} + 2H_3O^+ \text{ (aq)} \qquad (15.1)$$

In another sense, the overall reaction is an oversimplification of what actually occurs. Along with oxidation of reduced nitrogen, some of the element is assimilated into the bacterial protoplasm to form cells with an empirical formula of approximately $C_5H_7O_2N$. These are the bacteria that are actually catalysing the oxidation. This assimilation, however, accounts for only about 0.2% of the original nitrogen and therefore most of the nitrate produced is available for other purposes.

Example 15.6 shows how nitrification is an acid-generating step. When this occurs in large amounts and without other compensating reactions, it can lead to significant acidification of the water or soil.

---

### Example 15.6 Nitrification effect on pH

A consequence of the first step in nitrification is the concomitant release of hydrogen ions causing the local environment to be acidified. Suppose we are dealing with water containing alkalinity measured as 90 mg $L^{-1}$ $CaCO_3$, and with pH 6.8. At this pH, all of the alkalinity may be assumed to be due exclusively to the $HCO_3^-$ ion, and therefore its molar concentration would be $1.8 \times 10^{-3}$ mol $L^{-1}$.

If 6.0 mg $L^{-1}$ (as N) of $NH_4^+$ ($= 4.3 \times 10^{-4}$ mol $L^{-1}$) in the water is nitrified, then $8.6 \times 10^{-4}$ mol $L^{-1}$ of hydronium ion is produced. Before nitrification occurs,

$$K_{al} = \frac{a_{H_3O^+}[HCO_3^-]}{[CO_2]} = \frac{10^{-6.8} \times 1.8 \times 10^{-3}}{[CO_2]} = 4.5 \times 10^{-7}$$

Under these conditions, the equilibrium solubility of carbon dioxide $[CO_2]$ is calculated to be $6.3 \times 10^{-4}$ mol $L^{-1}$.

After nitrification, assuming a closed system, some of the hydrogen carbonate is converted to aqueous carbon dioxide. The final concentrations are then

$$[HCO_3^-] = (1.8 \times 10^{-3} - 8.6 \times 10^{-4}) = 9 \times 10^{-4} \text{ mol } L^{-1}$$

$$[CO_2] \text{ (aq)} = (6.3 \times 10^{-4} + 8.6 \times 10^{-4} \text{ mol } L^{-1}) = 15 \times 10^{-4} \text{ mol } L^{-1}$$

Under the new conditions, the pH of the water is calculated from

$$K_{al} = \frac{a_{H_3O^+}[HCO_3^-]}{[CO_2]} = \frac{10^{-pH} \times 9 \times 10^{-4}}{15 \times 10^{-4}} = 4.5 \times 10^{-7}$$

$$\{H_3O^+\} = 7.6 \times 10^{-7}$$

$$pH = 6.12$$

Therefore, the pH within this closed system has been reduced to 6.12. Taking into account carbon dioxide exchange with the atmosphere would alter the calculations.

---

## Uptake (assimilation)

After carbon, oxygen, and hydrogen, nitrogen is quantitatively the most important element required by plants and microorganisms for growth in water and soil. For many aquatic organisms and agricultural crops, the concentration of nitrogen in plant tissue averages about 1.5% by mass. Both ammonium and nitrate serve as a nutrient source. Ammonium is the preferred form but nitrate is the stable and common species in well-aerated environments. The first step of nitrate assimilation is ion exchange at the root or microbe surface and this is essentially an acid-neutralizing process. This is because the anion exchanged from the cell is usually an anion of a weak acid such as carbonate (reaction 15.29). Following release into the solution adjacent to the microbe or plant root cells, the carbonate is able to act as a proton acceptor (reaction 15.30).

$$\text{Root:}CO_3^{2-} + 2NO_3^- \text{ (aq)} \rightarrow CO_3^{2-} \text{ (aq)} + \text{Root:}(NO_3^-)_2 \tag{15.29}$$

$$CO_3^{2-} \text{ (aq)} + 2H_3O^+ \text{ (aq)} \rightarrow CO_2 \text{ (aq)} + 3H_2O \tag{15.30}$$

In a closed system then, acidity generated by nitrification is at least partially neutralized by assimilation. Therefore, nitrate uptake is a means of biologically immobilizing nitrogen species and, at the same time, neutralizing acidity.

## Abiotic ion exchange and adsorption

We have seen that most sediment and soil colloid surfaces are negatively charged, giving them the ability to act as cation exchangers. The positive ammonium ion can therefore be held immobile on the colloidal surface through reversible retention by clays, organic matter, and other materials.

$$\text{Soil}^-\text{:}K^+ + NH_4^+ \text{ (aq)} \rightarrow \text{Soil}^-\text{:}NH_4^+ + K^+ \text{ (aq)} \tag{15.31}$$

In contrast, nitrate, the more important inorganic nitrogen species under aerobic conditions, is an anion and does not interact with sediment and soil colloids in this way. Furthermore, compared to phosphate and other potential ligands, nitrate usually is able to form only weak complexes with most metals including those on the solid surface. Therefore, there is little tendency to take part in specific adsorption reactions with components of the solid phase. For these reasons, nitrate is a geochemically mobile species.

## Leaching

The net result of the biological and geochemical processes is that ammonium ion can be retained and immobilized by both biological and geochemical processes, whereas nitrate mobility is only subject to biological control. Nitrate therefore remains in the solution phase of water bodies. It is also readily leached out of soil into surface or groundwater under a number of particular environmental conditions:

- where there are few or no living plants, such as in a fallow field or a clear-cut forest;
- in situations where plants are dormant, as in wintertime. Storage of high nitrogen-content animal wastes outdoors in winter is a particularly serious problem. The nitrogen present in large concentration as nitrate may readily be leached from the biologically dormant manure, through soil and into water;
- in high-intensity agriculture where nitrate fertilizer exceeding plant requirements is applied.

There are several critical negative environmental consequences that arise due to excess nitrogen being leached into a water body.

Being an essential nutrient for aquatic microorganisms and plants, excess nitrate may be a cause of eutrophication in water bodies. While we noted previously that most often phosphorus not nitrogen is the limiting nutrient in lakes and ponds, there are many exceptions. In Florida in the southern USA, phosphate-containing minerals are found in the bedrock throughout much of that state. Much of the surface and groundwater is therefore relatively rich in dissolved and particulate phosphorus and increases in eutrophication depend on sufficient nitrogen compounds, particularly nitrate, being present.

Where nitrate is present in water from acid precipitation or as a result of nitrification of reduced nitrogen compounds, the associated acidity may overwhelm the alkalinity of the

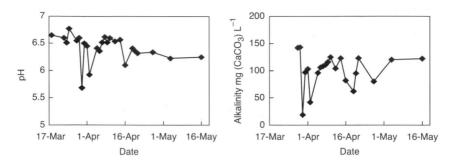

**Fig. 15.10** Changes in pH and alkalinity (mg $CaCO_3$ $L^{-1}$) at a depth of 1 m in Little Turkey Lake, Ontario, Canada over the time period between 19 March and 17 May 1982. Note the significant excursions to pH 5.5–6 and the greatly reduced alkalinity. (From Kwain, W. and J.R.M. Kelso, Risk to salmonids of water quality in the Turkey Lakes Watershed as determined by bioassay, *Can. J. Fish. Aquat. Sci.*, 45 suppl. 1 (1988), 127–35.)

water body. Figure 15.10 plots pH and alkalinity in the surface water of Little Turkey Lake, a small lake about 65 km north of Sault Ste Marie, Ontario. The data cover the period between 19 March and 17 May in 1982, a time that includes the spring thaw in early April. During this period with melting ice and snow, also a time when many fish are spawning, there is an increased flux of acidified water that reduces the background alkalinity and the pH of lakes and rivers in the catchment area. Recovery comes about as there is mixing of surface and less-affected deep waters and slow neutralization by reaction with soil and sediment minerals brings about a gradual return to 'normal' conditions.

A third problem relates to the toxicity of nitrate to humans and other mammals. In drinking water, maximum allowable concentrations have been set, usually between 10 and 50 mg $L^{-1}$, by many jurisdictions. Actually, nitrate itself is not toxic; rather, nitrite is a highly toxic species and is produced through reduction of nitrate by the bacteria *Escherichia coli* in the mammalian intestinal tract. The nitrite so produced reacts with hemoglobin causing severe oxygen deprivation, especially in children. Alternatively, it may react with secondary amines and amides to form carcinogenic N-nitrosamines.

Figure 15.11 summarizes most of the important reactions in the nitrogen cycle in all three compartments of the environment. Note that each of these is a natural process, but all are influenced or augmented (sometimes on a massive scale) by human interventions.

> **Main point 15.3** The global nitrogen cycle includes essential processes within and between all the compartments of the environment. Many of these processes are, in large part, biotic chemical reactions.

## 15.4 Microbiological processes—the sulphur cycle

Like that of nitrogen, sulphur chemistry has a major influence on processes in all the compartments of the Earth's environment. In the hydrosphere and in soil, sulphur is present in many inorganic and organic forms exhibiting oxidation states from –2 to +6. The most important reduced and oxidized mineral forms of the element are sulphides, such as pyrite ($FeS_2$), and sulphates, including gypsum ($CaSO_4\cdot2H_2O$). Minor amounts of sulphur are found as 'impurities' in many rocks. You will recall that some of the reduced species are released into air by several processes. Once in the atmosphere, oxidation reactions convert lower oxidation state species

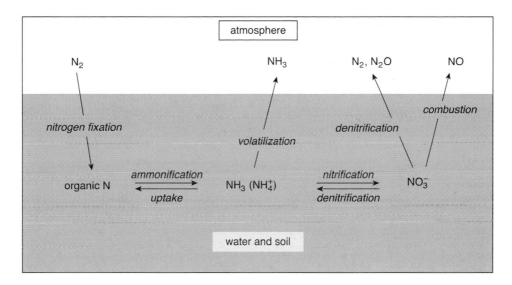

**Fig. 15.11** Nitrogen processes and relationships in water and soil, and their connection to the atmosphere

into sulphate, sulphate aerosols act as condensation nuclei for cloud formation, and sulphuric acid is one of the principal acidifying components in precipitation.

Sulphate concentrations in oceans around the world are close to 28 mmol L$^{-1}$. Variable, but averaging at 0.12 mmol L$^{-1}$, the concentration in fresh water is much smaller, but it is nonetheless one of the principal ionic species in lakes and rivers. In reducing environments, chemical species containing sulphur in the −2 state are formed. Like nitrogen, sulphur is an essential nutrient for microorganisms and plants, but the amounts required are approximately one order of magnitude smaller and sulphur is rarely the nutrient that limits biological growth. Nevertheless, the element plays an important role in a number of aquatic and terrestrial environmental processes.

The pE / pH diagram for sulphur was constructed earlier (Fig. 10.5) and shows how the principal inorganic forms of this element in the hydrosphere are sulphate under aerobic conditions and protonated or deprotonated hydrogen sulphide, under acidic and alkaline anaerobic conditions, respectively. Elemental sulphur is usually a rare and transient species in the aquatic and terrestrial environment.

The sulphur cycle is depicted in Fig. 15.12. Some aspects of reactions that produce gaseous species and of the atmospheric reactions themselves have been discussed in Chapter 5. We will now briefly consider three portions of the cycle that take place in water, sediments, and soils.

## Sulphur release during organic matter decomposition

In organic matter, sulphur exists as both carbon-bonded and oxygen-bonded forms:

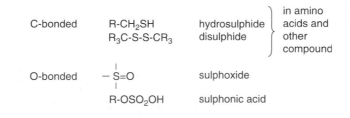

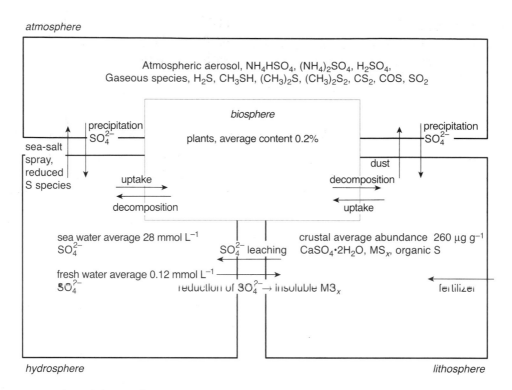

**Fig. 15.12** The sulphur cycle.

When organic matter undergoes microbial decomposition, the sulphur-containing groups within the organic compounds are simultaneously transformed. The carbon-bonded organic sulphur is present as a component of protein in the form of the amino acid cysteine and it is mineralized by a process (not a redox reaction) as shown in reaction 15.32. The enzyme responsible for the reaction is associated with a variety of bacterial species.

$$\text{HS}-\text{CH}_2-\overset{\displaystyle\overset{\text{COOH}}{|}}{\underset{\displaystyle\underset{\text{NH}_2}{|}}{\text{CH}}}+\text{H}_2\text{O}\xrightarrow[\text{desulphydrase}]{\text{cysteine}}\text{CH}_3\text{CO}-\text{COOH}+\text{HS}^-\,(\text{aq})+\text{NH}_4^+\,(\text{aq})$$

(15.32)

The $pK_{a1}$ of $H_2S$ is approximately 7. Therefore, in any reducing environment whose pH is below neutrality, a large proportion of the sulphide is present as the undissociated molecule, giving rise to the strong odours characteristic of this gas. This is commonly observed in some swamps and other wetlands. Hydrogen sulphide is also a toxic species.

For oxygen-bonded organic sulphur, microbial mineralization involves hydrolytic splitting of the S–O bond via sulphatase enzymes.

$$\text{ROSO}_2\text{O}^- + 2\text{H}_2\text{O} \rightarrow \text{ROH} + \text{SO}_4^{2-} + \text{H}_3\text{O}^+$$

(15.33)

Under environmental conditions sulphate remains in its deprotonated form, as the $pK_a$ value is 2.00.

## Sulphide oxidation

Under high-p$E$ aerobic conditions, sulphide is unstable and is easily oxidized via a variety of pathways. The various sulphide species may have been formed during decomposition of organic matter (reaction 15.32) or may have been present as sulphide minerals that were deposited at an earlier stage in the sediment or soil. Regardless of their origin, reaction 15.34 describes their oxidation to sulphate.

$$HS^- (aq) + 2O_2 + H_2O \rightarrow SO_4^{2-} (aq) + H_3O^+ (aq) \tag{15.34}$$

The principal chemoautotrophic bacteria responsible for sulphide oxidation include the common species *Thiobacillus thiooxidans*. These bacteria are present in most oxygen-containing waters, sediments, and soils and they multiply rapidly when supplied with a source of sulphide. The reaction simultaneously produces hydrogen ions and is thus an acidifying process.

These and other related microbial reactions are important in marine coastal environments and within deposits of sulphur-containing wastes such as sulphide mine tailings. Some specific cases will be discussed in later chapters.

## Sulphate reduction

In an organic-rich, reducing (low p$E$) aqueous environment, sulphate is readily reduced to species in the −2 or, less commonly, 0 oxidation states.

$$SO_4^{2-} (aq) + 2\{CH_2O\} + H_3O^+ (aq) \rightarrow HS^- (aq) + 2CO_2 + 3H_2O \tag{15.16}$$

$$2SO_4^{2-} (aq) + 3\{CH_2O\} + 4H_3O^+ (aq) \rightarrow 2S° + 3CO_2 + 9H_2O \tag{15.35}$$

Reaction 15.16 is the one we considered above to describe the oxidation of organic matter. It leads either to gaseous emissions of hydrogen sulphide (at low pH) or to production of aqueous soluble and insoluble sulphides. Many marine, estuarine, and fresh-water sediments as well as waterlogged soils contain sulphate-reducing bacteria. The most common of these is *Desulphovibrio desulphuricans*, an obligate anaerobe that grows at pH values above 5.5. The sulphide species generated ($H_2S$ or $HS^-$) are toxic to aquatic life; they may also react with metals present in the sediment / soil, such as iron (II) to produce insoluble sulphides that are components of marine shales.

Dimethyl sulphide can also be a product of reduction of sulphate in low-p$E$ environments. Production of this and other reduced sulphur gases is carried out by marine plankton and provides an important route for natural release of sulphur compounds into the atmosphere. Dimethyl sulphide is a particularly important reduced sulphur species and, within the atmosphere, is a key link in the global sulphur cycle.

**Main point 15.4** Sulphur chemistry, like that of nitrogen, is controlled through microbial mediation of processes—principally redox processes—in the hydrosphere and terrestrial environment.

The subjects we have discussed in this chapter have centred around natural microbial processes that occur in water and soil environments. Recently, there is growing interest in harnessing some of these natural processes for industrial purposes, such as for removing contaminants from industrial effluents. The literature link that follows provides a good example of this kind of approach, and relates to remediation of arsenic in the aqueous effluent from a lead and zinc mining operation in British Columbia, Canada.

**Literature link** Biological arsenic removal from contaminated water

Catriona Jackson
Researcher, Environmental Sciences Group (ESG)
Royal Military College, Kingston, ON

Arsenic is a common metalloid in the Earth's crust and can have toxic properties, depending on its chemical form. Fortunately, most geogenic arsenic is found as a component of sparingly soluble minerals, including orpiment ($As_2S_3$) and arsenopyrite (FeAsS), limiting the exposure pathway to humans and ecosystems. Natural weathering and human activity, however, can open up this exposure pathway by causing arsenic to dissolve in water. The most prominent anthropogenic source of arsenic is mining, which exposes minerals found underground, such as arsenopyrite, to oxygenated environments. Also, acidification is often used to extract valuable ores. These changes in pH and redox conditions convert pyrite ($FeS_2$) to sulphates ($SO_4^{2-}$) and iron oxyhydroxides (eqn LL15.1). This results in acid mine drainage (AMD, leachate) contaminated with dissolved forms of arsenic, including arsenate ($H_3AsO_4$) and arsenite ($H_3AsO_3$).

$$FeS_2 + 3.75O_2 + 3.5H_2O \longrightarrow Fe(OH)_3 + 2SO_4^{2-} + 4H^+ \qquad \text{(LL15.1)}$$

The large volumes of AMD contaminated with arsenic and other elements present a challenge for remediation. Anaerobic bioreactors may present a low cost solution as a result of their effectiveness in removing metals and metalloids from the dissolved state. Such bioreactors are used to foster environments conducive to the growth of sulphate-reducing bacteria (SRB), which thrive only in anaerobic conditions. These bacteria, which are also present in wetlands, use sulphate as an electron acceptor, reducing it to hydrogen sulphide ($H_2S$). Hydrogen sulphide is known to react with dissolved metals and metalloids to produce low-solubility metal-sulphide products (eqn LL15.2).

$$M^{2+} + H_2S \longrightarrow MS\ (s)\downarrow + 2H^+ \qquad \text{(LL15.2)}$$

In sulphidic environments, several arsenic sulphide complexes can be formed, depending on the sulphide concentration and the pH. These include thioarsenates ($H_3AsSO_3$) and thioarsenites ($H_3AsSO_2$), which are soluble arsenic sulphide compounds, as well as orpiment, which is a mineral. Amorphous orpiment is known to form in anaerobic bioreactor conditions, where dissolved forms of arsenic and sulphate exist in the presence of SRB. This creates a reducing sulphidic environment, allowing for a chain of reactions to occur that lead to the formation of amorphous orpiment, with soluble thioarsenates and thioarsenites as reaction intermediates (eqn LL15.3).

$$H_2AsO_4^- + H_2S \rightleftharpoons H_2AsO_3S^- + H_2O \rightarrow \rightarrow H_3AsO_2S\ (+H_2S) \rightarrow \rightarrow 3As_2S_3 + 3H_2S \qquad \text{(LL15.3)}$$

The soluble intermediates become the dominant arsenic species if the sulphide concentration is too low to form orpiment. Conversely, when the concentration of sulphide is too high, amorphous orpiment will dissolve (eqn LL15.4). This demonstrates the need to control sulphide concentrations within anaerobic bioreactors.

$$1.5As_2S_3 + 1.5H_2S \longrightarrow H_2As_3S_6^- + H^+ \qquad \text{(LL15.4)}$$

In Trail, British Columbia, Canada, a constructed wetland (CW) system removes dissolved arsenic, as well as other metalloids and metals contaminating leachate water from historic mine tailings landfills. The CW was built in 1997 by Nature Works Remediation Corporation and consists of two anaerobic bioreactor cells (in series) followed by a conventional wetland (Fig. LL15.1). The 800 L of water entering the CW each day can contain up to 600 mg $L^{-1}$ arsenic, and the concentration in the effluent is almost always less than 1 mg $L^{-1}$, which is the safe discharge criterion in Canada. Approximately 90% of the arsenic is removed in the anaerobic bioreactor cells, which are 800 m³ underground cells containing wood fibres from a pulp and paper mill lined with high density polyethylene.

The exact microbial and chemical mechanisms of arsenic removal inside the reactors are still not well understood and the reactors remain a major focus of research. X-ray absorption spectroscopy

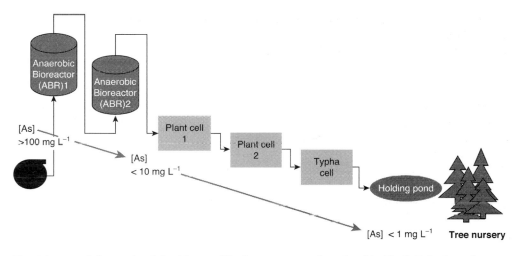

**Fig. LL15.1** Schematic of the Nature Works constructed wetland in Trail, BC, Canada.

(XAS) confirmed the presence of arsenic–sulphur compounds in the reactor matrix, indicating that arsenic–sulphide precipitation is a likely mechanism of arsenic removal. It is also possible that iron coprecipitates with sulphide to form FeS, which can adsorb arsenic to its surface.

While the reactors have been working effectively for more than a decade, further research is necessary to determine the long-term fate of this system. With a better understanding of the arsenic removal product and of the stability of that product under changing pH, redox and sulphide environments, this CW design could be used to treat water contaminated with arsenic in other areas of the world. The effluent water from the Trail system is below the safe discharge criterion for arsenic (1 mg L$^{-1}$) and is used to sustain a tree nursery. It should be noted, however, that the arsenic concentration in the effluent water is not below the safe drinking-water criterion (10 μg L$^{-1}$) and complementary remediation would be required to further reduce arsenic concentrations for drinking water purposes.

### Related research

Drury, W.J. Treatment of acid mine drainage with anaerobic solid-substrate reactors. *Water Environ. Res.*, **71** (1999), 1244.

Nature Works Remediation Corporation. http://www.nature-works.net/. © 2004. Accessed October 2009.

Neculita, C.M., G.J. Zagury, and B. Bussière, Passive treatment of acid mine drainage in bioreactors using sulphate-reducing bacteria: critical review and research needs. *J. Environ. Qual.*, **36** (2007), 1.

Newman, D.K., T.J. Beveridge, and F.M.M. Morel, Precipitation of arsenic trisulphide by *Desulphotomaculum auripigmentum*. *Appl. Environ. Microbiol.*, **63** (1997), 2022.

## ADDITIONAL RESOURCES

1. Paul, E.A. ed., *Soil microbiology and biochemistry*, 3rd edn, Academic Press, Oxford; 2007.

2. Tate, R.L., *Soil microbiology*, 2nd edn, John Wiley and Sons, Chichester; 2000.

3. Alexander, M., *Biodegradation and bioremediation*, 2nd edn, Academic Press; **San Diego, California**, 1999.

4. Rheinheimer, G., *Aquatic microbiology*, 4th edn, John Wiley and Sons, Chichester; 1992.

## PROBLEMS

1. Given that the Henry's law constant, $K_H$, for oxygen in water at 25°C is $1.3 \times 10^{-8}$ mol $L^{-1}$ $Pa^{-1}$, explain why a biological oxygen demand (BOD) value of $>5$ mg $L^{-1}$ is indicative of contaminated water.

2. Consider dissolved organic matter to have the generic formula $\{CH_2O\}$. For a 1 mg $L^{-1}$ (as C) aqueous solution of organic matter, calculate the mg of dissolved oxygen in the same volume required to oxidize it completely. Use this calculation to establish a relation between COD and DOC. Repeat the calculation using the generic formula for dissolved humic material (Fig. 12.3). Assume reaction of only the carbon and hydrogen in the humic material.

3. In the iron fertilization experiment to increase growth of phytoplankton in the oceans, what chemical species of iron would you expect to be present at equilibrium, if the iron was initially added as iron (III) chloride?

4. For an organic-rich soil, under controlled conditions in the laboratory at temperatures maintained in the mesophilic ranges, 20 to 30°C, the rate of carbon dioxide production commonly ranges from 5 to 50 mg $CO_2$ per kg soil per day. Using these data estimate the number of kg of carbon dioxide released from a 1 ha field under the same temperature conditions

5. 'Mineralization' refers to the process by which organic forms of an element are broken down and converted to inorganic species. In environmental situations this is most often a microbiological process. For nitrogen, indicate what forms of the element might be present in water or soil as reactants and products of the mineralization process.

6. Use the thermochemical tables to calculate $pE°$ values for the reductive half-reactions involving nitrate ($NO_3^-$ (aq)) in three different cases: when the product is; $NH_4^+$ (aq); $N_2O$ (g); and $N_2$ (g). Then calculate the $pE°$ (w) values for the same three reactions. What is the environmental significance of these results?

7. Evolution of ammonia is one process that can result in the transfer of nitrogen from an aqueous system to the atmosphere. Discuss environmental conditions that would favour such a process. Would you expect ammonia evolution to be significant: (a) in an acid bog; (b) in the oceans?

8. For water with pH 6.7 and alkalinity of 110 mg $CaCO_3$ $L^{-1}$, calculate the maximum concentration (mg N $L^{-1}$) of $NH_4^+$ that could be nitrified without the pH of the water falling below 6. Assume a closed system.

9. Why are reduced gaseous species of sulphur emitted from rice (paddy) fields, swamps, and the near shore borders of lakes, but not from open lakes—even though reduced sulphur compounds are found in the sediments in all these situations?

10. Would you expect gaseous sulphur emissions to be greater from the soil of a tropical savannah or of a tropical rainforest? What gaseous species are likely to be released?

11. Using data from Appendix B.2, calculate the $pE$ / pH equation for the $SO_4^{2-}$ / $SO_3^{2-}$ boundary. Plot it on the sulphur diagram. What is the significance of this plot in terms of the stability of the sulphite species in the hydrosphere.

12. The $pE$ of a groundwater sample is −1.2 and the pH is 8.83. The concentrations of $SO_4^{2-}$ and $HS^-$ are 2.29 and 0.003 mM, respectively. Is the system at equilibrium?

13. Some n-propanol (0.2 t) is accidentally discharged into an approximately circular waste lagoon with radius 50 m and average depth of 2 m. What is the increase in BOD and in COD in the water? Assume complete degradation to $CO_2$:

$$2C_3H_8O + 9O_2 \longrightarrow 6CO_2 + 8H_2O$$

14. When excess fertilizer is carried from soil in runoff water, it increases the BOD of the water. Calculate the mass of oxygen required to oxidize 10 kg of urea. Assume that nitrogen in the urea is oxidized to form nitrate.

# Chapter 16
# Water pollution and water treatment chemistry

The focus of this chapter is on understanding the nature of water pollution and the chemistry of water treatment for the protection of water resources. We will accomplish this by:

- defining what we mean by pollution
- reviewing the desired chemical characteristics of various types of water and corresponding water quality guidelines
- discussing wastewater treatment processes with emphasis on secondary and tertiary technologies.

At the outset of this book, we emphasized that our focus would be on the composition of the natural environment, the processes that take place within it, and the kinds of changes that come about as a result of human activities. With regard to many issues, we have discussed each of these aspects, including ways in which humans influence composition and processes in the environment. In these discussions, we have occasionally used the word 'pollution' but have not spent time to consider what that word really means. At this point, in the context of water chemistry, we have an appropriate place to consider this concept. After looking at how the term pollution can be defined, we will examine the ways in which the definition plays out in several different instances. Since wastewater that is discharged from urban centres is one of the principal contributors to water pollution, we will also spend some time discussing the chemistry of water treatment processes.

## 16.1 What is pollution?

### A definition based on purity

It is possible to define pollution in a variety of ways and it is worth thinking carefully about the different definitions. In terms of the hydrosphere, the ultimate definition could be to say that, unless water is 100% $H_2O$ (i.e. unless it is pure in the literal chemical sense), it is, at least to some extent, polluted. Obviously, such a definition is unrealistic and of limited utility. Every water sample in contact with the atmosphere contains dissolved gases including oxygen and carbon dioxide and every water sample in contact with sediment and rock contains other dissolved constituents such as species of silicon and calcium. These components are perfectly natural and, in fact, some are essential for the support of aquatic life.

**Table 16.1** Composition of some minor and trace elements in a sample of open-ocean sea water[a].

| Element | Concentration / $\mu g\ L^{-1}$ ± 95% confidence limit |
|---|---|
| Arsenic | 1.65 ± 0.19 |
| Cadmium | 0.029 ± 0.004 |
| Chromium | 0.175 ± 0.010 |
| Cobalt | 0.004 ± 0.001 |
| Copper | 0.109 ± 0.011 |
| Iron | 0.224 ± 0.034 |
| Lead | 0.039 ± 0.006 |
| Manganese | 0.022 ± 0.007 |
| Molybdenum[b] | 11.5 ± 1.9 |
| Nickel | 0.257 ± 0.027 |
| Selenium (IV) | 0.024 ± 0.004 |
| Uranium | 3.00 ± 0.15 |
| Zinc | 0.178 ± 0.025 |

[a] Sample is NRC Standard Reference Material NASS-2 from a depth of 1300 m at a location in the North Atlantic Ocean, SE of Bermuda.
[b] The relatively large concentration for molybdenum is not unusual, but is typical of values in oceans throughout the world.

As an indication of concentrations of dissolved species in clean natural water we may examine analyses of supposedly pristine waters. For example, in the copious literature on marine chemistry, we can find a variety of data on the composition of 'open ocean' water. The term 'open ocean' implies that the water has not been subject to significant additional human-derived inputs of chemicals such as you might find in coastal areas, and therefore represents the true composition of 'natural' sea water. Of course, in the limit, no such water exists. Average values for major elements in open ocean water are given in Table 9.1 and a more complete listing is in Appendix B.1. Table 16.1 reports analytical results for some trace elements in one well-characterized open ocean sample in the North Atlantic, south-east of Bermuda. Some of the values given there may be surprising, for example, the small concentration of iron and the much greater value for molybdenum. Compare these results with the average values given in the appendix to see if they are anomalous.

As is the case for sea water, there are also data available for 'pure' fresh waters from various locations; these too vary significantly from place to place. Average river-water concentrations are also listed in Appendix B.1. One might expect to find the purest water in far northern or southern latitudes or derived from glacial sources at high altitude—sources relatively remote from human habitations and activity. Even there, measurable small concentrations of some elements are found. Table 16.2 reports data for some anions and cations in snow samples from Terra Nova Bay in Antarctica. Clearly, in the natural environment there is no such thing as pure water in the chemically rigorous definition.

Just as absolute purity of water is impossible, so also is complete removal of all potential contaminants in a water supply or waste stream. While 'zero discharge' is a commendable goal, it is an ultimately unattainable one.

**Table 16.2** Concentration of some anions and cations in Antarctic snow[a].

| | Concentration / $\mu g\ L^{-1}$ | |
| --- | --- | --- |
| | Minimum | Maximum |
| Chloride | 25 | 40 100 |
| Nitrate | 8.6 | 354 |
| Sulphate | 10.6 | 4020 |
| Bromide | 0.8 | 49.4 |
| Phosphate | 1.8 | 49 |
| Fluoride | 0.1 | 0.2 |
| Sodium | 15 | 17 050 |
| Potassium | 3.1 | 740 |
| Magnesium | 2.7 | 1450 |
| Calcium | 12.6 | 1010 |
| Ammonium | 2.4 | 46.5 |

[a] From Udisti, R., S. Bellandi, and G. Piccardi, Analysis of snow from Antarctica: a critical approach to ion chromatography methods, *Fresenius J. Analyt.* Chem., **349** (1994), 289–93.

### A definition based on naturalness

A second possibility is to define water pollution as any concentration of chemical (or micro-organism) in water above the 'natural' (also called background or baseline) level—in other words, added concentrations due to anthropogenic inputs. Compared to our first definition, this one is somewhat more realistic, yet it too presents difficulties. But then, what is the natural level? For one thing, it is often impossible to distinguish between human and other environmental factors. In other situations, the natural level is itself sufficiently high to be toxic to humans, or is harmful in other ways. There are, for example, locations in East Africa and several other places around the world where groundwater contains elevated levels of fluoride. The high concentrations contribute to abnormal calcification of bones and teeth in humans and other animals. On the other hand, certain human inputs added deliberately or inadvertently may result in enhanced water quality. Adding calcium carbonate to an acid lake will certainly increase the calcium ion concentration above its previous 'natural' level but, at the same time, it should improve the quality of water by reducing its acidity and enhancing other environmental properties.

### A utilitarian definition

This brings us to a third possible definition of water pollution that is more utilitarian. One version of the definition is that a pollutant[1] is 'a substance or effect that adversely alters the aqueous environment by changing the growth rate of species, interferes with the food chain, is toxic, or interferes with health, comfort, amenities, or property values of people'. The definition implies a need to set standards or guidelines in order to indicate that water, whose chemical properties exceed the limits of the standards, may cause a particular environmental alteration or interference.

[1] The words pollutant and contaminant are used interchangeably in this book. Some persons make a distinction by defining a contaminant as any substance present in the environments above a 'natural' level whether or not it causes a detrimental effect, whereas a pollutant is a contaminant that causes adverse effects. (Porteous, A., *Dictionary of environmental science and technology*, John Wiley and Sons, Chichester; 1992.)

While we will make use of this last definition, we should recognize that it too has limitations. In order to define criteria, information is required concerning toxicity and other factors. Experimental work is carried out to establish standards and guidelines but there are always assumptions and incomplete data, and conclusions are partly subjective and often treat risk on a statistical basis.

Guidelines have been established by many jurisdictions and generally include:

- physical properties of temperature, colour, odour, and turbidity;
- general classes of chemical properties such as pH, total dissolved solids (TSS), salinity, hardness, biological oxygen demand (BOD), detergents, and petroleum residues;
- specific elements, complex ions, and organic compounds;
- radiological properties, that is, levels of radioactivity due to particular isotopes;
- microbiological properties, that is, counts of specific organisms and groups of organisms.

> **Main point 16.1** In terms of water, a pollutant can most usefully be thought of as a substance or effect that alters the aqueous environment, interfering in an adverse way with natural processes and / or causing inconvenience and ill-health to humans and other living things.

## 16.2 **Water quality**

### Drinking water

Clearly, decisions regarding the particular criteria will depend on the end use of the water. For drinking water, rather stringent requirements are in order, especially those related to toxicity but also including properties associated with aesthetics such as colour, odour, and taste. A partial list of drinking water standards set by the World Health Organization is given in Table 16.3.

### Toxicity

Central to any set of guidelines for drinking water are standards related to dissolved constituents that are potentially toxic. This requires that we examine the issue of how to define and assess the toxicity of these constituents.

A toxic substance is an element, compound, or microorganism that, upon exposure, creates a harmful effect on a living organism. Toxicity applies to plants, animals, and microorganisms but the degree to which any life-form is affected by a toxic agent depends on the species and other environmental factors. Serious cases of toxicity can lead to death of the organism either suddenly (acute toxicity) or after extended exposure (chronic toxicity). Toxicity is different from carcinogenicity (cancer-generating effects), although the consequences of both properties can obviously be extremely serious.

Measuring toxicity is a complex and highly developed science, usually carried out by means of laboratory studies using standard protocols. The fundamental experiment is one in which an organism is exposed to controlled amounts of the potential toxic agent under carefully defined conditions. After controlled exposure, the influence on the organism is measured in some appropriate way. This is called a *dose–response* study. In these experiments, the controlled variable is therefore the dose and the measured variable the response. There are many options that have been employed for determining both of these parameters.

- Dose may be the amount of the agent supplied in a single treatment, concentration supplied in a series of treatments, or length of exposure to a fixed concentration.
- The potential toxic agent (toxicant) may be a component of a medium in which microorganisms or plants are grown or it may be present in the atmosphere surrounding the

**Table 16.3** World Health Organization guidelines for drinking-water quality[a].

| Concentration / $\mu g\ L^{-1}$ | | Concentration / $\mu g\ L^{-1}$ | |
|---|---|---|---|
| Inorganics | | Organics | |
| As | 10 | Benzene | 10 |
| B | 500 | Benzo(a)pyrene | 0.7 |
| Cd | 3 | TCE | 40 |
| $CN^-$ | 70 | NTA | 200 |
| Cr | 50 | Pesticides | |
| Cu | 2000 | Atrazine | 2 |
| F | 1500 | Lindane | 2 |
| Pb | 10 | 2,4-D | 30 |
| Mn | 400 | DDT | 1 |
| Hg | 6 | Disinfection by-products | |
| $NO_3^-$ | 50 | Monochloramine | 3000 |
| $NO_2$ | 3 | Dichlorobromomethane | 100 |
| Se | 10 | Chloroform | 300 |

[a] From *Guidelines for drinking water quality*, 3d cdn, Chapter 8, WHO, Geneva, 2006, http://www.who.int/water_sanitation_health/dwq/gdwq3rcv/cn/index.html (accessed October 2009). Many other parameters are provided in this comprehensive online document along with an explanation of guideline values. Details are also provided about analytical procedures and treatment protocols. The guidelines are given in several languages.

organism; in these cases, dosage is a product of concentration and time. Alternatively, the test agent may be supplied in a form that can be ingested (in the case of animals) or injected into a specific tissue of an organism.

- There are options also in determining response. Most obviously, the death of the organism can be taken as the ultimate response, but more subtle non-lethal measures are also used. Growth rates, carbon uptake, reproductive performance, and changes in metabolic function of various kinds are among the commonly used response measures.

This kind of experiment typically leads to a *dose–response*[2] type of relationship as shown in Fig. 16.1. The figure describes a typical (but not universal) response pattern to increasing exposure to a toxic agent. At low dosages, the toxic agent generates no apparent effect, but above a particular dose the effect becomes evident on a small proportion of the population, with the proportion increasing as the dose increases until the total population shows response.

Parameters that are commonly reported as measures of toxicity are the $LD_{50}$ and $LC_{50}$. The $LD_{50}$ is defined using the dose–response curve as shown in Fig. 16.1 and is an estimate of the dose of contaminant that would be lethal to 50% of an infinitely large population of the test organism. The population being used to develop the dose–response curve may be comprised of any chosen organism, but it is usually a microorganism or small mammal such as the rat. The $LC_{50}$ is similarly defined as the concentration that has a lethal effect when exposure takes place over a specific time interval. This parameter is frequently used when discussing toxicity of substances in the ambient medium, for example, water for fish or air for humans. More difficult to

[2] In specific situations, response to doses of contaminants follows other patterns. A recent article (Renner, R., Redrawing the dose–response curve, *Environ. Sci. Technol.*, **38** (2004), 90A) examines the wide variety of dose–response curves that are observed and discusses the significances of the differences.

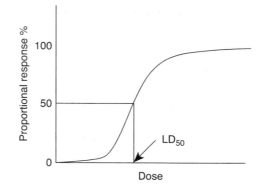

**Fig. 16.1** Typical dose–response curve obtained in a toxicity test. In the case where the measured response is the death of the organism, the dose that causes death of 50% of the test population is called the $LD_{50}$ (lethal dose 50).

define precisely are 'no observable adverse effects level' (NOAEL) also called the $LD_0$ or $LC_0$. Conceptually, these values are taken from the dose–response curve at a point just before the rapid rise in response is recorded.

The basic approach that is described here for measuring toxicity is widely followed and its fundamental validity is not questioned. At the same time, it is recognized that there are limitations to its application. A short list of some of these limitations includes the following:

- the fact that each organism responds differently; for obvious reasons, toxicity testing can rarely be done on human subjects and data related to humans are usually extrapolated from experiments obtained on other test animals;

- related to this, the toxicity of a chemical varies depending on the test species, and also on the stage of development of the species. A comprehensive test would have to take these issues into account;

- the need to take into account combined effects of other environmental factors that vary according to the individual situation;

- the effect of simultaneous exposure to multiple potential toxicants. Most tests are, however, done employing a single toxic agent;

- the simplest tests measure an acute response; it is much more difficult to quantitatively measure diverse chronic responses that are usually more subtle.

When creating guidelines or standards for drinking water or for other purposes, reference is made to the data that have been obtained from toxicity studies. For obvious reasons, the standards are set at amounts or concentrations much smaller (typically by factors of between 100 and 10 000) than those determined by the $LD_0$ or $LC_0$ measures. They also must take into account other qualitative issues such as appearance, taste, and odour. Setting values for standards is one area in which application of the precautionary principle should be a central feature.

Finally, it is important to note that *measuring toxicity* in the way described here is a different subject from determining the *mechanism of toxicity*. Mechanistic studies take into account details of the environmental form and behaviour of the toxicant as well as the metabolic processes where it is involved within the target organism. Development of behavioural models such as the biotic ligand model (BLM) described in Chapter 13 for metal-ion toxicity form a part of such a study. By pointing to specific molecular-level effects that can be used as response parameters, the mechanistic research is nevertheless helpful in establishing protocols for toxicity tests.

### Drinking water treatment

> **Sources of contamination of water bodies**
>
> - Point sources—include discharge from an industry or urban sewer, typically through a pipe. These streams of wastewater have the potential of being treated before release into the natural environment.
> - Diffuse sources—include runoff from large parking lots, agricultural fields and feedlots. The runoff is not localized into a stream or pipe and is therefore not amenable to simple treatment processes.
>
> In a different context, these two terms can also be applied to airborne emissions

Perhaps the most important point to be made about provision of safe drinking water is that the original source of the water must be protected. This requires that wastes that are discharged into natural water bodies be kept as clean as possible. For this reason, later in this chapter we will emphasize the need to apply high-quality wastewater treatment processes to waste streams from urban areas. Control of diffuse sources of pollution such as runoff and percolation from agricultural areas is also essential, but this centres largely on the need to prevent the waste effluents from reaching the nearby surface or ground water sources. To protect surface water from contamination by agrochemicals or animal wastes, a protected zone consisting of natural vegetation adjacent to the water body is frequently required.

Even well-protected sources will, however, usually require treatment in order to provide a safe supply. Drinking-water treatment has the goal of removing turbidity, chemicals such as those listed in Table 16.3 (this is a partial listing) and biological pathogens. Generally this requires two or more steps.

### Settling and filtration

The most basic treatment is to allow large particles to settle in a quiescent tank and then to pass the overlying water through a filter, typically a sand bed. Often a coagulant such as alum is added (see Section 16.2) as an aid for precipitating out some dissolved chemical species and also to enhance the production of a filterable 'floc'. Sometimes, activated carbon is also part of the filter bed. This has the ability to adsorb some of the organic impurities including soluble humic material.

### Disinfection

The filtration process offers only very limited protection against water-borne pathogens; disinfection is therefore almost always required as a final step in any drinking-water treatment plan. The prime requirement of the disinfection process is that it be able to efficiently eliminate any pathogenic organisms. A number of approaches can be effective in that regard—UV irradiation, treatment with ozone, with chlorine or with chlorine dioxide. There is a limitation in UV and ozone treatments, however, in that after application, no residual chemical agent remains that can inactivate organisms that may be picked up throughout the (often extensive) distribution system.

Chlorine is probably the most common disinfection agent, although it is not actually $Cl_2$ that is the active chemical. When chlorine gas is bubbled into water at near-neutral pH, it reacts rapidly to produce hypochlorous acid.

$$Cl_2 \text{ (g)} + 2H_2O \longrightarrow HOCl \text{ (aq)} + H_3O^+ \text{ (aq)} + Cl^- \text{ (aq)} \qquad (16.1)$$

The hypochlorous acid is the effective disinfecting agent and, instead of chlorine, its conjugate base, hypochlorite ion, may be added to the solution in the form of sodium or calcium hypochlorite. Because the deprotonated species is less efficient as a disinfectant than hypochlorous acid, the pH of the water to be treated should optimally be less than 7.

Besides being efficient and being capable of maintaining residual disinfecting properties, the other requirement for any water-purifying agent is that it not produce harmful by-products during the disinfection process. In this regard, chlorine / hypochlorous acid has some limitations. When added to water containing natural organic matter (NOM) in the form of soluble humic material, it is able to chlorinate the terminal groups of the humate molecules and thereby produce small quantities of a series of trihalomethane compounds ($CHX_3$). Chloroform is the most important of these. The disinfection by-products (DBPs) have been implicated in producing liver and bladder cancer in humans, and guideline values for their residual concentrations have been recommended (Table 16.3).

Chlorine dioxide gas is another useful disinfection agent, now used in many jurisdictions around the world. It is a free radical, analogous to the peroxy species that we came across in the smog-formation reactions. Like these, it is a powerful oxidizing agent and destroys pathogens by oxidizing their organic structures.

$$ClO_2 + 4H_3O^+ + 5e^- \longrightarrow Cl^- + 6H_2O \qquad (16.2)$$

Although it is a chlorine-containing species, chlorine dioxide does not add this element to organic compounds in the water and so it tends to produce a much smaller concentration of potentially dangerous DBPs compared to chlorine / hypochlorite.

Going back to the statement at the outset of this section, the most important and most sustainable solution to providing high-quality drinking water is to protect the source from original contamination. And, in preparation for any treatment protocol, knowledge of what contaminants or natural materials are present in the water is clearly a pre-requisite. The literature link that follows describes some research on methodology used to characterize the NOM that is present in a treated drinking water source.

### Literature link  The hydroxyl radical as a water treatment chemical

In recent decades, there have been several reports describing applications where the hydroxyl radical is employed as an oxidant to destroy residual organic chemicals in air or water. This requires the ability to generate the radical in an appropriate manner and at sufficient concentrations. One example is Venkatadri, R. and R. Peters, Chemical oxidation technologies: ultraviolet light / hydrogen peroxide, Fenton's reagent and titanium dioxide assisted photocatalysis, *Haz. Waste Haz Mat.* **10**, (1993), 107. When solar radiation is focused on a titanium dioxide semiconductor catalyst, the ultraviolet component of the sunlight activates the catalyst and the activated catalyst causes water to dissociate and form unstable hydroxyl radicals. These unstable radicals are formed when the titanium dioxide semiconductor promotes its electrons from the valence band (vb) to the conduction band (cb) where they are shared with water and oxygen molecules. Production of radical species occurs according to the following scheme.

$$TiO_2 + h\nu \rightarrow e(cb)^- + h(vb)^+ \qquad (LL16.1)$$

$$h(vb)^+ + OH^- \rightarrow \bullet OH \qquad (LL16.2)$$

$$h(vb)^+ + H_2O \rightarrow \bullet OH + H^+ \qquad (LL16.3)$$

$$e(cb)^- + O_2 \rightarrow O_2^- \qquad (LL16.4)$$

$$e(cb)^- + h(vb)^+ \rightarrow heat \qquad (LL16.5)$$

In this reaction sequence, $h(vb)^+$ represents the valence-band holes, $e(cb)^-$ represents the conduction-band electrons and $O_2^-$ is the superoxide ion. The efficiency of production of hydroxyl radicals depends in part on the crystallinity and surface area of the titanium dioxide.

The highly energetic radical species are able to oxidatively decompose some 'stable' chlorinated compounds such as the widely used solvents, trichloroethylene and perchloroethylene. The decomposition products are carbon dioxide, water, and hydrochloric acid. To remove soluble carbonate species, the water is previously acidified to pH 5, causing evolution of gaseous carbon dioxide $CO_2$ and, after irradiation, the water is subsequently neutralized to pH 7 with sodium hydroxide. In this way up to 400 000 litres of contaminated water could be treated per day.

Building on this research, new work has shown that hydroxyl radicals produced by a similar method have the potential to destroy NOM in surface water and could be applied as a step in the treatment of drinking water. (by Liu, S., M. Lim, R. Fabris, C. Chow, M. Drikas and R. Amal, $TiO_2$ photocatalysis of natural organic matter in surface water: impact on trihalomethane and haloacetic acid formation potential, *Environ. Sci. Technol.* **42**, (2008), 6218–6223.) The study from Australia examines the products formed during this free-radical-assisted degradation of NOM and addresses the question of whether these products are potential starting materials for the formation of halogenated compounds that are dangerous to human health.

In this study, the challenge of analysing natural water is enhanced by the need to be able to determine both the unreacted heterogeneous macromolecules of the humic substances family along with the medium and small-sized degradation products. This then calls for a combination of analytical techniques. To begin with, careful storage and pre-treatment protocols were required. Size-exclusion liquid chromatography with UV detection was used to show that much of the original NOM had molar masses in the range 250 to 4000 Da, and most of it was degraded to smaller molecules depending on conditions. Resin fractionation was also used before and after treatment to separate the soluble organic matter into very hydrophobic, slightly hydrophobic, charged hydrophilic and neutral hydrophilic fractions. Smaller degradation products, carbonyl compounds, were determined by a gas chromatography–mass spectrometry (GC–MS) procedure. In addition, to measure trihalomethanes, GC–MS was again employed, and other halogenated compounds were measured by a standard EPA procedure.

Like many similar studies, this research was carried out on a laboratory scale, using 500 mL samples. Do you see the potential for this treatment method to be upgraded to the large volumes of water encountered at the plant scale, perhaps dealing with a continuous flow of 100 000 $m^3$ per day? If so, would it replace any standard procedure now being used? At what stage in the treatment process should it be applied?

## Irrigation water

For industrial and irrigation purposes, quality demands are in some ways less severe than those for drinking water. The most important properties of water to be used for irrigation are the total concentration of soluble salts, the molar ratio of sodium to calcium and magnesium in the water, the concentration of potentially toxic elements especially boron, and the carbonate species concentration.

Table 16.4 details some properties set out for irrigation waters that are used in India. (These standards were, in part, based on ones established in the USA.) The criteria relate to the influence of water quality on sustainable plant growth as well as the safety of the (usually) food products being grown. They involve four major inorganic chemical parameters, but there are also (not shown) guidelines for potentially toxic chemicals and microorganisms. Criteria related to aesthetics are generally unnecessary, while some level of certain classes of mild pathogens may be acceptable in irrigation water.

> **Main point 16.2** Water quality requirements depend on the end use of the water. Water that is ultimately going to be used for human consumption requires strict regulations, while water that is to be used for agricultural or industrial purposes may have fewer, but different, restrictions.

**Table 16.4** Irrigation water standards for India.

| Total soluble salts[a] | |
|---|---|
| Conductivity / dS m$^{-1}$ (25°C) | Quality |
| <0.25 | Excellent |
| 0.25–0.75 | Good |
| 0.75–2.25 | Doubtful |
| >2.25 | Unsuitable |

| Sodium hazard[b] | |
|---|---|
| Sodium adsorption ratio (SAR) | Quality |
| <10 | Excellent |
| 10–18 | Good |
| 18–26 | Doubtful |
| >26 | Unsuitable |

| Boron[c] | | | |
|---|---|---|---|
| Boron concentration / mg L$^{-1}$ in crops that are | | | Quality |
| Sensitive | Semi-tolerant | Tolerant | |
| <0.33 | <0.67 | <1.0 | Excellent |
| 0.33–0.67 | 0.67–1.3 | 1.0–2.0 | Good |
| 0.67–1.0 | 1.3–2.0 | 2.0–3.0 | Doubtful |
| 1.0–1.3 | 2.0–2.5 | 3.0–3.8 | Unsuitable |
| >1.3 | >2.5 | >3.8 | Very toxic |

| Alkalinity[d] | |
|---|---|
| Carbonate alkalinity / meq L$^{-1}$ | Quality |
| <1.25 | Safe |
| 1.25–2.5 | Marginal |
| >2.5 | Unsuitable |

[a] Total soluble salts (salinity) are assessed by conductivity. Large concentrations of soluble salts are a contributing factor to soils themselves becoming saline. This reduces the productivity of most crops and, in the limit, makes the soil unsuitable for growth of most crops.
[b] The 'sodium hazard' is defined in terms of the sodium adsorption ratio, $C_{Na} / (C_{Ca} + C_{Mg})^{1/2}$, and is a measure of the potential of the water to saturate the exchange sites of the soil with sodium. The significance of this in terms of soil quality is described in Chapter 18.
[c] Boron is essential for normal crop growth, but is toxic when present in excessive concentrations. The sensitivity of crops is variable.
[d] Alkalinity of the water may contribute to harmful shifts in pH of the soil to the alkaline range.

## Wastewater

Wastewater that is recovered from domestic or industrial sources and is returned to the natural environment is subject to a different set of requirements. In general, if the water is to be discharged back into the hydrosphere, it should not contain dangerous levels of toxic chemicals or organisms, it should not supply excessive quantities of readily oxidizable (usually organic) compounds, and it should not be a source of nutrients that would support microbial growth. It is recognized that there will be some level of dilution when the wastewater is discharged into the receiving water body; for this reason, the requirements with respect to toxic components will be less stringent than those for drinking water.

Where wastewater is to be used for irrigation of productive lands, it is not necessary to remove nutrients or the benign dissolved and suspended organic matter. But the requirement to remove toxic chemicals and harmful microorganisms remains as an important issue. There are many industrial wastewater and treatment protocols that depend on what, if any, contaminants have been introduced during the industrial processes.

Discussion of aspects of the chemistry of many potential or actual contaminants in water has been or will be presented in other parts of this book. In the present section we will examine the principles of physical and chemical methods used to treat wastewater so as to remove the commonly occurring contaminants before it is returned to the environment.

Community wastewater (sewage) combines domestic wastes with industrial and other effluents. In most cases, the wastewater, whether or not it undergoes treatment, is discharged into a natural water body such as a river, lake, or the ocean. Increasingly, industrial effluents are under strict regulation to limit or eliminate discharge of known contaminants to the public system. The major components of community wastewater then derive from domestic and commercial sources and are made up of human wastes, solid and dissolved forms of food wastes, soaps and detergents, and soil residues. Typical measured properties of untreated sewage include:

- biological oxygen demand (BOD),    $250 \ mg \ L^{-1}$
- chemical oxygen demand (COD),    $500 \ mg \ L^{-1}$
- total solids (TS),    $720 \ mg \ L^{-1}$
- suspended solids (SS),    $220 \ mg \ L^{-1}$
- total phosphorus (TP),    $8 \ mg \ L^{-1}$
- total nitrogen (TN),    $40 \ mg \ L^{-1}$
- pH,    6.8

Throughout the day there are fluctuations in both the flow and contaminant concentration with higher values being characteristic of the morning and evening and lowest values occurring at night.

Of the above properties, BOD, SS, TP and TN are of greatest concern as they may all, in different ways, upset the normal balance of aquatic life. The natural levels of dissolved oxygen in water are sufficient to oxidize small amounts of animal and vegetative wastes via aerobic microbial reactions. In the process, the organic wastes are converted into simple organic and inorganic compounds, and oxygen into carbon dioxide. The carbon dioxide, under the influence of light, takes part in photosynthesis and oxygen is returned to the water. This cycle of self-purification is broken by the presence of excessive amounts of degradable organic matter (high BOD) producing anoxic conditions, by turbidity (high SS) that inhibits photosynthesis, and by unusually large concentrations of nutrients (most often, high TP) that stimulate plant and algal growth. For this reason, there may be government regulations to require that the effluent of a treatment facility meet specified criteria, such as BOD of $15 \ mg \ L^{-1}$, SS of $15 \ mg \ L^{-1}$, and TP of $1 \ mg \ L^{-1}$. In order to accomplish this, a variety of treatment procedures have been developed and applied on a large scale where up to $10 \ 000 \ m^3$ or more of wastewater a day may be processed. Wastewater treatment plants are one of the most common types of chemical plants in the world. There are many processes and designs, but some general features are frequently incorporated in any system. Treatment may include physical, chemical, and biological processes operating sequentially or simultaneously.

## 16.3 Wastewater treatment processes—primary and secondary methods

Wastewater treatment plants are often described as offering primary, secondary, and / or tertiary treatment. The primary processes are mainly physical ones. The waste stream is first subjected to pulverization to reduce large-sized solid material to smaller dimensions. It then passes through a screen and the water containing small particles flows into a clarifier where separation of oil, grease, and foams occurs by flotation with simultaneous sedimentation of heavier grit—sand, silt, and other solids. The settling process is often assisted by the addition of chemical coagulants such as alum. More will be said about this in a section below.

### The activated sludge process

The secondary processes are biological, where conditions are adjusted so that aerobic micro-organisms are able to thrive. A widely used method of secondary treatment is known as the activated sludge process (ASP). To create an aerobic environment, oxygen is provided by pumping coarse or fine air bubbles through an aeration tank or lagoon, or by vigorously mixing the surface of the wastewater in order to increase the area of contact between the water and the atmosphere. A further requirement is that the population of growing microorganisms be augmented by recycling some of the previously 'activated' biological mass. This biological material is called sewage sludge. Under these conditions, heterotrophic aerobic bacteria and protozoa grow and respire; in the process the organic matter of the raw sewage serves as a carbon source. The carbon is partially incorporated into the microbial biomass while another part of it is oxidized to produce carbon dioxide. At the same time a portion of the nutrients, phosphorus and nitrogen, is removed from solution as they too are taken up by the microbes. Metals that are present in wastewater interact with cellular material by passive adsorption through complex formation with extracellular functional groups. Some additional portion of the metals is retained by physical entrapment, and some is incorporated into the cellular structures. Further removal of all these elements and of other contaminants is accomplished by use of chemical coagulants.

> **Levels of wastewater treatment**
> - Primary—physical or chemically enhanced settling of suspended particles
> - Secondary—biological processes to convert organic matter into bacterial biomass and carbon dioxide, followed by settling of the products
> - Tertiary—a variety of advanced processes to remove specific contaminants or for disinfection of the water

After an appropriate residence time—on the order of 4 to 12 h—the water and suspended material (mixed liquor) from the aeration tank flows into the secondary clarifier where the biological floc is allowed to settle. Where a chemical coagulant, such as alum or ferric chloride, is added prior to the secondary clarifier, the chemical floc settles along with biological material. A small portion of the settled floc, now called sewage sludge, is returned to the clarifier as noted above. Most of the sludge is sent for further treatment or directly to waste. A schematic diagram of a typical ASP plant is shown in Fig. 16.2.

As a site for microbial growth, an alternative to the aeration tank is a trickling filter where the waste stream is allowed to drain slowly over a large surface-area medium—again in order to provide efficient contact between air and water.

A variety of other processes, used individually or in series, make up a tertiary system. These include filtration through a microscreen or sand bed, precipitation after chemical additions, adsorption on to granular activated charcoal (GAC), ion exchange, reverse osmosis, and disinfection by chlorination or ozonation. Some of these technologies are especially important for the

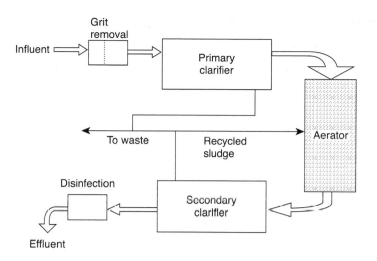

**Fig. 16.2** Schematic of an activated sludge wastewater treatment facility. The open arrows show the path of the wastewater stream, while the closed arrows follow the separated sludge.

treatment of specific kinds of industrial wastewater. The processes are chosen depending on the chemical nature of the wastewater stream and the quality requirements of the effluent water.

Typical before-and-after water quality values indicating the effectiveness of various treatment types (not including the tertiary methods) are given in Table 16.5. The three basic chemical quality parameters—biological oxygen demand (BOD), suspended solids (SS), and total phosphorus (TP)—are those most commonly monitored and regulated in a community wastewater-treatment facility.

## Chemical coagulants for turbidity removal

Chemical coagulant additions are frequently applied as a component of various levels of wastewater treatment. Coagulants serve two principal functions—to assist in the coagulation and flocculation processes, in order to maximize removal of very small solid particles of various compositions, and to react with and remove potential pollutant chemical species such as phosphorus in the water.

---

**Coagulation and flocculation**

- Coagulation is the process of destabilization of colloids, by altering surface properties in order to allow the individual particles to combine into larger ones
- Flocculation is the actual coming together or accumulation of the particles into a settleable mass

---

To understand how coagulants assist the removal of turbidity, we will once again make use of Stokes' law, this time to estimate settling rates in an aqueous medium. The previous application was concerned with the settling of airborne particulates as described in Chapter 6.

In the wastewater application, the Stokes–Cunningham slip correction factor is not required, and density and viscosity of water are used in place of those for air.

$$v_t = \frac{(p_p - p_w)g d_p^2}{18\,\mu} \tag{16.3}$$

where $v_t$ = terminal velocity of particles / m s$^{-1}$, $p_p$ = density of particle / kg m$^{-3}$, $p_w$ = density of water = $1.0 \times 10^3$ kg m$^{-3}$ at $P°$ and 20°C, $g$ = 9.8 m s$^{-2}$, $d_p$ = particle diameter / m, and $\mu$ = viscosity of water $1.0 \times 10^{-3}$ kg m$^{-1}$ s$^{-1}$ at $P°$ and 20°C.

**Table 16.5** Typical concentrations of wastewater contaminants during the wastewater treatment process[a].

| | Concentration / mg L$^{-1}$ | | | |
|---|---|---|---|---|
| | Raw influent | Primary | Secondary | |
| | | | Biological | Chemically assisted |
| BOD | 250 | 175 | 15 | 10 |
| SS | 220 | 60 | 15 | 10 |
| TP | 8 | 7 | 6 | 0.1–1 |

[a] BOD, Biological oxygen demand; SS, suspended solids; TP, total phosphorus.

---

**Example 16.1 Particle settling rate**

If we apply the Stokes relation to a spherical particle of sand, 0.1 mm ($10^{-4}$ m) in diameter, assuming a particle density of $2.65 \times 10^3$ kg m$^{-3}$, then

$$v_t = \frac{(2650 - 1000) \times 9.8 \times (10^{-4})^2}{18 \times 10^{-3}}$$

$$= 9 \times 10^{-3} \text{ m s}^{-1} = 9 \text{ mm s}^{-1}$$

Where the settling tank has a depth of 4 m, sand particles with these dimensions will settle in $4/(9 \times 10^{-3}) = 440$ s = about 7.4 min. A typical clarifier has been designed so that water takes an average of 1.5 h to pass through it (this is called the detention time).[3] Clearly, sand with particle diameters of 0.1 mm will have sufficient time to settle completely in a tank of the specified size.

We can repeat the calculation for a spherical particle of clay whose diameter is 0.002 mm.

$$v_t = \frac{(2650 - 1000) \times 9.8 \times (2 \times 10^{-6})^2}{18 \times 10^{-3}}$$

$$= 3.6 \times 10^{-6} \text{ m s}^{-1}$$

Within 1.5 h, clay particles of this size would therefore have settled to a depth of only

$$3.6 \times 10^{-6} \text{ m s}^{-1} \times 1.5 \times 3600 \text{ s} = 0.02 \text{ m} = 2 \text{ cm}$$

This calculation then shows that, for a settling tank with a detention time of 1.5 h, the 0.002 mm diameter particles will not settle to the bottom before moving out in the waste stream.

---

Turbidity in wastewater is due to a mixed collection of solids having a range of particle sizes, shapes, and densities. Much of the suspended material will, however, consist of colloidal material that settles too slowly to be removed within the detention time in the clarifier. One of the functions of coagulants is to assist and speed up the settling process.

Three coagulants are frequently employed in wastewater treatment. These are alum ($Al_2(SO_4)_3$, obtained as a concentrated solution containing about 5% aluminium or as a solid hydrate with 14 or 18 waters of hydration), ferric chloride ($FeCl_3$, usually a by-product of steel manufacturing and obtained in an aqueous solution form), and hydrated lime ($Ca(OH)_2$ in a solid form).

---

[3] The detention time is the time that wastewater spends in a particular part of the plant. For any tank it is readily calculated as follows: detention time (time units) = capacity (volume units) / flow rate (volume units / time units). Note how this definition bears similarity to the definition of residence time given in Chapter 1. In both cases the calculated time is given by amount divided by flux.

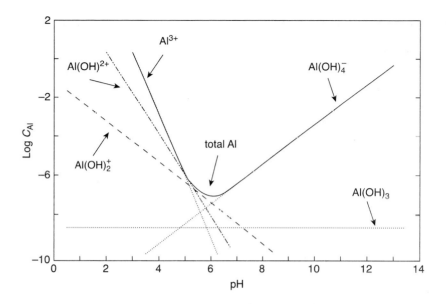

**Fig. 16.3** Solubility of individual aluminium species and total aluminium solubility in water as a function of pH.

Alum and ferric chloride both are sources of trivalent metal cations and their behaviour in wastewater treatment is somewhat similar. We will look at the case of alum. When it is added to any type of water containing sufficient alkalinity, the hydrated aluminium ion undergoes a stepwise series of hydrolysis reactions as indicated.

$$Al(H_2O)_6^{3+} (aq) + H_2O \rightarrow Al(H_2O)_5(OH)^{2+} (aq) + H_3O^+ (aq) \qquad (16.4)$$

$$Al(H_2O)_5(OH)^{2+} (aq) + H_2O \rightarrow Al(H_2O)_4(OH)_2^+ (aq) + H_3O^+ (aq) \qquad (16.5)$$

$$Al(H_2O)_4(OH)_2^+ (aq) + H_2O \rightarrow Al(H_2O)_3(OH)_3 (s) + H_3O^+ (aq) \qquad (16.6)$$

$$Al(H_2O)_3(OH)_3 (s) + H_2O \rightarrow Al(H_2O)_2(OH)_4^- (aq) + H_3O^+ (aq) \qquad (16.7)$$

The hydrolysis reactions involve successive deprotonation of the waters of hydration surrounding the central metal ion. Recall, as described in Chapter 13, that trivalent cations have a great tendency to undergo deprotonation reactions of this type. The extent of the reactions, however, also depends on solution conditions, particularly the availability of Brønsted bases to act as proton acceptors for the released hydronium ion. Notice that, except for $Al(OH)_3$, all of the deprotonated aluminium species are soluble ions.

A diagram showing the solubility of aluminium species and the total solubility versus pH is given in Fig. 16.3.

As we have seen earlier, a measure of the capacity of water to accept protons is the alkalinity and, almost always, community wastewater has a relatively high alkalinity—typically in the range from 100 to 300 mg L$^{-1}$ as CaCO$_3$. The pH of such water is usually between 6.5 and 7.5. These properties of the wastewater ensure that the equilibrium positions of reactions 16.4 to 16.7 are such that any added aluminium is largely converted to insoluble $Al(OH)_3$. Since the alkalinity is mostly due to hydrogen carbonate ion, the overall reaction of aluminium in wastewater may be approximated by

$$Al(H_2O)_6^{3+} (aq) + 3HCO_3^- (aq) \rightarrow Al(OH)_3(s) + 3CO_2 (g) + 6H_2O \qquad (16.8)$$

The above chemical processes occur in wastewater and play a role in the removal of small particles that cause turbidity. Two mechanisms, one essentially chemical and the other physical operate to clean up the water.

The aluminium hydroxide precipitate takes the form of a flocculated material having a very large surface area. Depending on conditions, the precipitate has a $pH_0$ value that is as high as 9. It therefore carries a net positive charge that acts to partially neutralize the negative charges of the suspended matter—the clays, particulate organic matter, and bacteria—all of which are characterized by having much lower $pH_0$ values. Neutralization of the double-layer charge allows the colloids to approach each other and come together into larger aggregates that readily settle in the clarifier. Secondly, a physical sweeping is also responsible for removal of turbidity, especially when a high concentration of coagulant is added to the waste stream. In this 'sweep floc process', the voluminous inorganic floc produced by reaction 16.8 physically entraps particles of clay and organic material as it settles, aiding in the clarification of the mixed liquor.

## Sludge digestion

Besides the treated wastewater, the other product from the treatment facility is digested sludge. The sludge is initially in the form of a slurry that consists mostly of microbiological material, undigested residual organic matter, and inorganic solids derived from the original wastewater as well as from added coagulant. Small concentrations of dissolved metals and inorganic species are also present. When removed from the clarifier, the solids content is only 0.1% or less, the rest being water. An effective treatment system subjects this slurry to anaerobic digestion in an enclosed reactor. In the oxygen-free environment, the complex microbial reactions that occur under these conditions may be summarized by the reaction (15.18) we described for the fermentation process in the previous chapter.

$$2\{CH_2O\} \rightarrow CH_4\,(g) + CO_2\,(g) \tag{16.9}$$

In order to maintain its temperature near the ideal of 35°C for mesophilic bacteria, part of the methane that is produced can be used as fuel to heat the reaction vessel. This ensures that the fermentation proceeds rapidly. In the early stages, the mixture is stirred but, as decomposition nears completion, stirring is stopped, allowing the solids to settle. During digestion, most of the pathogenic organisms are killed, and the product has very little objectionable odour. The process causes the sludge to become more concentrated, so that it finally contains about 5% solids. It may be further dewatered on open drying beds or by using a large centrifuge.

The options for final disposal of the sludge include placing it in a secure landfill or spreading it as an amendment to agricultural or forest soils. Two concerns regarding the sludge relate to metals such as cadmium and persistent pathogenic microorganisms that could remain in the digested material. We will discuss some uses, and problems associated with uses of this material in Chapter 19.

### A modern wastewater treatment plant

An overview of a typical modern wastewater treatment plant can be examined at the website listed at the end of this chapter under Additional Resources 5. The website describes the facilities and summarizes the processes used at the Houtrust plant at the Hague in the Netherlands. This plant employs chemical and biological treatments to generate water of acceptable quality. The by-product sludge is digested at the plant and this produces 30 000 $m^3$ of biogas, mostly methane, every day. The biogas is used to generate electricity, while the dewatered sludge is trucked to another site where it is incinerated. Besides treating the liquid and solids in the wastewater, an additional series of steps is applied to treat the air emanating from the wastewater- and sludge-treatment processes so that what is released to the atmosphere is free of unpleasant odours.

## 16.4 **Wastewater treatment processes—tertiary methods**

### Chemical coagulants for phosphate removal

The second function served by the alum coagulant is the removal of phosphorus from the waste-water stream. Phosphorus is of particular concern because of its role in promoting eutrophication, as we have seen in Chapter 14. In many jurisdictions there are strict limits set for maximum allowable phosphorus discharge; values of $1 \text{ mg L}^{-1}$ or lower concentrations in treated effluent are frequently specified. The biological processes described above serve to remove a portion of the phosphorus—typically about 20%—by incorporation within the insoluble biomass that is subsequently removed in the organic sludge. However, because the influent phosphorus concentration is of the order of 5 to $10 \text{ mg L}^{-1}$, the treated water still contains phosphorus in excess of the allowed, $1 \text{ mg L}^{-1}$ or lower, concentration. An advanced biological process that has the capability of effecting much more efficient removal of phosphorus will be briefly discussed later in the chapter.

As well as being an effective coagulant for removal of the turbidity associated with various colloids, alum is able to react with phosphate in order to precipitate out the highly insoluble aluminium phosphate. Usually, the reaction is shown simply as

$$Al^{3+} (aq) + PO_4^{3-} (aq) \rightarrow AlPO_4 (s) \tag{16.10}$$

While the reaction is described by the neatly balanced equation, what actually happens is almost certainly a much more complex process. When aluminium ion is added directly to a low-alkalinity waste stream containing phosphorus, some direct reaction of the type shown here may occur. However, in the presence of the large excess of high-alkalinity water, aluminium hydrolyses sufficiently rapidly that it is already converted to an insoluble hydrous oxide form before it is able to react with phosphate. Nevertheless, the solid is still able to remove phosphorus by way of specific adsorption on the active surface of the freshly precipitated floc; a possible reaction involves displacement of hydroxide ions by the partially protonated phosphate species.

$$Al(OH)_3 + HPO_4^{2-} (aq) + H_2O \rightarrow AlOH(HPO_4) \cdot H_2O (s) + 2OH^- (aq) \tag{16.11}$$

Following the surface reaction, phosphate slowly migrates to the interior parts of the aluminium colloid, leaving fresh surface to continue to adsorb more phosphate. As these processes continue, the gel-like hydrous aluminium oxide gradually converts more completely into aluminium hydroxyphosphate or aluminium phosphate, although complete conversion may not come about during the lifetime of the precipitate in a sewage treatment plant. That lifetime, incidentally, is not just the 4 to 12 h detention time in the aerator and clarifier, but much longer—perhaps 10 days—during which time most of the sludge is repeatedly recycled into the aerator from the clarifier. A smaller proportion of the sludge, perhaps 20%, is 'wasted' and collected for further treatment (see The final products after treatment of wastewater, at the end of this section). Recall that the principal purpose of the recycle is to supply to the fresh sewage a substantial quantity of microorganisms that will multiply and grow in the presence of the influx of nutrients in the influent. However, we can now see that a second useful function of the recycled sludge is that the precipitated alum is still capable of reacting with phosphorus by reaction 16.11, thus removing it from solution. In summary, the fresh alum added to a wastewater stream reacts in part with soluble phosphorus during or immediately after its hydrolysis; the 'recycled' hydrolytic precipitate of alum also acts to remove phosphorus even in the absence of new additions of coagulant.

Calculations of the amount of alum required in any given situation are usually based on the stoichiometry of reaction 16.10, assuming that, in the end, the precipitate contains a ratio of aluminium to phosphorus that is near 1:1. To ensure better nutrient removal, an excess of coagulant is often specified. Because of the cost associated with the large quantities of alum required to treat wastewater from a major community, it is advantageous to devote some effort to ensure optimum efficiency in its use. This is achieved by thorough mixing at the time of addition to maximize contact between aluminium and phosphorus during precipitation and equally good mixing during the time that the precipitate is present along with fresh influent.

It is possible to use iron (III) chloride or calcium hydroxide as coagulants in place of alum.

The iron (III) species behave in a manner similar to that described for aluminium (III)—hydrolysis, reaction with phosphate, simultaneous removal of suspended materials—but the best conditions for this coagulant to be used require a somewhat more acidic wastewater medium. A disadvantage of iron (III) chloride is that its solution is acidic and oxidizing, which means that it can be highly corrosive to pumps and other metallic parts of the delivery system.

Calcium hydroxide requires that the sewage mixed liquor pH be raised to at least 9.0 in order to ensure complete precipitation. Under these conditions, a discrete compound with formula $Ca_5OH(PO_4)_3$ has been identified and so the phosphorus removal reaction may be written as

$$5Ca(OH)_2 \text{ (aq)} + 3HPO_4^{2-} \text{ (aq)} \rightarrow Ca_5OH(PO_4)_3 \text{ (s)} + 6OH^- \text{ (aq)} + 3H_2O \qquad (16.12)$$

Because of the need to carry out precipitation in an alkaline solution, after settling and removal of the sludge but before discharge of the purified water, an additional pH adjustment is required to shift the pH downwards to a more acceptable value, usually between pH 6.5 and 7.5.

---

**Fermi question**

A city of 150 000 people, using on average 400 L of water per day per person, discharges wastewater without treatment into a nearby lake. The lake covers an area of 820 km² and has an average depth of 43 m. Predict the effect of establishing a secondary treatment system on the phosphorus levels in the water. The present phosphorus concentration in the lake is 22 $\mu$g L$^{-1}$ and its residence time in the water is 210 days. Use estimated data from this chapter to make the prediction. (This question includes some specific information making it somewhat different from the usual Fermi question. However, it still requires the development of other reasonable quantitative assumptions.)

---

## Chemical removal of nitrogen from wastewater

Because it is usually the limiting nutrient as far as eutrophication is concerned, phosphorus is in most cases the element of greatest concern in sewage treatment processes. Nitrogen, however, is also very important both because it also contributes to eutrophication and because, in the ammonium ion form, it can react with and reduce the concentration of dissolved oxygen in receiving waters. Therefore, there is increasing concern to develop methods to bring about its removal.

Unlike phosphorus, nitrogen has no easily produced insoluble forms. It therefore cannot be removed by a simple chemical precipitation procedure. When the nitrogen is present in the form of ammonium ion, there are, however, other relatively simple chemical procedures that can be employed for its removal. One method is to raise the pH of the water with calcium hydroxide (which precipitates out the phosphorus as described above) and then to pass the waste stream through a stripping tower where it is aerated in order to strip out the gaseous ammonia.

$$NH_4^+ \text{ (aq)} + OH^- \text{ (aq)} \rightarrow NH_3 \text{ (g)} + H_2O \qquad (16.13)$$

Subsequent readjustment of the pH to an acceptable value is then required. An alternative approach that does not involve pH modification is to remove ammonium ions from the neutral solution by ion exchange using the natural ion exchanger clinoptilolite or synthetic exchangers.

When disinfection by chlorine or bleach (NaOCl) is the final step in treatment of the water, almost complete removal of ammonium nitrogen is effected. As noted earlier, chlorine reacts with water to form hypochlorous acid.

$$Cl_2 \text{ (g)} + 2H_2O \rightarrow HOCl \text{ (aq)} + H_3O^+ \text{ (aq)} + Cl^- \text{ (aq)} \qquad (16.1)$$

Hypochlorous acid is the effective chlorinating species under treatment conditions, and reacts with ammonium ion to form mono-, di-, and trichloramines as follows.

$$NH_4^+ \text{ (aq)} + HOCl \text{ (aq)} \rightarrow NH_2Cl \text{ (aq)} + H_3O^+ \text{ (aq)} \tag{16.14}$$

$$NH_2Cl \text{ (aq)} + HOCl \text{ (aq)} \rightarrow NHCl_2 \text{ (aq)} + H_2O \tag{16.15}$$

$$NHCl_2 \text{ (aq)} + HOCl \text{ (aq)} \rightarrow NCl_3 \text{ (aq)} + H_2O \tag{16.16}$$

In the presence of carbon adsorption filters the chloroamines undergo a heterogeneous surface reaction that produces nitrogen gas as one of the products. Chlorination of wastewater is often the last step before it is discharged into the receiving water body. Besides functioning to convert ammonium into other forms, its principal purpose is as a disinfectant that kills most of the pathogenic organisms that have survived earlier steps in the treatment process.

## 16.5 Advanced microbiological processes

Biological treatment processes for removal of both phosphorus and nitrogen have been suggested as alternative methods to the chemical treatment systems thus far proposed. This requires specific design features of the treatment in order to remove one or both of the two nutrients. Of the proven biological phosphorus removal technologies, several include steps in which the wastewater passes through a series of reactors where the sequence of environments goes from anaerobic to anoxic to aerobic (Fig. 16.4). Recall that an anaerobic environment is one that contains no free oxygen ($O_2$), while an anoxic environment is devoid of oxygen in both free and combined ($NO_3^-$, $SO_4^{2-}$) forms. The purpose of the succession of zones is to allow for the development (growth) of microorganisms that have the needed specific traits. Bacteria of the genus *Acinetobacter* are particularly known for their ability to assimilate phosphorus during their growth.

### Microbiological phosphorus removal

The biological phosphorus removal process is a modification of the activated sludge process (Fig. 16.4). It begins with wastewater moving into two basins in series where anaerobic and then anoxic conditions are maintained. In the basins, the normal population of facultative decomposers causes the release of acetate and other fermentation products from soluble organic matter in the waste stream, by way of fermentation reactions such as those described in the previous chapter. The fermentation products are substrates that are favoured by *Acinetobacter* and other

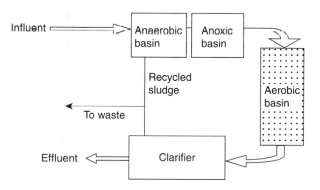

**Fig. 16.4** Design for advanced biological phosphorus removal. Note the sequence of anaerobic, anoxic, and aerobic treatments before settling of the phosphorus-enriched sludge in the clarifier. Primary and tertiary treatments can be additional components to this system. The open arrows show the path of the wastewater stream, while the closed arrows follow the separated sludge.

phosphorus-storing organisms, and they stimulate growth of these species compared with other microorganisms in the general population. Thus, the anaerobic conditions provide an environment that results in population selection and development of the phosphorus-storing species. Without the anaerobic phase, such organisms would be present in only very small amounts in the activated sludge.

The pre-stimulated waste stream then passes into an aerobic basin. Upon entry into this oxygen-rich zone (equivalent to an aerator in a conventional ASP), the selected microorganisms efficiently take up the soluble phosphorus from the wastewater stream. The phosphorus-enriched sludge is removed and treated in the usual way.

### Microbiological nitrogen removal

Nitrogen can be removed from wastewater using a microbiological 'nitrification–denitrification process', in which the two types of nitrogen reactions occur in series. In the first step, the environment is aerobic, favouring microbial nitrification to convert aqueous ammonium ion to nitrate.

$$NH_4^+ \text{ (aq)} + 2O_2 \text{ (aq)} + H_2O \xrightarrow[\text{Nitrobacter}]{\text{Nitrosomonas}} NO_3^- \text{ (aq)} + 2H_3O^+ \text{ (aq)} \qquad (16.17)$$

The second step requires anaerobic conditions. Denitrification occurs in the presence of denitrifying bacteria such as *Pseudomonas, Micrococcus, Serratia*, and *Achromobacter*. These bacteria mediate reduction processes in which sugars and carbohydrate and other organic compounds in the wastewater serve as electron donors and the soluble nitrate is converted into nitrogen gas (reaction 16.18). In many cases, smaller amounts of ammonium ion and volatile ammonia are produced as by-products.

$$4NO_3^- \text{ (aq)} + 5\{CH_2O\} \xrightarrow[\text{bacteria}]{\text{denitrifying}} 2N_2 \text{ (g)} + 4HCO_3^- \text{ (aq)} + CO_2 \text{ (aq)} + 3H_2O \qquad (16.18)$$

In reaction 16.18, the electron donor is represented by the generic organic form {CH₂O}. While wastewater biomass can act as the reducing agent for denitrification, in practice, methanol is often added as a supplement, because it ensures more rapid and complete reaction under anaerobic conditions.

A simple version of a biological treatment system based on the nitrification–denitrification process is shown in Fig. 16.5.

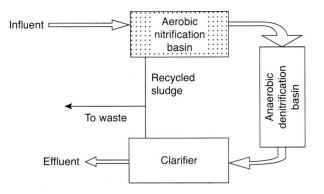

**Fig. 16.5** A schematic diagram of a biological nitrification–denitrification process for nitrogen removal from wastewater. Note that, in some ways, this system reverses the processes of the biological phosphorus removal system. The open arrows show the path of the wastewater stream, while the closed arrows follow the separated sludge.

Despite the fact that both phosphorus and nitrogen can be removed by microbiological processes, the two systems cannot easily be made compatible due to the sequence of environments required. It is, however, possible that a combined chemical / biological system for sequential phosphorus and nitrogen removal could be achieved. The introduction of an anaerobic denitrification basin (required in the N process) following the aerobic basin from the phosphorus process (in Fig 16.4) would create a system that allows for nitrification / denitrification to take place, following the earlier phosphorous treatment. However, removal of the two nutrients using these and other biological wastewater treatment plants requires very careful design and control of the operating conditions. Even in a biological system, it is frequently necessary to add alum (or another coagulant) at an appropriate location within the plant, in order to enhance phosphorus and solids removal.

One of the major drawbacks to the biological systems is the requirement for several large tanks where the various environments and processes are established. There are, therefore, substantial construction and operational costs as well as the cost of land for the complete facility.

Recent advances in design of the nitrogen biological system have allowed for the nitrification and denitrification processes to proceed simultaneously in an essentially aerobic single reactor.[4] When proper coagulation conditions are maintained, oxygen-consuming nitrification (reaction 16.17) occurs in the well-aerated outer layer of the floc at a rate sufficient to limit penetration of oxygen into the centre of the floc. The interior anaerobic zone then becomes a site where heterotrophic bacteria can carry on their denitrifying activity (Fig. 16.6).

---

**Main point 16.3** Wastewater treatment systems are commonly referred to as primary, secondary, or tertiary, depending on the type of processes employed during the treatment. Organic matter, phosphorus, and nitrogen are constituents that need to be removed from urban wastewater in order to reduce the impact of the effluent on the receiving water body and the surrounding ecosystem. Chemical and biological processes for removal of these contaminants have been developed.

---

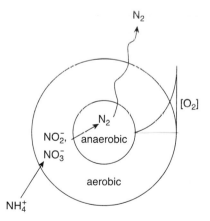

**Fig. 16.6** Diagrammatic representation of a floc showing an aerobic outer shell where ammonia may be nitrified to produce nitrite and nitrate, and an anaerobic core where denitrification takes place. Note the decline in oxygen concentration moving into the centre of the floc.

---

[4] Zeng, J.H., Z. Yuan and J. Keller, Improved understanding of the interactions and complexities of biological nitrogen and phosphorus removal processes, *Rev. in Environ. Science & Bio / Techn.* 3 (2004), 265.

## 16.6 The final products after treatment of wastewater

There are two endproducts of wastewater treatment. Water itself is the principal product and, if the treatment has been successful, much of the suspended solids, dissolved organic matter, nutrients and harmful compounds have been removed. Other soluble inorganic chemicals are, at least to some degree, also removed. Before final release into a receiving water body or for irrigation, chlorination or another disinfection step is employed in order to kill organisms that could be toxic to aquatic life or ultimately to humans and other animals.

There are other chemicals that are increasingly being found in municipal wastewater and that are only partially removed by the conventional processes described here. An increasingly ubiquitous number of these chemicals are in the pharmaceuticals and personal care products (PPCP) category. Shampoos, for example contain siloxanes, compounds that are persistent and have been found to have subtle effects on the development of fish embryos. These and many others will require increasing attention in order to protect the health of the natural surroundings and of humans themselves.

Specialized treatment processes are required for wastes from most manufacturing industries and, increasingly, there is a requirement that this treatment be carried out before the industrial water is discharged into the municipal system.

As noted above, the second endproduct of wastewater treatment, the solid sewage sludge, must also be disposed of, usually on land. It is important to ensure that the sludge does not contain harmful chemicals, especially metals (which will not be destroyed during treatment). When its safety can be assured, sludge can be a useful resource for application to soil. When well mixed with the surface soil, it can increase the organic matter content and provides a source of a small amount of nitrogen and other nutrients that are important for plant growth.

An alternative to disposal on land is to incinerate the dewatered sludge, producing useful heat and / or electricity from the energy generated. It is clearly important to ensure proper control of the atmospheric emissions from this operation.

> **Main point 16.4** There are two products resulting from wastewater treatment—the water itself, which has been treated to render it suitable for discharge, and a solid sludge that must undergo further disposal.

---

### ADDITIONAL RESOURCES

1. Binnie, C. and M. Kimber, *Basic water treatment, 4th edn*, IWA Publishing, London; 2009.

2. Newman, M.C. and M.A. Unger, *Fundamentals of ecotoxicology*, Lewis Publishers, Boca Raton, Florida; 2002.

3. Wright, D.A. and P. Welbourn, *Environmental toxicology*, Cambridge University Press, Cambridge; 2002.

4. United Nations Economic and Social Commission for Western Asia, *Wastewater treatment technologies—a general review*. http://www.escwa.un.org/information/pubdetails.asp, accessed October, 2009.

5. Veolia Environment, *A site visit of the Hague wastewater treatment plant*, http://www.wastewater treatment-hague.veoliaenvironnement.com/technologies/air-treatment.aspx, accessed November, 2009.

## PROBLEMS

1. A city's drinking water supply has a pH of 7.2 after filtration and chlorine is to be added as the agent for disinfection. Calculate the ratio of HOCl to $OCl^-$ in the solution after chlorine is added. What can be done to improve conditions for disinfection?

2. Estimate the total organic carbon (TOC) concentration of wastewater whose chemical oxygen demand (COD) is 500 mg $L^{-1}$ ($O_2$). What fraction of the total (dissolved and particulate) solids content of 720 mg $L^{-1}$ is then made up of organic material? Assume the organic fraction can be represented as {$CH_2O$}. Of what might the remaining solids consist?

3. An influent wastewater stream contains 330 mg $L^{-1}$ organic matter (both suspended and soluble) and 27 mg $L^{-1}$ ammonium ion (as N). Calculate the total BOD. What assumptions is it necessary to make?

4. A wastewater contains 7.2 mg $L^{-1}$ phosphorus and is treated with 15 mg $L^{-1}$ aluminium in the form of an alum solution. Assume that phosphorus is precipitated out by reaction 16.10 and that any excess added aluminium forms $Al(OH)_3$. Calculate the mass of inorganic sludge produced in one day in a plant treating 20 000 $m^3$ of wastewater.

5. After wastewater treatment in the activated sludge process, nitrogen is mainly in the form of ammonia and ammonium ion. Plot the fraction of nitrogen that is in the ammonia form (and therefore strippable by air purging) as a function of pH (at 25°C) over the pH range 6 to 10.

6. A settling tank treating $5.5 \times 10^6$ L of water per day has dimensions as follows: length, 12.2 m; width, 7.0 m; depth, 3.5 m. Calculate the detention time for water in the tank. Calculate the minimum particle size (expressed as diameter of spheres) that could settle in the tank.

7. Use reaction 16.12 to calculate the daily volume of $Ca(OH)_2$ solution of concentration 6.4 g $L^{-1}$ Ca, required to treat pH-adjusted wastewater containing 6.1 mg $L^{-1}$ phosphorus. The treatment plant processes 27 000 $m^3$ of water each day. Assume a safety factor (fractional excess) of 2.

8. Nitrification of ammonium ion is one of the steps during biological nitrogen-removal processes. In wastewater whose pH and alkalinity are 7.2 and 156 mg $L^{-1}$ (as $CaCO_3$), respectively, a concentration of 7.8 mg $L^{-1}$ (as N) ammonium ion is present before the process begins. Calculate the pH and alkalinity after nitrification has gone to completion, assuming this to be the only reaction that affects the pH.

9. A wastewater treatment plant produces sludge containing 1800 kg of dry organic solids each day. Assuming the generic formula {$CH_2O$} for the solids and complete anaerobic digestion by reaction 16.9, calculate the fuel value of the generated methane in joules, barrels of oil, and kilowatt hours.

# Part C
# The terrestrial environment

How little I know of this world
Deeds of men, cities, rivers,
Mountains, arid wastes,
Unknown creatures, unacquainted trees!
The great Earth teems
And I know merely a niche.
*Rabindranath Tagore, 1913*

# Chapter 17
# The terrestrial environment

The focus of this chapter is on the terrestrial environment, the land portion of the Earth, including the inorganic and organic materials from which the soils are formed. Here, we will discuss:

- the physical and chemical soil formation processes
- the important role played by soil organic matter in general, and more specifically its role in carbon sequestration in forests
- the significant nature of soil as a complex three-phase system.

In the public consciousness, environmental issues usually focus on subjects of water or air, which we ingest into our bodies every day. But a complete study of environmental chemistry must also include important subjects concerning the terrestrial environment. Keep in mind that humans live on land and we obtain most of our food through land-based agriculture. And beyond agriculture, the integrity of the Earth's total environment depends on the terrestrial compartment. There is an intimate connection between the solid material and the air and water that surround it and are within it.

The terrestrial environment is comprised of rocks and soil and the living mater associated with these. Rocks and soil together are referred to as the lithosphere, and it is this compartment of the environment that is of concern to us in this section of the book. The land area makes up 29% of the total area of the Earth's surface; the terrestrial environment is subdivided into the categories shown in Table 17.1.

The reactivity of solid materials that make up the lithosphere is, in large part, dependent on the particle size of these materials. Soils—finely divided materials with a relatively large surface area exposed to air or water—react much more readily with environmental agents than do massive rocks. This is one reason why, in the short term, the chemistry of the terrestrial environment must be concerned mostly with soils. Furthermore, soils (defined in a broad sense) cover approximately 80% of the Earth's land mass and the next largest fraction is made up of snow and ice in the Arctic and Antarctic regions. Exposed rocks make up only about 5% of the terrestrial surface.

Over the long period of the history of science, there have been many reasons for studying soils, but two stand out as being of direct and practical importance to humans and other living species. One of these is that soils are the principal plant growth medium and form the basis of agriculture and forestry. Especially over the last century, the subject of soil science has developed around these interests. Soil scientists have been particularly concerned with nutrient cycles and the relations between elements and compounds in the soil and their uptake by plants. They are also interested in other agronomic factors such as the interconnections between composition of soil materials, particle size, soil texture, and the resultant soil physical properties. Soil science

**Table 17.1** The Earth's terrestrial environment[a].

|  | Area / $10^6$ km$^2$ | %of total |
|---|---|---|
| Total land area | 148 | 100 |
| Ice-covered land | 17.2 | 12 |
| Arable land | 14.1 | 10 |
| Pasture and meadow | 33.8 | 23 |
| Forest | 39.5 | 27 |
| Other[b] | 43.4 | 29 |

[a] Data are taken from the *FAO production yearbook*, Vol. 39, Food and Agriculture Organization, Rome; 1986, with updated values for arable land and pasture from the *FAO statistical yearbook*, 2007–8, and for forest cover from the *FAO global forest resources assessment, 2005*.
[b] The 'Other' category includes mountainous land, deserts, and some land that is potentially available for pasture or direct food production.

of this type—related to plant production—is a highly developed science with a body of knowledge and practice that has increased in volume and sophistication over more than a century. We should also not forget the contribution of farmers, agricultural practitioners around the world, to our knowledge of the behaviour, including chemical behaviour, of soils.

Another separate but connected reason for studying soils is more recent and relates to the fact that soils play a major role as an environmental agent. Key links in the global carbon, nitrogen, phosphorus, and sulphur cycles, as well as in many others, involve soil chemical processes. Organic matter decomposition, nitrification, denitrification, phosphorus fixation, and sulphide oxidation are just a few of these processes. There are two broad environmental implications related to such reactions. On the one hand, soil chemical processes affect the nature and amount of elements that are released to the hydrosphere and atmosphere. In turn, soils are the locus of inputs from other compartments of the environment and are themselves affected by processes there. For example, rainfall chemical composition is altered when rain percolates through the soil, perhaps draining into rivers and lakes, or maybe reaching the water table and becoming part of the groundwater reservoir. Through the interactions, the soil properties are also altered by their encounter with the rain.

Soil reactions involved in the global element cycles have been occurring over long periods of geological time, although human activities in recent years have perturbed some of them to a significant extent. There are other specific types of chemical reactions that only recently have come to be played out in the soil environment. A good example is related to the application of pesticides in agriculture. Pesticides are used to control insects, weeds, or pathogenic microorganisms on growing crops. These chemicals degrade over time and their movement and rate of degradation are, in part, determined through their interactions with the soil. Another example is the disposal of waste materials—municipal garbage, mine tailings, sewage sludge, sometimes even known toxic materials—in the soil environment. In other words, soils are an important environmental agent and a study of the environmental properties of soils is important along with studies of their agronomic properties.

**Main point 17.1** Soils cover a significant fraction of the Earth's terrestrial environment. They provide a supporting medium for many forms of life and are the basis of agriculture and forestry. Soils are also an important environmental agent, acting as a 'filter' for aqueous and solid inputs including rain, municipal wastes, pesticides, and other chemicals.

## 17.1 Soil formation

### Soil mineral matter

A starting point for examining properties of soils is to consider the natural processes by which they are formed from exposed rocks on the Earth's surface. These processes have been going on throughout the history of the Earth and continue to operate in the present time.

Table 17.2 lists elements in order of abundance in the Earth's crust—the crust being defined as the approximately 32-km thick layer on the surface of the planet. Information about concentrations of other elements is given in Appendix B.1. Exposed soil material is a thin surface layer covering part of the crust; the great mass of crustal material consists of the underlying igneous and metamorphic rocks.

The complex processes by which rocks on the Earth's surface are transformed into soils are collectively known as weathering and we can consider these processes under two general headings.

### Physical weathering

- Physical weathering results in the breakdown of massive rock materials into smaller aggregates, eventually fine enough and sufficiently well developed that they are considered to be soils.[1] This is brought about in several ways.

- Freeze–thaw in temperate regions leads to enhancement of fracturing along the natural fracture planes of rocks. Further enhancement occurs when water enters into cracks and expands (by 9%) upon freezing.

**Table 17.2** Percentage abundance of the predominant elements and their oxides[a] in the Earth's crust.

| Element | Abundance / % | Oxide | Abundance / % |
|---------|---------------|-------|---------------|
| O  | 46.6 |          |      |
| Si | 27.7 | $SiO_2$    | 58.2 |
| Al | 8.13 | $Al_2O_3$  | 15.2 |
| Fe | 5.00 | $Fe_2O_3$  | 7.2  |
| Ca | 3.63 | $CaO$      | 5.1  |
| Na | 2.83 | $Na_2O$    | 3.8  |
| K  | 2.59 | $K_2O$     | 3.1  |
| Mg | 2.09 | $MgO$      | 3.5  |

[a] Note that geologists and other Earth scientists often report elemental analyses in terms of oxides, even though most are not found predominantly in the oxide form. Besides silicon, the only non-metal present in significant concentrations in many rocks is oxygen. Therefore the sum of concentrations of all metal oxides should be close to 100%. In the listing here, we have accounted for approximately 96% of the total mass.

---

[1] Soil can be defined as the layer of unconsolidated particles that are derived from weathered rock and organic material, and that contains water and / or air in the void space. Soil covers the upper surface of much of the Earth and supports plant life. This is not the only definition of the term 'soil', and an excellent discussion of different concepts is given in Chapter 1 of Additional Resources 1.

- Fire causes expansion of rocks but, due to their low thermal conductivity, the surface expands much more rapidly than the rock below. This causes strains that are released by fracturing. The same effect occurs to a lesser extent due to daily temperature variations.

- Deposition of salts in already fractured rocks may occur. If these salts have a thermal expansion coefficient higher than that of the surrounding rock, diurnal and seasonal temperature changes can cause pressure-induced breakage. Similarly, some minerals deposited in cracks, especially clays like montmorillonite, undergo major expansion on hydration. This can also result in cleavage of massive material.

- Abrasion due to erosion by wind or water, but particularly under the influence of glaciation, further subdivides rocks into smaller aggregates.

- Once finely divided material has formed on a rock surface it is subject to transport by wind, water, or ice. The pressure release resulting from erosion causes expansion in the vertical plane at right angles to the normal horizontal fracture planes, and this results in further cleavage.

- Where plants, particularly trees, are growing, penetration of roots into crevices generates pressure sufficient to cause breakage.

The net effect of these and other processes is that massive rocks are broken down into material with smaller particle size and correspondingly larger specific surface area. We have seen that particle size is very important in terms of both physical and chemical behaviour of solids in any environment.

## Chemical weathering

Occurring simultaneously with the above and other *physical processes*, a wide range of *chemical reactions* also takes place. Some chemical reactions are associated with the activity of micro- and macroorganisms, while others are purely abiotic.

### Hydrolysis reactions

Hydrolysis is a general term for processes in which water is an essential reactant. Various types of hydrolysis reactions play a major role in the weathering of rocks and minerals, many of which are some type of aluminosilicate compound. A typical example related to the igneous mineral, orthoclase feldspar, leads to the production of the clay mineral kaolinite. In this reaction, orthoclase is referred to as a *primary* mineral, while its weathering product, kaolinite, is a *secondary* mineral.

$$2KAlSi_3O_8 \text{ (s)} + 2H_3O^+ \text{ (aq)} + 7H_2O \rightarrow Al_2Si_2O_5(OH)_4 \text{ (s)} + 4H_4SiO_4 \text{ (aq)} + 2K^+ \text{ (aq)} \quad (17.1)$$

orthoclase                                    kaolinite                    silicic acid

In the reaction, silicon is released as silicic acid from the soil at the same time as the clay mineral is produced. Further hydrolysis causes additional desilication, with the end result being the production of aluminium hydroxide in the form of the mineral gibbsite.

$$Al_2Si_2O_5(OH)_4 \text{ (s)} + 5H_2O \rightarrow 2Al(OH)_3 + 2H_4SiO_4 \text{ (aq)} \quad (17.2)$$

kaolinite                    gibbsite        silicic acid

This particular weathering sequence is especially important in the humid tropics due to the abundant rainfall and elevated temperatures. At pH values between 2 and 9, silicic acid ($pK_{a1} = 9.7$) remains in the fully protonated form and has an aqueous solubility of approximately 150 mg L$^{-1}$. Its prolonged solubilization and the accompanying mineral transformations lead to formation of red soils depleted in silica but rich in kaolinite and hydrated aluminium (and also iron) oxides. Depending on their specific properties, such soils are referred to as laterites, latosols, or oxisols.

We can be more general, and consider primary aluminosilicate minerals as a group. They are weathered to produce any one or more of the secondary clay minerals as summarized in reaction 17.3

$$\text{aluminosilicate (s)} + H_3O^+ \text{(aq)} + H_2O \rightarrow \text{clay mineral (s)} + H_4SiO_4 \text{(aq)} + \text{cation (aq)}$$

$$(17.3)$$

Both the specific and the general reaction show that water and hydronium ions are the agents of weathering. There are several natural sources of hydronium ion including carbon dioxide released in the soil by microbial respiration, and low molar mass acids that are products resulting from the decomposition of organic matter in the soil. In some situations, the natural sources of hydronium ion are augmented by anthropogenically derived acids, most especially nitric acid from fertilizers, and sulphuric and nitric acids that are present in rainfall in certain regions of the Earth.

Because hydrolysis involves uptake of hydrogen ion and release of alkali and alkaline earth cations by the mineral, it is not surprising that the pH of a slurry (called the abrasion pH) of secondary weathering products tends to be somewhat lower than that measured on the corresponding finely divided primary minerals (Table 17.3). For the unweathered primary minerals, pH is largely controlled by the small solubility of the metal cations in the ambient solution. During weathering, these cations are removed and the slightly acidic properties of the clay mineral products tend to control the acid–base properties of the slurry.

### Acid–base reactions

Besides silicate minerals, major areas of the Earth's solid surface, especially those with exposed sedimentary deposits, are made up of carbonate rocks, principally limestone ($CaCO_3$) and dolomite (($Ca, Mg)CO_3$). In areas of carbonate bed rock, physical weathering releases finely divided particles of the same composition and these are incorporated in the soil.

Through chemical weathering processes, the carbonate minerals act as a carbon dioxide sink and, at the same time, moderate the associated acidity of any water that percolates the soil.

$$(Ca, Mg)CO_3 + H_2O + CO_2 \longrightarrow Ca^{2+}(Mg^{2+}) + 2HCO_3 \qquad (17.4)$$

As is evident from the equation, this process is another type of hydrolysis reaction.

### Chelation reactions

Beyond simple hydrolysis, chelation contributes substantially to chemical weathering. Iron and aluminium are two major elements whose solubility in uncomplexed forms is extremely small (see Example 13.1). Yet evidence of substantial dissolution of these elements has been observed in many soils. This has been shown to be due to the formation of soluble organic complexes. For example, in forested areas of the temperate zone, translocation of iron and aluminium from

**Table 17.3** Abrasion pH of minerals.

| Mineral | Type[a] | Abrasion pH |
|---------|---------|-------------|
| Olivine | P | 10–11 |
| Augite | P | 10 |
| Oligoclase | P | 9 |
| Orthoclase | P | 8 |
| Quartz | P | 6–7 |
| Kaolinite | S | 4–7 |

[a] P, Primary minerals; S, Secondary minerals

upper to lower layers of the soil is associated with solubilization by chelation with ligands that are derived from soil organic matter. Calculations and measurements on soil solutions both show that more than 90% of the soluble iron and aluminium is present in the form of organic complexes. The ligands that form complexes with metal ions include inorganic species, but more important are ones derived from biological sources often associated with decomposition of dead plant and microbial material. The anions of acids like citric acid complex strongly with iron and aluminium and, in addition, are sources of hydrogen ions. Humic and fulvic acid are also important multidentate ligands capable of complexing with metals and accelerating chemical weathering (Section 13.3).

Consider a rock weathered by air and water in the surface environment. Table 17.4 lists concentration ratios calculated as the concentration of particular elements in the altered material on the weathered rock surface divided by the concentration of the same element in a freshly cut interior portion. Where these ratios differ greatly from one, it indicates a major accumulation (large ratio) or loss (small ratio) of the element during weathering. For bare exposed rock, the weathering occurs to minimal thickness, and the chemical changes are relatively small—due largely to abiotic hydrolytic processes as indicated above. For rock covered with lichen, weathering occurs to a much greater depth and there is loss of most of the calcium and an accumulation of iron in the weathered layer. Lichens are microorganisms that have both algal and fungal components. The fungus attaches to the rock or soil surface and extracts nutrients that are used by the algal component. The algae are then able to carry out photosynthesis and produce carbohydrates and other organic molecules, some of which have chelating properties. These chelating agents enhance the rate of chemical weathering. It is evident that the chemical changes brought about by chelating reactions are different from those associated with hydrolysis alone.

### Redox reactions

Oxidation / reduction is a third important chemical weathering process. Oxidation occurs when primary minerals containing oxidizable elements in low oxidation states are exposed to the atmosphere. The resulting increase in oxidation state disturbs the charge balance of the mineral, and loss or gain of other elements in the compound may occur to maintain neutrality. The result is the formation of secondary minerals with different properties. Oxidation of iron in

**Table 17.4** Concentration ratios for Hawaiian basalt that had erupted in 1907[a]. Exposed rock was free of any lichen, while part of the same deposit was covered with the lichen *Stereocaulon volcani*.

|  | Exposed rock | Lichen-covered rock |
| --- | --- | --- |
| Weathering rind thickness / mm | <0.002 | 0.142 |
| Element mass concentration ratio: weathered / fresh |  |  |
| Fe | 1.21 | 6.36 |
| Al | 0.47 | 0.58 |
| Si | 1.20 | 0.21 |
| Ti | 0.97 | 0.27 |
| Ca | 1.24 | 0.004 |

[a] From Jackson, T.A. and W.D. Keller, A comparative study of the role of lichens and 'inorganic' processes in the chemical weathering of recent Hawaiian lava flows, *Am. J. Sci.*, **269** (1970), 446–66.

the primary mineral biotite produces the 2:1 layer clay mineral vermiculite. Accompanying the oxidation of iron (II) to iron (III), potassium is lost. An idealized conversion is shown.

$$K_2(Mg, Fe(II))_6(AlSi_3O_{10})_2(OH)_4 \rightarrow Mg_{0.84}(Mg_{5.05}, Fe(III)_{0.9})(Si_{2.74}Al_{1.26}O_{10})_2(OH)_4 \quad (17.5)$$

biotite                                                   vermiculite

Other iron-containing minerals provide us with a further example. The redox behaviour of hydrous iron oxide is responsible for many changes in mineral chemistry. Under oxidizing conditions, the stable form of iron oxide is $Fe_2O_3$ in the form of haematite or in hydrated forms, given the simple formula $FeOOH$, and known as goethite or limonite. These minerals are highly insoluble, but under reducing conditions they may dissolve as iron (II) species, only to be redeposited in the same or some other location under an oxidizing regime. When considering $pE$ / pH diagrams, we observed that this behaviour is thermodynamically favoured (Fig. 10.P.1).

In most discussions about redox chemistry in soils, the focus is centred on iron because of its abundance (Table 17.2) in the Earth's crust. However, the chemistry of many other elements (e.g. manganese, arsenic, chromium) has important redox components too. Even if a particular element is not directly subject to oxidation or reduction, its environmental behaviour can be affected indirectly by changes in the form of a major element such as iron. Frequently the iron is present in an amorphous form as hydrous iron (III) oxide, and other metals as well as non-metals are adsorbed or coprecipitated as impurities in the solid. If conditions change, and the iron is reduced to a soluble form, the coprecipitated elements are released and can simultaneously go into solution.

### Other reactions

Hydration reactions contribute to physical weathering, as shown above, but also bring about chemical alteration of certain minerals. The hydration reactions between haematite and goethite and between gypsum and anhydrite are cases in point.

$$Fe_2O_3 + H_2O \rightarrow 2FeOOH \quad (17.6)$$
haematite              goethite

$$CaSO_4 + 2H_2O \rightarrow CaSO_4 \cdot 2H_2O \quad (17.7)$$
anhydrite             gypsum

Ion-exchange reactions alter the nature of 'available' elements at the surface exchange sites of soil colloids. Actual structural changes can also be associated with ion exchange. These are particularly important with respect to clay minerals, where replacement of one interlayer ion by another alters the interlayer spacing and therefore the chemical and physical properties of the clay. The 2:1 clay mineral illite is very similar to montmorillonite, except that much of the isomorphous substitution is due to $Al^{3+}$ replacing $Si^{4+}$ in the tetrahedral layer (Section 14.6). Furthermore, the interlayer cation, $Ca^{2+}$, is replaced by $K^+$ that is bonded strongly with the adjacent tetrahedral layer. These chemical changes give rise to a different type of clay with significantly altered properties—having much reduced cation-exchange capacity (CEC) and resistance to physical expansion associated with wetting.

All of the physical and chemical weathering processes we have discussed can occur simultaneously or in overlapping sequences and, as they take place, they influence one another. The net result of the weathering of rocks over extended periods is the production of a finely divided material that is classified as a soil. The rate of soil formation can be as high as 1 or 2 cm depth of new soil (1300 to 2600 t ha$^{-1}$) in 100 years under warm, moist climates, but is much less in temperate dry regions. This is in contrast to rates of erosion that may be hundreds of times greater. It is not unusual for wind erosion to carry away several centimeters of surface soil over one year and large amounts may be lost in a single catastrophic water event.

The processes of soil formation and development are continuous and on-going. The 'endproduct' is not a stable material but is itself subject to further change, due either to natural or anthropogenic factors. The kinds of changes that are occurring at the present time are a subject of great interest to those chemists who study the terrestrial environment.

## Soil organic matter

Beginning with rocks alone, weathering reactions produce an essentially inorganic mineral soil and this mineral matter does, in fact, usually make up the larger proportion of actual soils. Soils are, however, not purely inorganic materials. A second component is organic matter, which often makes up < 1 to 5 per cent by weight (weight%) of the soil mass. Much higher concentrations of organic matter occur in important types of soils such as those derived from peat and the surface layers of forest soils. In contrast, desert soils are almost purely inorganic. In any case, even when the proportion of organic matter is small, it plays a disproportionately significant role in many soil physical and chemical processes.

The primary sources of organic matter are plant tissue—roots of growing and dead plants and litter such as leaves and branches that have fallen on the surface. All these components are found in various stages of decay ranging from fresh material to a product that is highly decomposed by microbial and chemical processes creating a structure and appearance that is vastly different from that of the parent material. The biomass of soil microorganisms—bacteria, fungi, actinomycetes, and protozoa—is itself an important contributor to the organic fraction of soil and the millions of microorganisms typically make up between 0.05 and 0.5% by mass (dry weight) of the top 15 cm of soil in a natural setting. Other small soil animals, principally earthworms, contribute typically one-quarter of this fraction to the biomass. While earthworms are only a minor component of the biomass, they are important because of their ability to enhance aeration and water movement, and to translocate organic matter within the surface soil.

A typical breakdown of the composition of fresh plant residues is summarized in Fig. 17.1. The carbohydrate in most plant tissues is largely in the form of cellulose, hemicellulose, and, to a lesser extent, starch. All are polymers of (different forms of) glucose.

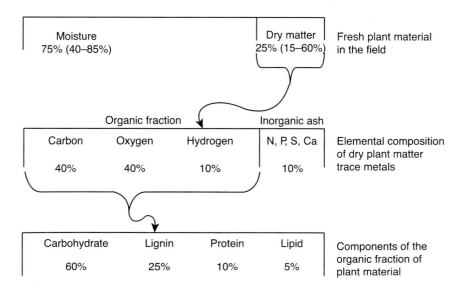

**Fig. 17.1** Composition of fresh plant materials. The figure shows components of fresh material, of dry matter, and of the organic fraction, and typical concentrations of each.

### Decomposition processes

Decomposition is a complex chemical and microbiological process. Considering all the classes of chemical compounds in plant residues, decomposition occurs at varying rates, depending mostly on the suitability of each compound as a food source for microorganisms.

The rate of disappearance of organic substrates can be approximated by first-order kinetics.

$$dM/dt = -kM \qquad (17.8)$$

where $M$ is the mass of organic material and $k$ is the value of the rate constant (in units of day$^{-1}$). The rate constant depends on the nature and state of subdivision of the substrate material as well as on environmental conditions, particularly temperature and moisture. Typical values (Additional Resources 2) for temperate climates are given below.

| | | $k/d^{-1}$ |
|---|---|---|
| Sugars, starches, simple proteins | rapidly decomposed | 0.2 |
| Complex proteins | | |
| Hemicellulose | | 0.08 |
| Cellulose | | |
| Lignins, lipids | slowly decomposed | 0.01 |

Heterotrophic organisms incorporate carbon compounds into their own biomass. Degradation and resynthesis produce new soluble or insoluble organic compounds. Inorganic elements from the plant residues are mineralized as species such as $NH_4^+$, $NO_3^-$, $H_2PO_4^-$, $SO_4^{2-}$, $Ca^{2+}$, and $K^+$ and / or immobilized in the microbial structure. Simultaneously, some of the original organic carbon is evolved by respiration as $CO_2$. The release of $CO_2$ by decaying plants to some extent offsets removal of this greenhouse gas by photosynthesis. There is considerable interest and, in some cases, controversy about the relative significance of these two opposing processes in various ecosystems.

The microbial processes of decomposition are represented in the following sequence that relates to a situation where a soil is supplied with a fresh input of undecomposed organic matter. Examples of this would include a forest soil receiving leaf fall during the autumn season and a field where a 'green manure' crop is grown and then ploughed under.

| | |
|---|---|
| Soil | stable microbial population |
| ↓ | |
| Soil + fresh organic matter | zymogenous microorganisms multiply causing decomposition of the new supply of OM; simultaneous degradation to smaller molecules and synthesis of humic material; $CO_2$ evolution |
| ↓ | |
| Soil + microbial biomass + partially decomposed OM | as fresh plant material is consumed, development of OM substrate deficiency; reduced microbial activity due to death of microorganisms; return to autochthonous population |
| ↓ | |
| Humus | relatively stable, chemically and microbiologically |

The product resulting from these complex processes is soil humic material, called humus, whose chemistry we have examined earlier. This is a relatively stable material, but over years is itself slowly decomposed to produce carbon dioxide as a final product. In a tropical region, where deforestation is brought about prior to use of the land for agriculture, loss of organic matter may be particularly rapid for the first few years—anywhere from 20 to 60% of the original amount may be lost each year. Once it reaches a 'normal' baseline value, further losses occur at a much slower rate.

In Section 12.3, it was pointed out that the carbon content of humic material is frequently found to be approximately 60%. Fresh plant material, depending on source, typically contains 0.5 to 5% nitrogen giving it a carbon to nitrogen (C:N) ratio between 12 and 120. Table 17.5 contains examples of %N and C:N ratios for a few selected materials. (Additional C:N ratios are given in Table 19.3.)

Soil microorganisms incorporate nitrogen and carbon into their bodies as proteins and structural materials. Upon death of the microbes, the incorporated elements remain in the soil. However, some additional organic carbon serves as a respiratory substrate, and is released to the atmosphere as carbon dioxide. As a consequence, the organic carbon content of the soil OM decreases while the nitrogen content remains the same. Therefore, the C:N ratio declines as decay proceeds and ultimately reaches a stable value in the humus between 10 and 13.

---

### Example 17.1 **Nitrogen content in humic materials**

What is the relation between nitrogen and humus (organic matter) content in a soil?

Using a typical C:N ratio of 12 for humus and an OM:C ratio of 1.7 (100 / 60), the ratio of organic matter to nitrogen (OM:N) is

$$OM:N = 12 \times 100 / 60 = 20 \text{ (or } 100 / 5)$$

This means that there is approximately 5% nitrogen in the humic material.

---

The 5% value is typical of good-quality humus as, for example, that studied in the southern San Joaquin Valley, California, which contains about 5.0 to 5.5% nitrogen. (For reference see footnote to Table 17.5.)

Being a source of nitrogen, an important plant nutrient, is one of several essential contributions that the organic matter makes to the soil, even though it may constitute only a few per cent or less of the total mass. Other important contributions, physical, chemical and biological, will be made clear in subsequent chapters.

**Table 17.5** Nitrogen contents and C:N ratios for various materials.[a]

|  | % N | C:N |
|---|---|---|
| Plant tissues—herbaceous |  |  |
| Clover or bean leaf | 2.5–5.0 | 8–16 |
| Grass shoots (young) | 3–4 | 12–15 |
| Grass shoots (mature, yellow) | 0.5–2.0 | 20–80 |
| Animal and microbe |  |  |
| Insects, mammals | 6–12 | 5–10 |
| Fungi (grown on leaf) | 3–4 | 11–16 |
| Bacteria | 4–12 | 5–14 |

[a] Data in table reproduced from: Peacock B., *Balancing the nitrogen budget*, The University of California Cooperative Extension, Tulare County, http://cetulare.ucdavis.edu/pubgrape/ng296.htm, accessed November, 2009.

## 17.2 **Carbon dioxide sequestering in forests**

Issues of organic matter decomposition and soil formation are closely connected with the important subject of carbon dioxide sequestering during growth of forests. Of the approximately 8.5 Gt of carbon that are released into the atmosphere each year by anthropogenic processes, about 3.7 Gt remain in the atmosphere, while the other 4.8 Gt are removed (sequestered) into the oceans (2.3 Gt) and terrestrial systems (2.5 Gt). The principal terrestrial removal process is thought to occur by forest growth. In some countries, active efforts are being made to reduce greenhouse gas build-up in the atmosphere by promoting extensive reforestation, along with other improved forestry and agricultural practices.

While this is a laudable effort, the extent to which it contributes to carbon dioxide removal depends on knowing the total $CO_2$ budget—uptake by photosynthesis, release by respiration, decay and fire—in order to determine the net sequestering effect. Reliable estimates of uptake values in various ecosystems have been made (Appendix A.3) and they show that the greatest accumulation rates are achieved where there is a plentiful supply of nutrients and water. For example, a tropical forest covering 100 000 ha has the capability of accumulating 0.83 Mt of carbon each year, while the rate of uptake in a boreal forest is less than half this value.

One of the many unknowns is the rate of decay of biomass that has accumulated in the ground underlying the forest. The terrestrial biomass may make up about 20% of the total biomass in a forest system. Except for freshly deposited leaves and other litter, the organic matter is mostly a relatively stable humic material that decomposes very slowly. Yet, in situations where a forest has been cleared or burned over, the exposed organic-rich soil is subject to more rapid rates of decomposition. As a consequence, an area that was formerly a sink for carbon dioxide becomes a source. In general, a sustainable practice is to replant a forested area soon after tree harvesting.

## 17.3 **Soil formation as a complex process**

While we have considered the processes of genesis of inorganic and organic components of soil in separate categories, it is important to realize that there are interactions between the components. The effects of individual processes and properties are not necessarily additive. For example, iron (III) and aluminium (III) oxide minerals in soils are able to specifically adsorb and therefore immobilize some metals that are present in the soil solution. However, there are situations where soil humic material is strongly associated with the same oxides. Coatings of humic substances on the oxide surfaces inhibit the specific adsorption reactions, and mobility of metal ions is therefore greater than in soils containing uncoated minerals. Therefore, a detailed knowledge of the amount and mineralogy of oxide minerals in soils is alone insufficient to predict the degree of metal retention by the soil. We will come across other cases where interactions produce anomalous effects of this sort.

The complex combinations of various physical and chemical (abiotic and biotic) processes acting on parent rocks and organic residues produce a finely divided material that we call a soil. Reflecting the origin and nature of the parent materials and the geological environment within which they were weathered and transported, soils have a wide range of properties. The variation is evident on both the horizontal and vertical scale, and over time there are continuing changes as well. Common features of all soils include:

- soil is a finely divided, heterogeneous, porous material made up of mineral and / or organic matter;
- pore spaces are filled with air and / or water, depending on moisture conditions.

Figure 17.2 shows the principal features common to most soils.

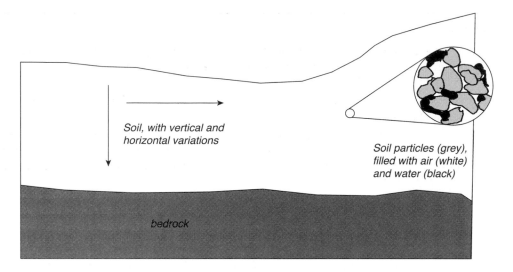

*Soil, with vertical and horizontal variations*

*Soil particles (grey), filled with air (white) and water (black)*

*bedrock*

**Fig. 17.2** Soil, the surface layer of much of the terrestrial environment. A three-phase mixture, it consists of finely divided organic and inorganic particles and pore spaces filled with water and / or air. The soil is highly heterogeneous in both the vertical and horizontal dimensions.

Soils are therefore a three-phase mixture, and a complete description of their role in environmental reactions and cycles requires a consideration of the interactions involving all three phases.

We pointed out earlier that formation processes are on-going and therefore soils are continuously being generated and altered around the Earth. But we must underline the fact that losses of soil—by erosion, by being covered with hard surfaces due to urbanization, and by degradation—occur much more rapidly than formation. The following two chapters will describe in more detail some of the ecological contributions of soils—contributions that are essential for the support of life as we know it on the planet.

> **Main point 17.2** Soil is formed over periods of geological time by a combination of physical and chemical (including biotic) processes. The processes act on consolidated rocks to produce the mineral component and on plant and animal material to produce the organic portion of the soil. The product of these weathering processes is a finely divided material that contains air and / or water in the pore spaces. The nature and composition of any soil continues to change and can be significantly affected by inputs from human activities.

## ADDITIONAL RESOURCES

1. Brady, N.C. and R.R. Weil, *The nature and properties of soils*, 14th edn., Prentice Hall, Upper Saddle River, New Jersey; 2008.

2. Paul, E.A., ed., *Soil microbiology, ecology and biochemistry*, 3d edn., Academic Press, San Diego, California; 2006.

3. Blume, W.E.H., Functions of soil for society and the environment, *Rev. Environ. Science Bio / Techn.*, **4** (2005), 75–79.

4. McBride, M.B., *Environmental chemistry of soils*, Oxford University Press, New York; 1994.

## PROBLEMS

1. Consider the terrestrial portions of the global nitrogen cycle (Fig. 15.9). What human activities have perturbed aspects of this cycle, and in what ways would other processes have to adjust to keep the cycle in balance?

2. The weathering of some clay minerals containing potassium is affected by growing plants. Explain why this might be so.

3. The surface soil in the humid tropics is often depleted in silica and enriched in iron and aluminium oxides. In contrast, the surface mineral layer of a forest soil in a temperate region may be devoid of significant iron and aluminium minerals and have a high concentration of silica. Suggest an explanation.

4. In an experimental forest, a sample of leaves is weighed over a one-year period and is found to lose 83% of its mass. Assuming first-order kinetics, what is the rate constant for their decomposition.

5. Using the rate constants provided in the text for degradation of various components of soil organic matter, compare the time required for degradation to a 10% residual, of tree leaves (mostly cellulose and hemicellulose) with small twigs (mostly lignin).

6. Estimate the total organic nitrogen in 1 tonne of soil that contains 1.5% organic matter (as humus).

7. What are the possible feedback processes, related to the forest carbon budget, resulting from increasing mixing ratios of carbon dioxide in the atmosphere?

8. The concentration of copper in the surface organic layer of a forest soil is 37 ppm, and in the underlying mineral layer is 17 ppm. The bulk densities of these two layers are 0.36 and 1.22 g mL$^{-1}$, respectively. The very low density of the surface material occurs because it consists in large part of partially degraded organic material and there is very little of the heavier mineral matter. Which of these two layers has the higher copper concentration per unit volume?

9. The concentration of cadmium in the top 15 cm of soil in a field (often called the plough layer) can be estimated by taking a representative sample, dissolving the soil, and analysing it by atomic absorption spectroscopy with electrothermal atomization. The concentration is found to be 0.78 ppm. Suppose that dewatered (solid) sewage sludge containing 22 ppm cadmium is added at the rate (mass per area) of 3 t ha$^{-1}$. Assuming sludge is well mixed within the plough layer, calculate the new average concentration of cadmium within this part of the soil. The bulk density of the soil is 1.1 g mL$^{-1}$.

# Chapter 18
# Soil properties

The focus of this chapter is on the properties of soils, especially those properties that relate to their environmental significance. We will accomplish this by:

- investigating the fundamental physical and chemical properties of soils
- introducing the concept of soil profiles and examining three specific profiles, a Spodosol, Alfisol, and Vertisol
- noting the environmental services provided by soils and discussing how a soil's ability to provide these services relates to its properties
- discussing problems related to perturbed soil chemistry.

Everyone recognizes the importance of soil for the sustenance of human and other living communities on the Earth. Soils are the fundamental resource supporting agriculture and forestry, as well as contributing to the aesthetics of a green planet. They are also a reservoir from which minerals are extracted and onto which solid wastes are sometimes disposed. In addition, soils act as a medium and filter for the collection and movement of water. And, critically for life as we know it, by supporting plant growth soil is a major determinant of atmospheric composition and therefore of the Earth's climate. All these contributions may be summarized by saying that soils provide a variety of essential environmental services. For these and many other reasons, it is of greatest importance to maintain the integrity of this essential resource. In this chapter, we will spend time examining soil properties in the environmental context.

Following our approach in discussing soil genesis, it is helpful to consider the properties of soil in terms of both their physical and chemical characteristics. We will come to realize that there is a close and overlapping relation between these two aspects.

> **Main point 18.1** Besides being a medium for the production of food, fibre, and fuel, soils are also an important environmental agent. Soils interact with chemicals produced within or added to soils in ways that depend on both their physical and chemical properties.

## 18.1 Physical properties

### Particle size

Every soil is made up of many particles that have varying chemical composition and size. Using established physical separation methods, a sample of soil can be subdivided into fractions on the basis of particle size. Particle size is a primary physical property and there are several classification schemes used to assign names to each fraction. Figure 18.1 uses the scheme recommended by the International Society of Soil Science (ISSS).

By this convention, soil is arbitrarily defined as material with particle sizes less than 2.0 mm. In fact, many soil analysis protocols begin with a sieving (2.0 mm) step to separate soil from the larger-sized aggregates. Within the material defined as soil, there are three primary size categories: in order of decreasing particle size—sand, silt, and clay. Keep in mind our earlier classification of colloids as particles smaller than 10 μm in size. Soil particles in the clay and small silt fractions are therefore in the colloid category. Keep in mind also that these colloidal particles often have surfaces that have high chemical reactivity.

The physical and to some degree chemical properties of soils are highly dependent on particle size.

Sand is 'light', easily worked, has good drainage (but poor water retention), and is readily aerated. Chemically, the most important components of sand are usually primary minerals such as quartz and feldspars; these are relatively inert and poor sources of nutrients. Soils rich in clay are 'heavy', difficult to work, and have poor drainage and aeration (but retain water strongly). Recall again the material on clay minerals in Section 14.6. Be careful to note the distinction between *clay minerals*, with their unique chemical properties and *clay-sized particles*. Clay-sized material may be comprised of some combination of the clay minerals themselves, organic matter, and finely divided primary minerals and hydrous iron and aluminium oxides.[1] All of these fine materials have large surface areas and take part in ion exchange and / or adsorption reactions. Therefore, the clay-sized particles are able to interact with and retain nutrients; in this way, they may be productive plant growth media.

## Texture

Soil texture is a collective term that defines a real soil by the proportion of different particle-size components. Texture nomenclature is based on triangular diagrams such as the one shown in Fig. 18.2.

To use the diagram, consider a case where a soil contains 35% clay, 30% silt, and 35% sand-sized material. Beginning at 35 on the clay axis, a line is drawn parallel to the sand axis. Similarly, a line at 30 on the silt axis is drawn parallel to the clay axis. The point of intersection is in the region known as clay loam and the soil is so named.

Soils that are desirable from an agricultural perspective often fall in the middle region of the diagram. Such soils have the beneficial physical properties of lighter soils and tend to be quite easy to work, but also have moderate moisture-retaining ability, and chemical reactivity due to the contribution of the clay-sized materials. Note that the region named 'clay' is large and

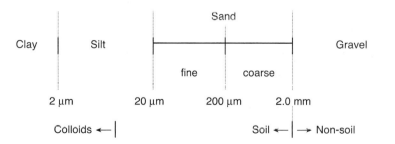

**Fig. 18.1** Soil particle-size classification according to the International Society of Soil Science.

[1] Note that in soil terminology, clay may refer to either a size category or a class of minerals. To distinguish between these two usages, it is best to speak of clay-sized materials and clay minerals, respectively.

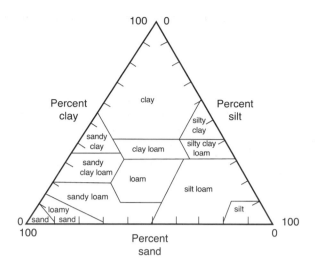

**Fig. 18.2** Soil texture triangle.

includes soils that may contain even less than 50% clay. This is because finely divided material has a dominating effect on the behaviour of many soils.

## Density

The density of soil depends on the minerals or organic components that make up its composition. The density of individual particles—called *particle density*—is considerably less than 1 g mL$^{-1}$ for organic matter, and greater than 5 g mL$^{-1}$ for some metal oxides or even > 7 g mL$^{-1}$ for less-common minerals such as metal sulphides. Many widely distributed soil minerals, including quartz, the feldspars, and clay minerals, have densities that fall within the approximate range from 2.5 to 2.8 g mL$^{-1}$ and this may be taken as a good estimate of the particle density of most mineral soils.

The *bulk density* is the density of the soil as it is found in the field. Bulk density takes into account the pore spaces between particles, and is therefore smaller than the particle density. In the case of mineral soils situated within a depth of 1 m or so from the surface, the bulk density is often between 1.2 and 1.8 g mL$^{-1}$ for materials containing a good proportion of sand. For soils with higher clay content, the density falls in a slightly lower range from 1.0 to 1.6 g mL$^{-1}$. When both particle and bulk densities are known, the pore space of a soil may be readily calculated.

---

### Example 18.1  **Pore space in soil**

Consider a silt loam soil that has a particle density of 2.65 g mL$^{-1}$ and a bulk density of 1.50 g mL$^{-1}$. Calculate the pore space in the soil.

$$\text{Pore space (\%)} = 100 - \frac{\text{bulk density}}{\text{particle density}} \times 100$$

$$= 100 - \frac{1.50}{2.65} \times 100 = 43\%$$

---

Sandy soils have pore space values, ranging from 35 to 50%, those of fine textured soils range from 40 to 60%, and soils very high in organic matter may have more than 60% pore space. Subsoils—soils at depth—tend to be more highly compacted than surface soils and usually have low organic matter contents so that the pore space is often only about 25% of the total volume in the field.

## Structure

A fourth physical property of soils is structure. Structure is a term used to describe the way in which individual particles are aggregated together to form larger units. These units are readily observed in the field when digging or ploughing the soil. Because organic matter has the ability to act as a cementing agent, it plays a key role in developing and maintaining structure in soils. Other chemical entities and the nature of the parent material are also determining factors. Structure is a field characteristic and is usually destroyed when a soil sample is collected and brought to a laboratory for analysis. An abbreviated and self-explanatory set of definitions of common structures is given in Table 18.1.

## Permeability

Permeability, also called hydraulic conductivity, is a measure of the ability of soil to allow the flow of water vertically and horizontally. For example, when it rains heavily, a permeable soil transports the water rapidly downwards. The rate of lateral movement of groundwater in the subsoil is also determined by permeability. Many soils have downward permeabilities in the range 1 to 5 cm h$^{-1}$; rates smaller than 0.5 cm h$^{-1}$ are considered to be very low, while rates greater than 15 cm h$^{-1}$ are high.

For soils with little or no structure, permeability is largely a function of soil texture. The coarse desert sands of the Arabian Peninsula have very high permeabilities, while some fine-grained alluvial soils in Iraq have permeabilities in the very low categories.

Structure is also a major determinant affecting permeability, especially in the case of clays. Soils with well-developed structure are more permeable than structureless ones. The volume of space created by structural boundaries may be small compared with the total pore volume of the soil, but it can be the major route for water transport. The greater volume of pore space between the fine-grained particles *within* structures then plays only a minor role in movement of water. Water also moves via cracks that have been created by drying, by earthworms, or by roots of plants.

Permeability tends to decrease with depth in a soil, but the changes are often irregular and discontinuous. *Hardpans* are regions of impermeable layers in the soil profile. In agricultural soils, a *hardpan* or *ploughpan* is frequently observed just below the depth (typically about 15 cm) to which the plough or cultivator extends, and is due to the effect, over many years, of heavy machinery or animals passing over the same area of land. It has been observed that soils with a normal bulk density of 1.2 to 1.5 g mL$^{-1}$ may be compacted to densities greater than 2 g mL$^{-1}$ in this way. Occasional deep ploughing or, alternatively, the use of no-till

**Table 18.1** A partial listing of soil structure types.

| | |
|---|---|
| Structureless | No observable aggregation or no definite orderly arrangement around natural lines of weakness |
| Granular | Individual particles held together to form crumb-like aggregates |
| Block-like | Soil particles arranged around a point and bounded by flat or rounded surfaces—includes blocky to granular–spheroidal forms |
| Plate-like | Soil particles arranged around a horizontal plane and bounded by flat horizontal surfaces |
| Prism-like | Soil particles arranged around a vertical axis and bounded by relatively flat vertical surfaces |

technologies are ways employed to minimize the problem of interrupted water percolation in such situations.

In studying soil chemistry it is important to realize that there are many ways by which the physical properties of soils influence their environmental (including chemical) behaviour. Permeability is probably the most significant single factor in this regard. Soils with low permeability can become waterlogged. Oxygen is consumed in the stagnant water and air exchange is limited, eventually creating a reducing environment. Likewise, permeability affects transport of chemicals through the soil. Any complete description of movement of chemicals in the soil requires knowledge both of the distribution of the chemical between soil and water (as described in Chapter 14) and the soil hydrological properties. The latter properties are determined by the various physical attributes of the soil.

> **Main point 18.2** Soil texture and structure affect the flow of water through the soil, but also determine the extent of solid surface available to react with chemicals moving through the pores.

## 18.2 Chemical properties

We can now move on to look at some chemical properties of soils that are relevant to our overall study of environmental chemistry.

### Total elemental composition and organic matter content

A logical starting point is to look at a total analysis of the elements. Ranges of values for the major and some minor elements present in the mineral portion of soils are included in Table 18.2. The chemical composition is determined by the nature of the starting materials from which the soil was formed and by the processes that it has undergone over time.

For many purposes, compositional data are not particularly useful because they do not indicate whether the elements are found as components of the mineral lattice or are associated with surface adsorption phenomena. In the former case, especially for silicate minerals, such

**Table 18.2** Ranges of values for major and minor element composition of the mineral component of soils[a].

| Major elements[b] | | Minor elements[c] | |
|---|---|---|---|
| Element | % | Element | mg kg$^{-1}$ |
| Si | 30–45 | Zn | 10–250 |
| Al | 2.4–7.4 | Cu | 5–15 |
| Fe | 1.2–4.3 | Ni | 20–30 |
| Ti | 0.3–0.7 | Mn | ~400 |
| Ca | 0.01–3.9 | Co | 1–20 |
| Mg | 0.01–1.6 | Cr | 10–50 |
| K | 0.2–2.5 | Pb | 1–50 |
| Na | tr–1.5 | As | 1–20 |

[a] The average crustal abundance of all the elements is influenced by concentrations in near surface rocks and is given in Appendix B.1.
[b] tr, Trace.
[c] A more complete list of minor elements in whole soil is given in Table 18.8.

elements would be substantially 'insoluble' except over geological time, and therefore would not play a significant role with respect to plant growth or in terms of most environmental processes. Those elements that occupy ion-exchange sites on soil particles or those that are weakly adsorbed are, however, much more available to be taken up by plants and other organisms or to be transported through the soil dissolved in water. In fact, the terms 'available' and 'extractable' are widely used by Earth scientists and there are operational definitions and analytical procedures for determining the fraction of elements that is present in available forms. We will look at this later.

Organic matter is a relatively small component (by mass) in most soils; usually it is intimately mixed into and, in some cases, chemically associated with the mineral soil. Although there are exceptions, a general trend whereby the organic content decreases with depth is usually observed. Table 18.3 reports values for organic *matter* (OM) in a variety of near-surface soil situations. Analytical methods usually give a measure of organic carbon (OC) content. Based on a widely held assumption that soil organic matter is about 60% carbon, multiplying by a factor of $100 / 60 = 1.7$ is used to convert OC to OM values.

Soil organic matter is broadly classified into humic material (HM) and non-humic material. When we considered some characteristics of HM in Chapter 12 we noted that the HM is derived from either terrestrial or aquatic sources and that there are some general differences between the two types. Terrestrial HM comes primarily from plant residues—fallen leaves and branches from trees in a forest, dead grass and other meadow plants after a dry season or winter, or crop residues left from harvest and perhaps ploughed into the surface layers of the soil. The partially decomposed and resynthesized matter is relatively stable and contributes to good soil structure and adds to the cation-exchange capacity of the soil.

The non-humic organic fraction of the soil is made up of many compounds, among them the complex polysaccharides, cellulose, hemicellulose, and pectin. It also includes smaller carbohydrate molecules that have been released as these organic polymers decompose. Because the simple monosaccharides are readily soluble and also serve as a favoured food source for micro flora and fauna, their concentration in soil is not great. High molar mass polysaccharides, on the other hand, are relatively stable. Like the HM macromolecules they act as cementing agents, contributing to soil structure by binding individual particles together into larger units. Proteins and amino acids are also an excellent food source for soil microorganisms, but small amounts can be measured at any time in the soil. Lipids are a minor component, but are relatively resistant and long-lived in the soil environment.

Much more resistant to decomposition are the lignins, complex phenolic polymers that are responsible for the toughness of plant parts. Lignins are a major component of the woody components of plants. Tannins are also plant-derived polyphenols that are, in some cases, highly resistant to degradation. For this reason lignins and tannins, like HM, can make up a significant portion of the organic matter especially in forest soils.

Over time, some of the non-humic material is incorporated into the HM fraction and thus, in some ways, the non-humic compounds have a transient existence.

**Table 18.3** The range of organic matter content in most soils.

| Soil type | Organic matter content (%) |
|---|---|
| Temperate agricultural soils | 1–5 |
| Tropical agricultural soils | 0.1–2 |
| Forest soils (surface horizons) | >10 |
| Peat soils | >20 |

**Table 18.4** Percentage of total metal extracted from soil using two extractants[a].

| Extractant | Percentage extracted from soil | | | | |
|---|---|---|---|---|---|
| | Co | Ni | Cu | Pb | Cr |
| NH$_4$OAc (pH 7) | 0.3 | 0.86 | 1.1 | 0.51 | 0.37 |
| DTPA | 1.3 | 2.7 | 5.0 | 8.7 | 0.11 |

[a] From McLeod, S.E. and G.W. vanLoon, A study of elemental contamination in Orchard Park, Kingston, Ontario, *Ontario Geography*, **17** (1981), 91–104.

Although it makes up only a relatively small proportion of the total soil mass, organic matter is the key component in carrying out important soil processes. These include regulation of water retention and flow, development of soil structure through binding of individual particles, being a source of carbon and energy for microorganisms as well as a source of nitrogen and other nutrients for both microbes and plants, and providing exchange sites where other nutrients reside and are available for uptake by growing plants. Refer back to the description of humic acid in Chapter 12, where you will see that humate molecules have charged sites that can interact with nutrient ions in the soil solution. This last process is what we will discuss next.

## Available elements

As noted above, the available element is the portion of the element in the soil that can take part in a range of chemical and biological reactions. To some extent, the terms *available* and *extractable* element are synonymous. To measure extractable elements, a soil is shaken in an aqueous solution containing chemicals chosen to displace that portion of the element that is supposed to be readily available for uptake by growing plants. Depending on the nature of the soil and other environmental circumstances, many extractants and conditions have been recommended for this purpose. Most methods make use of the law of mass action in order to displace the cations that are present on the exchange sites (as when a 1 mol L$^{-1}$ ammonium acetate solution is used for extraction) and / or involve chelating agents that combine with the immobilized element to assist its dissolution (as in extraction with DTPA (diethylenetriaminepentaacetate) solution). The use of complexing agents tends to dissolve a larger fraction of the transition metals (Table 18.4).

In an agricultural context, measuring concentrations of available elements is used to assess their potential to provide plant nutrition. In environmental situations, such data are used to predict mobility of the element and possible toxicity or other adverse consequences.

## Exchange capacity

### Cation-exchange capacity

In Chapter 14, we showed that components of sediments and soils have the ability to electrostatically adsorb positive ions on to their surface. The clay minerals and organic matter are particularly important in this regard. A quantitative assessment of the ability to interact with cations is the cation-exchange capacity (CEC)—one of the most commonly measured properties of soils. Example 18.2 illustrates a calculation of cation-exchange capacity in a soil. The example makes use of one of the many standard methods[2] for determination of exchange cation concentration. It involves extraction by aqueous ammonium chloride at a concentration of 1 mol L$^{-1}$ and with pH adjusted to 4.5.

[2] Nommick, H., Ammonium chloride–imidazole extraction procedure for determining titratable acidity, exchangeable base cations and cation exchange capacity in soils. *Soil Sci.*, **118** (1974), 254.

### Example 18.2 **Soil cation-exchange capacity**

Suppose 1.00 g of soil is extracted with 100 mL of the ammonium chloride solution in order to displace the exchange cations from the soil. After filtering, the dissolved calcium, magnesium, potassium, and sodium are determined by atomic absorption spectroscopy. The hydrogen ion is determined by titration.

The concentrations of the principal cations found in the extractant solution are as follows:

| | |
|---|---|
| Ca | $30.3\ \mu g\ mL^{-1}$ |
| Mg | $3.2\ \mu g\ mL^{-1}$ |
| K | $2.2\ \mu g\ mL^{-1}$ |
| Na | not detected |
| $H_3O^+$ | $2.60\ mmol\ L^{-1}$ |

To determine the cation-exchange capacity, it is necessary to calculate the total positive charge associated with these ions. In most cases, no other cation would be present in significant amounts. All the cations in solution were originally present on exchange sites and therefore their positive charge must be equivalent to the number of negative sites in 1 g of soil.

For calcium, the positive charge associated with the exchange complex is determined as follows. The 'concentration' of positive charge due to this divalent cation is equal to

$$2 \times 30.3\ \mu g\ Ca\ mL^{-1} / 40.1\ g\ mol^{-1} Ca = 1.51\ \mu mol\ mL^{-1}$$

This concentration is multiplied by 100 mL to give the total positive charge due to exchangeable calcium that had been extracted from 1 g of soil. The final result is, by convention, expressed in terms of cmol ($10^{-2}$ mol) of positive charge per kg (cmol (+) $kg^{-1}$) of soil.

$$1.51\ \mu mol\ (+)\ mL^{-1} \times 100\ mL \times \frac{1000\ g}{kg\ soil} \times \frac{10^{-4}\ cmol}{\mu mol}$$

$$= 15.1\ cmol\ (+)\ kg^{-1}\ soil$$

Note that this standard unit of cmol (+) $kg^{-1}$ has the same value as the traditional and still widely used unit of meq $(100\ g)^{-1}$.

A similar calculation can be done for the other elements that were extracted from the exchange complex, and the corresponding values are, for Mg, 2.6 cmol (+) $kg^{-1}$ soil and for K, 0.6 cmol (+) $kg^{-1}$ soil.

For $H_3O^+$ the charge on 1 g of soil is calculated by multiplying the concentration of hydronium ion (mol $L^{-1}$) by the volume of solution, 0.100 L. This is then converted into the cmol (+) $kg^{-1}$ as shown here.

$$\frac{2.6 \times 10^{-3}\ mol\ L^{-1}}{1.00\ g\ soil} \times 0.100\ L \times \frac{1000\ g}{kg} \times \frac{10^2\ cmol}{mol}$$

$$= 26.0\ cmol\ (+)\ kg^{-1}\ soil$$

The cation-exchange capacity (CEC) is then the sum of the values for each ion.

$$CEC = (15.1 + 2.6 + 0.6 + 26.0)\ cmol\ (+)\ kg^{-1}$$

$$= 44.3\ cmol\ (+)\ kg^{-1}\ soil$$

The actual cations that occupy exchange sites depend on the nature of the soil particles as well as on other environmental circumstances. The two alkaline-earth elements and two alkali metals used in the example are almost always the four most important exchange metal cations, and for many soils the quantitative order of importance is Ca > Mg > K > Na. Only in particular circumstances do other metal cations contribute significantly to the CEC value.

Aluminium ion is prominent on exchange sites of some acidic soils. Being trivalent, it is tightly bound to the negatively charged sites. Aluminium is not a plant nutrient and in fact excessive amounts may be toxic to plants.

Hydronium ion makes a variable contribution to the CEC. Under acid conditions, whether natural or anthropogenic, a large proportion of the cation-exchange sites may be occupied by hydronium ions, while in neutral and alkaline soils their contribution is negligible. Where a considerable fraction of the exchange sites is taken up with hydronium ion, the nutrient-supplying capacity is diminished. Furthermore, such soils have reduced capacity to neutralize additional acidity. A measure of proportion of metals (compared with hydronium ions) on exchange sites is the base saturation, defined as

$$\text{Base saturation} = \frac{\text{number of exchange sites occupied by Ca + Mg + K + Na}}{\text{total number of exchange sites}} \times 100\%$$

A large base saturation value is usually considered to be a desirable feature. Soils that have a small CEC and / or a small base saturation value are therefore especially susceptible to acidification by either natural or anthropogenic inputs.

---

### Example 18.3 **Base saturation of cation exchange sites**
Using the data from Example 18.2, base saturation = sum of the concentrations of $Ca^{2+}$, $Mg^{2+}$, and $K^+$ on exchange sites in the soil (the value for $H_3O^+$, 26.0 cmol (+) $kg^{-1}$ soil, is not included), which is then divided by the CEC and multiplied by 100%.

$$\text{Base saturation} = \frac{15.1 + 2.6 + 0.6}{44.3} \times 100\%$$
$$= 41\%$$

---

Widely ranging CEC values occur. Sandy soils, low in organic matter, have very small CEC values, often less than 5 cmol (+) $kg^{-1}$, while soils high in certain clays and / or organic matter may have a CEC of more than 100 cmol (+) $kg^{-1}$. The two factors that dominate the magnitude of the CEC are the nature and content of clay minerals and the content and degree of decomposition of the organic matter. Both clays and OM have negative sites in their structures and these are the source of most of the cation-exchange capacity in soils. This is clear from the Table 14.6 showing CEC values for individual constituents of soil and sediment. Some typical values of whole soil cation-exchange capacity are given in Table 18.5. In some cases the soil name itself enables prediction of a very approximate CEC value.

**Table 18.5** Cation-exchange capacity (CEC) values / cmol (+) $kg^{-1}$ for a variety of selected surface soils.

| Surface soil | CEC / cmol (+) $kg^{-1}$ |
| --- | --- |
| Kentville sandy loam (Nova Scotia, Canada) | 10 |
| St. Quintin peat (Quebec, Canada) | 155 |
| Darlington clay loam (Manitoba, Canada) | 50 |
| Baker Lake sand (NWT, Canada) | 18 |
| Sassafras sand (New Jersey, USA) | 2 |
| Lipa clay loam (Luzon, Philippines) | 36 |
| Nabha silt loam (Punjab, India) | 9.8 |
| Gezirach clay (Sudan) | 52 |

Finally, it is important to be aware that there is a very extensive literature on methods for determining CEC, and their significance. One should not embark on a study that interprets and compares CEC values without becoming familiar with some of this literature.

### Soils of variable charge

What has been described in the previous section implies that the CEC (or the surface charge) has a fixed value for a given soil. In some cases this is true—particularly for soils of temperate regions, which tend to contain clay minerals such as montmorillonite. Most of the charge on these clays is fixed charge—that is, the charge is independent of the environment in which the clay is found. Tropical soils, however, frequently contain other minerals in the clay-sized fraction, particularly the hydrous oxides of iron and aluminium and these minerals are characterized by having a pH-dependent variable surface charge. In Chapter 14, we discussed the basis of this variable-charge phenomenon. As we showed there, an important defining characteristic of these soils is the pH at which positive and negative charges are balanced. This is called the $pH_0$.

The cation-exchange capacity of variable-charge soils depends, not only on the soil properties themselves but also on the pH of the surrounding environment. When the ambient pH is lower than $pH_0$ of the soil component, then that material is in protonated form, and its net surface charge is positive. By this means, the soil develops a certain amount of anion-exchange capacity. Conversely, when the ambient pH is greater than $pH_0$, the net surface charge of the variable-exchange material is negative and contributes to the cation-exchange capacity of the soil. There are many important implications of this. It means that natural and anthropogenic factors (for example, acid rain or nitrification of ammonium ion, both of which add acid to the soil environment) are able to alter the exchange properties of variable-charge soils in the field. It also means that measurements of exchange capacity should be carried out under conditions of pH that are similar to those of the intrinsic soil environment. For example, determining the cation-exchange capacity of a variable-charge soil whose field pH is 5.6 using pH 7.0 aqueous ammonium acetate as extractant would enhance the surface negative charge, resulting in an overestimate of the CEC value.

The pH is not the only environmental factor affecting the charge properties of variable-charge soils. The presence of other species that are specifically adsorbed (covalently bonded) to the soil minerals can substantially change the surface charge. Perhaps the most important example is phosphate, which is strongly retained by iron- and aluminium-rich soils. If the phosphate displaces water molecules from the hydrous oxides, it serves to increase the negative charge of the mineral (Fig. 18.3), adding to the CEC and also giving it a higher $pH_0$ value. For this reason, addition of phosphate may be a beneficial agronomic practice in order to increase the cation-exchange capacity of a highly weathered tropical soil. However, it may not have the expected effect of providing plant-available phosphorus nutrient, since this anion is covalently and tenaciously bonded to the soil and is not readily extracted by plant roots.

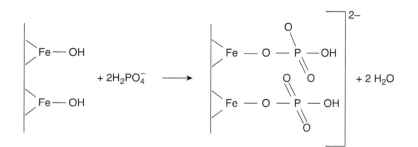

**Fig. 18.3** Specific adsorption of phosphate on a hydrous iron oxide surface. The negative charge and CEC are increased by this reaction.

It should be clear from this discussion that the charge properties of soils are one very important feature that influences their behaviour as environmental agents. Positive species present in the soil / water environment tend to be retained by the negative charges on the surface of many soil components, while negative species are more likely to be mobile, and therefore leachable. Importantly, however, charge is only one of the properties that affects retention. As is illustrated by Fig. 18.3, species like phosphate are able to form covalent bonds with specific soil components and these reactions are, to a large degree, independent of surface charge.

## Soil pH

The soil pH depends on the nature and history of the soil. Soils rich in carbonate minerals are usually somewhat alkaline. On the other hand, those containing large amounts of humic material are often, but not always, acidic. The acidity arises from microbial decomposition of the organic matter producing organic acids as a metabolic product, as well as from carbon dioxide released during respiration. Soils containing adsorbed iron and aluminium are also acidic as a consequence of the hydrolysis of the iron (III) and aluminium (III) cations. Prolonged leaching of the principal exchangeable metal cations and their replacement by hydronium ion also contributes to increasing acidity. In many insances, there is a fairly close positive correlation between the base saturation and the pH of the soil.

| **Terminology commonly used to describe the acid–base status of soils:** | | |
|---|---|---|
| pH | less than 4 | strongly acid |
| | 4 to 5 | moderately acid |
| | 5 to 6 | slightly acid |
| | 6 to 8 | neutral |
| | 8 to 9 | slightly alkaline |
| | 9 to 10 | moderately alkaline |
| | greater than 10 | strongly alkaline |

Soil pH is affected by changes in redox status. If a soil containing hydrous iron oxide is submerged, oxygen is consumed by the degradation of organic matter. Recall the sequence in Section 15.2. In the ensuing anaerobic environment, the iron (III) oxide is reduced as represented by the half-reaction shown.

$$Fe(OH)_3 \text{ (s)} + 3H_3O^+ \text{ (aq)} + e^- \rightarrow Fe^{2+} \text{ (aq)} + 6H_2O \tag{18.1}$$

Because hydronium ions are consumed in the reduction, there is an accompanying rise in pH of the soil. Flooded soils therefore tend to exhibit higher pH values than their upland counterparts.

**Main point 18.3** The chemical nature of soils is determined by the combination of mineral and organic matter that makes up the soil. Particularly important are the surface characteristics of soils, since these play a major role in determining the types of interaction—either retention or release of species—with the interstitial water.

## 18.3 Soil profiles

The description of soil's physical and chemical properties given in the previous sections does not take into account the fact that soils in the field are not a monolithic mass of unchanging composition. Rather, they are characterized by large spatial variability in both the horizontal and vertical dimensions. As one scans a landscape, there are obvious variations across the Earth's surface—differing landforms, visible changes in soil colour, and changes in land-use patterns. Although less visible, important variations in soil properties also exist on a much smaller scale of area. Likewise, there are important changes in physical and chemical properties with depth.

If one digs a pit, typically about 1 m or more deep, it is usually observed that there is a series of horizontal layers of changing colour and / or texture. Such layers are referred to as *horizons*, and the assemblage of soil horizons is called a *soil profile*. Numerous classification systems have been developed to categorize various types of profiles. Every system is based on a description of the chemical and physical properties of the horizons, but assumptions about the soil-forming (pedogenetic) processes that the soil has undergone may also be taken into account. One of the widely accepted systems, the Comprehensive Soil Survey System, was developed by the Soil Survey Staff of the US Department of Agriculture. Unlike most other taxonomic systems, this one is based exclusively on the observed properties of the soil in the field. It makes few or no assumptions about the processes that gave rise to those properties. The Comprehensive System has been widely adopted, but its adoption does not preclude using general and simpler terminology from other methods of classification. We cannot go further into this important and extensive subject here.

It is helpful, however, to be aware of some of the widely (but not universally) accepted terminology used to describe certain features of soil profiles. To do this, we will examine three 'typical' examples from very different environmental situations. We will describe some of the relevant features, introduce basic terminology, and indicate aspects of the environmental properties of these profiles.

## A Canadian Shield Spodosol

The first example is of a Canadian Shield Spodosol, also often called a Podzol. Spodosols are typical of humid, temperate regions, usually under forest cover; they have formed on relatively acidic parent material and are often characterized by being coarse-textured. A profile of such a soil from the boreal forest region of Montmorency, Quebec, in central Canada is shown in Fig. 18.4.[3]

Organic horizons[4] overlie the mineral soil. Two such horizons, named O1 and O2, are distinguished in the present example. The former contains largely undecomposed litter (leaves, needles, twigs) from the forest canopy, while the latter consists of the same material in a form that has been partially or completely humified so that the original morphological features of the litter may not be recognizable.

The processes of decomposition of organic matter release organic acids and generate carbon dioxide—both processes contribute to acidification of the soil. In this Spodosol, the pH is in the highly acidic category and, importantly, the acidity is due entirely to natural processes. Rainfall, even neutral rainfall, is thus acidified as it percolates through the coarse material. Moving downward into the upper horizon of mineral soil—the A horizon—it leaches out readily soluble components. Iron and aluminium solubility is greatly increased by the pore-water acidity, while silicon solubility is unaffected by pH in the acid region. Therefore, the upper horizon of mineral soil becomes depleted in iron, aluminium, and other soluble elements, while most of the silica remains behind. The effect is clearly visible in the soil profile in that the top mineral layer has a characteristic grey ashy appearance in contrast to the more highly coloured organic soil above and mineral soil below. The dark red-brown colour of the deeper mineral soil is, to a large extent, due to iron oxides as will be explained below. Horizons where the main process is leaching are called *eluvial* horizons—symbolized as E horizons. Note that the pH of the E horizon in this Spodosol is still low, although somewhat higher than in the overlying organic layers.

---

[3] Bentley, C.F. (ed.), *Photographs and descriptions of some Canadian soils*, The University of Alberta, Edmonton; 1979.

[4] Terminology used to define soil horizons is a complex and controversial subject that we cannot be concerned with here. In the occasional assignments of soil types employed in this book, we will use the US Soil Taxonomy system (Soil Survey Staff, *Keys to soil taxonomy*, AID, USDA, SCS, SMSS Technical Monograph No. 19, 5th edn, Pocahontas Press, Inc., Blacksburg, Virginia; 1992.)

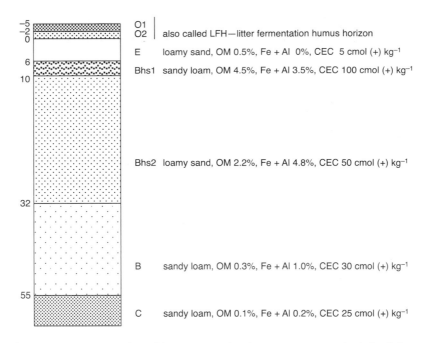

**Fig. 18.4** The Montmorency soil profile. Horizon depth in cm given to the left of the profile. Forest floor at 0 cm.

Underlying the E horizon is an accumulation or depositional layer, which is called the B horizon. In this layer, a portion of the compounds transported from above is precipitated from solution due to a variety of causes including a rise in pH, a decline in microbial activity, and adsorption. In a Spodosol, iron and aluminium are deposited as hydrous oxides in the B horizon along with organic matter. These compounds together give this part of the profile a characteristic dark brown or red-brown colour that contrasts with the overlying light-coloured layer. In the particular soil being described here, the horizon is named Bhs where the 'h' refers to humic material and the 's' designates iron and aluminium sesquioxides[5] that have accumulated in the region. Below the Bhs horizon, the B horizon continues but is much less affected by depositional processes.

The C horizon is a relatively unaltered subsoil that overlies and grades into the bedrock.

In a general sense, then, a Spodosol is typically found in temperate forested regions and is characterized by an organic layer overlying a mineral soil, the surface of which is depleted in iron, aluminium, and organic matter. These constituents are deposited below. The pH of a Spodosol is in the acidic to very acidic range but rises somewhat with increasing depth.

### A tropical Alfisol

The second soil profile example is taken from south India[6] and is in the general class of soils called Alfisols. The soil, named Tyamagondalu, is located near Bangalore in Karnataka state and is typical of many red soils in the southern Deccan Plateau. These soils have developed on weathered gneiss and are deep, clayey, and moderately acid to neutral. They support extensive cultivation of millet, some legumes, and rice. The profile is shown in Fig. 18.5.

---

[5] The prefix 'sesqui' means one and a half, indicating that there are 1.5 oxygens for each metal in iron (III) and aluminium (III) oxides.
[6] Murthy, R.S., L.R. Hirederur, S.B. Deshpande, and B.V. Venkata Rao (eds.), *Benchmark soils of India*, National Bureau of Soil Survey and Land Use Planning (ICAR), Nagpur; 1982.

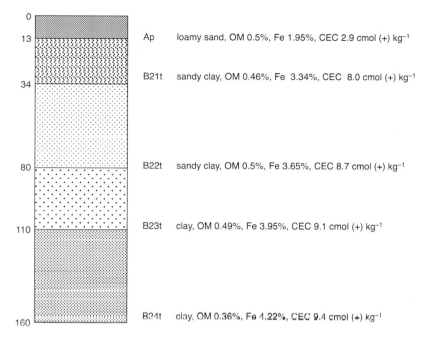

| | | |
|---|---|---|
| Ap | loamy sand, OM 0.5%, Fe 1.95%, CEC 2.9 cmol (+) kg$^{-1}$ |
| B21t | sandy clay, OM 0.46%, Fe 3.34%, CEC 8.0 cmol (+) kg$^{-1}$ |
| B22t | sandy clay, OM 0.5%, Fe 3.65%, CEC 8.7 cmol (+) kg$^{-1}$ |
| B23t | clay, OM 0.49%, Fe 3.95%, CEC 9.1 cmol (+) kg$^{-1}$ |
| B24t | clay, OM 0.36%, Fe 4.22%, CEC 9.4 cmol (+) kg$^{-1}$ |

**Fig. 18.5** The Tyamagondalu soil profile. Horizon depth in cm is given to the left of the profile.

The surface horizon is a loamy sand with pH about 6.8 and organic matter content of less than 1%. In the tropical environment of south India, the principal eluviated species are not iron and aluminium (these are essentially insoluble at neutral pH) as in Spodosols but rather silicon, whose solubility is quite high in the neutral pH range. As a result, over a lengthy time period, eluviation generates a soil that, to considerable depth, is low in silica and contains relatively large proportions of iron and aluminium oxides along with the clay mineral kaolinite. The small content of organic matter is characteristic of soils in the arid tropics since the limited yearly input of fresh crop residues is readily oxidized under the dry, high-temperature conditions. The surface horizon is designated as Ap where the suffix 'p' indicates a compacted plough layer—a soil that has been altered by human activities, particularly cultivation.

Underlying the Ap horizon is a series of accumulation horizons designated as B21t, B22t, B23t, and B24t. B horizons are subdivided into a sequence B1, B2, and B3 where B1 indicates a transition between A and fully expressed B material, B2 is the true B horizon, and B3 is a transitional soil to that in the C horizon. In the present profile, all the soil in this horizon is in the B2 category. The second digits in this system are to denote morphological differences within the B horizon observed in the field or by laboratory testing. The suffix 't' symbolizes that the transported material that has accumulated is a silicate clay mineral, in this case kaolinite.

Therefore, the Tyamagondalu soil contains a plough layer of relatively coarse material that is underlain by a series of layers where clay minerals accumulate. The entire profile is rich in iron and aluminium oxides and kaolinite.

### A subtropical Vertisol

The third soil profile example is a Vertisol from the central Transvaal Plateau of South Africa. This is a subtropical area characterized by hot summers and mild dry winters. The total precipitation is about 600 to 800 mm y$^{-1}$. The topography is generally flat or a series of gently undulating ridges with the dominant vegetation consisting of grasses and several varieties of trees (such as acacia) and bushes. Agriculture is centred on maize, sunflower, and fodder under dry land conditions; wheat and tobacco when irrigated.

The soil has developed on basic igneous rocks such as norite. The Hartbeespoort soil[7] profile is shown in Fig. 18.6.

The A11, A12, and A13 horizons are heavy black clay soils distinguished from one another only by subtle structural differences and root content that declines from A11 to A13. Because the dominant clay mineral is montmorillonite, the soil has a large cation-exchange capacity. Organic-matter content is considerably lower than 1% and thus it makes little contribution to the CEC. (Note that the black colour of this and many similar soils in the tropics and subtropics is due to the nature of the parent material, not to the presence of organic matter.) The large content of montmorillonite also causes the soil to undergo a high degree of swelling and shrinking as the soil is wetted and dried. The cracks that form on drying provide a route by which surface material can slough off and fall down to a deeper part of the profile, resulting in a natural cycling and mixing of the soil.

While Vertisols can be chemically rich, they are difficult to work because of the nature of the clay that makes them sticky when wet and very hard in the dry season. In the Hartbeespoort soil, there is some free calcium carbonate throughout the profile and this is reflected in a soil pH that is in the slightly alkaline range. The type of clay does not change with depth but the clay content declines, and this leads to smaller cation-exchange capacity values in the B and C horizons. They do, however, remain at a moderately (and surprisingly) high value.

## Soil orders

The above three examples illustrate how physical and chemical properties of the horizons in a soil profile are used in order to describe and classify individual soils. National soil surveys have been carried out in most countries, providing soil maps of varying detail. A general soil map of

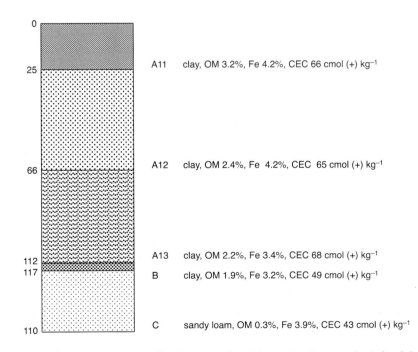

| | | |
|---|---|---|
| A11 | clay, OM 3.2%, Fe 4.2%, CEC 66 cmol (+) kg$^{-1}$ |
| A12 | clay, OM 2.4%, Fe 4.2%, CEC 65 cmol (+) kg$^{-1}$ |
| A13 | clay, OM 2.2%, Fe 3.4%, CEC 68 cmol (+) kg$^{-1}$ |
| B | clay, OM 1.9%, Fe 3.2%, CEC 49 cmol (+) kg$^{-1}$ |
| C | sandy loam, OM 0.3%, Fe 3.9%, CEC 43 cmol (+) kg$^{-1}$ |

**Fig. 18.6** The Hartbeespoort soil profile. Horizon depth in cm is given to the left of the profile.

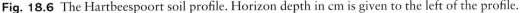

[7] Dudal, R., *Dark clay soils of tropical and subtropical regions*, Food and Agriculture Organization of the United Nations, Rome; 1965.

**Table 18.6** Soil orders according to the USDA system of soil taxonomy[a].

| Soil order | Principal characteristics |
| --- | --- |
| Alfisols | Soils with grey to brown surface horizons, medium to high exchangeable cation supply, and subsurface horizons of clay accumulation; usually moist but may be dry during warm seasons |
| Aridisols | Mineral soils in dry areas; may have a horizon where calcium carbonate, gypsum, or other salts accumulate |
| Entisols | Recent soils without pedogenic horizons |
| Histosols | Organic soils of various types; greater than 30% organic matter to a depth of greater than 40 cm |
| Inceptisols | Young soils formed by alteration of parent materials. Little evidence of accumulation; soils are usually moist |
| Mollisols | Soils with nearly black, humic-rich surface horizons and good supply of exchangeable, mostly divalent, cations |
| Oxisols | Highly weathered soils with a depositional horizon rich in hydrous oxides of aluminium and iron |
| Spodosols | Mineral soils with an eluvial layer underlain by a depositional horizon having accumulations of amorphous organic materials and aluminium and iron hydrous oxides |
| Ultisols | Soils that are usually moist with horizon of clay accumulation and low base saturation; formed in areas of warm, moist climate, usually under forest |
| Vertisols | Soils with high content of swelling clays and wide deep cracks at dry seasons |

[a] From Soil Survey Staff, *Keys to soil taxonomy*, AID, USDA, SCS, SMSS Technical Monograph No. 19, 5th edn, Pocahontas Press, Inc., Blacksburg, Virginia; 1992.

the world is provided in Reference 2 of the Additional Resources list. Table 18.6 lists the soil orders defined by the USDA Soil Taxonomy system and notes the principal features characteristic of each order.

> **Main point 18.4** As determined by the local geology and their environmental history, soils develop a variety of changing physical and chemical properties with depth. A description of these features is called a soil profile.

## 18.4 Soils and ecosystem services

Aside from being a medium that provides physical support, water and nutrients for plants, soils also contribute to the global environment in several other ways. The ability of soils to provide a number of these *ecosystem services* cannot be overemphasized. One helpful description of the services assigns the following categories:

- Cultural services—non-material services such as the provision of attractive landscapes, like an old-growth forest or a well-maintained diverse agricultural area.
- Provisioning services—agriculture, forestry and biomass production, which provide products required by humans. Soils also act as a gene reservoir, containing more numbers and types of microorganisms than all other above ground sources put together.
- Regulating services—regulation of water movement in the ecosystem and a key link in the global carbon cycle.

- Supporting services—the fundamental properties of the soil and the on-going pedogenetic processes that maintain these properties, thus enabling the three other types of services to continue.

These combined services are absolutely essential to maintain the overall global ecosystem in balance. Problems arise when the soil is degraded in such a way that there is diminished ability to provide the various services,

## Environmental problems associated with soils

Many of the environmental problems associated with soil are substantially physical ones. Soil erosion due to growing crops like corn every summer season over a period of many years is a widespread phenomenon. In the *Corn Belt* across the midwestern United States, average losses of topsoil have been estimated to be around 25 t ha$^{-1}$ each year. Serious erosion occurs to varying degrees in many other locations throughout the world.

Too little or too much water is also a problem in terms of sustaining good productivity of soils. About one person in seven lives in arid or semi-arid regions and desertification is spreading in parts of North Africa, the Middle East, and the Indian subcontinent. On the other hand, waterlogging due to canal irrigation is a major issue near some of the world's great rivers and inland seas—the Nile, the Tigris–Euphrates, and the Indus Rivers and the Caspian and Aral Seas.

While problems of erosion or water imbalance have an apparently 'physical' basis, there are chemical aspects to these issues as well. Besides, there are other soil issues that more obviously are related to soil chemistry. The following sections discuss these issues and show some of the connections between soil and other compartments of the global environment.

## Nutrient loss from soil by leaching

Intensive land use has encouraged the application of chemical inputs to maximize the yield of crops and forests. Care is required to ensure that any chemical used for this purpose is not leached into groundwater or washed away as run-off. In Chapter 20, we will consider in some detail the fate of organic biocides when applied to a crop or the soil. Here, we will briefly examine the behaviour of inorganic nutrients used in agriculture—chemicals like ammonium sulphate or calcium phosphate.

The flux of water (rainfall or irrigation), soil texture, and nature of plant cover play major roles in determining the extent of leaching. Downward movement of chemical species is inhibited geochemically by being bound to soil-particle surfaces and biologically through uptake by the plant. Therefore, leaching occurs to a smaller extent where the soils are fine textured with reactive surfaces, and where plants are actively growing.

The physicochemical interactions between nutrient and soil are major determinants with respect to leaching. Cationic macronutrient species include ammonium, potassium, calcium, and magnesium ions added in the form of ammonium salts, potassium chloride (potash), calcium carbonate (limestone), and calcium magnesium carbonate (dolomite). The positive ammonium and metal ions tend to be associated with the negatively charged exchange complex comprised of clay minerals and organic matter. Soils with a large cation-exchange capacity therefore hold cations and prevent their leaching. At the same time, species held on a soil are in equilibrium with their dissolved counterparts, and so at least a small portion is present in the associated water. When there is competition with high concentrations of other cations (including hydronium ion) in the soil solution, additional nutrient ions are released from the solid phases and are then subject to leaching.

Ammonium is a special case because, under high p$E$ conditions, it is readily oxidized to nitrate. Being an anion, the nitrate is not strongly bound to solid particles in most soils. It is then readily available to be taken up by plants. If, however, it is not utilized by plants or microorganisms, it then remains in solution.

To maximize production from high-yielding varieties of grains and other crops, heavy applications of nitrogen fertilizers are widely used. It has frequently been observed that a significant fraction of nitrate added or produced from inorganic fertilizers is not taken up by plants and is readily leached from the rooting zone, sometimes into groundwater. There is considerable variability in the amounts of nitrates leached from soils depending on the nature of the soil, amount of water, crop, tillage practices, and rates, types, and times of application of fertilizer. Nitrate from agricultural sources reaching surface or groundwater is a very serious environmental issue. In a surface water body it can contribute to eutrophication, although, as we have noted, more frequently phosphorus is the limiting nutrient. Nitrate also is toxic, particularly to young children, and drinking water standards have been set in various jurisdictions (Section 16.2).

### Nitrate pollution from agriculture in the United Kingdom

Around the world, nitrate pollution of water is being recognized as a serious problem associated with agriculture. Being a major nutrient, and one produced at significant economic and environmental cost, losses from the terrestrial environment into the hydrosphere mean reduced availability for its principal purpose—that of enhancing crop growth. The common aquatic and terrestrial forms of nitrogen are ammonium and nitrate and both of these forms can be utilized by plants. Ammonium is frequently the initial inorganic nitrogen form released into a soil–water system because it is produced by mineralization of the organic nitrogen in plant and animal residues and because the element is added as an ammonium salt, liquid ammonia, or urea. In temperate regions, there are situations where animal manures are spread on fields during cold winter weather. Chemical and microbial reactions are slow, and the ammonium compounds are not used or transformed in any way. When the spring snowmelt occurs, large amounts of ammonium may then run off or percolate through the soil. In countries like the Netherlands and other European and North American countries where livestock are raised in large numbers within a limited area, disposal of manure on land is a major issue, especially in the wintertime.

During a normal warm growing season, in all but low-pE environments, nitrification rapidly occurs and, at these times, nitrate is the more common nitrogen species. Where there is more nitrate than is required by the growing crop, the excess can run off or be leached through the soil. It is interesting to consider the features that lead to high nitrate loadings in streams and other surface waters. The situation in Britain is a good example. A map (Fig. 18.7) showing average levels in surface waters indicates a marked gradient of increasing nitrate concentration running from northwest to southeast. The highest values, typically from 5 to 9 mg N $L^{-1}$ or more, occur in the areas in and around East Anglia. Not surprisingly, low values are found in streams located in areas having high elevations where run-off is rapid, with high rainfall, which creates a dilution effect, and limited agricultural activity. In contrast, elevated levels are found in the drier and intensively cultivated lowlands.

Ammonium increases the possibility of fish asphyxia and, in public water supplies, it reacts with chlorine to produce the chloramine compounds, reducing chlorine's effectiveness as a disinfectant. A maximum acceptable level of ammonium is 0.5 mg $L^{-1}$. On the other hand, nitrate, the principal nitrogen species in an oxygenated aqueous environment, can also be toxic to humans. While it is rapidly excreted by adult kidneys, it accumulates in infants younger than 6 months and is reduced to nitrite where it combines with haemoglobin in the blood to form methaemoglobin, a form of the protein that is unable to carry oxygen. The result is the so-called *blue baby syndrome*, a sometimes fatal type of oxygen deprivation. Maximum acceptable nitrate levels in water are frequently set at 50 mg $L^{-1}$.

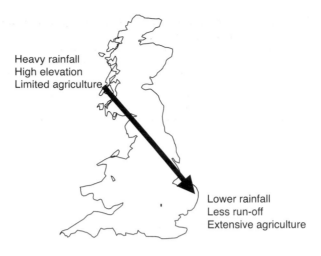

**Fig. 18.7** Gradient in nitrate concentrations in rivers of England and Scotland. Concentrations increase in a southeasterly direction from levels that are typically $<1$ mg L$^{-1}$ in the northeastern highlands to $>5$ mg L$^{-1}$ in the heavily agricultural southeastern plains. From Goudie, A.S. and D. Brunsden, *The environment of the British Isles: an atlas*, Clarendon Press, Oxford; 1994.

### Nutrient loss by soil erosion

The environmental behaviour of the second important plant nutrient, phosphorus, is quite different from that of nitrogen. We have seen (Chapter 14) that phosphorus, present in aqueous solution as the family of orthophosphate species, is strongly and specifically bound to particular soil minerals including aluminium and iron oxides and 1:1 clays. The fixation can be so strong that soluble phosphorus in run-off or leachate is almost negligible. It is when erosion occurs that losses of this nutrient become significant.

In many other instances too, erosion can be a major contributor to loss of nutrients and other chemicals from the land. Species sorbed to soil particles are carried away in the runoff water where they can later be released in soluble forms when they reach rivers or lakes. Table 18.7 gives concentrations of both soluble and particle-bound calcium and phosphorus measured in run-off from eroded forested areas developed on Alfisols in western Nigeria. The amount of nutrient associated with suspended material in run-off can far outweigh the soluble nutrient and is especially strongly dependent on slope.

### Reactions with acids and bases

As we noted earlier, mineral composition is the principal factor determining the intrinsic pH of a soil. Where carbonate minerals are present, soil pH is usually in the range 7.5–8. On the other hand, soils containing significant quantities of exchangeable aluminium (III) and iron (III) or high in organic matter are generally somewhat acidic. Soil pH is also affected by acid inputs from a variety of sources—many of these associated with acid-generating reactions that occur naturally. One acid source involves microorganisms whose respiration produces carbon dioxide. Microbial respiration therefore enriches the soil atmosphere in carbon dioxide resulting in reduced pH of the soil solution. Another contribution to reduced pH in soil solution comes from the production of organic acids by microbes that degrade soil biomass. Each of these factors is influenced by climate and setting and the net result is that there is a range of possible 'natural pH' values for different soils.

There is also a variety of anthropogenic sources of acidity, but two stand out as being quantitatively especially important. Acid precipitation at various locations in the world supplies a large amount of hydronium ion to the soil. A mean annual precipitation pH of 4.2 is not unusual in

**Table 18.7** Nutrient concentrations in water and soil particles in run-off from a forested Alfisol in Western Nigeria[a].

| Slope / % | Concentration / $\mu g\ L^{-1}$ | | | |
|---|---|---|---|---|
| | Soluble metal | | Particle-bound metal | |
| | Calcium | Phosphorus | Calcium | Phosphorus |
| 1 | 2.7 | 0.1 | 475 | 3.9 |
| 5 | 2.6 | 0.1 | 725 | 5.5 |
| 10 | 1.6 | 0.4 | 790 | 8.2 |
| 15 | 1.4 | 0.6 | 1135 | 14.7 |

[a] From Lal, R., Soil erosion problems on an Alfisol in Western Nigeria and their control, Monograph No. 1, IITA, Ibadan, Nigeria; 1976.

some parts of northern Europe and eastern North America (Fig. 5.2). Since the 1990s, due to the expansion of coal-fired electricity production large areas of China are also subject to significant acid rain.

Consider a situation where the yearly rainfall with average pH of 4.2 is 1000 mm (1 m) and the area is 1 ha. The total number of moles of acid being added to the 1 hectare area of soil is then

$$M_{H_3O^+} = 1\ m \times 10\,000\ m^2 \times 1000\ L\ m^{-3} \times 10^{-4.2}\ mol\ L^{-1} = 630\ mol$$

A second very common source of acid input into soil occurs due to the addition of nitrogen-containing fertilizers, where the added nitrogen is in the reduced form and is subject to nitrification. As an example, consider a situation in which 200 kg ha$^{-1}$ nitrogen is added in the form of solid ammonium sulphate. This corresponds to 200 000 g / 14 g mol$^{-1}$ = 14 000 mol of nitrogen in the form of ammonium ion.

The initial nitrification reaction is

$$2NH_4^+\,(aq) + 3O_2 + 2H_2O \xrightarrow{\text{Nitrosomonas}} 2NO_2^-\,(aq) + 4H_3O^+\,(aq) \qquad (18.2)$$

According to the stoichiometry of this equation, after the reaction is complete 28 000 mol of hydronium ion have been generated in situ in the soil within the 1 ha area. The further oxidation of nitrite to nitrate does not lead to the formation of any additional acid or base. The obvious but often neglected conclusion is that use of nitrogen fertilizers can make a much greater contribution to soil acidification than does acid rain.

There are many reactions potentially available to neutralize acidity that has been added to the soil or generated within it. Factors that determine which one, if any, is operative include the nature of the particular soil, the presence of vegetation, and the anions associated with the input acidity. We can divide the types of reactions that neutralize acidity into two categories—those that are geochemical and those that are biological.

### Geochemical reactions that neutralize acidity

A soil that contains carbonate minerals neutralizes acid by dissolution of the solid carbonate species. For calcite (limestone) the reaction is

$$2CaCO_3\,(s) + H_2SO_4\,(aq) \rightarrow 2Ca^{2+}\,(aq) + 2HCO_3^-\,(aq) + SO_4^{2-}\,(aq) \qquad (18.3)$$

When this reaction occurs, the soil solution is enriched in calcium and hydrogen carbonate ions and depleted in hydronium ion. Soil pH, which is initially in the neutral or mildly alkaline ranges, is usually little affected by these reactions unless the supply of carbonate minerals is very small.

For soils containing little or no carbonate species, neutralization occurs due to cation-exchange reactions of the type shown

$$\text{Soil:K}^+ + \text{H}_3\text{O}^+ \text{ (aq)} + \text{NO}_3^- \text{ (aq)} \rightarrow \text{Soil:H}_3\text{O}^+ + \text{K}^+ \text{ (aq)} + \text{NO}_3^- \text{ (aq)} \qquad (18.4)$$

Such reactions serve to neutralize the solution acidity, while the base saturation of the exchange complex declines. Depletion of important nutrient cations like calcium and potassium occurs when they are displaced from exchange sites at a rate that is faster than they can be replaced by weathering of primary minerals. The extent to which the exchange reaction occurs depends on the original exchange capacity and base saturation of the soil. Because every soil has some exchange capacity, it has been suggested that the magnitude of the CEC along with the base saturation is a reasonable measure of the acid-neutralizing capacity of soils—equivalent to alkalinity for a natural water body. Although CEC is important in many situations, other physical and chemical factors also come into play and must be considered in a detailed evaluation of soil resistance to acidification.

Soils containing iron and aluminium hydrous oxides are capable of removing sulphate (and with it hydronium ion) by specific complexation. One possible way to describe this process is

$$\text{Al(OH)}_3 \text{ (s)} + \text{SO}_4^{2-} \text{ (aq)} + 2\text{H}_3\text{O}^+ \text{ (aq)} \rightarrow \text{AlOHSO}_4 \text{ (s)} + 4\text{H}_2\text{O} \qquad (18.5)$$

The reaction is a specific one between sulphate (but not nitrate) and solid aluminium or iron hydroxide and, while taking up sulphate, it simultaneously eliminates hydronium species. The net result is diminished soil solution acidity and also diminished ionic concentration through removal of both positively and negatively charged species. The simple equation, however, describes only one possible means of interaction. An alternate removal process is through specific adsorption of the sulphate on to the amorphous hydrous oxide surface. Hydronium ion is then retained as a counterion. Either process can be an important way of removing both sulphate and acidity simultaneously in tropical Oxisols and other red soils, as well as in the depositional horizons of Spodosols where aluminium and iron oxides accumulate. In many cases, these will be soils having an intrinsic low pH and these reactions provide a way in which acidic soils can neutralize some additional acidity.

The AlOHSO$_4$ is not very insoluble and there are situations where the soil solution is sufficiently dilute that sulphate and aluminium are released back into the aqueous phase. Depending on the solution pH and other factors, the aluminium ion goes into solution as $\text{Al}^{3+}$, $\text{AlOH}^{2+}$, or is complexed with some other ligand such as $\text{F}^-$. When redissolution occurs, the elevated aluminium concentrations can have a serious detrimental effect on the growth of plants. If water carrying soluble aluminium species reaches lakes or rivers, the metal can also be toxic to a variety of aquatic species.

The three types of neutralization reactions described here occur sufficiently rapidly to cause the pH of an acidic solution to rise substantially during the time it takes the water to pass down the soil column. There are other reaction types, including some of the weathering reactions described in Chapter 17, that consume hydronium ions much more slowly. The reaction with orthoclase feldspar is a good example.

$$2\text{KAlSi}_3\text{O}_8 \text{ (s)} + 2\text{H}_3\text{O}^+ \text{ (aq)} + 7\text{H}_2\text{O} \rightarrow \text{Al}_2\text{Si}_2\text{O}_5(\text{OH})_4 \text{ (s)} + 4\text{H}_4\text{SiO}_4 \text{ (aq)} + 2\text{K}^+ \text{ (aq)} \qquad (18.6)$$

While this reaction does occur, it takes place extremely slowly—too slowly to be of any significance—during an individual rainfall event.

### Biological processes that neutralize acidity

Besides being under geochemical control due to the processes that we have just looked at, acid inputs are simultaneously subject to biological control. Most importantly, acidity associated with nitric acid is influenced by the fact that nitrogen in the form of nitrate is a macronutrient for all plants. Therefore, where plants are growing, the nitrate is taken up into the roots by reactions that begin with an anion-exchange process at the root surface. As shown in the following important reaction, the exchanging anion is usually considered to be carbonate.

$$\text{Root}:CO_3^{2-} + 2NO_3^- \text{ (aq)} + H_3O^+ \text{ (aq)} \rightarrow \text{Root}:(NO_3^-)_2 + HCO_3^- \text{ (aq)} + H_2O \quad (18.7)$$

The reaction shows that, as nitrate is taken up, hydronium ion is neutralized by the released carbonate weak base, and thus the soil pH is buffered. As a result, the large amount of acidity generated by nitrification of ammonium fertilizers is consumed when applied at the time when plants are growing. Obviously, where acid is added to or generated in a fallow field or clear-cut forest, the biological control processes are limited.

Clearly, there are several possible abiotic and biotic reactions related to neutralization of acids that can occur within the soil. In all cases, both the soil and the soil solution are altered in the process. The effect may be to temporarily enhance the growth of plants by supplying nutrients such as nitrate or soil-derived ones such as calcium and magnesium. In other situations, potentially toxic elements like aluminium are dissolved and adversely affect plant growth.

Looking at the other side of the issue, we see that the soils that are most likely to be unable to neutralize acid inputs are those having no free carbonate minerals, small CEC values, small content of iron and aluminium oxides, and ones where no vegetation is growing.

---

### Example 18.4 **Removal of incoming acidity by two soils**

Consider the following situations and comment on the mechanism of neutralization.

Rain with an average pH of 4 falls on two fields. Soil pore-water data are obtained during a period when no plants are growing.

- The first field is a black soil, containing 0.7% OM, and having a CEC of 71 cmol (+) $kg^{-1}$. The soil pore water has a pH of approximately 7.3. Cations in the pore water are, in order of decreasing concentration, $Ca^{2+}$, $Mg^{2+}$, and $K^+$, and the total concentration of positive ions is $1.2 \times 10^{-4}$ mol $L^{-1}$.

- The second field is a red soil, containing 1.1% OM and having a CEC of 6.3 cmol (+) $kg^{-1}$. The soil pore water has a pH of approximately 6.5. Cations followed the same order of concentration, but small amounts of $Al^{3+}$ are also present. The total concentration of positive charge is $3.9 \times 10^{-5}$ mol $L^{-1}$.

In the first case, the large CEC must be associated with a clay mineral like montmorillonite, as the OM concentration is small. Neutralization of the acidity is efficient and is due to exchange of hydronium ions with metal ions on the exchange sites. The metal ions replace hydronium ions in solution and the total positive charge does not decrease; the small increase is probably due to dissolution of soluble salts from the soil.

In the second case, the fact that the soil is red is suggestive of the presence of hydrous iron (and aluminium) oxides. The small CEC is indicative of a clay mineral like kaolinite. Neutralization is partly due to cation exchange as above, but the total positive charge concentration remaining in the pore water is only about one-third of that in the rain. This indicates that the sulphuric acid component has been retained by the metal oxides as summarized in reaction 18.5.

---

## Acid sulphate soils

Acid sulphate soils are characteristic of marine coastal plains in areas rich in organic matter. They are associated with brackish water, especially in areas covered by mangrove swamps. In Vietnam over 2 million ha of these types of soil are developed on pyrite-rich former marine deposits.

In submerged areas, organic debris from the mangroves accumulates, producing a dark-coloured organic muck and a highly reducing environment. The sea water provides a continuous source of sulphate that, under the anaerobic conditions, acts as an electron acceptor for the oxidation of the organic matter as has been shown in Section 15.4. The reaction is mediated via the sulphur-reducing bacteria *Desulphovibrio* and, depending on pH, produces hydrogen sulphide whose odour is often detectable in such coastal areas.

$$SO_4^{2-} (aq) + 2\{CH_2O\} + 2H_3O^+ (aq) \rightarrow H_2S + 2CO_2 + 4H_2O \qquad (18.8)$$

The low $pE$ conditions also favour solubilization of iron (III) minerals by reducing the metal to iron (II), which then reacts with the reduced sulphur to form pyrite ($FeS_2$). A possible overall representation of the complex set of reactions is

$$4Fe(OH)_3 + 8SO_4^{2-} (aq) + 16H_3O^+ (aq) \rightarrow 4FeS_2 + 15O_2 + 30H_2O \qquad (18.9)$$

Over time this reaction proceeds and builds up a plentiful supply of pyrite that acts as a reservoir for the reduced sulphur.

The waterlogged soils have a pH that is usually in the slightly acidic region but a major drop in pH occurs if the soil is allowed to dry. By the reverse of reaction 18.9 very large quantities of sulphuric acid are released and the pH of the 'reclaimed' soil may fall as low as 1.5 or 2. The deposited iron (III) compounds give the fine-grained organic soil a yellow mottling and it is frequently called a 'cat clay'. The low pH results in high concentrations of soluble aluminium and these two factors alone make such soils completely unsuitable for agriculture. Manganese toxicity and phosphorus deficiency have also been observed.

For these reasons, coastal soils that have the potential to become acidic should not be drained. In the Mekong Delta, areas have been kept continually submerged and used for lowland rice culture. Previously drained soils can be treated with lime if the acidity is not too great. Where the pH is less than 3.5, reclamation requires uneconomically large quantities of lime and the soils are frequently abandoned.

## Salt affected soils

We saw in Chapter 5 that precipitation always contains small concentrations of many elements. When rain percolates freely through a well-drained soil, some of these dissolved ionic species are retained at various depths by interaction with soil particles. At the same time, weathering and leaching can cause dissolution of elements from the soil. Soil pore-water composition therefore is determined by a combination of removal and dissolution reactions. Overall, the drainage water usually contains only a very small concentration of ionic species, and there is no significant accumulation of salts in any part of the soil profile.

In contrast, where there is limited precipitation along with a high rate of evaporation, the downward movement of water may be insufficient to leach out all the salts that accumulate near the soil surface. Salt affected soils are therefore common in arid and semi-arid regions of the world including parts of Tunisia, Iraq, Sudan, Pakistan, and Australia. Salinity and related problems arise when, over extended periods of time, evaporation and evapotranspiration from the soil exceed the downward percolation of rainfall or irrigation water. When the input water itself contains relatively high concentrations of salts, the possibility of the soils accumulating salts is enhanced. Typically, loss of water by upward movement away from the surface is greater than that by downward movement when the water table is closer than 1 to 1.5 m below the surface.

Salt affected soils are classified with respect to their pH, electrical conductivity (EC) of the saturation extract, cation exchange capacity, exchangeable sodium percentage (ESP), and sodium absorption ratio (SAR). The latter two terms require some explanation.

Exchangeable sodium percentage is the fraction, expressed as a percentage of exchangeable sodium ions $[(Na^+)_E]$ compared to total exchangeable ions (CEC),

$$ESP = \frac{[(Na^+)_E]}{CEC} \times 100 \qquad (18.10)$$

A related parameter is the sodium adsorption ratio (SAR); it is used to characterize the sodium content of soil solutions and is defined as follows.

$$\text{SAR} = \frac{C_{Na^+}}{(C_{Ca^{2+}} + C_{Mg^{2+}})^{1/2}} \qquad (18.11)$$

The concentrations (mmol $L^{-1}$) $C_{Na^+}$, and $C_{Ca^{2+}}$, $C_{Mg^{2+}}$ are measured on aqueous extracts of the soil. Units of SAR are (mmol $L^{-1})^{1/2}$.

Electrical conductivity is a measure of total ionic concentration of the soil solution and is determined on a saturated extract.

- 'Normal' soils have EC values[8] less than 4 dS $m^{-1}$ and ESP values below 15%.

- Saline soils have EC values greater than 4 dS $m^{-1}$ with an ESP less than 15%. Because the soluble salts in saline soils are mostly neutral—made up of cations such as $Ca^{2+}$ and $Mg^{2+}$ and the anions $Cl^-$ and $SO_4^{2-}$—the pH of saline soils is generally under 8.5.

- Sodic soils have EC values less than 4 dS $m^{-1}$ but ESP greater than 15%. The concentration of neutral salts is small and salts such as sodium carbonate are important. In water, because the carbonate ion hydrolyses producing hydroxyl ion, the pH of these soils is typically high, lying between 8.5 and 10.

- Saline–sodic soils are a fourth category of salt-affected soils. These soils have both high EC (>4 dS $m^{-1}$) and high ESP (>15%). The dominating presence of neutral salts usually maintains the pH value at less than 8.5.

A high concentration of accumulated salts, especially in sodic or sodic–alkaline soils, leads to several environmentally significant consequences. Fine-grained soils have a large clay content and the clay minerals become dispersed or 'peptized', meaning that the individual particles remain as separate units. This occurs because the large (hydrated radius) of the sodium ion does not efficiently neutralize the negative charge of the small clay particles. They therefore disperse themselves when wet and form dense structures when dry. The result is that the soil loses any structure that it previously exhibited, and becomes highly impervious to the movement of water. With regard to plant growth, the elevated salt concentration requires an expenditure of energy as a physiological response by the plant in order to maintain a constant water potential gradient between the root and soil solution. As a result, plant growth is inhibited. Furthermore, high concentrations of particular ions such as sodium in the soil solution can create a nutrient ion imbalance—for example, calcium deficiency is frequently observed in high-sodium soils. In some cases, toxic levels of certain elements—boron is a common example—may be reached in salt affected soils. Where soils are alkaline, the hydroxyl ion contributes to toxicity along with the other factors.

Depending on the situation and the specific properties of the salt-affected soil in question, there are a number of technologies that can be applied for their reclamation.

- Flushing the salts from a saline soil is perhaps the simplest remediation procedure. This requires using water that is itself low in salts and ensuring adequate drainage of the leached water. By flushing, neutral soluble salts are removed from the rooting zone and the conductivity moves toward a normal value. Because of low hydraulic conductivity, simple flushing cannot be used when the ESP of the soil is initially large.

- For saline–sodic soils, a similar flushing procedure can be used, but the input water should contain a high concentration of calcium and / or magnesium ion in order to increase the soil permeability. As a result, exchangeable sodium ions are replaced by those of the divalent alkaline-earth metals. Subsequent flushing using water with low ion concentrations can then be done without dispersion of the clays, in order to bring the conductivity into the normal range.

---

[8] The traditional units used to measure electrical conductivity are mmho $cm^{-1}$ (a mho is a reciprocal ohm). The newer SI unit, decisiemens per metre (dS $m^{-1}$), has an identical numerical value.

- Because sodic soils have a large concentration of sodium ion and are also alkaline as a result of the carbonate ion, it is necessary to remove both these species. One treatment is to add calcium sulphate (gypsum) to the soil while maintaining continuously moist conditions. The calcium sulphate reacts with sodium carbonate to produce insoluble calcium carbonate and soluble (neutral) sodium sulphate. At the same time, a large fraction of the sodium on the exchange sites is replaced by calcium. Using low-conductivity water, subsequent flushing of the sodium sulphate from the soil is then possible.

- An alternative treatment procedure makes use of elemental sulphur, which is worked into the surface soil where it is oxidized microbially to produce sulphuric acid. The acid again serves to convert sodium carbonate into sodium sulphate, which is then washed out of the soil.

It is important to note that soils, especially those in the sodic categories, can be irreversibly affected by excess salt concentrations. Such soils may have become so impervious to water movement that flushing is almost impossible. Attempts to enable some movement of water have been effected by reducing the bulk density through intensive tilling. This allows water carrying calcium ions to penetrate into the macropores resulting in a limited degree of leaching.

## Soil problems related to metals

### Metal contamination of soils

In considering the issue of salt-affected soils, we were concerned with the accumulation of metal ions in the form of soluble salts near the soil surface. For the most part the metal ions were sodium, potassium, calcium, and magnesium. In different environmental circumstances, other metal ions can also accumulate in the soil, although the amounts are usually much smaller than those of the alkali and alkaline-earth families. Some of these may be type B metals with toxic properties and their availability for plant uptake or to be leached into groundwater is an important environmental issue.

Excessive concentrations of metals can be associated with naturally occurring ore deposits, but more usually are the result of human activities. In the next chapter, we will examine some specific cases where waste materials containing metals are disposed of on land. Here we will review the general principles that determine the fate of metal ions in the soil. With soil, we are always dealing with soil / water relationships, so all of the principles discussed in Section 13.1 are applicable in the present case as well.

With respect to water, we emphasized that all samples, even samples that are considered to be 'pure', have trace quantities of 'impurities'. Not surprisingly, this situation is even more dramatically expressed in soils, where it is observed that there is always a background concentration of less-common metals and other elements as well. The observed levels are variable all over the Earth's surface and depend on many factors. Table 18.8 gives some background concentration ranges and mean values (in all cases using units of $\mu g\ g^{-1}$ (ppm)) that have been measured by various researchers on soils that are not considered to be 'contaminated' in any special way. There is no universally accepted definition of contamination that can be applied to soils. Frequently, however, a contaminated soil is described as one that has a metal concentration more than 3 (or some other factor) times greater than a recommended average value.

Where do the trace elements come from? To some extent, they are derived from the original minerals that were subject to weathering and produced the mineral soil. This does not necessarily mean that they reflect the composition of the associated bedrock, since many soils have been transported from other locations to their present site by wind, water, or ice. Likewise, the organic components of the soil contain small amounts of many metals and they too may be derived from either local or distant sources. To these important 'original' sources of metals in the soil, other inputs are added. One of these additional sources is deposition from the atmospheric

**Table 18.8** Concentrations ($\mu g\ g^{-1}$) of minor metals, including some semi-metals in uncontaminated soils.

| Element | Concentration / $\mu g\ g^{-1}$ | | | |
|---|---|---|---|---|
| | World[a] Mean (range) | World[b] Mean (range) | USA[a] Mean | Canada[a] Mean (range) |
| Arsenic | | 6 (0.1–40) | | |
| Cadmium | 0.5 | 0.06 (0.01–7) | | <1 |
| Chromium | 200 (100–300) | 100 (5–3000) | 53 | 43 (10–100) |
| Cobalt | 8 (10–15) | 8 (1–40) | 10 | 21 (5–50) |
| Copper | 20 (15–40) | 20 (2–100) | 25 | 22 (5–50) |
| Lead | 10 (15–25) | 10 (2–200) | 20 | 20 (5–50) |
| Manganese | 850 (500–1000) | 850 (100–4000) | 560 | 520 (100–1200) |
| Mercury | 0.01 | | 0.071 | 0.059 (0.005–0.1) |
| Nickel | 40 (20–50) | 40 (10–1000) | 20 | 20 (5–50) |
| Selenium | 0.01 | 0.5 (0.1–2) | 0.45 | 0.26 (0.03–2) |
| Strontium | 350 | 300 (50–1000) | 240 | 210 (30–500) |
| Zinc | 50 (50–100) | 50 (10–300) | 54 | 74 (10–200) |

[a] Data reported in McKeague J.A. and M.S. Wolynetz, Background levels of minor elements in some Canadian soils, *Geoderma*, **24** (1980), 299–307.
[b] Data reported in Allaway, W.H., Agronomic controls over the environmental cycling of trace elements, *Advan. Agronomy*, **29** (1968), 235–74.

aerosol. A detailed inventory[9] of the origin of metals in the atmosphere shows that important sources include wind-blown soil particles (dust), volcanoes, and volatile and particulate organic matter derived from forested areas and from the sea. The biogenic sources are especially important for many metals (accounting for more than 30% of the annual releases to the atmosphere) including most of those listed on Table 18.8. Industrial emissions are responsible for much of the flux of lead, cadmium, and zinc.

In soil, the metals are present in a variety of forms. In some cases they are structural components of soil minerals or minor constituents incorporated into soil organic matter. Metals can also be deposited by specific sorption processes onto surfaces of pre-existing minerals of various types. An additional fraction of soil trace metals is accounted for by electrostatic ion retention on mineral or organic-matter exchange sites. Finally, a (usually very small) concentration of metal species is found in the pore water of any soil.

Environmental issues regarding soil trace metals often centre around their mobility and this is related to the form in which the trace metal is present as well as the environmental situation. The most important environmental factors are the amount, chemical nature, and movement of water through the soil. Metal associated with the original inorganic material may not be readily available for uptake or leaching. Especially in the case of silicate species, such metals are resistant to weathering, tend to persist in the solid structure, and therefore do not achieve the mobility associated with being in solution. Aluminium present in aluminosilicate minerals such as the feldspars is essentially inert on a time-scale of years or decades. It is only over much longer timescales that the weathering alterations we noted earlier become significant.

[9] Nriagu, J.O., A global assessment of natural sources of atmospheric trace metals. *Nature*, **338** (1989), 47–9.

### Redox chemistry of soil metal species

There are other associations involving the mineral phase, in particular, metals chemically and specifically bonded to surfaces of various minerals. As has been frequently mentioned, the hydrous oxides of iron and aluminium are particularly active sorbents. Mobilization of metals in these associations can occur as a result of acid conditions or changes in the redox status of soils. Reducing conditions are especially significant in this context. When a soil that has previously existed in an oxidizing environment—under well-aerated conditions—is flooded, oxygen movement is restricted and the pre-existing oxygen is used up in microbial aerobic reactions. Once the oxygen is depleted, other species can act as electron acceptors in conjunction with facultative anaerobes in the soil. We saw earlier that sulphate and nitrate were two such species. Solid mineral phases can serve the same purpose: manganese (IV) oxide and iron (III) oxide, both in hydrated forms, are two important solid electron acceptors. Studies under controlled conditions have shown that a sequence of reduction events occurs when a soil is flooded.[10] The order and approximate pE range over which event occur are

$$O_2 \text{ (aq)} + 2H_3O^+ \text{ (aq)} + 2e^- \rightleftharpoons H_2O_2 \text{ (aq)} + 2H_2O \quad pE = 6.3 \text{ to } 5.8 \quad (18.12)$$

$$H_2O_2 \text{ (aq)} + 2H_3O^+ \text{ (aq)} + 2e^- \rightleftharpoons 4H_2O \quad (18.13)$$

$$2NO_3^- \text{ (aq)} + 12H_3O^+ \text{ (aq)} + 10e^- \rightleftharpoons N_2 \text{ (g)} + 18H_2O \quad pE = 4.2 \text{ to } 3.6 \quad (18.14)$$

$$MnO_2 \text{ (s)} + 4H_3O^+ \text{ (aq)} + 2e^- \rightleftharpoons Mn^{2+} \text{ (aq)} + 6H_2O \quad pE = 3.6 \text{ to } 3.1 \quad (18.15)$$

$$Fe_2O_3 \text{ (s)} + 6H_3O^+ \text{ (aq)} + 2e^- \rightleftharpoons 2Fe^{2+} \text{ (aq)} + 9H_2O \quad pE = 1.9 \text{ to } 1.4 \quad (18.16)$$

$$SO_4^{2-} \text{ (aq)} + 8H_3O^+ \text{ (aq)} + 8e^- \rightleftharpoons S^{2-} \text{ (aq)} + 12H_2O \quad pE = -0.7 \text{ to } -1.4 \quad (18.17)$$

The reactions are reversible, although oxidation occurs over a pE range that is approximately 0.8 units higher than for the corresponding reduction. The consequence of reactions 18.15 and 18.16 is that hydrated manganese and iron oxide minerals in soils are subject to reduction, a process that mobilizes the metals in the form of 2+ species.

Manganese and especially iron are quantitatively important soil constituents, but trace elements are subject to similar mobilization / immobilization phenomena. Some of the phenomena are second-order effects related to the behaviour of major elements. Arsenic is a good example.[11] Under high-pE conditions, arsenic is present as $H_2AsO_4^-$ and $HAsO_4^{2-}$ species (see the literature link in Chapter 15 and also Chapter 10, problem 9) and these are strongly adsorbed by hydrous iron oxides in the soil. Under reducing conditions, solubilization of iron by reaction 18.16 releases, along with iron, arsenic that has been sorbed on the iron oxide surface. The reducing situation also converts the As (V) species into As (III) in the form of the weak acid, under most environmental conditions. Another factor controlling metal solubility is the precipitation of metal sulphides. When pE is low enough so that sulphate is reduced to sulphide (reaction 18.17), the $S^{2-}$ forms highly insoluble compounds with several metals including copper, nickel, and zinc, so that their already low solubility in soil pore water is suppressed even further. If the soil matrix reoxidizes, the trace metals are then released back into the associated solution.

### Organic matter and mobility of metals in soil

Organic matter can act either to mobilize or to immobilize metals in soil. The solubility of metals that are structural components of organic matter or that form strong complexes with it is determined by the solubility of the associated organic matter. Often, decomposition to form products

---

[10] Patrick, W.H., Jr. and A. Jugsujinda, Sequential reduction and oxidation of inorganic nitrogen, manganese and iron in flooded soil. *Soil Sci. Soc. Am. J.*, **56** (1992), 1071–3.
[11] Masscheleyn, P.H., R.D. Delaune, and W.H. Patrick, Jr., Arsenic and selenium chemistry as affected by sediment redox potential and pH, *J. Environ. Qual.*, **20** (1991), 522–7.

that are smaller and more soluble is an important factor in increasing the solubility of such metals. In temperate forest regions, the mineral soil is overlain by an annually renewed layer of organic litter. The litter is subject to degradation and the products include small molecules like organic acids as well as a stable insoluble humic material. The organic acids act as ligands for many metals enhancing movement in percolating water down through the soil profile. In part, this explains the formation of an eluvial layer in a Spodosol.

Some of the low molar mass fulvic acid material is itself soluble, giving the soil pore water a characteristic yellow or light brown colour. It is interesting that the solubility of soil humic material is larger when the pore water is neutral or mildly alkaline rather than acidic. Metals complexed with this soluble material are, of course, themselves made soluble in the high-pH situation. In this respect, metal solubility may not follow the usual pattern of being greater in acid conditions.

The opposite effect is also possible. There are situations in which organic matter—the more insoluble, high molar mass humic fraction—acts as a reservoir to reduce the mobility of metals. Examples of the ability of humus to inhibit metal leaching in soils are well documented. Lead, released from combustion of leaded gasoline in vehicle engines, is deposited in soils on roadsides, and it has been observed that the metal accumulates in the surface of the soil in association with organic matter. Over many years, in most locations, only very limited movement downward has been observed.

The metals associated with the exchange complex, both mineral and organic, are a small fraction of the total metal content in most instances. Yet this fraction is readily available for plant uptake—a positive consequence when the metal is a required micronutrient, but negative if it exerts a toxic effect. The exchangeable metals are the most readily mobilized. They can be displaced by other ions that are present in large concentration in the soil solution, including the hydronium ion. It is this portion of soil metals that is made more soluble by acid inputs, as we noted in an earlier section.

It is clear, then, that many factors control the ability of a metal to move down the soil profile in the soil water. Added to the chemical issues, the actual flux of water also plays a major role. Leaching of metals is therefore favoured where there is high rainfall and where soils are coarse textured allowing water to move rapidly downwards. This was a factor explaining the development of highly leached tropical soils such as Oxisols, but it is also a factor that determines leaching of contaminant metals under many other environmental conditions.

**Literature link** Metals in the soil of an abandoned industrial site

We have seen that speciation, or the form in which particular elements exist, is a feature that defines their behaviour in the environment. An investigation in Southern Italy (*Assessing the origin and fate of Cr, Ni, Cu, Zn, Pb, and V in industrial polluted soil by combined microspectroscopic techniques and bulk extraction methods* by Roberto Terzano, Matteo Spagnuolo, Bart Vekemans, Wout de Nolf, Koen Janssens, Gerald Falkenberg, SaverioFiore, and Pacifico Ruggiero (*Environ. Sci. Technol.*, **41** (2007), 6762–6769) is a study of several trace metals found in the soil at a location of an earlier industrial site. The mobility of the metals, and thus the possibility that they could migrate into water courses or into plant tissue, depends on the particular geochemical form of the metal. It was for this reason that two very different types of analytical methods were employed in order to identify species of chromium, nickel, copper, zinc, lead and vanadium within the soil matrix.

The first of these methods is an indirect one that makes use of single and sequential extraction procedures—operational procedures where samples of the soil are shaken with various aqueous extractants that are able solubilize different portions of each metal. The elements are then determined in the extractants using standard atomic spectroscopy methodology. The solubility of the metals in different solutions allows one to infer their association with components of the solid material. Using EDTA as an individual extractant allows for assessment of the available element, the fraction that is potentially available to be taken up by growing plants. The sequential extraction method employs several solutions used in sequence, namely water, dilute acid, a reducing solution, and then an oxidizing solution. In this consecutive order of extractants, water-soluble metal, metal associated with carbonate minerals, metal adsorbed onto iron and manganese oxides, and metal associated with organic matter can be determined. Any residual metal that remains unextracted is considered to be a structural component of a primary mineral and therefore can be considered to be essentially immobile both geochemically and biologically.

To a large extent in the Italian study, most of the metals remained unextracted after treatment with all the solutions and therefore were taken to be present as primary minerals. A smaller but significant portion of the zinc, copper, vanadium and lead, however, also appeared in several of the extractant solutions. Most commonly, the reducing solution dissolved a significant fraction of the metal taken to indicate association with amorphous iron and manganese oxides.

You can see that sequential extraction procedures can provide limited but useful information about the metal ion relations of the bulk soil.

The direct methods employed to identify mineral species in the soil included three microscopic X-ray techniques (using synchrotron-generated radiation) that allowed for examination of individual mineral grains. Note that these methods contrast with the extraction procedures that require a larger sample of soil and therefore measure some kind of average composition. μ-XRF (X-ray fluorescence) analysis gave percentage composition of elements in the mineral species of individual grains. μ-XRD (X-ray diffraction) and μ-XANES (X-ray absorption near-edge structure) provided unique spectra that could be compared with standards to enable identification of specific minerals or of associations with other soil components. Using these methods, all six elements were found to exist as primary, sometimes complex, oxide minerals, most in the spinel family. In this sense, they were present in the soil in substantially insoluble geochemical forms and, therefore, the X-ray data confirmed the broad picture revealed by the extraction data.

This, however, was not true in every case. While the extraction method found a large portion of lead to be in the reducible fraction, attributed to association with iron and manganese oxides, the X-ray methods could not detect any such association. Rather, it was found to exist as the mineral minium ($Pb_3O_4$). Minium is an insoluble mineral, but reducing conditions could favor the reduction of $Pb(IV)$ to $Pb(II)$ a more soluble form of the element.

As a general conclusion, the authors state, "The microscopic techniques adopted proved to be very effective to identify, in the case of rather unusual mineral phases, the more stable metal, while the more labile ones were better assessed with the help of bulk extraction methods." Besides providing information about the mobility and availability of these trace elements in the soil, the identification of particular mineral species enabled a tentative ascription of the industries that caused the pollution: polyvinyl chloride and cement-asbestos productions.

---

**Main point 18.5** Nutrient leaching, acidification of soil, various types of salinity, and metal contamination are environmental issues associated with soils. All are related to the physical and chemical properties of soils, as well as to external environmental pressures.

---

## ADDITIONAL RESOURCES

1. Tan, K.H., *Environmental soil science, 3rd edn*, CRC Press, Boca Raton; 2009.

2. Sparks, D.L., *Environmental soil chemistry, 2nd edn*, Academic Press, San Diego 2003.

3. Rengel, Z., *Handbook of soil acidity*, Marcel Dekker Inc., New York, 2003.

4. Palm, C., P. Sanchez, S. Ahamed and A. Awiti Annu, *Soils: A contemporary perspective*, The Annual Review of Environment and Resources 32:99–129, (2007). Available online at http://environ.annualreviews.org.

See also Additional Resources in Chapter 17.

## PROBLEMS

1. The permeability of the Black Cotton Soils (high in montmorillonite) of the Deccan Plateau in central India is very high (up to 20 cm h$^{-1}$) at the beginning of the monsoon season, but soon becomes much less (below 1 cm h$^{-1}$) as the rains continue. Suggest an explanation.

2. Discuss the types of water-contamination problems that are possible when septic tanks for sewage disposal are located in sandy or in clayey soils.

3. Explain why clay-rich soils have desirable physical and chemical properties for use as liners for landfill sites.

4. Consider the following data for a forest soil.

| Horizon | Bulk density / g mL$^{-1}$ | Particle density / g mL$^{-1}$ |
|---|---|---|
| O ($-5$ to 0 cm) | 0.19 | 1.78 |
| E (0 to 8 cm) | 1.08 | 2.61 |
| B (42 to 66 cm) | 1.52 | 2.65 |

   Comment on reasons for the differences in values and their significance in terms of porosity and permeability in each horizon.

5. Salts are commonly spread on highways during winter to prevent build-up of ice. Sodium chloride is most commonly used, but calcium magnesium acetate $(Ca_{0.3}Mg_{0.7})(C_2H_3O_2)_2)$ has been recommended as an alternative because it biodegrades, is less toxic to aquatic life, and is less corrosive. There is concern, however, that it might increase the mobility of trace metals in roadside soils. What metals would be of concern and how could these two salts affect their mobility in soils?

6. With reference to an ion-exchange medium, selectivity refers to the thermodynamic tendency to retain a particular species. The order of selectivity for alkali-metal cations by most clay minerals is

$$Cs^+ > K^+ > Na^+ > Li^+$$

   Explain this in terms of the aqueous solution chemistry of these ions.

7. In soils of the eastern United States, the CEC (cmol ($+$) kg$^{-1}$) has been described as related to the OM (concentration as %) by the equation

$$CEC = 4.83 + 3.87\,OM$$

   with $r = 0.73$ and $N = 57$ ($r$ is the correlation coefficient and $N$ is the number of data points). How does this relation correspond to the generalities presented in this chapter?

8. 10 g of soil were shaken with 100 mL 1 M NH$_4$OAc, allowed to stand for 2 h, and filtered. 10 mL of the filtrate was taken and found to have concentrations of 220 ppm Ca$^{2+}$, 180 ppm Mg$^{2+}$, and 270 ppm mg K$^+$. What is the CEC of the soil?

9. A fraction of soil organic matter was found to have a molar mass of 5000 g mol$^{-1}$. The organic matter contained 85 carboxyl groups mol$^{-1}$ (p$K_a$ = 4.8) and 16 phenolic groups mol$^{-1}$ (p$K_a$ = 6.0). At a pH of 7.0, what is the CEC of the organic matter?

10. Logging of a forest by removing all the mature trees is a controversial forestry practice. Aside from issues such as the effect on species biodiversity and erosion, clear-cutting can alter chemical processes in the soil and even in the global environment. Explain how this practice could lead to increased nitrification and denitrification—and how this may affect soil acid–base properties and the stratospheric ozone concentration.

11. The following are chemical properties of two Venezuelan surface soils. Predict their relative sensitivity to acidic inputs from rain or fertilizer and give reasons for your prediction.

| Soil from | pH | OC | N | Clay | Ca | Mg | K | Na | Al | CEC cmol (+) kg$^{-1}$ | BS % |
|---|---|---|---|---|---|---|---|---|---|---|---|
| | | | | Percentage in soil of | | | | | | | |
| Machiques | 6.0 | 0.75 | 0.08 | 7.2 | 0.3 | 1.4 | 0 | 0.01 | 0.1 | 3.7 | 44 |
| Barinas | 5.6 | 1.91 | 0.17 | 32.6 | 6.6 | 0.9 | 0.5 | 0.1 | 1 | 8.4 | 96 |

12. In one study it has been shown that soil farmed using conventional tillage is a source to the atmosphere of 39 kg C ha$^{-1}$ while, with minimum tillage, the same area is a sink for 11 kg C ha$^{-1}$. Suggest possible reasons for these differences.

13. Traditional agricultural methods are used throughout the Zaire river basin and have been described.[12] Comment on the significance of the following practice in terms of soil chemistry and other related properties.

    (a) Land with the thickest vegetation is cleared (by fire) before planting sorghum or manioc.

    (b) Several crops are grown together or in sequence in the same area, so that the land is covered with vegetation over an extended period of time.

    (c) Household waste, sod, and dry grass are incorporated into the soil before or after planting. Sometimes the compost is incinerated in mounds, later used for planting root crops such as yams.

    (d) In some areas, nomadic communities are invited (sometimes even paid) to set up temporary animal corrals after which the site is used for cropping.

14. The planting of high-yielding hybrid varieties of grains is widespread and has contributed to increasing the global grain supply. A co-requirement is the application of large amounts of fertilizer to support the high levels of growth. For example, corn (maize) yields of greater than 20 t ha$^{-1}$ are obtainable under good agronomic conditions. This might require the addition of 500 kg ha$^{-1}$ of nitrogen in the form of urea. Calculate the amount (kg ha$^{-1}$) of limestone (CaCO$_3$) that should be added to the soil just to balance the acidity generated by this fertilizer, assuming complete nitrification of the urea.

15. Elemental sulphur is sometimes used for neutralizing the alkalinity in sodic soils. Write the equation for the oxidation of sulphur to its stable aerobic form, and calculate the amount of sulphur required to 'neutralize' every kg of sodium carbonate in the soil.

16. Soils in the Mekong Delta of South-East Asia are maintained in a submerged state and are used for growing rice. These soils formed on marine deposits that are rich in pyrite. What chemical changes would be expected when these soils are drained in order to grow a different crop?

[12] Miracle, M.P., *Agriculture in the Congo Basin*, University of Wisconsin Press, Madison, Wisconsin; 1967.

# Chapter 19
# The chemistry of solid wastes

**The focus of this chapter is on the chemistry of solid wastes and how best to make use of them in creative ways that cause little adverse impact on the natural environment. Here, we will discuss:**

- the various types of bulk solid wastes and their potential to be useful resources
- the chemical issues related to managing wastes from mining and metal production
- uses of organic wastes, especially as compost and as an energy source
- the processes that can be used to manage mixed urban wastes such as landfilling and incineration, and the related environmental considerations.

In any human society, bulk solid wastes are produced as a by-product of the normal and fundamental activities of living. These wastes can be as rudimentary as food scraps, ash from fires, and excreta from humans and animals. In a modern, highly industrialized society, however, the wastes go much beyond these fundamental materials both in quantity and variety. The amounts of waste produced by intensive agriculture and by modern industry are staggering, to say nothing of waste generated by ordinary citizens in a wealthy consumer-oriented urban setting. Table 19.1 compares bulk solid waste products associated with various human activities.

The term *bulk wastes* distinguishes these large-volume waste products from specific chemicals such as pharmaceuticals, hospital wastes, dyes, and chemical additives that are sometimes discarded, but over which detailed control protocols should be followed.

We use the word waste with some hesitation. Almost any substance that is discarded and is therefore designated as waste can also be thought of as a potential resource. Throughout human history, societies have found ways in which to use waste—organic residues as fertilizers, animal dung as fuel, inert materials as landfill. In the present era too, there are efforts to discover new uses for materials that have served their primary purpose. This has given rise to the concepts of reuse and recycling. Alongside these efforts are ones to minimize the production of waste by-products so that a reduced quantity of residual material remains to be discarded. This is one of the concerns of production according to the principles of *green chemistry*, which we briefly discuss in Chapter 21.

With or without the reduction / reuse / recycling efforts, the need to discard some solid materials is inevitably an endproduct of many activities.

**Table 19.1** Bulk solid waste materials associated with various human activities.

| Human activity | Bulk solid waste materials |
|---|---|
| Food production | Plant residues, animal manures |
| Food manufacturing and consumption | Packaging materials (paper, plastics of many types, aluminium, steel, glass), food wastes, sewage sludge |
| Shelter construction and use | Construction materials (stone, brick, concrete, metals, wood) soil and other landscaping wastes, ash from heat production |
| Transportation | Scrap metal, rubber and other polymers (natural and synthetic), road and rail construction wastes (concrete, asphalt, steel) |
| Consumable products | Metals, glass, ceramics, polymers, paper, and other fibres |

## Disposal of bulk solid wastes

In the same way that liquid (aqueous) wastes from homes, factories, and agriculture are often discharged into water bodies, so also many solid waste materials are disposed of on land. The solid wastes cover a wide range of types—everything from animal manures, to municipal garbage, to waste steel from manufactured objects, to mine tailings. The waste materials ultimately interact chemically and physically with environmental components, but the time-scale over which such reactions occur may be short or long depending on the type of materials and their degree of exposure to environmental forces. Two important environmental components that play a central role in many reactions are oxygen and water and the solid wastes undergo various kinds of 'weathering' processes, the same processes that are responsible for soil formation. The resulting products from weathering can include gases or liquids that become part of the atmosphere or hydrosphere. In most cases, residuals and reaction products remain with the solid phase of the waste itself or migrate into the soil on which the waste has been disposed.

Certain types of solid wastes are deliberately spread out over a large area of land in order to maximize the degree of mixing and interaction with the soil. Composted organic residues and manures are examples of this. Mixing is done so that the beneficial effects of the added solid are assimilated by the soil that receives it.

In contrast, for some wastes the degree of interaction is made as small as possible. Mixed urban waste, for example, is often compacted and then encapsulated in a *sanitary landfill* by using synthetic membranes to isolate the garbage from the atmosphere, land, and groundwater. By minimizing free movement of air and water from outside the landfill, reaction rates are minimized and the reactions that occur within the entombed material are different from those of the typical oxidative hydrolytic processes. There are persons who practice a type of 'archaeology' on old landfills, and situations are reported where newspapers buried 80 years ago have been unearthed and are perfectly readable.[1]

Because there is a wide variety of solids that have been disposed of on land, it is best (once again) to consider the subject of solid wastes in terms of general principles. We will use some of the most common examples to illustrate these principles.

**Main point 19.1** Creative research is required to minimize and find uses for solid waste. Where, as a last resort, such bulk solid wastes must be discarded, they are often placed on or in the land for disposal. In some cases, effort is made to isolate them from the soil; in other cases they are deliberately incorporated into the soil.

[1] Rathje, W.L. and C. Murphy, *Rubbish: the archeology of garbage*, HarperCollins, New York; 1992.

## 19.1 **Solid wastes from mining and metal production**

In total volume and mass, tailings are the major waste products from many mining operations. Throughout the world, these deposits cover hundreds of thousands of hectares of land (Fig. 19.1). Tailings are the rejected rock produced during upgrading of a mined ore. In many mineral deposits, the original concentration of the ore mineral itself is very small—a few parts per million in the case of gold, for example—and much of the associated rock must be rejected. A common method of upgrading or *beneficiation* is by flotation. This process requires that the entire mass of ore that is mined be ground to small particle size and separated by gravity, aided by surfactants in an aqueous slurry. The *concentrate*, containing most of the desired mineral, is then sent for further processing and the *tailings*, which comprise the major portion of the originally mined rock, are piped out into a large receiving pond. Besides the flotation method, other physical and chemical processes are used for upgrading the ores, depending on the nature of the mineral and the form in which it is found. The waste materials are variously referred to as tailings, gangue or spoils.

Drainage in relatively coarse-grained tailings is rapid, leaving a dry sandy deposit that either fills a depression in the landscape or is built up into an artificial hill. Fine-grained deposits and those that have an affinity for water may not easily become dry and so stay partially wet or sometimes remain totally under water. Some types of tailings are deliberately kept submerged in a tailings pond.

**Fig. 19.1** The Quebrada Honda tailings deposit in Peru will be up to 130 m deep and will extend over a distance of 3.9 km. The capacity is for 530 000 m$^3$ of tailings from the Cuaione and Toquepala copper mines.

The physical and chemical properties of tailings are highly variable, depending on the nature of the ore mineral itself and of the host rock. Problems associated with the disposal of tailings depend on the particular properties. Three commonly encountered situations will be discussed briefly here.

## Benign tailings deposits

In some cases, the gangue minerals that form the bulk of the tailings are relatively inert. The diamond mines of South Africa represent a case in point. There, the diamonds are found as clusters in funnel-shaped 'pipes', which are intrusions of kimberlite into the host rock. The kimberlite is variable in composition but usually consists of resistant ultrabasic igneous minerals including olivine ($(Mg,Fe)_2SiO_4$), biotite ($K(Mg,Fe)_3AlSi_3O_{10}(OH)_2$), garnet (e.g. $Fe_3Al_2(SiO_4)_3$), and ilmenite ($FeTiO_3$). The igneous intrusion contains sporadic inclusions of fossil-containing shales derived from earlier sedimentary deposits. There are many of these diamond-rich pipes around the country, with major sites being located near Kimberley and Pretoria.

Tailings from such mines are made up of finely ground rock fragments whose mineral composition reflects that of the kimberlite host rock. These mostly silicate minerals undergo only very limited and very slow chemical reactions after exposure to air and water. Because they are relatively inert, leaching of metals from them is minimal. As a result, chemical contamination of associated water bodies occurs only rarely.

### Stabilization and reclamation problems

There are problems, however. If the tailings remain untouched as a massive deposit of inert material, they are unsightly and wind-blown dust is an aesthetic problem as well as a potential health hazard. For these reasons, major efforts have gone into stabilizing the deposits. The most appropriate method of stabilization usually involves covering them with a suitable type of vegetation. To do this, the physical and chemical properties of the tailings as plant growth media must be considered.

Most tailings have relatively homogeneous particle-size distributions, with particle-size median values in the sand or large silt-size regions (Fig. 18.1). Because organic matter concentrations are close to zero, there are no agents for binding particles and so the deposit has little 'soil' structure. As a consequence, permeability is high, drainage is rapid, and there is very limited water-holding capacity. Where the tailings are dark-coloured, their albedo is small and they absorb solar radiation efficiently. Therefore, daytime temperatures in the surface material can be extremely high. Taken together these factors are responsible for a dry, hot medium that is not in any way a hospitable site for growth of plants.

There is a second problem associated with trying to establish vegetation on benign tailings deposits. Tailings such as those from the kimberlite diamond mines do not contain significant natural sources of the major plant nutrients such as nitrogen, phosphorus, potassium, and calcium. Furthermore, the fact that the mineral components are inert means that release of minor and trace nutrients into available forms is not sufficient to support the growth of most plants.

One solution to the combined physical and chemical problems is to incorporate organic matter (OM) into the surface of the tailings. Along with the organic amendment, chemical fertilizers are added at the time of planting and / or afterwards. The OM serves to improve the nutrient- and water-holding capacity and can be a source of some nutrients, while fertilizers further supplement the nutrient supply. Without the OM, the cation exchange capacity of the coarse-grained, unreactive material is so small that any soluble cations not immediately taken up by plants are leached from the deposit. Complexing components of the OM provide cation-exchange sites for retention of positively charged nutrient species. Furthermore, a portion of the organic complexing components dissolve and thus enhance the rate of chemical weathering of the tailings minerals. Once a growth cycle has been established, natural processes of litter fall and decay at least partially maintain the organic content of the 'soil'. The processes of soil formation then proceed at an accelerated rate. Organic matter of various sorts, for example, composted municipal waste or sewage sludge, can serve as an appropriate amendment. This then becomes a good example where combining two waste materials can produce a useful product.

## Tailings from sulphide ore deposits

Tailings that contain several per cent sulphur as sulphide minerals present additional problems for reclamation. This situation is frequently encountered in the mining industry. Sulphide minerals, especially pyrite, $FeS_2$, are present in the host rock associated with several important ore minerals including copper, nickel, lead, zinc, and sometimes gold. Sulphides are also present in coal deposits and significant concentrations are found in the spoils associated with coal mining. Table 19.2 gives the composition of sulphide tailings associated with gold mining in South Africa and in spoils from a coal mine in Pennsylvania.

The gold and coal mine tailings suffer from similar problems to spoils defined above as benign; compared with good soils, they are very coarse-grained and are low in nitrogen, phosphorus, potassium, and other nutrients. These properties alone make them a challenge with respect to establishing a vegetative cover. In addition, a major problem associated with the presence of sulphur in the tailings is the internal generation of sulphuric acid, giving them a very low pH.

### Acid-generating reactions

The reactions that produce acidity, most of which have a microbiological component, begin with pyrite or other sulphide minerals. The reactions are related to those discussed in Chapter 15. Some that have been clearly identified are the following:

The sulphur in pyrite is initially oxidized by oxygen in the air.

$$2FeS_2 + 7O_2 + 6H_2O \rightarrow 2Fe^{2+} (aq) + 4SO_4^{2-} (aq) + 4H_3O^+ (aq) \tag{19.1}$$

The acid-tolerant iron-oxidizing bacterium *Thiobacillus ferrooxidans* then causes oxidation of the iron (II) that was released in the initial step.

$$4Fe^{2+} (aq) + O_2 + 4H_3O^+ (aq) \rightarrow 4Fe^{3+} (aq) + 6H_2O \tag{19.2}$$

The iron (III) that is formed also acts as an oxidizing agent and further reacts with solid pyrite, without the need of molecular oxygen.

**Table 19.2** Properties of waste material from two mines containing associated sulphide minerals[a].

| Composition of waste material | Gold mine tailings, Witwatersrand, South Africa | Coal spoils, Pennsylvania, USA |
| --- | --- | --- |
| Sand and gravel / % | 50 | 67 |
| pH | 2.5–3.1 | 3.3 |
| Available elements / $\mu g\ g^{-1}$ | | |
| Mg | 110 | 127 |
| Ca | 1400 | 102 |
| K | 15 | 199 |
| P | 13 | 3 |
| N total / % | 0.02 | 0.003 |
| S total / % | 1.5–3.5 | Several % |

[a] From Additional Resources 1.

$$FeS_2 + 14Fe^{3+} (aq) + 24H_2O \rightarrow 15Fe^{2+} (aq) + 2SO_4^{2-} (aq) + 16H_3O^+ (aq) \qquad (19.3)$$

For reactions 19.2 and 19.3, the rate-determining step is the oxidation of iron (II) to iron (III). The sum of these two reactions gives reaction 19.4, which is the same as reaction 19.1. Therefore, the three processes, taken together, describe a self-accelerating sequence of oxidation of pyrite.

$$2FeS_2 + 7O_2 + 6H_2O \rightarrow 2Fe^{2+} (aq) + 4SO_4^{2-} (aq) + 4H_3O^+ (aq) \qquad (19.4)$$

At the low pH generated in this way within the tailings, the iron (II) remains in solution and can be leached from the deposit. When the leachate moves away from the tailings and is exposed to the atmosphere and less acidic surroundings, the iron (II) is oxidized to iron (III) that readily loses protons (reaction 19.5). The solid product, amorphous hydrous iron oxide is often given the approximate formula $Fe(OH)_3$. Deposits of this orange-red solid are frequently observed as coatings on the sediments where the leachate has moved downstream from the highly acid tailings.

$$4Fe^{2+} (aq) + O_2 + 18H_2O \rightarrow 4Fe(OH)_3 + 8H_3O^+ (aq) \qquad (19.5)$$

Other mechanisms for pyrite oxidation have also been suggested.[2] One of these begins with the dissolution of iron and the oxidation of sulphide to neutral sulphur.

$$2FeS_2 + O_2 + 4H_3O^+ (aq) \rightarrow 2Fe^{2+} (aq) + 2S_2^0 + 6H_2O \qquad (19.6)$$

This is followed by oxidation of iron (II) (reactions 19.2 or 19.5) and of elemental sulphur, the latter reaction mediated by *Thiobacillus thiooxidans* bacteria.

$$S_2^0 + 3O_2 + 6H_2O \rightarrow 2SO_4^{2-} (aq) + 4H_3O^+ (aq) \qquad (19.7)$$

Each of these reactions (except reactions 19.2 and 19.6) is an acid-producing process. Similar reactions can be written for other sulphide minerals such as pyrrhotite (FeS) and chalcopyrite ($CuFeS_2$) that invariably are found in association with pyrite. Over a period of years, the sulphide minerals all oxidize, generating vast quantities of acidity and of sulphate ion. The amount of acid produced is so great that leachate emanating from sulphide tailings deposits can have pH values of 1 or even lower. The sulphate-rich pore water and leachate is referred to as acid mine drainage. Being a highly acidic solution, it solubilizes other metals so that concentrations of metals like copper, zinc, and lead can also be very high. All these factors add to the hostile environment for plant, animal, and 'normal' microbial growth in the tailing deposits themselves and in nearby soil and water affected by the leachate.

Reclamation strategies for sulphide-bearing tailings must therefore include all the issues related to their physical properties as well as taking into account the essential requirement for reducing acidity of the environment. Liming is one possible amendment, but the limestone treatment must be repeated at regular intervals, as the production of acid is an ongoing process. Keeping the tailings submerged in a pond is also possible as this maintains the sulphide minerals in a stable reduced form. In spite of the difficulties, reclamation and revegetation of sulphide tailings have been carried out and self-supporting grasslands and tree plantations have been established at some sites.

It is interesting to note the similarities between sulphide tailings and the acid sulphate soils described in Chapter 18—two distinct local environments that arise in totally different settings.

[2] Kittrick, J.A., D.S. Fanning, and L.R. Hossner (eds.), *Acid sulphate weathering*, SSA Special Publication Number 10, Soil Science Society of America, Madison Wisconsin; 1982.

**Literature link** Establishing vegetation on mine tailings

The issues involved in establishing a vegetative cover on tailings deposits are reviewed by Mendez, M. O. and R. M. Maier, *Phytoremediation of mine tailings in temperate and arid environments* in Reviews in Environmental Science and Biotechnology, **7** (2008), 47.

The term 'phytoremediation' refers to the improvement of the tailings matrix using plants of various types. The authors distinguish between two approaches taking into account specific characteristics of plants, where the focus is on tolerance of the plants to, or uptake of problem elements (mostly metals):

1. *phytoextraction* in which the plants take up the problem elements, including ones that may have toxic properties, followed by plant harvest and disposal in an appropriate manner;
2. *phytostabilization*, where growth of a vegetative cover is a containment strategy that assists in immobilizing elements within the tailings.

General requirements for the establishing of vegetation on tailings include pH control, most often liming to reduce the acidity, addition of organic amendments and major nutrients. The plants that are appropriate for phytoextraction are ones that are specialized to tolerate very high levels in their tissues; the specific location of high concentrations varies from species to species. Plants used in phytostabilization prevent uptake of metals in the region around the roots (rhizosphere) or translocation of the metals into above ground parts.

Some specific properties of appropriate plant species are shown to be:

| General plant characteristics | Phytoextraction | Phytostabilization |
|---|---|---|
|  | High biomass productivity; rapid growth rate | Large canopy or good ground cover; perennial; deeply rooting |
| Bioconcentration factor (BCF) | >1 | <1 |
| Translocation factor (TF) (ratio of shoot:root concentration) | >1 | <1 |
| Shoot metal concentrations (mg kg$^{21}$) |  |  |
| As | >1000 | <30 |
| Cd | >100 | <10 |
| Cu | 1000–5000 | <40 |
| Mn | 1000–10 000 | <2000 |
| Pb | 1000–10 000 | <100 |
| Zn | >10 000 | <500 |

## Phytoextraction

A common problem encountered in the phytoextraction approach is that growth rates and hence biomass production are very low. When coupled with modest bioconcentration factors, this means that removal of metals down to acceptable levels could require hundreds of growth / harvesting cycles over an equivalent number of years. Some partially successful attempts have been made to engineer transgenic strains that have the ability to take up larger amounts of the target metals. The approach used was to insert genes that stimulated production of either γ-glutamylcysteine synthetase (ECS) or glutathione synthetase (GS). These are enzymes that affect production of thiol peptides, phytochelatin and glutathione, respectively. As we saw in Chapter 13, it is these sulphur-containing peptides that have an affinity for transition and Type B elements, whose removal is being sought. Even with these and other improvements, however, the time required and costs of production stand in the way of the widespread application of phytoextraction.

### Phytostabilization

Hardy native plants that grow prolifically and have an extensive root system are most suitable for use in phytostabilization. Growth of such vegetation may encourage sorption of metals onto root surfaces, and / or metal accumulation in root tissues. An important requirement is that the plants have a small translocation factor so that elevated concentrations are not present in above-ground parts that will be eaten by wildlife or harvested. After establishing appropriate species that are tolerant to the tailings and adapted to the local climatic situation, the initial plant cover should provide conditions for a successional development that ultimately leaves the site with a diverse, self-sustaining vegetative cover. The cycling of plant biomass encourages a heterogeneous microbial population and assists in minimizing water erosion and leaching. On balance, the relative ease of application and lower costs make phytostabilization a more promising approach to remediation than phytoextraction

The study provides examples of attempts at revegetation carried out in laboratory and field experiments around the world. The authors conclude by emphasizing the potential for further use of phytoremediation, but also its current significant limitations.

## Red mud

The 'red mud' problem[3] is in some ways the opposite of that associated with sulphide tailings. Red mud is a particular type of waste connected with the production of alumina ($Al_2O_3$) from bauxite (a highly impure form of $Al_2O_3$)—an initial step in aluminium production. In the Bayer process, aluminium is solubilized (as $Al(OH)_4^-$ also written as $AlO_2^-$) by treatment of finely milled crude bauxite with concentrated aqueous sodium hydroxide under controlled high temperature and pressure conditions.

$$Al_2O_3 + 2OH^- + 3H_2O \longrightarrow 2Al(OH)_4^- \qquad (19.8)$$

Alumina is later recovered after neutralizing the highly alkaline solution. The residual solids contain clay minerals, quartz, and insoluble iron and titanium oxides. This waste material, called red mud, has a chemical composition that is typically about 15% iron, 15% aluminium, 8% titanium, and 5% silicon and contains variable amounts of sodium and calcium, mostly as hydrated oxides. The amount of red mud produced is between 0.5 and 2 tonnes per tonne of alumina produced and, throughout the world, the total annual production of red mud is about 70 million tonnes.

The red mud is pumped as a slurry from the bauxite plant. The viscous slurry contains 20–40% solids having a wide range of particle sizes, but a substantial fraction is in the silt to clay size range so that complete settling occurs only very slowly. The large sodium content ensures that the clay does not form aggregates and remains in a highly dispersed and impermeable form. Furthermore, the suspended solids are very hygroscopic and have considerable ability to retain the slurry water. These factors lead to one of the problems associated with red muds—they do not dry out quickly when disposed of on land. One technique used to speed up the process is to spread a thin layer of the mud over a large area and allow it to dry before repeating the process. While drying, a crust of hardened material forms on the surface, inhibiting evaporation of the deposit underneath. To promote more complete drying, the crust is broken up periodically to expose the wet material below. Settling of the solid particles can also be enhanced by addition of flocculants, typically organic polymers containing anionic functional groups such as carboxylate. A different method used to promote settling can be applied to situations where the bauxite wastes are generated near the ocean. Mixing about one part red mud slurry with

---

[3] The Red Mud Project: http://www.redmud.org/home.html, accessed October, 2009.

five parts seawater results in partial neutralization of the slurry and this causes precipitation of calcium hydroxycarbonate minerals along with a cocktail of iron-, aluminium-, titanium- and silicon-containing solids.

Another problem associated with the muds is their highly alkaline nature. Unlike acid mine drainage, leachate from alumina wastes does not solubilize most metals; however because of its amphoteric nature, aluminium itself is an exception. As we have seen, aluminium is soluble in basic solution as the anionic hydroxy ($Al(OH)_4^-$) species (Fig. 16.3). The concentration of the element in water associated with red mud can therefore be very high and this, along with the alkaline nature of the solution, contributes to the toxic nature of the leachate, a problem that is not contained if it moves into surface or groundwater. Using sea water to promote neutralization to about pH 9.5 has the added benefit that it greatly suppresses the solubility of aluminium. Again, consult Fig. 16.3.

The combined physical and chemical properties of red muds make disposal in tailing ponds a problem, and alternative disposal methods and uses for the material are being sought. Revegetation is possible only if the deposit is covered with a thick layer of soil or dredged sediment. Attempts to use sewage sludge as an amendment before growing plants have met with some success. Some red mud can be used for making bricks or ceramic products. It has also been shown to function somewhat effectively as a dephosphorizing agent in wastewater treatment. None of these applications can cope with the millions of tonnes of material that is produced each year and research is on-going for better methods of reclamation.

> **Main point 19.2** Depending on the nature of the materials, disposal of mine and mineral processing wastes gives rise to a number of physical and chemical problems. Many tailings deposits are stabilized by treatment with nutrients and other amendments that allow for revegetation of the deposit. The production of acid is a particular problem associated with tailings containing sulphide minerals, while alkaline conditions are associated with the processing of aluminium ores.

## 19.2 Organic wastes

### Direct disposal of animal wastes on land

Worldwide, a very common farming practice is the application of animal manure on fields as an amendment to improve soil quality. Manure is a good source of organic matter, and also contains nutrients, especially nitrogen, phosphorus, and potassium. The composition of manure is variable, depending, in the first instance, on the species of animal, but also on diet and other conditions. Considering the common livestock species—cattle, goats, sheep, horses, pigs, and poultry—there is a wide variation in daily output of faeces and urine, but the average is often taken to be approximately 60 kg per 1000 kg mass of animal (see Table 19.3). Again, depending on species and diet, percentages of components in the excreta are approximately: total solids 10%; nitrogen 0.5%; phosphorus 0.2%; and potassium 0.3%. Concentrations are similar for most types of livestock, with the notable exception of poultry for which the nutrient values are somewhat higher than for the mammals. It is the organic matter and nutrients that make the manure a particularly good soil amendment. Using it for this purpose also responds to the objective of returning to the soil some of the components that have been removed by grazing or harvesting of plants.

### Fate of nutrient elements

The spreading of animal manure brings up aesthetic issues and there are major health concerns associated with the presence of pathological microorganisms. There are also environmental issues connected with the fate of the chemical components of the manure. Before the nutrient

**Table 19.3** Typical values for daily output of manure from livestock.

| Livestock type | Weight / kg | Total manure / kg | Nitrogen / g | Phosphorus / g |
|---|---|---|---|---|
| Cattle | 450 | 30 | 140 | 50 |
| Hogs and pigs | 100 | 15 | 100 | 30 |
| Chickens | 1.8 | 0.32 | 4.4 | 1.4 |

elements can be utilized by growing plants, the organic matter must undergo decomposition, releasing the inorganic forms that are available for uptake. To encourage degradation, a well-oxygenated environment is preferable. Even in optimum cases, however, it is likely that some of the nutrients remain in an unmineralized form beyond the growing season, and delayed release of the plant-available nutrient species may occur at times when there are no plants to take up the nutrient and no other means by which it can be immobilized. Depending on the element, the hydraulic conductivity and chemical properties of the soil, and other environmental conditions, nutrients can move through the soil into surface or groundwater.

Consider the major nutrients. Potassium is retained as $K^+$ on the cation-exchange sites of soil materials that have adequate CEC values. Phosphorus is retained in various forms by its association with iron and aluminium hydrous oxides or as insoluble calcium species. Many problems occur, however, with nitrogen. In an oxygenated soil, ammonium that is released by mineralization undergoes nitrification, yielding nitrate as the thermodynamically stable species. As we have seen, most soils have only a small anion-exchange capacity and therefore very little ability to interact with the negatively charged nitrate species. Specific adsorption of nitrate is also not significant. Therefore, nitrate tends to be mobile, and excess nitrogen released from manure frequently ends up in water bodies. A particular situation that leads to high levels of nitrate leaching occurs where manure is stored outdoors in unprotected piles during winter prior to spreading at the beginning of the growing season. Nitrate leached from the pile is not retained by the soil, and there are no plants to act as a biological control. It can therefore move downwards below the water table; toxic levels in groundwater are sometimes observed in this situation. Similar nitrate problems can arise after the manure is spread on the field. As described in the previous chapter, nitrate pollution of surface waters in the UK is an example of the consequences of poor nitrogen management.

While leaching of phosphorus and potassium is less common, these elements can be mobilized by erosion and transport of dried manure particles. We have already noted that non-point-source contamination, especially by phosphorus, makes a major contribution to eutrophication of some surface water bodies.

### Nutrient management

Proper management can minimize the chemical contamination of water from application of animal wastes. Proper management includes taking into account factors such as the type of soil, slope, drainage, and relation to natural water courses and aquifers. The amount of manure that is spread over a given area must be limited, typically to about 30 t ha$^{-1}$ each year. How this is applied, in wet or dry forms, is also important. Thirty t ha$^{-1}$ is equivalent to supplying one hectare of land with fresh waste from two or three cattle or five or six hogs. Furthermore, to ensure maximum uptake, crops should be planted soon after application of the manure. Where heavy feeders like maize are used, there is reduced chance of leaching or run-off losses. Detailed 'nutrient-management plans' are an important planning component where modern, high-intensity agriculture is being practiced. It is not uncommon for farms to have thousands of hogs or tens of thousands of poultry kept in a small confined area. The amount of waste produced in such operations can be equivalent to that of a small town, and disposal without proper planning can have disastrous consequences.

There are salinity problems associated with animal wastes as well. The total salt (chlorides of the alkali and alkaline-earth elements) content of manure can be quite large, anywhere from 1 to 10% on a dry mass basis. Concentrations in leachate from the manure can therefore be substantial, and in some cases they are excessive, also having a deleterious effect on plant growth or on the groundwater quality. The relative concentrations of the four major cations are in the order calcium > potassium > magnesium > sodium. This order is similar to that frequently observed for cations on the exchange complex and therefore it does not upset the normal ionic balance. It is the total salt concentration that creates the salinity hazard. As might be expected, these problems are more intense in arid regions of the Earth.

Finally, when manure is stored in lagoons or piles for periods of time before being spread on fields, a major concern is loss of nitrogen from the material. Significant in this regard are losses to the atmosphere. Most of the nitrogen is originally present in reduced forms as urea, uric acid or other forms of organic nitrogen all of which are rapidly hydrolysed to form ammonia / ammonium ion. Under stored conditions, a significant fraction of the ammonia can be volatilized and this contributes to the formation of aerosol particulates such as ammonium sulphate.

Other gaseous species of nitrogen can also be generated through the reactions discussed in Chapter 15. When conditions are aerobic, nitrification occurs to produce nitrite and nitrate. If anaerobic conditions then ensue in pockets of saturation, denitrification follows with both nitrous oxide and dinitrogen gases being produced. Depending on conditions, up to 15% of the nitrogen can be lost in this way, although the amount in different cases is highly variable.

## Production and use of compost

For the disposal of bulk organic solid wastes, composting offers a number of attractive features that provide environmental benefits[4]. Composting is a process in which organic solids undergo decomposition to produce a relatively stable humus-like material. The process is microbiological, involving a suite of aerobic microorganisms including bacteria, actinomycetes, and fungi. It is often carried out on an individual or family unit scale by placing the wastes in a pile or in a specially designed container. Alternatively, composting can be carried out on an industrial scale under carefully controlled conditions. With proper attention to quality control, the product is suitable for use as a soil amendment that acts as a source of organic matter as well as of small quantities of essential nutrients. A prime requirement is that there be an appropriate ratio of carbon:nitrogen (C:N) in the compostable materials. It is also essential that the final product does not contain harmful components such as elevated concentrations of potentially toxic metals, organic chemicals, and pathogens.

---

**Composting benefits**

- a source of micronutrients;
- supplies substrates that support a diverse population of heterotrophic organisms;
- improves soil structure and water holding capacity;
- keeps sequestered carbon within the terrestrial environment.

---

While the production process can be simple, careful control of conditions is necessary to ensure rapid and efficient synthesis of the stable product. The factors to be considered are:

- the nature of the organic material to be used in producing compost;
- the need to control environmental conditions for composting—temperature, aeration, and water supply;
- the time required to produce a mature and stable product.

[4] Termorshuizen, A.J., S.W. Moolenaar, A.H.M. Veeken and W.J. Blok, The value of compost, *Rev. Environ. Science Bio / Technol.*, **3** (2004), 343.

### Composting reactions

The composting reactions are essentially aerobic, analogous to the activated sludge process in wastewater treatment. In these reactions, aerobic microorganisms grow and produce carbon dioxide and microbial biomass. In the process, other components of organic matter are converted into more stable species akin to humic materials. The microorganisms require water and oxygen and so compost must be kept moist and, at the same time, must be well aerated. This is done by supplying water when necessary, while preventing the mass of material from becoming too compacted and / or by mechanically agitating it to allow the entry of air. As a consequence of the degradation reactions being exothermic, there is a temperature increase in the compost. The temperature should be allowed to rise to 50 or 60°C so that decomposition occurs at a reasonably rapid rate. The high temperature also ensures that pathogenic organisms are destroyed. While aeration is essential, excessive aeration by agitation must be avoided so as not to release too much internally generated heat. Besides affecting the rate of decomposition, a final temperature of at least 60°C should be reached and maintained for several hours or days to ensure that pathogenic microorganisms and enzymes are permanently inactivated.

The carbon:nitrogen ratio of the raw materials is critical for controlling the rate and extent of composting reactions. A large ratio means that there is insufficient nitrogen for optimum microbial growth. When this occurs, decomposition is incomplete and the compost will be immature with partial degradation products—acetic, propionic, and *n*-butyric acids, all phytotoxic—remaining in the dormant material. Furthermore, when immature compost is added to soil as an amendment, further decomposition takes place. The heterotrophic microorganisms that bring about the decomposition take up nitrogen from the soil, immobilizing it and therefore rendering it unavailable to plants. On the other hand, if the C:N ratio is too small, the excess nitrogen present is released by ammonification reactions as potentially toxic ammonia. Furthermore, the ammonia uses oxygen as it undergoes nitrification and this, along with the intrinsically large organic carbon content, can result in a reducing (low p$E$) and acidic (low pH) environment.

The optimum carbon:nitrogen ratio is around 30 and it is best controlled by careful selection of proportions of high ratio (sawdust, straw, paper) and low ratio (food scraps, manures) solids. Soil or inorganic nitrogen fertilizers can also be added, if it is necessary, to decrease the C:N ratio. Table 19.4 shows the C:N ratio for a variety of materials used in composts.

**Table 19.4** Carbon to nitrogen ratios in organic waste materials[a].

|  | C:N |
|---|---|
| High nitrogen content materials | |
| Grass clippings | 19:1 |
| Sewage sludge (digested) | 16:1 |
| Food wastes | 15:1 |
| Cow manure | 20:1 |
| Horse manure | 25:1 |
| High carbon content materials | |
| Leaves and foliage | 40–80:1 |
| Bark | 100–130:1 |
| Paper | 170:1 |
| Wood and sawdust | 300–700:1 |

[a] Data obtained from Cornell Cooperative Extension, T. Richard http://compost.css.cornell.edu/Factsheets/FS2.html, accessed November, 2009.

## The nature of composts

Typical ranges and median elemental composition of composted products are shown in Table 19.5. Organic matter makes up most of the mass of the compost. Much of the OM is in the form of nitrogen-rich humic-like materials. Functional groups, especially carboxylic acids, give compost a high potential for ion-exchange sites. The sites are occupied by the usual cations but minor elements are also present. The trace-metal content is often much larger on a mass fraction basis than in most soils (compare Table 19.5 with Table 18.8). Many of the trace metals are minor nutrients and their presence can add to the fertilizer value of compost. In excessive amounts, however, these and other metals could be toxic to plants or to animals that consumed the plants.

As a soil amendment, compost has several virtues. These virtues also apply to its use as an amendment in mine-tailings reclamation. Most important is that the large organic matter content can improve the physical properties of sand-rich or clay-rich soils. It is also a source of small but substantial amounts of essential nutrients, including trace elements. At the same time, the organic matter increases the exchange capacity of the soil. Having pointed out the virtues, the possibility of toxicity due to trace metals, or in particular cases due to organic contaminants, must always be considered.

## Sewage sludge as a soil amendment

In our discussion of wastewater treatment, we saw that the process ends with two products—the treated water that is released, usually to a lake or river, and sewage sludge. Sludge has been disposed of in a variety of ways including disposal at sea, in landfills, and through incineration, but a particularly common and important means of disposal is to use it as a soil amendment on productive (agriculture or forestry) land.

### The nature of sewage sludge

Sewage sludge is initially a slurry made up of solid material, often about 1% of the total mass, with high residual water content. It is principally organic in composition, but also contains inorganic components derived from the original sewage as well as from any metal-based coagulants added during treatment. Depending on the sources, sewage sludge composition is highly variable. Typical values for major and minor elements in some sludge samples are given in Table 19.6. Several of the elements listed in the table are nutrients and these, along with the organic matter, suggest that sewage sludge, like compost, can serve as an excellent soil amendment. Nitrogen

**Table 19.5** Composition of selected composts on a dry mass basis.

|  | C | N | P | K | Ca | Mg | Na | S |
|---|---|---|---|---|---|---|---|---|
| Median / % | 31 | 1.1 | 0.27 | 0.26 | 4.0 | 0.3 | 0.3 | 0.2 |
| Range / % | 27–40 | 0.51–1.8 | 0.15–0.6 | 0.07–0.97 | 1.2–7.5 | 0.08–0.6 | 0.2–0.67 | 0.2–0.6 |
| Number of samples | 4 | 6 | 6 | 6 | 5 | 5 | 4 | 5 |
|  | Cu | Ni | Mn | Zn | Hg | Pb | Cd | Cr |
| Median / mg kg⁻¹ | 230 | 110 | 500 | 930 | 4.5 | 600 | 7 | 220 |
| Range / mg kg⁻¹ | 100–630 | 0.76–190 | 400–600 | 500–1650 | 4–5 | 9–900 | 0.04–100 | 2–270 |
| Number of samples | 6 | 3 | 3 | 6 | 2 | 4 | 5 | 4 |

[a] Reported in He, X-T., S.J. Traina, and T.J. Logan, Chemical properties of municipal solid waste composts, *J. Environ. Qual.*, 21 (1992), 318–29.

**Table 19.6** Median values for major and minor element content of dried sewage sludge from seven American states[a].

| Element | Content / % | Element | Content / mg kg$^{-1}$ |
|---------|-------------|---------|------------------------|
| Organic C | 30.4 | Al | 4000 |
| Total N | 2.5 | Cu | 850 |
| $NH_4^+ - N$ | 0.13 | Ni | 190 |
| $NO_3^- - N$ | 0.019 | Mn | 200 |
| Total P | 1.8 | Zn | 1800 |
| Total S | 1.1 | Pb | 650 |
| K | 0.24 | Cr | 910 |
| Na | 0.12 | Cd | 20 |
| Ca | 3.8 | Hg | 6 |
| Mg | 0.46 | Fe | 8000 |

[a] Elliott, L.F. and F.J. Stevenson (eds.), *Soils for management of organic wastes and wastewaters*, Soils Science Society of America, Madison, Wisconsin; 1977.

is present in both free and combined forms with about one-quarter to one-half available as ammonia plus nitrate. The remainder is present as structural components of organic molecules. The macronutrient content is somewhat greater than that of most composts. Sludge also builds up the soil organic matter content in the same manner as compost or animal or green manures. It therefore improves soil structure and water retention, while it supplies small but significant amounts of important nutrients to the soil.

Unfortunately, many sludges originating from urban wastewater sources contain relatively large concentrations of non-nutrient metals. These too are, in most cases, present in greater concentrations than they are in compost. Some of the elements are potentially toxic to plants growing on soil treated with sludge and / or to animals (including humans) who might consume the plants. Industrial effluents can be a source of elevated concentrations of toxic metals, although, in many parts of the world, technological and legislative developments have diminished this problem. Households can be a major source as well; lead, copper, and zinc are all leached from domestic pipes, especially if the water is soft. Many consumer products such as phosphate-based detergents supply small amounts of metals to the wastewater stream.

Aerobic sludge is that which has been obtained directly from an aerobic system such as from the aerator of an activated sludge plant. The prolonged aerobic digestion results in auto-oxidation of the synthesized biomass. Anaerobic sludge is obtained after the aerator sludge has undergone further reaction in the absence of air in order to produce methane. These processes were described in Chapter 16.

### Application of sludge to soil

Sludges are applied either as a slurry containing only about 1% solids or more usually in a dewatered, partially dried form where the water content is much smaller. In applying sludge to agricultural land, there are obvious precautions, similar to those observed in the use of animal manures, that must be followed. Among these are:

- consideration of proximity to residences and wells;
- maintaining a safe distance from surface water bodies—depending on slope and soil permeability;
- considering depth of the groundwater table—depending on soil permeability;

- avoiding application on shallow soils, when bedrock is near the surface;
- observing a waiting period after application of sewage sludge before planting certain vegetable crops, because some pathogenic organisms may have survived the treatment process.

The precautions related to potentially toxic metals require that sludge not be used on organic soils, on soils with low pH, and on land already containing high concentrations of the metals. Recommended applications take into account the metal content of the sludge. One set of recommendations specifies that there should be sufficient nitrogen available in the sludge compared to the amount of potentially toxic metal. This is expressed as a minimum that the concentration ratio

$$\frac{\text{nitrogen in the form of ammonia plus nitrate}}{\text{specified metal}}$$

must meet before the sludge is considered acceptable.

---

### Example 19.1 **Acceptable sludge characteristics**

Suppose a sludge contains 0.28% nitrogen (ammonia plus nitrate), and 3.3 mg kg$^{-1}$ selenium. The minimum acceptable N:Se ratio is set at 500. Is the sludge acceptable according to this one guideline?

$$0.28\% - 2.8 \text{ g kg}^{-1} - 2800 \text{ mg kg}^{-1}$$

Therefore the calculated ratio for this sludge is:

$$2800 \text{ mg} \div 3.3 \text{ mg} = 850$$

Since $850 > 500$, the sludge is acceptable for use according to the specified criterion for this single element.

---

Another type of criterion assigns relative toxicity factors to the most abundant trace elements found in sludges. In the UK, the weighting is based on zinc equivalents, with zinc, copper and nickel factors of 1, 2 and 8, respectively. In approximately 75% of cases, these three elements have been found to be the most likely to contribute to toxicity. The amount of the three metals added to a soil should not exceed 560 kg ha$^{-1}$ (measured in zinc equivalents) over a 30-year period.

---

### Example 19.2 **Acceptable sludge application rates**

Consider a sludge that contains 920 mg kg$^{-1}$ zinc, 540 mg kg$^{-1}$ copper, and 60 mg kg$^{-1}$ nickel. Determine the amount of sludge that may be applied to a soil.

$$\text{Zinc equivalent concentration} = 1 \times 920 + 2 \times 540 + 8 \times 60$$

$$= 2480 \text{ mg kg}^{-1}$$

Note that the unit mg kg$^{-1}$ is equivalent to g t$^{-1}$. Thus, the sludge contains 2480 mg kg$^{-1}$ = 2480 g (Zn eq) t$^{-1}$ (sludge).

The recommended application limit of zinc or zinc equivalents is 560 kg (Zn eq) ha$^{-1}$ = 560 000 g (Zn eq) ha$^{-1}$. Thus,

$$\text{Recommended application limit for sludge} = \frac{5.60 \times 10^5 \text{ g (Zn eq) ha}^{-1}}{2480 \text{ g (Zn eq) t}^{-1} \text{ (sludge)}}$$

$$= 225 \text{ t (sludge) ha}^{-1}$$

According to this criterion, over a 30-year period, a maximum of 225 t of sludge could be applied to 1 ha of land.

---

As would be expected, most of the metals found in sludge are originally present in organic forms, although some may be associated with mineral phases originating from soil materials that found their way into wastewater. In soils, the trace metals are present in a wide variety of associations; with organic matter, with carbonate minerals, with iron and aluminium oxides, and incorporated into the silicate lattice. Metal associated with organic matter in sludge is released when the organic material decomposes in the field and most of it then combines with some other solid phase in the soil. The metal is therefore quite immobile, and usually remains in the top layers of the soil where it can be taken up by plants or accumulate over time.

Salinity problems are not commonly associated with the application of sewage sludge. Much of the soluble salt content has been dispersed in the aqueous effluent from the treatment plant and the sludge itself does not have a large soluble ionic concentration.

### Waste reclaiming waste

An attractive concept is to use two waste materials together in a way that is an environmental benefit to both, as we have shown in the example of the use of sewage sludge as an amendment for mine tailings. We emphasized that tailings are frequently deficient in water-holding capacity, nutrients, and exchange sites to hold the nutrients. Sludge, on the other hand, is rich in organic matter and contains at least small amounts of many nutrients. Therefore, its application to tailings (usually some additional major nutrients such as nitrogen and phosphorus are required as well) provides many of the needs of the tailings. This approach has been recommended for treatment of copper spoils.[5] Native shrubs and grasses were used in the revegetation protocol. Growth was greatly enhanced on spoils where sludge was added. This creative idea is, however, not without problems. Uptake of some of the important trace elements in the tailings and in the sludge gave high values in the shrub leaves. This indicates that, in this and many other cases, the revegetated spoils could not be used to produce agricultural crops for human or animal consumption.

## Small-scale biogas synthesis

Similarly to the anaerobic digestion of sewage sludge, a process that is carried out on an industrial scale, *biogas* can be synthesized at a domestic level in a relatively simple process. Biogas synthesis makes use of a substrate of biological waste material, particularly animal manures, to produce two products—a combustible gas and a residual organic slurry. The gas is a mixture of methane and carbon dioxide that makes a convenient clean-burning fuel with many applications, principally for cooking and heating. In Europe, where farms are sufficiently large, the gas is frequently used to generate electricity on site; the electricity is used for the farm and for household operations or is sold to the district grid. The sludge material that is left behind retains many of the desirable properties of the original manure, which makes it suitable as an amendment for soils. A design of a simple biogas generator widely used in rural areas around the world is shown in Fig. 19.2.

### Biogas production processes

Biogas generation is a complex anaerobic process, one that can be conceptually divided into three steps, all of which involve bacterial mediation. The first step is solubilization and hydrolysis, whereby solid organic matter made up largely of polysaccharides, proteins, and lipids is broken down into the component sugars, amino acids, glycerol, and fatty acids. Once in solution, the small molecules act as substrates for reactions producing acetic acid, carbonate species, and hydrogen—the latter two compounds being partially partitioned into the gas phase according

---

[5] Sabey, B.R. R. L. Pendelton, and B.L. Webb, Effect of municipal sewage sludge application on growth of two reclamation shrub species in copper mine spoils, *J. Environ, Qual.,* **19** (1990), 580–6.

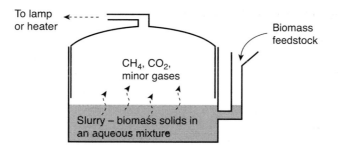

**Fig. 19.2** A simple biogas generator, isolated from the atmosphere so as to generate an anaerobic environment.

to Henry's law. Smaller amounts of other substances like sulphide species and ammonia are also produced and relatively biorefractory[6] compounds including aromatics and fatty acids remain unreacted at this stage. Finally, methanogenesis occurs through reactions in which ammonia and hydrogen take part to further reduce the acetate and carbonate species.

Using the generic simple formula for carbohydrates and the simplest carboxylic acid, a summary description of a possible reaction species is as follows.

$$2\{CH_2O\} + H_2O \rightarrow CH_3COO^- \text{ (aq)} + H_3O^+ \text{ (aq)} \tag{19.9}$$

$$CH_3COO^- \text{ (aq)} + H_2O \rightarrow CH_4 + HCO_3^- \text{ (aq)} \tag{19.10}$$

$$HCO_3^- \text{ (aq)} + 4H_2 + H_3O^+ \text{ (aq)} \rightarrow CH_4 + 4H_2O \tag{19.11}$$

The hydrogen in reaction 19.11 derives in part from reactions such as:

$$3\{CH_2O\} + 2H_2O \rightarrow CH_3COO^- \text{ (aq)} + CO_2 + 2H_2 + H_3O^+ \text{ (aq)} \tag{19.12}$$

In the acidic environment, some of the carbonate species are converted into carbon dioxide through the normal equilibrium reaction

$$HCO_3^- \text{ (aq)} + H_3O^+ \text{ (aq)} \rightarrow CO_2 + 2H_2O \tag{19.13}$$

The end result is a mixture of gases containing methane and carbon dioxide as major components, but also including small amounts of hydrogen, ammonia, and hydrogen sulphide. The methane and carbon dioxide mixing ratios are typically in the ranges 60–70% and 30–40%, respectively, with the heating value of the gas depending strongly on their relative amounts. If the feedstock is pre-dominantly cellulosic, relatively high ratios of carbon dioxide are produced, while lipids and proteins favour methane.

The reaction conditions must be carefully controlled in order to optimize production of high heat content biogas. The pH of the reaction mixture should be in a range 6.5–8.5. The allowed alkalinity is variable but may be as high as 14 000 ppm calcium carbonate. Temperatures should be in the range 20–60°C—the optimum temperature for mesophilic bacteria is about 35°C and for thermophiles about 55°C. In winter, or to operate at higher temperatures, some of the produced biogas may be required as supplemental heat for the digester, thus reducing the net usable production efficiency. Nitrogen is required to support the microbiological activity and so the carbon:nitrogen ratio of the feedstock is critical. As was mentioned earlier in this chapter the optimum value of the ratio for composting is about 30; a similar or somewhat lower value is appropriate for biogas production. Substances whose ratios are close to the optimum are shown

---

[6] Biorefractory means resistant to biological degradation processes.

in Table 19.4. Wood, even when finely chopped, is not suitable because it has a high lignin (carbon) content and a relatively small active surface area and, depending on the type of wood, its C:N ratio ranges from 50 to 400. Urine has a C:N value of 0.8. A near-ideal mixture consists of animal wastes diluted by straw as is found in cattle bedding mixtures.

The production of biogas is environmentally attractive because a clean-burning fuel is produced. Sulphur emissions remain low and, compared with direct burning of biomass, the problem of particulate emissions is nearly eliminated. The residual sludge contains most of the nitrogen, phosphorus, potassium, and other nutrients of the original feedstock. The organic content is still substantial, being about 30% of the original value—most of it biorefractory lignin-related materials as well as bacterial cellular components. When added to soil, the clean-smelling sludge supplies nutrients as well as being a source of organic matter that enhances the physical and chemical exchange properties of the soil.

The process represents a compromise between using wastes exclusively for fuel or exclusively as a soil additive. Using typical figures, the extent of the compromise may be summarized as in Example 19.3.

---

**Example 19.3 Disposal methods for agricultural biomass and their environmental benefits**

- When added directly to soil, using average values for biomass composition, one tonne of composted biomass (dry weight) provides the following approximate amounts of nutrients: 6 kg nitrogen; 1.5 kg phosphorus; and 3 kg potassium.

- When burned, using data from Table 8.6, one tonne of biomass produces approximately $1.5 \times 10^{10}$ J of energy.

- When converted to biogas and sludge, depending on conditions, one tonne of biomass produces about 200–700 $m^3$ of biogas . For a 65 : 35 ratio of methane:carbon dioxide in the gas, the heating value is two-thirds that of pure methane (Table 8.6) = $2/3 \times 3.7 \times 10^7 = 2.4 \times 10^7$ J $m^{-3}$. The total energy content of the gas then ranges between $4.8 \times 10^9$ and $1.7 \times 10^{10}$ J. The lower number would be closer to the actual value in many cases. A slurry of sludge containing 300 kg of the original solid biomass remains to be applied to the soil.

---

**Main point 19.3** Organic residues of plant and animal origin have value both as soil amendments and as energy sources. Two environmentally desirable options are available for their disposal—composting and then application to land, or conversion to gaseous fuel. In the latter case, the residue can be applied to the land.

## 19.3 **Mixed urban wastes**

The material that is discarded by urban society is a complex mixture of many substances, and the nature of the mixture is highly diverse depending on the situation. In high-income countries, the traditional dump (euphemistically called a 'sanitary landfill') has contained a wide range of organic wastes, paper, glass, metals, and plastics. Table 19.7 gives ranges of composition, determined in five studies, of material discarded from urban areas in the USA.

All of the materials in the table have some value in terms of nutrient content, energy content, and / or as a resource for recovery of metals or other commodities. Partly for this reason, in recent years wealthy societies have sought ways to make use of the discarded materials. Paper, cardboard, glass, steel, aluminium, and various types of plastics are removed from the waste stream prior to landfill; then are processed in various ways, and then recycled in new forms. Recycling is not a novel phenomenon in low-income countries. There has traditionally been a market for many of the materials, and systems have long been in place to enable collection for reuse and recycling.

**Table 19.7** Composition of the solid waste stream as determined in several studies in the USA[a].

|  | Median / % | Range / % |
|---|---|---|
| Food waste | 8.8 | 8.0–22.8 |
| Yard waste (grass, weeds, leaves, etc.) | 19.8 | 11.0–28.2 |
| Other | 8.9 | 7.8–16.2 |
| Total organic | 39.0 | 28.8–50.8 |
| Newsprint | 5.2 | 3.1–7.3 |
| Cardboard | 7.0 | 6.2–10.8 |
| Mixed paper | 21.8 | 5.6–32.7 |
| Total paper | 35.0 | 14.9–48.7 |
| Ferrous metal | 5.1 | 3.2–7.5 |
| Aluminium | 1.2 | 0.3–1.6 |
| Other metal | 0.2 | 0.2–0.7 |
| Total metal | 6.0 | 4.9–9.0 |
| Plastics | 7.3 | 7.0–9.4 |
| Glass | 5.0 | 2.4–8.3 |
| Other inorganic materials | 5.3 | 1.9–12.2 |
| Total plastics + glass + other non-metal inorganic materials | 18.0 | 16.1–27.3 |

[a] Data reported in O'Leary, P. and P. Walsh, Introduction to solid waste landfills, Course material for C240-A180 Solid Waste Landfills, a correspondence course from the Solid and Hazardous Waste Education Center, University of Wisconsin, Madison.

## Landfilling

Disposing of mixed urban waste in a *sanitary landfill* is one means by which urban garbage is handled. Landfilling may follow removal of components for composting or recycling, or it may be used for disposal of ash after an incineration process. In many cases, waste disposal is a one-step process in which all the collected material is directly placed in a dump. For public health and aesthetic reasons, a properly sited landfill should not be located in the immediate vicinity of population centres. Balancing this is the need to take into account the economics related to transport of the garbage away from the location where it has been generated. To maximize the amount of garbage that can be disposed of in any site, the waste is usually compacted, often with heavy machinery, into a dense mass. What happens within this highly concentrated mixture of solid materials?

### Degradation reactions

Most garbage sites contain substantial amounts of degradable organic substances—food wastes, wood, paper, and other fibres—in addition to more inert materials. Some organic materials will be present even when an efficient composting and / or separation process precedes the landfilling operation. Degradation of the OM is again a microbial process and, as long as oxygen is present to act as an electron acceptor, the usual oxidation reactions proceed through various intermediates eventually producing carbon dioxide and water as the major products. Nitrogen is converted to nitrate. However, in a typical landfill the compact nature of the deposit inhibits air transfer into the bulk of the mass, and the local environment soon becomes anaerobic. The first stages of anaerobic decomposition lead to the production of small molar mass organic acids along with additional carbonate species and hydrogen gas. The decomposition processes will include reactions such as those that apply in the case of biogass production (reactions 19.9 to 19.13). At this stage, the pH of leachate in the landfill drops to around 4.5, sufficiently low that some metals are solubilized. Calcium and magnesium are the most common metallic ions found in leachate at this point. Under the low-pH conditions, methane-producing bacteria do not function efficiently and little fermentation takes place. Eventually, however, methanogenic bacteria become dominant and the principal degradation process is fermentation leading to the formation of methane and carbon dioxide in a 1:1 molar (and volume) ratio.

$$2\{CH_2O\} \rightarrow CH_4 + CO_2 \tag{19.14}$$

As in the case of domestic biogas production, the proportion of methane is again higher for degradation of proteins and fats. The 'average' landfill generates these gases in an approximately equimolar ratio.

Assuming that the average degradation reaction is represented by reaction 19.14, 370 m³ of methane would be produced from 1 t of waste. Because a typical mixed waste contains about 75% decomposable organic matter and because the decomposition is never complete, actual amounts of methane produced are often found to be nearer to 25% of this value (approximately 100 m³). Nevertheless, this is a valuable energy resource that can be used to heat adjacent buildings, as an energy source for a factory, or for other purposes. Methane is, of course, a potent greenhouse gas, providing another reason why it should not be allowed to leak into the atmosphere without being collected and used as fuel. Engineered collection systems (Fig. 19.3) involve sealing the landfill to prevent escape of the gas into the surrounding soil, and installing pipes to transport the gas to the location where it is stored or used. The seal usually requires a synthetic polymer liner and capping system.

### Leachate

The degradation does not proceed cleanly to only gaseous products. A liquid leachate is also produced whose composition varies over the stages of decomposition described above. Figure 19.4 goes beyond the simple reactions 19.9 to 19.13 and shows a sequence of this variation along with changes that take place in gas composition and production over a 2-year period. After the aerobic phase of decomposition is complete, the initial anaerobic processes produce a leachate with a large organic matter content as shown by the chemical oxygen demand (COD) plot in the figure. This COD is due in large part to organic acids produced by the anaerobic decomposition. The low pH solution at this point can dissolve some metals, including those that are toxic, and it can also contain a wide variety of refractory organic species. It is important, therefore,

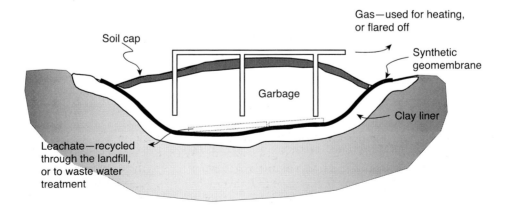

**Fig. 19.3** A well-designed urban waste disposal site.

to contain the leachate. The synthetic liner serves this purpose, and along with it a thick clay underlayer is provided. A liquid-collection system directs the leachate to the lowest point within the site where it can be pumped to the surface. There, it is either recirculated through the landfill or removed for further treatment. In some cases, the treatment is done at a wastewater treatment plant where landfill leachate is added as a component of the influent stream.

> **Fermi question**
> Estimate the area of land required to accommodate all the solid waste generated in your city (or one nearby) over the next 50 years. Assume that landfilling is the only disposal option.

## Incineration

Incineration of urban waste materials is an alternative to landfilling. There are several attractive features to this alternative. One is that it provides an efficient means for energy recovery from much of the waste, and it is a means of disposal that can be set up in an area immediately adjacent to a population centre. Secondly, it reduces the volume of waste considerably so that much less land is required for final disposal. Furthermore, it eliminates problems associated with methane generation and leachate that emanate from a landfill site. In spite of its considerable advantages, there are large environmental problems to consider as well.

### Energy value of wastes

It is useful to begin an examination of the environmental consequences by looking at the energy-production issue. We noted that, in a typical landfill, about 100 m³ (4500 mol) of methane could be recovered from 1 t of waste. When the methane is burned, the reaction, with enthalpy is

$$CH_4 + 2O_2 \rightarrow CO_2 + 2H_2O \quad \Delta H = -890.3 \text{ kJ mol}^{-1} \tag{19.15}$$

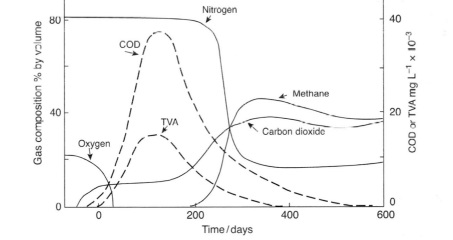

**Fig. 19.4** Changes in gas composition (solid line) and leachate composition (dashed line) over a 2-year period in a landfill. COD, Chemical oxygen demand; TVA, total volatile acids. COD is measured in terms of acetic acid. The total volume of gas evolved from the landfill reaches a maximum at about 380 days. (Based on a figure in Pohland, F.G., J.T. Derien, and S.B. Ghosh, Leachate and gas quality changes during landfill stabilization of municipal refuse, In *Proceedings of the 3rd International Symposium on Anaerobic Digestion* (ed. R.L. Wentworth), pp. 185–201, Dynatech, Cambridge, Massachusetts; 1983.)

Therefore, from 1 t of waste, it is possible to obtain 4500 mol $\times$ 890 300 J mol$^{-1}$ = 4.0 $\times$ 10$^9$ J of energy, assuming an efficient collection and recovery system.

If, alternatively, the 1 t of waste is burned directly, typically about 85% is combustible (the combustible value is somewhat higher than that used for decomposition, because it includes plastics that break down in a landfill only very slowly). Again, using the generic carbohydrate formula, the combustion reaction can be written as

$$\{CH_2O\} + O_2 \rightarrow CO_2 + H_2O \quad \Delta H = -440 \text{ kJ mol}^{-1} \quad (19.16)$$

In this situation, 1 t of waste = 1 $\times$ 10$^6$ g / 30 g mol$^{-1}$ = 3.3 $\times$ 10$^4$ mol of $\{CH_2O\}$
This produces 3.3 $\times$ 10$^4$ mol $\times$ 0.85 $\times$ 440 000 J mol$^{-1}$ = 1.2 $\times$ 10$^{10}$ J of heat energy.

There is therefore about a threefold gain in maximum energy recovery when using direct combustion compared with recovering methane from a landfill and using it as fuel.

### Combustion products

There are, however, other environmental consequences to consider. During incineration, most of the organic matter is converted into carbon dioxide and water. Depending on the nature of the waste and on the combustion conditions, other gases are also produced—sulphur dioxide, the NO$_x$ compounds, polyaromatic hydrocarbons, and chlorinated organic substances—of which some are of particular environmental concern. We will say something about some of the particularly hazardous chlorinated organics later.

While incineration reduces the mass of the mixed urban waste considerably, it does not eliminate it entirely. There are also residual solids called ash. Some of the ash (usually <1%) is emitted through the stack as fly ash. An incinerator that treats 20 t h$^{-1}$ of waste can release 2 to 10 kg h$^{-1}$ of fly ash. A much larger fraction of the solid component (>99%) is present as bottom ash that remains as residue in the combustion chamber after incineration is complete. Recovery of a good portion of the fly ash is technically possible using technology described in Chapter 6. This residue and the bottom ash must be disposed of and both components present their own unique problems.

### Gaseous emissions and fly ash

With regard to metals, at least a portion of the volatile metals such as cadmium, lead, and mercury are vaporized and emitted from the stack, while more refractory elements tend to remain in the bottom ash. The fate of metals is not described in a simple way, however. Volatility depends on the form of the element, the nature of other materials in the waste stream, and the operating conditions of the incinerator.

Take the example of lead, an element that is present in waste materials in paints and discarded batteries, among other things. These should never be incinerated but, without great care to remove them, it is almost inevitable that some will be included in the waste stream. The metal (Pb$^0$) begins to volatilize significantly at around 1000°C; this temperature is usually achieved in the afterburner chamber of an incineration unit. Therefore, at least some of the lead is released in elemental form into the vapour phase. If the furnace is operating under oxygen-limited pyrolysis (reducing) conditions, sulphur will be converted to sulphide (S$^{2-}$) that reacts with lead to generate lead (II) sulphide. This compound also becomes significantly volatile at around 1000°C. Under oxidizing conditions—an air-rich combustion mixture—some of the lead is converted to lead (II) oxide. The oxide is more volatile than either lead metal or lead (II) sulphide. A major effect on lead release occurs when there are chlorine-containing compounds in the waste stream; a common source is the widely used polymer, polyvinyl chloride (PVC). The reaction to produce lead (II) chloride is rapid and the compound is highly volatile so that losses of the metal are very rapid and complete. Other metals too have their own characteristic reaction possibilities, but in most cases the presence of chlorine enhances metal vaporization.

After leaving the combustion unit and moving into the stack, the gases cool and further transformations of vaporized metals take place. One possibility is that metal species will homo-geneously form nuclei that coagulate and grow to become a stable aerosol. A second possibility is that the metal, in free or combined form, will condense on the surface of other ash particles. Incorporation in the aerosol may also occur by a third mechanism wherein the metal *reacts at* rather than *condenses on* the surface of another particle. In each of these cases, the metal becomes part of the fly ash, and will be expelled from the stack unless a removal process has been engineered into the system.

It may be possible to take advantage of the last (reaction) mechanism in order to design a means by which metals are efficiently removed from the stack gases. Uberoi and Shadman[7] tested various solid sorbents for their ability to capture lead emitted in a simulated flue gas in a laboratory reactor. The bed of sorbent was kept at an elevated temperature to ensure that the lead was in the vapour phase when it passed through this material. Alumina and kaolinite, among others, were found to be especially efficient in retaining lead, by reactions of the type

$$Al_2O_3 + PbCl_2 + H_2O \xrightarrow{SiO_2} PbO \cdot Al_2O_3 + 2HCl \tag{19.17}$$

Lead deposited by chemical reaction on such sorbents is not extractable by water, in contrast to metal that condenses rather than reacts to form an aerosol.

Retention of fly ash from municipal incinerators and other combustion sources is a very important issue. Much of the ash is well within the $PM_{10}$ (fraction of small particles smaller than 10 $\mu$m in diameter) size range, and it can have a very high concentration of metals. The more volatile metals including cadmium, copper, lead, and zinc are especially concentrated. Lead and zinc have been found at levels between 1 and 10% in the particulates. Moreover, because these metals are condensed on the surface of the aerosol, they have been found to be readily available in physiological reactions and with respect to solubilization. For this reason, fly ash has been classified as a hazardous material in some jurisdictions and its disposal in landfills is proscribed. But again, one can think of possibilities for employing fly ash as a resource rather than a waste. If amounts and availability of the trace metals are carefully controlled, it may be possible to use it as a soil amendment. One such approach, at least at an experimental stage,[8] is to use small amounts of fly ash in combination with sewage sludge. Both these materials can supply essential micronutrients for growing plants, and the organic matter in the sludge is a source of chelating agents that moderate the mobility and availability of the metals. As with all such technical 'solu-tions' to environmental problems, a cautious approach is essential before applying such ideas to the production of consumable products.

### Toxic organics

We have emphasized the fact that metal concentrations are a major problem in the gaseous emis-sions and in fly ash. Organic contaminants are also very important. The compounds of great-est concern are chlorinated organic substances including chlorobenzenes (CB), chlorophenols (CP), polychlorinated biphenyls (PCB), furans (PCDF), and dioxins (PCDD). Some examples of the chlorinated compounds are shown in Fig. 19.5. Each of these examples represents a class of compounds having varying degrees and positions of chlorine substitution. This is indicated using the generic structure for PCBs. In total, there are 209 possible congeners and, depending on the source of the PCBs, various congeners pre-dominate. The PCB compounds are somewhat

---

[7] Uberoi, M. and F. Shadman, Sorbents for the removal of lead compounds from hot flue gases. *Am. Inst. Chem. Eng. J.*, **36** (1990), 307–9.

[8] Sajwana, K.S., S. Paramasivam, A.K. Alva, D.C. Adriano and, P.S. Hooda, Assessing the feasibility of land application of fly ash, sewage sludge and their mixtures, *Adv. Environ. Res.*, **8** (2003), 77–91.

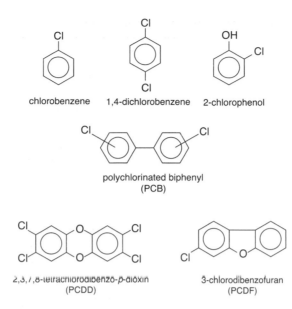

**Fig. 19.5** Some chlorinated organic compounds that have been found in fly ash. A generic formula for PCBs is given. The 2,3,7,8-tetrachlorodibenzoparadioxin is considered to be a reference compound for dioxins in general.

toxic and may be weakly carcinogenic; the more highly chlorinated PCBs have greater toxicity. Polychlorinated biphenyls have been an important industrial chemical with several uses, especially as liquid insulators for heavy electrical equipment. Because of their toxicity, production has been curtailed but much of the produced material is still in use. The PCBs can be destroyed by various means but, when they find their way into landfills or conventional incinerators, uncontrolled release as leachate, in gaseous form, or in fly ash can occur.

Table 19.8 lists some properties of PCBs having 1, 4, or 8 chlorines in their structures. It is instructive in that it shows that vapour pressure, aqueous solubility, and reactivity all decrease substantially with increasing chlorine content.

Also shown in Fig. 19.5 are examples of the dibenzo-*p*-dioxins (PCDDs) and dibenzofurans (PCDFs). Again, there are many congeners of these compounds and, as a group, they are loosely referred to as dioxins and furans. They are not an industrial product as such, but are produced and released in small amounts during combustion of chlorine-containing organic materials such as polyvinyl chloride (PVC) plastics. PCBs are themselves a precursor and the dioxin product is much more toxic than the original PCB. Unlike metals, which can only be released if they are already present in the waste material, many organic compounds can be formed in the incinerator from other organic materials. Chlorine is obviously a requirement for formation of chlorinated organics. It is available in various inorganic forms and also in some polymers such as the commodity product polyvinyl chloride.

On a mass basis, concentrations of the organic compounds in ash are much smaller than those of the metals. For the chlorinated substances maximum concentrations in fly ash may be as high as 1 µg g$^{-1}$ for the individual species. Because of their intrinsic volatility, the amounts are substantially less in bottom ash. The compounds are initially released in the vapour phase or are adsorbed on to smoke and dust particles. If they are not removed in the stack, they are carried into the atmosphere where they may condense or associate with other components of the aerosol. Eventually, they settle as dry deposition or are washed out by rain into water or soil or on to the surface of vegetation.

**Table 19.8** Properties of PCBs having variable chlorine content[a].

| Number of Chlorines | %Cl | $P_v$ / Pa | $S(aq)$ / g L$^{-1}$ | $t_{1/2}$ | | |
|---|---|---|---|---|---|---|
| | | | | Air | Water | Soil |
| 1 | 18.8 | 0.9 to 2.5 | 1.2–5.5 | 1 week | 8 months | 2 years |
| 4 | 48.6 | 0.002 | 0.010–0.043 | 2 months | 6 years | 6 years |
| 8 | 68.8 | 0.00002 | 0.000001 | 6 years | 6 years | 6 years |

[a] $P_v$, Vapour pressure at 25°C; $S(aq)$, aqueous solubility; $t_{1/2}$, approximate half-life in the various environments.

Dioxins and the other chlorinated compounds have very limited solubility in water. The octanol / water partition coefficient ($K_{OW}$) for 2,3,7,8-TCDD is approximately $10^6$–$10^7$ and the $K_{OM}$ value is $1.4 \times 10^4$. The compounds therefore have a tendency to partition into soil, especially when the soil has a substantial organic content. Leaching tests indicate negligible solubility, although very small amounts of smaller and / or more polar compounds such as chlorophenols and chlorobenzenes are dissolved. If ash is deposited on the surface of the ground, the chlorinated compounds usually remain in the top 15 to 30 cm. In most cases, any vertical or lateral movement in pore water would occur via soluble organic matter or dispersed soil colloids rather than in true solution.

As we have noted in Chapter 6, the polynuclear aromatic hydrocarbons (PAHs) are another category of organic compounds emitted from combustion processes including incineration of solid waste. PAHs are present in both fly ash and bottom ash in concentrations that are typically 100 to 1000 ng g$^{-1}$. Like the chlorinated compounds, they too have a strong affinity for organic matter as reflected in their $K_{OW}$ values that fall in the range from $10^4$ to $10^7$. This means that the PAH compounds also tend to associate with the soil phase, the tendency being greater when soil organic matter concentrations are large. Very little PAH dissolves directly in water but, where the water contains dissolved or particulate organic matter, there is a competition for association of the PAH between the aqueous-phase organic matter and solid phase organic matter. The limited transport that does occur in soil is through the latter association.

One set of specifications for municipal incinerator operations[9] recommends that maximum emissions from an urban waste incinerator be 11 mg Rm$^{-3}$ for total particulates, 7 µg Rm$^{-3}$ for cadmium, 76 µg Rm$^{-3}$ for lead, 56 µg Rm$^{-3}$ for mercury, and 0.14 ng Rm$^{-3}$ for TCDD. The unit Rm$^{-3}$ is for a reference cubic metre, meaning that the conditions have been normalized to 25°C, $P^o$, and 11% oxygen. In order to ensure compliance with emission standards, furnace design and operation must be carefully evaluated. In the particular study referred to, it was necessary to ensure that the combustion temperature consistently be maintained at greater than 1100°C and never allowed to fall below 1000°C, that the residence time of gases in the furnace be at least 1 s, that there be good turbulence, and that oxidizing conditions are maintained by keeping at least 6% residual oxygen after burning.

### Bottom ash

In almost all incineration situations, the amount of solid residue at the bottom of the furnace is much greater than that of fly ash and depends on what, if any, prior removal of waste material has taken place. Typically, 10–20% of the total mass remains after combustion of mixed urban wastes; it consists of 'siftings' that have fallen through the gratings in the furnace as well as the

[9] Anonymous, National Incinerator Testing and Evaluation Program: environmental characterization of mass burning incineration technology at Quebec City, Report EPS3 / UP / 5, Environment Canada; 1988.

bottom ash. Assuming that efficient combustion has occurred, the slag-like material contains $SiO_2$, $Al_2O_3$, $CaO$, $Fe_2O_3$, $Na_2O$, $K_2O$, and $MgO$ as its principal components. Small amounts of other elements are associated with the non-combustible solids and are often incorporated in a silicate matrix. The high temperature in the furnace volatilizes most of the low boiling metals and organic materials leaving much smaller concentrations than are present in fly ash. (The PAH compounds are an exception in that their concentration in bottom ash is comparable to or greater than that in fly ash.)

Even when present only in low concentration, leachability of undesirable chemicals from the residue is still a possibility. To test for this, standard leaching tests have been developed such as the Sequential Batch Extraction Procedure (ASTM Standard No. D4793–88). In this test, the ash is mixed with water in a mass ratio of 20 parts water to 1 part ash. The mixture is shaken for 18 h, separated, the aqueous phase saved, and the leaching procedure repeated for a total of five cycles. The combined leachate is then analysed for content of metal and organic contaminants. In most cases, for municipal solid waste (MSW), chlorinated organic compounds of concern are extracted at levels near or below the detection limit. Most metals too are extracted in only very small concentrations. The bottom ash is therefore a much more inert and benign product than is fly ash, and it can usually be used as construction material (for example for rail or road beds) or disposed in a landfill. Ferrous and non-ferrous metals are recovered from some incinerator facilities, especially in Europe.

One final point to make about incineration as a waste-disposal measure arises from the need to have combustible materials to burn. These are, of course, the same materials (food wastes, paper, plastics, etc.) that are most appropriate for recycling or composting. As a result, incineration is a disincentive for these other, environmentally more desirable means of diversion of the waste stream away from landfilling. Furthermore, because the incinerator requires a constant source of fuel it actually can encourage the production of a steady supply of combustible waste.

> **Main point 19.4** Urban wastes comprise a complex mixture of materials. Disposal must take into account their resource and energy values and the environmental consequences of the various management strategies. Landfilling, composting, and incineration all have environmental consequences associated with them.

## ADDITIONAL RESOURCES

1. Hargreaves, J.C., M.S. Adl, P.R. Warman, A review of the use of composted municipal solid waste in agriculture, *Agric. Ecosys. Environ.*, **123** (2008), 1.

2. Steinfeld, H. and T. Wassenaar, The role of livestock production in carbon and nitrogen cycles, *Ann. Rev. Environ. Res.*, **32** (2007), 271.

3. Wong, M.H. and A.D. Bradshaw, eds., *The restoration and management of derelict land: modern approaches*, World Scientific Publishing Co. Pte. Ltd., Singapore, 2002.

4. Salomons, W. and U. Forstner (eds.), *Chemistry and biology of solid waste*, Springer-Verlag, Berlin; 1988.

## PROBLEMS

1. The 'Trelogan' tailings[10] are associated with a calcareous lead / zinc deposit in Britain and have the properties shown.

$$pH = 7.0; \quad CEC = 2.8 \text{ cmol } (+) \text{ kg}^{-1}; \quad conductivity = 2.3 \text{ dSm}^{-1}$$

[10] Data adapted from Additional Resources 1.

Describe the nature of any expected physical and chemical problems associated with revegetation of these tailings, and how the problems might be overcome.

| Particle size / mm | Percentage of total |
|---|---|
| >2 | 6.7 |
| 2.0–0.2 | 14.6 |
| 0.2–0.02 | 29.7 |
| 0.02–0.002 | 24.8 |
| >0.002 | 24.2 |

| Total analysis of tailings | | | | | | | |
|---|---|---|---|---|---|---|---|
| Element | $\mu g\ g^{-1}$ | Element | $\mu g\ g^{-1}$ | Element | $\mu g\ g^{-1}$ | Element | $\mu g\ g^{-1}$ |
| N | 126 | Ca | 138 500 | Cu | 205 | Pb | 39 800 |
| P | 160 | Mg | 1500 | Cd | 267 | Zn | 95 000 |
| K | 1070 | F | 185 | Ni | 87 | | |

2. Use the data in Table 19.3 to estimate the C : N ratio in cattle manure. Compare the value with that given in Table 19.4.

3. Arsenic is often associated with sulphide minerals. Because of this, significant amounts of arsenic may be discarded along with the tailings from mining operations involving these minerals. Refer to the material provided in Chapter 10, Problem 9, and indicate what species of arsenic would initially be present in the tailings, and what species might be generated over time.

4. The nickel $pE$ / pH diagram is shown in Fig. 19.P.1. Included in setting up the diagram is a concentration of $Ni^{2+}$ of $10^{-4}$ mol L$^{-1}$ and a total S concentration of $10^{-3}$ mol L$^{-1}$. Predict the

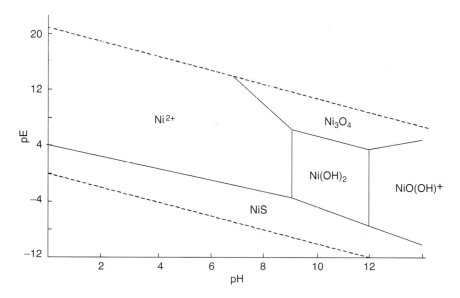

**Fig. 19.P.1** Nickel $pE$ / pH diagram.

environmental fate of Ni discharged as the mineral NiS if it is disposed into a well-aerated tailings pond or, alternatively, if it is placed at the bottom of a deep lake.

5. If the acidic leachate from sulphide tailings flows in a stream along the surface of the land, it is frequently observed that a red-brown deposit is found downstream on the rocks or sediment in the stream bed. Discuss the chemistry that might explain this observation.

6. Use values of the acid dissociation constants for aluminium aquo complexes to calculate the solubility of aluminium in red-mud leachate that has a pH of 10.3.

7. Describe the significance of the following properties in assessing the quality of compost: concentration of carboxyl groups; carbon:nitrogen ratio; total phosphorus concentration; concentration of trace metals.

8. Consider two cases regarding a landfill.

    (a) There is no collection system and the landfill gases are allowed to escape into the atmosphere.

    (b) The gases are collected and flared off, with 83% efficiency.

    Using data from Table 8.4 compare the relative global-warming potentials (GWPs) of the gases released in the two systems.

9. At the United Nations Environment Program (UNEP) website, there is a graphical representation showing composition of municipal solid waste in seven European countries and 7 large Asian cities. Compare the composition of waste in Norway with that in Kathmandu (Nepal). Discuss likely reasons for the differences and what strategies might be most appropriate for managing these two types of waste. (http://maps.grida.no/go/graphic/municipal_solid_waste_composition_for_7_oecd_countries_and_7_asian_cities, accessed November, 2009).

    Use the compositional data (above) and consult the Environment Section of the Eurostat yearbook (http://epp.eurostat.ec.europa.eu/portal/page/portal/publications/eurostat_yearbook, accessed November 2009) to determine the total amount of metal, glass, plastic and paper waste in Norway and the total amount of each recovered.

10. A possible classification of metals in incinerator ash is based on their leachability.[11]

| Ion exchangeable | Readily leached |
| --- | --- |
| Metals associated with oxides or carbonates | Can be leached under acidic conditions |
| Metals associated with iron and manganese | Can be leached under reducing conditions |
| Metals bound to sulphide or organic matter | Can be leached under strongly oxidizing conditions |
| Metals in the silicate lattice | Not leachable |

Use your knowledge of metal association with solids to explain this classification system.

---

[11] Fraser, J.L. and K.R. Lum, Availability of elements of environmental importance in incinerated sludge ash, *Environ. Sci. Technol.*, **17** (1983), 52–4.

# Chapter 20
# Synthetic organic chemicals

The focus of this chapter is on synthetic organic chemicals—especially those that are toxic or can generate other adverse effects within the environment. We will investigate their nature and environmental properties by:

- defining what we mean when we use the terms *biocide* and *xenobiotic compound*
- investigating the nature of their chemistry and stability; redox properties and photoactivity
- discussing the means by which their transport in the environment occurs, primarily in soil and water systems
- introducing the concepts of leachability and a simple function that predicts their potential for harm.

In this chapter, we are introducing another very large topic—a topic on which much research has been focused and that is the subject of thousands of papers and books. It is again appropriate to begin with some definitions and a description of the scope of our discussion. The subject is synthetic organic chemicals, but we are principally concerned with chemicals that can have a harmful effect on living organisms. Sometimes such chemicals are referred to as biocides whose literal meaning comes from Greek *bio* = life, and Latin *cide* (*caedere*) = to kill. An alternative name with a similar broad significance is *xenobiotic compounds* (Greek *xeno* = alien, strange, and *biotic* = to life).

Included in the general listing of synthetic organic chemicals are substances made for the deliberate purpose of targeting and killing specific organisms (category 1 in Table 20.1). Pesticide is a general name for this particular subclass of biocides. Within the subclass, specific types of pesticide may be identified—insecticides, bactericides, fungicides, and herbicides, for example. When the pesticide is designed to eliminate all types of living organisms, it is called a fumigant or sterilant.

Category 2 synthetic organic chemicals are ones that are produced either deliberately for a useful purpose—for example petroleum products—or are a by-product of some other process—such as dioxins and furans that are released during combustion of some materials like municipal and medical waste.

**Table 20.1** Examples of biocides.

| Function | Chemical class | General formula of typical compound | Specific example | $LD_{50}$ oral value[a] / mg kg$^{-1}$ |
|---|---|---|---|---|
| **Category 1** | | | | |
| Insecticide | Organophosphorus | | Parathion | 6–15 |
| Insecticide | Organochlorine | Various | Methoxychlor | 5000–7000 |
| Herbicide | Triazine | | Metribuzin | 2200 |
| Herbicide | Carbamate | | Aldicarb | 0.93 |
| Herbicide | Chlorophenoxy | | 2,4-D | 375 |
| Herbicide | Pyridylium | | Paraquat | 150 |
| **Category 2** | | | | |
| Petroleum residue | Hydrocarbon | $C_nH_{2n+2}$, $C_nH_{2n}$, etc. | Isooctane | 3300 |
| Unwanted by-product | Dioxin | | 2,3,7,8-TCDD | 0.022–0.10 |

[a] $LD_{50}$ values are oral (rat) values in mg kg$^{-1}$ and are taken from various sources. The significance of the $LD_{50}$ is discussed in the text below and in Section 16.2.

**Some types of biocides**

- acaricide—lethal to spiders and mites;
- algicide—lethal to algae;
- fungicide—lethal to fungi;
- herbicide—lethal to plants;
- insecticide—lethal to insects;
- pesticide—often used as a synonym for biocide, although sometimes it excludes the herbicide category;
- fumigant (sterilant)—lethal to a broad spectrum of organisms.

An overlapping category of synthetic organic chemicals of importance in an environmental context is the list of *persistent organic pollutants (POPs)*, defined through the Stockholm Convention of the United Nations Environment Program. The POP lists include some pesticides, but also a number of organic compounds released into the environment intentionally or unintentionally through industrial and other processes. The characteristic feature of all POPs is that they are stable and persist in the natural environment for long periods of time—years or even decades. Twelve compounds have been designated as the 'dirty dozen' and include eight pesticides (aldrin, chlordane, DDT, dieldrin, endrin, heptachlor, mirex, and toxaphene); two industrial chemicals (polychlorinated biphenyls and hexachlorobenzene); and the two incineration by-products (dioxins and furans). Note that most of these compounds have chlorine in their structures; as a general rule, building chlorine into an organic compound adds to its stability. A useful web-based guide concerning the properties and requirements for manufacture and handling of POPs is referenced as Additional Resources 1.

For simplicity and constancy, we will continue to use the terms biocide and xenobiotic compounds to cover all kinds of synthetic organic compounds that find themselves in the environment and that can be harmful to living organisms. The substances range from relatively benign petroleum residues leaked from storage tanks to highly toxic dioxins, small amounts of which form and are released during combustion and during manufacture of chlorophenol chemicals, as well as when chlorine-based bleaches are used in the pulp and paper industry.

You can see that we are proposing a very broad definition of biocides. It includes obviously toxic compounds, but also ones that are not acutely harmful. Nevertheless, petroleum products and other 'relatively benign' materials are aesthetically unpleasant and, at least in the long term, are somewhat toxic to plants, animals, and microorganisms of various types. You are reminded that one of the measures of toxicity is the $LD_{50}$ value, which gives the dose that causes death to 50% of an experimental population after exposure through various defined routes such as oral ingestion or contact through the skin (Chapter 16). Where exposure to the chemical takes place by inhalation of air or by aquatic species through living in contaminated water, the toxicity parameter is $LC_{50}$, referring to the concentration of the chemical in the ambient air or water. In both cases, a small $LD_{50}$ or $LC_{50}$ is characteristic of a very toxic compound. A suggested rating system for various biocide lethal dose values is given in Table 20.2.

**Table 20.2** A suggested classification of toxicity associated with biocides[a].

| Rating | EPA[b] class | Exposure | $LD_{50}$[c] or $LC_{50}$[d] |
|---|---|---|---|
| Very high | I | Oral | 0–50 |
| | | Dermal | 0–200 |
| | | Inhalation | 0–0.2 |
| High | II | Oral | 50–500 |
| | | Dermal | 200–2000 |
| | | Inhalation | 0.2–2 |
| Medium | III | Oral | 500–5000 |
| | | Dermal | 2000–20 000 |
| | | Inhalation | 2–20 |
| Low | IV | Oral | >5000 |
| | | Dermal | >20 000 |
| | | Inhalation | >20 |

[a] From Briggs, S.A., *Basic guide to pesticides: their characteristics and hazards*, Taylor & Francis, Washington, DC; 1992.
[b] EPA, Environmental Protection Agency (USA).
[c] The $LD_{50}$ values apply to the oral and dermal exposure routes and are a total dose of chemical per unit body weight (mg kg$^{-1}$).
[d] LC in $LC_{50}$ stands for 'lethal concentration'. $LC_{50}$ values apply to inhalation and are the concentrations of the biocide in the ambient medium (mg L$^{-1}$).

> **Fermi question**
> Select a pesticide and estimate the total quantity that you might inadvertently ingest each year. Compare your annual intake with the $LD_{50}$ value. Choose a pesticide that is commonly used in your area, and use concentrations that are at the maximum guideline levels.
>
> As an example of data that can be employed in this calculation, maximum levels set for the herbicide 2,4-D by the USA Environmental Protection Agency are 0.1 ppm in water and range from 0.1 to 0.5 ppm for most food products.

The voluminous literature on all these substances includes material on their synthesis, mode of action, agricultural or industrial applications, and toxicological properties. However, in line with the goals set out at the beginning of the book, we limit ourselves to a discussion of what happens during the time they are in the environment—air, water, and soil—after the compounds have been synthesized but before they are taken up by living organisms. Given the very large number of compounds and the infinite range of soil, water, and atmospheric properties, there is still a great deal to say even within this defined area and we will restrict ourselves to general principles, as illustrated by some specific examples. While the focus is on biocide behaviour in the soil part of the terrestrial environment, many of the principles apply equally to behaviour in sediment and in water itself.

### Stability and mobility

The general question we wish to examine is 'What happens to synthetic organic compounds once they are released into the general environment?' A specific example of this question would be 'What is the fate of the herbicide glyphosate,[1] once it has been sprayed on a field as a broad-spectrum weed killer?' Aside from its toxicological effect on the target organism (assuming there are target organisms in the general case), there are two aspects involved in answering such questions. One is the *chemical stability* of the biocide—inversely related to how and at what rate it degrades. The other is the *mobility* of the biocide—the mechanisms by which and rates at which it is transported through various compartments of the environment. The two aspects overlap. If degradation is rapid, then mobility usually becomes less of an issue. If transport is fast, then different degradation mechanisms may operate as the pesticide moves to a new environment. It is important to remember that chemical breakdown of the pesticide into other products does not always mean loss of toxicity. Degradation products can also have biocidal properties—even enhanced ones in some cases—and their transport properties are therefore also important. Keeping in mind the interconnections, we will look at the two issues, stability and mobility, separately and in sequence.

> **Main point 20.1** Organic biocides include many types of synthetic compounds foreign to the soil / water environment. Some are materials inadvertently discharged into the environment, while others, of which pesticides are the most common example, have been manufactured and applied because of their particular toxicological properties. In both cases, we are interested in their persistence and mobility in the environment.

---

[1] Glyphosate, $HOOC-CH_2-CH_2-NH-CH_2-PO(OH)_2$, is a recently developed non-selective herbicide that is absorbed by plant foliage, effectively destroying many deep-rooted annuals, biennials, and perennials. Building in resistance to glyphosate by genetically modifying a crop like soybeans or canola is a widely used strategy in modern plant breeding. Using this approach, the field can be sprayed while the crop is growing, killing all other plants but leaving the crop itself untouched. Clearly, there are complex environmental implications associated with this strategy.

## 20.1 Chemical stability of organic compounds

### Degradation and toxicity

The ultimate decomposition products of an organic molecule are simple stable species like carbon dioxide and water along with compounds like hydrochloric acid or ammonia that incorporate other elements in the original molecule. Because these products are inorganic species, the overall process is called mineralization. In the course of mineralization, however, there will usually be intermediates of varying stability. In many cases, the intermediates have lower toxicity than the parent compound. One such example where this is true is in the degradation of the POP dichlorodiphenyltrichloroethane (DDT). Dichlorodiphenyltrichloroethane degrades very slowly in water and soil by dechlorination reactions, and one of the products found under reducing conditions is dichlorodiphenyldichloroethene (DDE; reaction 20.1).

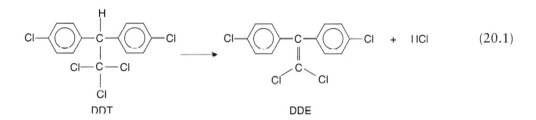

$$\hspace{12cm} (20.1)$$

The product is less toxic than the parent compound. In fact, removal of chlorine from organochlorine molecules nearly always has a detoxifying effect.

Less frequently, such as with some organophosphorus compounds, degradation products exhibit enhanced toxicity. Organophosphorus compounds like parathion are properly called phosphorothioates because of the presence of the P=S group. Inside cells, oxygenating enzymes convert this group to P−O (reaction 20.2).

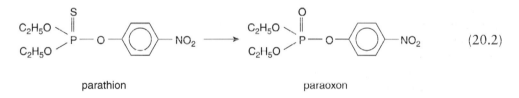

$$\hspace{12cm} (20.2)$$

The phosphorus atom in the oxygenated product, paraoxon, then reacts with the acetylcholinesterase enzyme (AcE) forming a stable complex that inhibits the ability of AcE to catalyse acetylcholine hydrolysis during transmission of nerve impulses. This interference with the neurotransmission accounts for the activity of paraoxon as a pesticide and also as a serious potential health hazard to other organisms including humans. Importantly, then, it is the oxygen-containing species that is the effective agent in killing target insects or other organisms. But this highly toxic form is produced only inside the cell.

Both paraoxon and parathion are subject to degradation via hydrolysis reactions, in which water acts as a nucleophile. These reactions are catalysed in organisms by various phosphatase enzymes (reaction 20.3). However, the presence of a sulphur attached to phosphorus in parathion instead of the oxygen in paraoxon reduces the tendency for reaction with the nucleophile (in addition to water, the nucleophile can also be a nucleophilic part of AcE) and therefore the sulphur analogue is less toxic. Consequently, it becomes an environmental issue of concern when the more toxic oxygen-containing forms of these organophosphorus compounds are produced by oxygenation outside the target organism. For this reason, it is important to consider the fate, including transport, of the reaction products as well as of the biocide itself.

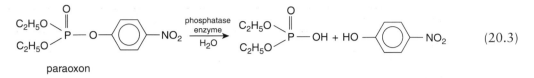

(20.3)

For the thousands of organic compounds considered as environmental biocides, there are as many possible sets of reactions that could occur. It is convenient to document some of the more common processes in the context of the tendency of different functional groups to undergo specific types of transformation reactions. We will consider these under two headings—photolytic reactions and non-photolytic reactions. Non-photolytic reactions include the more usual chemical reactions of hydrolysis, oxidation, and reduction, involving components of soil and water.

## General classes of degradation reactions

### Photolytic degradation

Photochemical degradation of organic compounds is a possibility only when the chemical is exposed to sunlight. In practice, this means that reactions occur during daytime and it is necessary that the chemical be in the gas phase, in or on atmospheric aerosol particles, in surface waters, or on the exposed surface of plants or soil. Even when these conditions obtain, not every compound undergoes photolytic decomposition. The additional requirements are that the molecule be capable of absorbing some portion of the radiation from the solar spectrum to produce an excited state and that the quantum yield for subsequent decomposition be large compared to yields for other deactivation pathways (see Photochemical reactions in Section 2.3).

Table 20.3 lists some approximate mean bond enthalpies in organic molecules. Also tabulated are the approximate wavelengths and associated energies for maximum absorption by common chromophoric groups. Recall that the solar spectrum at the Earth's surface has a low-wavelength cut-off at about 285 nm because more energetic radiation has been almost completely absorbed by ozone in the stratosphere (Fig. 3.2).

Looking at these two sets of data, we could rule out the possibility of extensive photodegradation of an alkane, since very little radiation is absorbed by the sigma bonds between carbon–carbon and carbon–hydrogen. These bonds absorb only very energetic radiation (wavelengths <200 nm). We could also predict only limited degradation of naphthalene. Even though the molecule absorbs strongly at 286 and 312 nm, the energy required to break an aromatic C–C or C–H bond is greater than that taken up from the solar radiation. On the other hand, we might predict that the seed and soil fungicide *fenaminosulph* would be susceptible to photolysis because of the ability of the azo group to absorb and because of the relatively weak C–N bonds. This prediction is, in fact, borne out in practice.

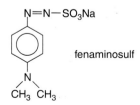

fenaminosulf

Another example of direct photolysis is the photochemical degradation of *trifluralin* in an alkaline aqueous solution leading to the formation of N-dealkylated products (reaction 20.4).

**Table 20.3** *Top part*: Mean bond enthalpies for bonds in organic molecules.[a]
*Bottom part*: Approximate wavelengths and associated energies of radiation absorbed by
chromophoric groups. Some of these groups also absorb at other wavelengths below 270 nm[b].

| Bond in polyatomic molecule | Mean bond enthalpy / kJ mol$^{-1}$ |
|---|---|
| Carbon-carbon (single) | 348 |
| Carbon-carbon (double) | 613 |
| Carbon-carbon (aromatic) | 518 |
| Carbon-hydrogen (alkane) | 412 |
| Carbon-hydrogen (alkene) | 440 |
| Carbon-hydrogen (aromatic) | 431 |
| Carbon-fluorine | 484 |
| Carbon-chlorine | 338 |
| Carbon-bromine | 276 |
| Carbon-iodine | 240 |
| Oxygen hydrogen | 463 |
| Nitrogen-hydrogen | 388 |
| Carbon-oxygen (single) | 360 |
| Carbon-oxygen (double) | 743 |
| Carbon-nitrogen (single) | 305 |
| Nitrogen-nitrogen (double) | 409 |

| Chromophore | Wavelength / nm | Energy / kJ mol$^{-1}$ |
|---|---|---|
| Carbonyl | 285 | 332 |
| Nitro | 280 | 427 |
| Nitroso | 300, 665 | 399, 180 |
| Nitrate | 270 | 443 |
| Azo | 340 | 351 |
| Phenol | 270 | 443 |
| Naphthalene | 286, 312 | 419, 384 |

[a] Values reported in Atkins, P., *Physical chemistry*, 6th edn, W.H. Freeman and Company, New York; 1997.
[b] The following chromophores absorb radiation only at wavelengths less than 270 nm (equivalent to an energy of 443 kJ mol$^{-1}$): alkane; alkene; alkyne; benzene; carboxyl; alcohol; ether; ester; amino; amido; nitrile.

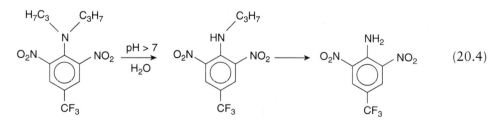

$$(20.4)$$

These two cases are examples of direct photochemistry; this refers to the situation where a
molecule absorbs radiation and the captured energy is sufficient to cause dissociation or some
other kind of chemical reaction. For this and other direct photolytic reactions, the reaction rate
obeys first-order or pseudo-first-order kinetics

$$\text{rate} = f\,[\text{trifluralin}] \tag{20.5}$$

where $f$, the photochemical rate constant, is a property of the solar flux as well as of the molecule itself, as was shown in eqn 2.17.

Another type of photochemical process is *indirect photolysis*. In this situation, a sensitizer molecule (not the target compound) is radiatively excited. If the excited species is sufficiently long-lived, it is able to transfer energy or an electron, hydrogen atom, or proton to a different receptor molecule (the biocide). In this way, without absorbing radiation directly, the receptor molecule can be activated so as to take part in a subsequent chemical reaction. There are specific examples of sensitized photodegradation such as the case of rotenone, a natural product extracted from the derris root that is susceptible to excitation by sunlight. The rotenone in the excited state is able to transfer its surplus energy to *aldrin* and other organochlorine compounds leading to their degradation. In a more general sense, a number of important environmental species, including some mineral surfaces and humic material, are capable of acting as sensitizers. The rate of an indirect photolytic reaction is given by

$$\text{rate} = f\,[\text{sensitizer}][\text{receptor}] \tag{20.6}$$

but, when the sensitizer is present at a much higher concentration than the receptor, the reaction becomes pseudo-first-order.

$$\text{rate} = f_1[\text{receptor}] \tag{20.7}$$

where $f_1 = f\,[\text{sensitizer}]$.

The ability to take part in indirect photochemical reactions extends the possibility of photolysis to compounds that would by themselves be stable to solar radiation.

### Non-photolytic reactions

Degradation in the soil / water matrix by a variety of conventional thermal reaction processes also occurs. When these processes are purely chemical in nature they are referred to as being abiotic; as long as the required reactants are present, such reactions can take place in a sterile environment. Others are chemical reactions that are mediated by the autochthonous or zymogenous microorganisms present in the particular environmental situation. The microorganisms may degrade the compound as a primary substrate from which they derive energy and / or nutrient requirements. Alternatively, the compound may be cometabolized along with the principal substrates required by the autochthonous population. Both such processes are called biotic processes. The importance of biotic processes in other environmental circumstances has been more fully described in Chapter 15. A number of reactions take place both biotically and abiotically and, depending on circumstances, one or other mechanism pre-dominates. It is sometimes difficult to distinguish between mechanisms and to determine which one is operative in a given situation.

Some of the most important classes of chemical reactions involving organic compounds in soil and water are listed here and in each case one or more examples are provided.

### Hydrolysis

Hydrolysis is a general term to describe a nucleophilic reaction where water interacts with a substrate molecule replacing a portion (the leaving group) of the molecule with OH. A general equation for hydrolysis is

$$\text{RX} + \text{H}_2\text{O} \rightarrow \text{ROH} + \text{HX} \tag{20.8}$$

This common type of reaction proceeds by either purely chemical or microbiological mechanisms, but abiotic hydrolysis is often very slow unless the rate is augmented by a catalyst. Following are seven examples.

**1.** Ethers (reaction 20.9), esters (reaction 20.10), and thioesters (reaction 20.10, in cases where C = S replaces C = O) undergo hydrolysis as follows.

$$R - O - R' + H_2O \rightarrow R\,OH + R'OH \tag{20.9}$$

$$R - \overset{\overset{O}{\parallel}}{C} - R' + H_2O \longrightarrow R - \overset{\overset{O}{\parallel}}{C} - OH + R'\,OH \tag{20.10}$$

An important instance of this occurs with 2,4-dichlorophenoxyacetic acid (2,4-D), one of the most widely used herbicides. It contains both carboxylic acid and ether functional groups. Hydrolysis takes place as shown, involving ether C–O bond scission.

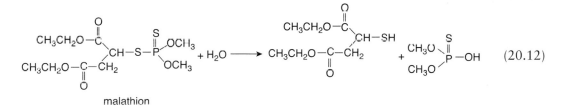

$$\tag{20.11}$$

*Parathion* and *malathion* (reaction 20.12) are potent organophosphorus insecticides that undergo hydrolysis at the phosphorus centre.

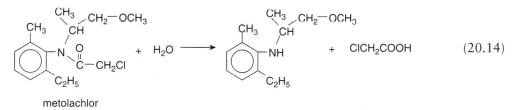

$$\tag{20.12}$$

malathion

Note that further hydrolysis of the alkyl carbon groups could occur.

**2.** Amides are hydrolysed producing an acid and an amine. Depending on the pH of the solution, the products may exist as a free acid and amine (ammonium) salt or a free amine and carboxylate salt or an equilibrium mixture of all these species.

$$R - \overset{\overset{O}{\parallel}}{C} - NH - R' + H_2O \longrightarrow R - COOH + R'NH_2 \tag{20.13}$$

*Metolachlor*, a selective pre-emergence herbicide, undergoes this type of abiotic hydrolysis process.

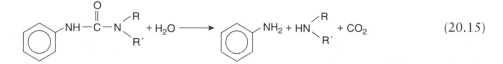

$$\tag{20.14}$$

metolachlor

**3.** Phenylurea compounds are hydrolysed to give two amines: an aniline and an aliphatic amine.

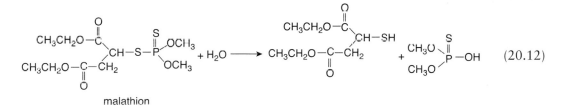

$$\tag{20.15}$$

An example of this reaction is the pesticide *fenuron*, a herbicide that is used to control a broad spectrum of weeds including deep-rooted grasses and woody plants.

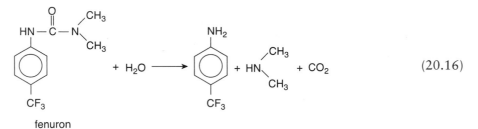

fenuron

$$(20.16)$$

4. Nitriles are hydrolysed to produce, in succession, an amide and then a carboxylic acid.

$$R\text{—}CN + H_2O \longrightarrow R\text{—}\overset{O}{\overset{\|}{C}}\text{—}NH_2 \xrightarrow{H_2O} RCOOH + NH_3 \qquad (20.17)$$

The selective herbicide *ioxynil* (and its bromine analogue *bromoxynil*) undergoes such hydro-lytic processes.

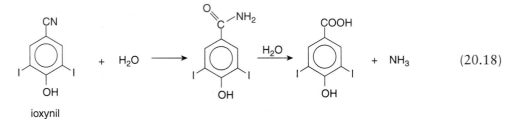

ioxynil

$$(20.18)$$

5. Carbamates hydrolyse to generate an amine, an alcohol, and carbon dioxide.

$$R\text{—}NH\text{—}\overset{O}{\overset{\|}{C}}\text{—}O\text{—}R' + H_2O \longrightarrow RNH_2 + R'OH + CO_2 \qquad (20.19)$$

The hydrolysis reaction for *carbaryl*, a very widely used horticultural insecticide, occurs as follows. In this case, the aromatic alcohol 1-naphthol is formed.

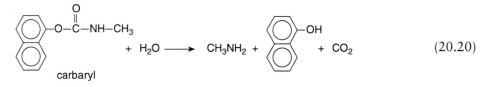

carbaryl

$$(20.20)$$

6. The thiocarbamate family likewise react to form amines, thiols, and carbon dioxide.

$$R\text{—}NH\text{—}\overset{O}{\overset{\|}{C}}\text{—}S\text{—}R' + H_2O \longrightarrow RNH_2 + R'SH + CO_2 \qquad (20.21)$$

This is illustrated by the hydrolysis of *benthiocarb*, a selective pre-emergence herbicide used especially against annual grasses and broad leaf weeds in rice fields.

$$Cl\text{—}\langle\bigcirc\rangle\text{—}CH_2\text{—}S\text{—}\overset{O}{\overset{\|}{C}}\text{—}N\overset{C_2H_5}{\underset{C_2H_5}{<}} + H_2O \longrightarrow HN\overset{C_2H_5}{\underset{C_2H_5}{<}} + Cl\text{—}\langle\bigcirc\rangle\text{—}CH_2\text{—}SH + CO_2 \qquad (20.22)$$

benthiocarb

**7.** A well known example of hydrolysis occurs with the triazine herbicides.

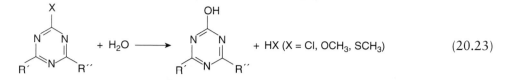

$$(20.23)$$

A case in point is the hydrolysis of *simazine*.

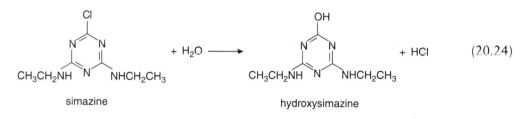

$$(20.24)$$

simazine                    hydroxysimazine

## Oxidation

Oxidation reactions are extremely important in terms of the total degradation of organic species in the environment. It is usually through oxidation that the final mineralized products are produced. For oxidation to occur, an appropriate electron acceptor must be available and the nature of these oxidants depends on the environmental circumstances. In surface soils and water, where the p$E$ is high (see Chapter 10), there is a plentiful supply of oxygen and, in addition, more powerful oxidants produced by photochemical processes may be present. These include the hydroxyl radical, hydrogen peroxide, ozone, and singlet oxygen ($O_2[^1D]$). While the concentration of each of these species is normally small compared with that of ground-state dioxygen, they can play a dominant role in bringing about oxidation of otherwise refractory molecules. Under the variety of anaerobic conditions, other less-powerful oxidants like nitrate and sulphate are potential electron acceptors for the oxidation of biocides. When all oxidants are depleted, reductive degradation becomes operative. Most oxidation reactions are mediated by microorganisms and, again, it may be helpful to refer back to Chapter 15 for further discussion of environmental conditions affecting these processes.

Six of the types of oxidation reactions are as follows:

**1.** Alkanes or aliphatic hydrocarbon substituents are oxidized, often at the terminal group, to produce alcohols, aldehydes, and then carboxylic acids.

$$RCH_3 \rightarrow RCH_2OH \rightarrow RCHO \rightarrow RCOOH \qquad (20.25)$$

Various microorganisms using unique mechanisms are capable of bringing about parts or all of this sequence. Subterminal oxidation is also possible. Larger molecules tend to be more resistant to degradation.

**2.** Alkenes are oxidized through a number of processes resulting in the production of ketones, alcohols, and carboxylic acids. When attack occurs at the double bond, the following reaction sequence occurs from alkene to epoxide to 1, 2-diol to a-hydroxyacid and finally to a carboxylic acid that has one less carbon than the original alkene.

$$R-CH_2-CH=CH_2 \longrightarrow R-CH_2-CH-CH_2 \longrightarrow R-CH_2-CH-CH_2$$

$$\longrightarrow R-CH_2-CH-COOH \longrightarrow R-CH_2-COOH + CO_2 \qquad (20.26)$$

3. Branched-chain hydrocarbons, while somewhat more stable than straight chain molecules, undergo similar types of biological oxidation reactions. Alicyclic hydrocarbons are highly resistant to oxidative degradation in most situations.

4. Aromatic hydrocarbons are also highly resistant to oxidation reactions, as was observed in our earlier discussion of polyaromatic hydrocarbons in the atmosphere. Nevertheless, oxidation does occur with the rate and extent strongly influenced by the nature of substituents on the molecule. Halogens and nitro and sulphonate groups have a stabilizing effect compared with hydroxy, methoxy, and carboxylate groups. The number and positions of substituents are also important.

One mechanism for the oxidation of benzene itself involves the initial formation of an epoxide or oxirane, which is subsequently converted to the diol with concomitant rearomatization of the benzene ring.

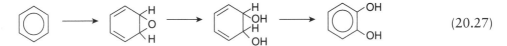

$$(20.27)$$

Further oxidation can lead to ring fission with the production of a dicarboxylic acid.

Oxidation of petroleum hydrocarbons in gasoline (petrol) and oil spills is an important example where oxidative degradation of aromatic compounds comes into play. The rate of degradation of these compounds is of considerable current interest in situations like leakage from gasoline tanks and oil spills associated with extraction or transport of crude petroleum.

Many of the generalizations in 1 to 4 above regarding hydrocarbon decomposition have been found to apply in a study[2] of the microbial degradation of petroleum drilling cuttings. The cuttings, containing fuel oil, calcium salts, and a small amount of emulsifier, had a pH of 9.1, an organic matter content of 12.4%, and a carbon: nitrogen ratio of 103. They were applied to a silty clay agricultural soil of pH 5.1, organic matter content of 2.62%, and carbon:nitrogen ratio of 8.4. If the soils had been sterilized, there was little or no degradation over a 9-month period but, in non-sterilized soils, about 75% of the original mass of hydrocarbon was degraded during this time. Amongst the classes of hydrocarbon, there were very large differences in decomposition rate, with essentially all the low molar mass alkanes (those with less than 27 carbons) being degraded within 16 days. There was slower decomposition of branched chain and aromatic fractions. The residual unreacted material was an unresolved complex mixture of mostly high molar mass hydrocarbons.

5. Particular functional groups of biocide molecules are also subject to oxidative processes, although biotic hydrolysis reactions more often initiate a degradative sequence. A thioether sulphur atom can be oxidized in two steps, first to a sulphoxide and then to a sulphone (reaction 20.28).

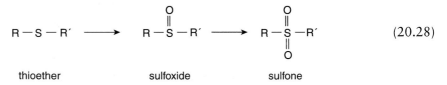

$$(20.28)$$

thioether              sulfoxide              sulfone

*Aldicarb* is a potent and broad-spectrum insecticide, acaricide (active against spiders), and nematicide (active against nematodes) that is susceptible to such oxidation (reaction 20.29). As in the case of parathion, the oxidized analogue is more toxic than the original compound.

[2] Chaineau, C-H., J-L. Morel, and J. Oudot, Microbial degradation of fuel oil hydrocarbons from drilling cuttings, *Environ. Sci. Technol.*, **29** (1995), 1246–54.

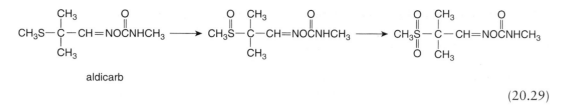

aldicarb

$$(20.29)$$

**6.** Another very important oxidative reaction is $\beta$-oxidation of fatty acid side chains, through a ketone intermediate to an acid with two fewer carbons in the chain.

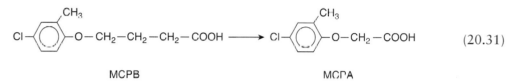

$$R-CH_2-CH_2-COOH \longrightarrow R-\overset{\overset{\displaystyle O}{\|}}{C}-CH_2-COOH \longrightarrow R-COOH \qquad (20.30)$$

An example of this reaction is the conversion of *MCPB* (4-(4-chloro-2-methylphenoxy) butanoic acid), a herbicide used for post-emergence weed control in cereals and pasture, to MCPA (4-chloro-2-methyl-phenoxyacetic acid) as shown.

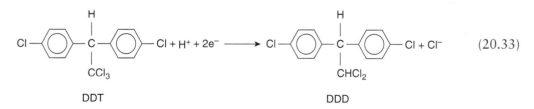

MCPB $\qquad$ MCPA

$$(20.31)$$

## Reduction

Reduction reactions occur under low-p$E$ environmental situations such as those we observed to occur in anoxic groundwater and flooded soils. Some important and well-defined biotic degradation processes take place in these situations. Four examples are included.

**1.** Dehalogenation is a principal process of degradation of refractory halogenated organic compounds. While oxidative decomposition of such compounds is slow, the reduction reactions are frequently faster. Two general mechanisms of dehalogenation have been identified. Hydrogenolysis takes the form

$$R-X + H^+ + 2e^- \rightarrow R-H + X^- \qquad (20.32)$$

Dichlorodiphenyltrichloroethane (*DDT*), the well-known insecticide, is reduced to *DDD* (dichlorodiphenyldichloroethane) in this way. (Compare with reaction 20.1.)

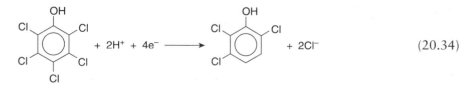

DDT $\qquad$ DDD

$$(20.33)$$

*Pentachlorophenol*, a common wood preservative, is similarly reduced to trichlorophenol.

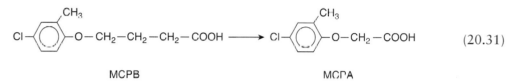

$$(20.34)$$

Advantage is taken of the increased rate associated with reducing conditions in some waste water treatment processes in order to enhance the degradation of halogenated compounds present in a waste stream. Reduction is also a possible means by which halogenated compounds are detoxified in a sanitary landfill. Zero-valent iron may be used as an electron-donating reductant to increase the rate of reaction. Besides iron metal, many organic hydrocarbons in a landfill or in soils and sediments will act as electron-donating compounds to bring about such reductive dechlorination reactions.

2. A second type of reductive dehalogenation reaction is called vicinal dehalogenation because the halogen-leaving groups are arranged vicinally (i.e. 1,2) in the compound.

$$R-\overset{\overset{\displaystyle X}{|}}{C}H-\overset{\overset{\displaystyle X}{|}}{C}H-R'+2e^- \longrightarrow R-CH{=}CH-R'+2X^- \tag{20.35}$$

This mechanism has been observed for *lindane* in flooded soils and in anaerobic sludge.

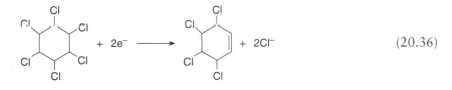

(20.36)

lindane

3. Reduction of nitro groups is a multistep process that, if carried to completion, has an overall reaction represented by

$$R-NO_2 + 6H^+ + 6e^- \rightarrow R-NH_2 + 2H_2O \tag{20.37}$$

The insecticide *fenitrothion* and the herbicide *trifluralin* undergo such transformations under reducing conditions.

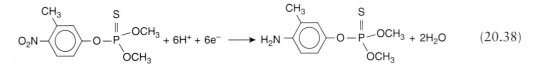

fenitrothion

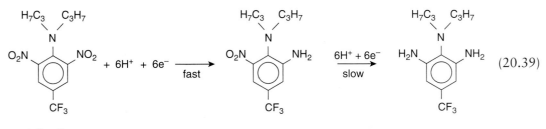

trifluralin

4. A fourth type of reduction involves dealkylation or dealkoxylation.

$$\underset{\text{(X=O, S, NH)}}{R-X-R'} + 2H^+ + 2e^- \rightarrow \underset{\text{(R'=(CH}_2)_n\text{CH}_3 \text{ or O(CH}_2)_n\text{CH}_3)}{R-XH+R'H} \tag{20.40}$$

This reaction is operative in the case of *carbofuran*, a carbamate insecticide, and *methoxychlor*, a close analogue of DDT.

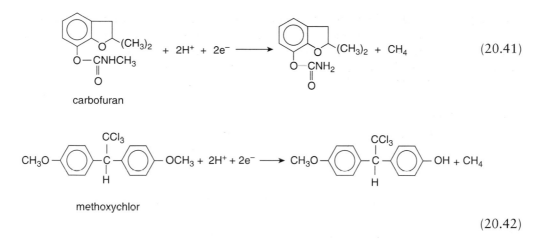

$$\text{(20.41)}$$

$$\text{(20.42)}$$

## Rates of degradative reactions

While all of the above reactions do occur and therefore bring about the degradation of bio-cides in the environment, the rate at which the transformations occur is also a critical feature in determining whether or not the original compound can cause harm to humans or other non-target organisms. The chemical nature of the substance is obviously of key importance (as molecular stability is inversely related to rate of degradation), but rate also depends on the availability of other reactants as well as the general environmental situation. Photolytic, biotic, and abiotic thermal processes can all contribute to breakdown of the molecule. The extent to which each of these occurs varies greatly in each environment, even between micro-environments, and so specific studies are required for individual situations. In this context, great care is needed in transferring information from laboratory experiments to that which would apply in the field. Some of the important considerations that need to be included are discussed now.

### The nature of the biocide

Very broad conclusions regarding the persistence of xenobiotic classes may be drawn from present data. Table 20.4 summarizes some of this information as it is related to several important classes of pesticide compounds. In the table, persistence $(t_{3/4})$ is defined loosely as the approximate time taken for 75% of the pesticide to degrade in the field. Persistence, as defined, is equivalent to two half-lives.

**Table 20.4** Persistence of pesticides in the environment[a].

| Pesticide class | Persistence $(t_{3/4})$ / months[b] |
|---|---|
| Chlorinated hydrocarbons | 16 to >24 |
| Urea, triazine | 3 to >18 |
| Benzoic acid, amide | 3–12 |
| Phenoxy, toluidine, nitrile | 1–6 |
| Organophosphorus | 0.2–3 |
| Carbamate, aliphatic acid | 0.5–3 |

[a] From Edwards, C.A., *Persistent pesticides in the environment*, 2nd edn, CRC Press, Cleveland, Ohio; 1973.
[b] The $t_{3/4}$ value is the time it takes for 75% of the pesticide to degrade in the field.

### Temperature

For any abiotic thermal reaction, temperature affects the rate of degradation according to the usual Arrhenius relation—typically a doubling of degradation with every 10 degree rise in temperature. A similar situation obtains in the case of biotic reactions, but each microorganism has its own optimum temperature. Below 0°C, most microbiological processes essentially cease, as they also do at extremely elevated temperatures. High-temperature situations pertain to environments such as dark-coloured soil under intense solar radiation where surface soil temperatures can be high enough (>50°C) to kill off some of the mesophilic population.

### Moisture

Moisture is essential for many abiotic and most biotic reactions. Almost all microorganisms require an abundance of moisture to flourish, and degradation rates increase with increasing moisture in soil up to the maximum water-holding capacity in the field. At this stage, called *field capacity*, the soil is moist, but much of its volume remains filled with air. In the presence of even more moisture, the soil becomes saturated and pores are then completely filled with water. Under these conditions oxygen transfer from the atmosphere is limited, and anaerobic, reducing conditions set in with accompanying microbial switch-over to facultative and anaerobic species. The actual degradation mechanisms may then change in a substantial way.

### Soil / water properties—pH

Either acids or bases can enhance hydrolysis by nucleophilic substitution and therefore hydrolytic degradation rates are subject to a strong pH dependence (Fig. 20.1). The example shown here is a specific one but is common for many biocides. The degradation of *atrazine* is rapid in acidic and basic solutions but is very slow at intermediate pH values such as are found in most environments. At pH 6 and 25°C, the pseudo-first-order rate constant of approximately $10^{-10}$ s$^{-1}$ corresponds to a half-life ($= \ln 2 / k$) of over 300 y. On the other hand, at pH 4, the rate constant is close to $10^{-7}$ s$^{-1}$ with a half-life of only around 3 months.

The situation illustrated by Fig. 20.1 describes data obtained in laboratory experiments. Outside the laboratory, however, hydrolysis of atrazine does occur to a significant extent even under

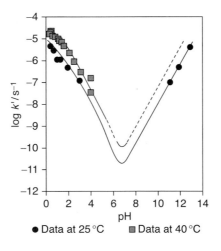

● Data at 25 °C   ▧ Data at 40 °C

**Fig. 20.1** The pH / rate constant profile for the triazine pesticide, atrazine. (The figure is redrawn from some of the experimental and interpolated data provided in Plust, S.J., J.R. Loehe, F.J. Feher, J.H. Benedict, and H.F. Herbrandson, Kinetics and mechanism of hydrolysis of chloro-1,3,5-triazines. Atrazine, *J. Org. Chem.*, **46** (1981), 3661–5.)

apparently neutral conditions. In part this is because the bulk pH of the water body or soil solution does not indicate the true availability of protons or hydroxyl ions to the pesticide. Furthermore, other organic or inorganic species can catalyse the reaction and so the microenvironment provided by soluble species or solid phases must also be taken into account.

### Soil / water properties—organic matter

Through dissociation of acidic groups, natural organic matter (NOM) is a source of hydrogen ions and is therefore one environmental component that determines the bulk pH as well as the local availability of protons. In solution or as a solid, NOM can interact strongly with many other organic compounds, as shown in Chapter 12. The complexed species then show altered— either enhanced or diminished—reactivity. Probably the most significant effect of NOM on degradation is with respect to biotic processes. A soil rich in organic matter usually has a large and diverse microbial population. Heterotrophic organisms, by definition, require pre-synthesized organic compounds, and as they make use of the NOM they can also cometabolize the foreign compound. Zymogenous species that make use of the particular biocide chemical also contribute to degradation.

### Soil / water properties—inorganic species

Dissolved metal ions are known to catalyse some nucleophilic substitution reactions. By forming a complex through an electron-donating atom like oxygen, electrons are withdrawn from the molecule making it more susceptible to attack by a nucleophile such as water. Reaction 20.43, the hydrolysis of an amide, is an example.

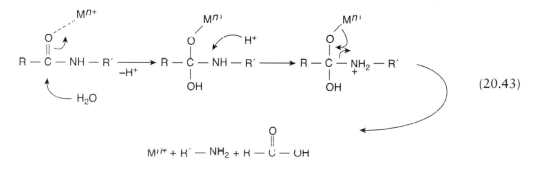

$$(20.43)$$

Note again that, in the sequence shown, the final products may be present in varying degrees of protonation depending on bulk or local pH.

Surface-bound metals can effect the same enhancement. For example, it has been shown that organophosphorus compounds hydrolyse more rapidly in the presence of hydrous iron and aluminium oxides such as are found in abundance in some tropical soils as well as in the depositional horizon of Spodosols and other forest soils.

Besides being available to form complexes with organic biocides, certain soil minerals, including hydrous iron (III) oxide (as limonite or goethite), are oxidizing agents, enhancing the oxidation of organic material in the soil. On the other hand, minerals like iron sulphide (pyrite) provide sites for reduction at the interface between the solid and the associated water. As was noted above, natural organic matter, especially relatively undecomposed material with a small O:C ratio, is also essentially a reducing agent that can contribute to anoxic conditions and a reducing environment.

### Kinetic calculations

In calculating the rate of degradation of organic compounds and the kinetic parameters associated with a particular reaction, it is common to assume that the process is first-order. In many cases this is a reasonable assumption because the organic compound (such as a pesticide) is usually present in a very small concentration compared with other reactants (water and organic matter, for example) in the soil environment.

---

### Example 20.1  **Rate of degradation**

An example of a quantitative assessment of pesticide degradation is taken from a study of the spontaneous decomposition in water of diuron, a commonly used soil sterilant.[3]

The abiotic degradation mechanism, resulting in a single product, 3,4-dichloroaniline, occurs through the following steps:

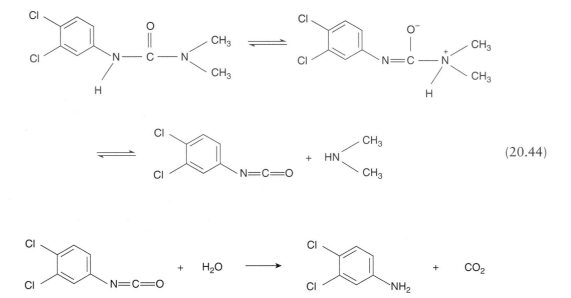

(20.44)

The reaction begins with the formation of the zwitterion by general acid–base catalysis. The zwitterion is in equilibrium with the phenyl isocyanate derivative, which hydrolyses to produce the aniline derivative.

The loss of diuron and the simultaneous build-up of the final product can be followed simultaneously using liquid chromatography with dual-wavelength detection. Figure 20.2 shows one result obtained at 70°C, pH 7.2 maintained by a phosphate buffer, and with an ionic strength of 1 mol $L^{-1}$.

Using data such as those shown in Fig. 20.2, and assuming that the degradation follows pseudo-first-order kinetic behaviour ($C_t = C_0 e^{-kt}$), the value of the rate constant can be calculated from a plot of $\ln(C_t / C_0)$ vs. $t$ (Fig. 20.3). The straight line behaviour is confirmation of first order kinetics (actually pseudo-first-order, because of the high 'concentration' of the water solvent).

For the plot shown, the value of $k_{obs}$ (= −slope) is 0.390 and the half-life (= $\ln2 / k_{obs}$) is 1.77 days.

In this paper, the authors determined $k_{obs}$ values under several buffer concentrations and over a range of pH. From this they calculated values of $k_0$, the rate constant at zero ionic strength.

After obtaining results at several temperatures and with high buffer concentrations, a linear Arrhenius plot was also constructed.

$$\ln k = -E_a / RT + \ln A, \quad \text{the natural logarithm form of the equation } k = Ae^{-E_a / RT}$$

---

[3] Salvestrini, S., P. Di Cerbo, S. Capasso, Kinetics of the chemical degradation of diuron, *Chemosphere*, **49** (2002), 69.

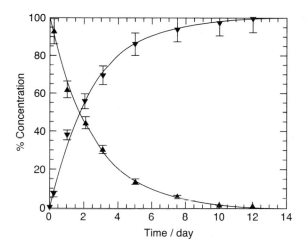

**Fig. 20.2** Time profile for the degradation of diuron. $T = 70°C$, pH = 7.2, 0.38 M phosphate buffer, ionic strength = 1 mol L$^{-1}$ (KCl). ▲ diuron reactant, ▼ 3,4-dichloroaniline product. Figure from Salvestrini, S., P. Di Cerbo, S. Capasso, Kinetics of the chemical degradation of diuron, *Chemosphere* 49 (2002), 69.

From this, the value for ln $A$ was determined to be 43 ($A$ has units d$^{-1}$) and $E_a$ was 127 kJ mol$^{-1}$. Using these values, you can see that the rate of reaction is very slow at more 'normal' temperatures. Outside the neutral pH range, degradation of diuron was found to be significantly catalysed by both hydronium and hydroxyl ions, as was observed for the case of atrazine, shown above.

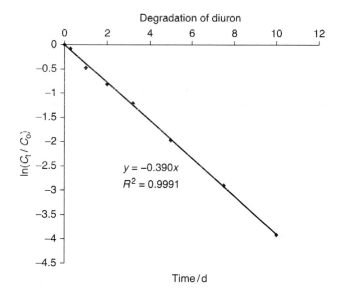

**Fig. 20.3** Plot of lnC$_t$ / C$_o$ vs. time in days for the degradation of diuron.

It is important to emphasize again that, while laboratory kinetic experiments provide useful information about reaction mechanisms and factors affecting rates, the actual rate constants obtained must be used with extreme caution when considering the situation in the field.

The great variability of local environments and especially the potential role of microorganisms will always have a large effect, different from place to place, on how rapidly a particular biocide degrades.

> **Main point 20.2** Biocides degrade by either abiotic (including photolytic) or biotic processes. Their stability depends on the chemical structure, the availability of reactants and microorganisms required for decomposition, and a variety of environmental factors. The degradation products have their own unique toxicological properties.

## 20.2 **Mobility of organic compounds**

The second factor related to environmental behaviour of organic compounds is their ability to move through the compartments of the environment. Pesticides, as one class of examples, are usually sprayed over planted fields or fields with already growing plants. The pesticide is then in contact with the soil, and rainfall or irrigation water can carry it downward into the soil, or laterally into streams and other water bodies. Likewise, other organic compounds that are spilled or buried will remain in place or be moved by water that is on or in the soil. A number of mechanisms are responsible for their migration, or attenuation, at the surface and in the subsurface of the soil (Fig. 20.4).

### Aqueous transport

Organic compounds move through the soil within both the gas and liquid phases. In most cases aqueous phase transport is the most important process. In water, chemicals are carried downward by percolating rain and irrigation waters, upward by capillary action in poorly drained arid situations, and laterally on slopes and in groundwater aquifers.

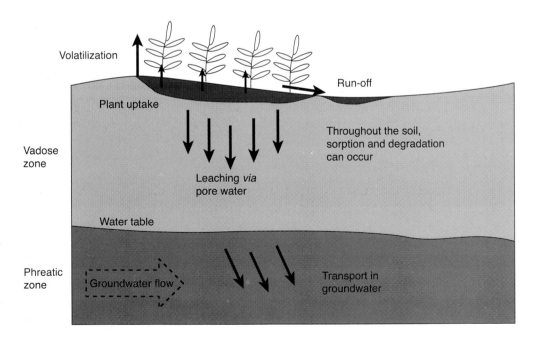

**Fig. 20.4** Mechanisms responsible for biocide migration and attenuation in the soil and associated environments.

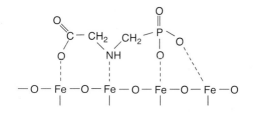

**Fig. 20.5** Bonding between glyphosate and an iron oxide mineral surface.

Whether the compound moves readily or tends to remain fixed in association with the soil, depends on the combined properties of the particular chemical and the soil. In this regard, a frequently calculated property is the octanol:water partition coefficient ($K_{OW}$) that was described in Chapter 14. As was shown there, this parameter is empirically related to two other closely connected coefficients, $K_{OM}$ and $K_{OC}$, describing the distribution between organic matter and water and organic carbon and water, respectively. These latter parameters are themselves related to the fundamental distribution coefficient $K_d$, which is the ratio of the concentration of the compound in soil to its concentration in the associated water, (eqns 20.45 and 20.46),

$$K_d = K_{OM} f_{OM} \qquad (20.45)$$

$$K_d = K_{OC} f_{OC} \qquad (20.46)$$

where $f_{OM}$ and $f_{OC}$ are the fractions of organic matter and organic carbon, respectively, in the soil. Recall that the important assumption here is that the organic matter fraction of the soil is solely responsible for sorption and retention of the compound. This is not true in cases where specific interactions between soil minerals and certain species play an important role in retention. The strong bonds formed between the carboxylate, amine, and phosphonate groups of the herbicide glyphosate (Fig. 20.5) with iron in hydrous oxide minerals provide an example of retention by the inorganic components in soil. In spite of exceptions like this one involving glyphosate, the relations described above are widely used and often approximately valid.

Consider the three distribution constants that we have described. $K_d$ is used to determine the distribution of the xenobiotic compound between soil and water in a particular situation. On the other hand, $K_{OW}$ is a fundamental property of the biocide itself.

The intermediate parameter, $K_{OM}$ (or $K_{OC}$), is particularly useful; while it is a property of an individual chemical, it can be readily converted to a $K_d$ value if the organic matter content of the soil is known.

For example,

$K_d = K_{OM} \times f_{OM}$ from eqn 20.45

for a $K_{OM}$ of 50 mL g$^{-1}$ (a quite small value), when the soil contains 1% organic matter,

$K_d = 50$ mL g$^{-1}$ OM $\times 0.01$ g (soil) g$^{-1}$ (OM) $= 0.50$ mL g$^{-1}$ (soil)

A further parameter sometimes used to measure mobility is the $R_f$ function,[4] borrowed from chromatography. It is a measure of the fractional transport of the compound compared with the water solvent. It too can be derived using the $K_d$ along with knowing the porosity and bulk density of the soil.

$$R_f - \frac{\text{rate of movement of solute}}{\text{rate of movement of aqueous phase}} \qquad (20.47)$$

---

[4] $R_f$ is the retardation factor, a term used in describing the movement of a solute in a thin-layer chromatography experiment. It is defined as the ratio of the distance that a solute moves to the distance that the solvent moves in a given time, during a chromatographic experiment.

Assuming equilibrium, and considering a particular volume of the soil / water system, the definition of $R_f$ is equivalent to

$$R_f = \frac{\text{amount of solute in the aqueous phase}}{\text{total amount of solute}}$$

$$= \frac{x_m}{x_m + x_s} \qquad (20.48)$$

where $x_m$ = the amount of solute in the aqueous phase and $x_s$ = the amount of solute sorbed to the soil *at any particular time*.

$$\frac{1}{R_f} = \frac{x_m + x_s}{x_m} = 1 + \frac{x_s}{x_m} \qquad (20.49)$$

The ratio $x_s / x_m$ is the ratio of the amount of solute sorbed on soil to that in the aqueous phase *in a given volume of the column*. This fraction is related to the $K_d$ value by taking into account the density of soil particles ($d$) and the fractional porosity (p) of the soil.

$$\frac{x_s}{x_{aq}} = K_d \times \frac{\text{mass of soil}}{\text{mass of water}} = K_d \times \frac{d \times \text{volume of soil}}{1.0 \times \text{volume of water}} = K_d \times d \times \left(\frac{1-p}{p}\right) \qquad (20.50)$$

and

$$R_f = \frac{1}{1 + K_d \times d \times \left(\frac{1-p}{p}\right)} \qquad (20.51)$$

Using the definition of $R_f$ in eqn 20.47, the rate of movement of the solute = $R_f$ × rate of movement of the aqueous phase.
For a $K_d$ value of 0.50 mL g$^{-1}$ as calculated above, and assuming soil porosity of 45% and particle density of 2.5 g mL$^{-1}$,

$$R_f = \frac{1}{1 + 0.50 \times 2.5 \times \left(\frac{1-0.45}{0.45}\right)} = \frac{1}{2.53} = 0.40$$

Small values of each of the equilibrium $K$ parameters and large values of the $R_f$ indicate little affinity for the organic solvent or soil and therefore a strong tendency to be present in the aqueous phase. When $K_{OM} = 0$, there is no interaction with soil and $R_f = 1$. Consequently, the compound moves freely with water. When $K_{OM}$ is very large, $R_f$ approaches 0, signifying that the compound is completely immobilized.

Table 20.5 lists ranges of $K_{OM}$ values and corresponding $K_d$ (assuming 1% organic matter) and $R_f$ values.

**Table 20.5** Distribution coefficients and mobility properties of various classes of organic compounds in soil. (assuming soils contain 1% organic matter)

| $K_{OM}$ / mL g$^{-1}$ | $K_d$ / mL g$^{-1}$ | $R_f$ | Mobility | Class (typical) |
|---|---|---|---|---|
| 0–50 | 0–0.5 | 1–0.4 | Very high | Aliphatic acids |
| 50–150 | 0.5–1.5 | 0.4–0.2 | High | Carbamates |
| 150–500 | 1.5–5 | 0.2–0.07 | Medium | Benzoic acids |
| 500–2000 | 5–20 | 0.07–0.02 | Low | Triazines |
| 2000–5000 | 20–50 | 0.02–0.01 | Slight | Organophosphates |
| >5000 | >50 | <0.01 | Immobile | Organochlorines |

### Example 20.2  **Rate of movement of a biocide in groundwater**

Consider a situation where groundwater is moving through the subsoil at a rate of 2.3 cm h$^{-1}$ in a soil with fractional porosity 0.27, consisting of particles whose density is 2.6 g mL$^{-1}$. The water contains a xenobiotic organic compound whose $K_d$ value is 10 mL g$^{-1}$.

$$R_f = 1 / (1 + 10 \times 2.6 \times 0.73 / 0.27) = 0.014$$

Therefore, the rate of movement of the biocide is

$$0.014 \times 2.3 = 0.032 \text{ cm h}^{-1}$$

What are the factors contributing to transport properties?

- The organic compound itself. Structural features of the molecule define its relative tendency to remain associated with soil or sediment particles or to move with the aqueous phase. Molecular factors contributing to retention are evidenced by large $K_{OW}$, $K_{OM}$, and $K_d$ values and include having substantial hydrocarbon components, halogenation, being non-ionizable or cationic, and having the ability to form covalent bonds with soil minerals. All these factors will favour interaction with soil organic or inorganic matter, thereby preventing the compound from being transported by water. On the other hand, hydrophilicity (small $K_{OW}$, $K_{OM}$, and $K_d$) and therefore mobility are associated with the molecule having limited hydrocarbon and halogen components but containing ionizable and polar groups as features of the structure.

- The soil. We have emphasized that, to a large extent, it is organic matter that is responsible for retention and this is the reason for the use of the coefficients $K_{OM}$ (or in some cases the related term $K_{OC}$). Retention is in part due to the hydrophobic nature of much of the organic matter complex, and its interaction with hydrophobic structures in the xenobiotic organic compound. Usually, to a lesser extent, retention is also due to interaction between the compound and mineral surfaces. Because most soils have significant cation exchange capacity, interaction with positively charged biocide molecules can be very significant.

## Vaporization

Mobility of pesticides has another aspect besides transport through porous media in the aqueous phase. Transport in the gas phase is a second means by which biotic compounds move vertically or horizontally. Gas-phase transport in soil does occur, but is usually not quantitatively as significant as transport via water. What can be very significant, however, is transport across the landscape, in the atmosphere above the soil surface.

The principal molecular property defining the ability of a pure compound to vaporize is its vapour pressure. For neutral, covalent compounds, this parameter depends in large part on the molar mass and polarity of the material; small molar mass and a nonpolar structure favour volatility. For many xenobiotic compounds, the vapour pressure at 25°C ranges from 10$^{-6}$ to 1 Pa. (For comparison, recall that the vapour pressure of water is about 3000 Pa at the same temperature.) Exceptions with higher vapour pressures include low molar mass hydrocarbons such as some petroleum residues. The fact that the vapour pressure of many compounds of interest is very small does not mean that evaporative losses are necessarily negligible. In an open environment with even minimal air movement, the atmospheric partial pressure of the compound is almost zero except in the microlayer just above the surface of the compound. Therefore, while the position of the equilibrium for vaporization may lie strongly to the left, the very small concentration of the gaseous form may ensure that continuous evaporation occurs.

$$\text{Biocide (liquid or solid)} \rightleftharpoons \text{Biocide (gas)}$$

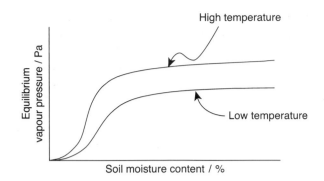

**Fig. 20.6** Equilibrium partial pressure of an organic compound in soil with varying water content. At low values of soil moisture, retention of the compound is due to interaction with the soil. The plateau values are characteristic of the vapour pressure of the solute from a saturated aqueous solution.

Environmental factors affect the rate of vaporization—most importantly temperature, but also placement of the biocide (i.e. whether on the surface or incorporated into the soil), air currents (wind), and the nature of the surface on which the biocide is deposited. Chemical interactions between a biocide and soil, the same ones that inhibit transport in water, can greatly reduce the equilibrium vapour pressure and also retard the rate of solid or liquid to vapour transition.

If a plot of vapour pressure against soil moisture content is made, assuming a constant concentration of biocide in the soil, there is an increase in vapour pressure until a constant value is reached (Fig. 20.6). Where water content is small, the forces retaining the condensed phase biocide on the soil particles act to reduce the extent of evaporation but, as moisture content increases, water molecules occupy adsorptive sites, releasing the biocide so that it may more readily vaporize. The plateau region occurs when monolayer coverage of all active sites has been achieved, and the compound is freed from the influence of the soil matrix. For sufficiently soluble species, this is close to a situation where the biocide is vaporizing from water itself, a situation that in equilibrium terms is described by Henry's law (Chapter 11).

When considering vaporization from a solid surface as in the soil example, the vapour pressure of the pure chemical is the most important molecular property. However, it is frequently the case that evaporation takes place from an aqueous solution as when herbicides are used to control water weeds or applied in wetland paddy culture. For soluble species, relations defined by Henry's law again control the equilibrium vapour pressure.

With biocides sprayed on plants, sorption is due to retentive chemical forces on the plant surface and also to absorption into the epidural layer of the plant tissue.

### The octanol / air coefficient

Recently,[5] it has been postulated that evaporation from plant surfaces and also from the organic fraction of soil is best described by a dimensionless octanol/air coefficient $K_{OA}$, exactly analogous to the octanol / water coefficient, $K_{OW}$. The reasoning behind this is again that plant material has both hydrophilic and hydrophobic character, well represented by the amphiphilic solvent octanol. Values of $K_{OA}$ for chlorinated biocides including chlorobenzenes, PCBs, and DDT range from $10^4$ to $10^{12}$. Such large values indicate a strong tendency for these compounds to

[5] Harner, T. and D. Mackay, Measurement of octanol—air partition coefficients for chlorobenzenes, PCBs and DDT, *Environ. Sci. Technol.*, **29** (1995), 1599–606.

remain associated with growing crops in the terrestrial environment. This is consistent with the observed accumulation of organochlorines in plants and soil. Between $-10°C$ and $+20°C$, the value of $\log K_{OA}$ increases linearly with $1 / T$. This means that as the temperature increases, the compound has a greater tendency to evaporate from the plant surface. The major increase in volatility with temperature is critical in determining their airborne mobility in different parts of the world. We described the 'grasshopper effect' earlier, whereby chemicals like DDT, when released in tropical or temperate (during the warm growing season) regions, vaporize to some extent, are carried elsewhere by global air currents, and then condense in cold areas. Even in the absence of local sources, significant concentrations of organochlorines have been measured in the polar regions.

> **Main point 20.3** Mobility in air and water is related to the distribution of the chemical between the different phases. Like chemical stability, distribution also depends on the structure of the compound as well as physical and chemical properties of the soil and air.

## 20.3 Leachability

Leachability is described in a way that combines aspects of chemical stability and mobility in water. The leachability of a compound is a measure of its tendency to remain undegraded while moving in the aqueous phase to a new location. This is an important parameter because leachable compounds can potentially move to other perhaps more sensitive parts of the environment. Situations where drinking water has been contaminated by leached pesticides have been observed in many places. Taking chemical stability and mobility together, attempts have been made to characterize the leachability of organic chemicals in a simple manner. An example of this approach is the groundwater ubiquity score,[6] *GUS index*, defined in the following way.

$$\text{GUS} = \log_{10}\left(t_{1/2}^{\text{soil}}\right) \times (3.77 - \log_{10}(K_{OM})) \tag{20.52}$$

where $t_{1/2}^{\text{soil}}$ is the half-life in days for degradation in the soil and $K_{OM}$ is the sorption coefficient in mL g$^{-1}$ as defined above. As we have emphasized, it is important that the $t_{1/2}^{\text{soil}}$ be based on field experiments, as any measure of degradation rates is highly influenced by experimental conditions. As such, the value of $t_{1/2}^{\text{soil}}$ includes loss by biotic and abiotic degradation, as well as by vaporization. The $t_{1/2}^{\text{soil}}$ is the stability term; a large value means that the compound is stable under environmental conditions and is therefore potentially subject to leaching. Because of the great variability in environmental settings and conditions, values of $t_{1/2}^{\text{soil}}$ that are reported for a particular pesticide can sometimes be highly variable and must always be treated with considerable caution. $K_{OM}$ is the mobility term and a small value indicates little interaction with the soil, therefore favouring leachability.

The values used in Table 20.6 are averages from a number of situations. The suggestion is that a GUS index of less than 1.8 characterizes a species that is not prone to leaching (either because it degrades rapidly or is strongly retained by the solid matrix), while a score of greater than 2.8 is characteristic of highly leachable compounds. Table 20.6 lists selected pesticides along with associated $K_{OM}$ and $t_{1/2}^{\text{soil}}$ and GUS values.

---

[6] Gustafson, R.L., Groundwater ubiquity score: a simple method for assessing pesticide leachability. *Environ. Toxicol. Chem.*, **8** (1989), 339–57. This paper provides the following groundwater ubiquity score equation: $\text{GUS} = \log_{10}(t_{1/2}^{\text{soil}}) \times (4 - \log_{10}(K_{OC}))$. The description provided in the text above has modified the original derivation so as to use the $K_{OM}$ value in the calculation rather than $K_{OC}$.

**Table 20.6** Properties related to the leachability of selected pesticides[a].

| Pesticide | Class | $t_{1/2}^{soil}$ / days | $K_{OM}$ / mL g$^{-1}$ | GUS[b] |
|---|---|---|---|---|
| Picloram | Picolinic acid | 206 | 15 | 6.00 |
| Atrazine | Triazine | 74 | 63 | 3.68 |
| Carbofuran | Carbamate | 37 | 32 | 3.55 |
| Metolachlor | Amide | 44 | 58 | 3.30 |
| Simazine | Triazine | 56 | 81 | 3.25 |
| Aldicarb | Carbamate | 7 | 10 | 2.34 |
| Oxamyl | Dithiocarbamate | 8 | 15 | 2.34 |
| Carbaryl | Carbamate | 19 | 250 | 1.75 |
| Toxaphene | Organochlorine | 9 | 56 000 | −0.93 |
| Trifluralin | Dinitroaniline | 83 | 4700 | 0.19 |
| Chlorpyrifos | Organophosphate | 54 | 3600 | 0.37 |
| Heptachlor | Organochlorine | 109 | 7600 | −0.23 |
| Dieldrin | Organochlorine | 934 | 7100 | −0.24 |
| Chlordane | Organochlorine | 37 | 11 000 | −0.43 |
| DDT | Organochlorine | 38 000 | 120 000 | −6.00 |

[a] Data summarized from Gustafson, R.L., Groundwater ubiquity score: a simple method for assessing pesticide leachability. *Environ. Toxicol. Chem.*, **8** (1989), 339–57.
[b] GUS, Groundwater ubiquity score. Pesticides in this table with GUS values from 6.00 to 3.25 (i.e. picloram to simazine) are considered to be 'readily leached', whereas those with GUS values from 1.75 to −6.00 (i.e. carbaryl to DDT) are considered to have 'limited leachability'.

---

### Example 20.3 **GUS values and leachability**

The herbicide picloram (a derivative of picolinic acid) has the following GUS index.

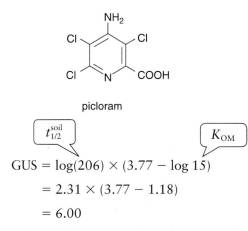

$$GUS = \log(206) \times (3.77 - \log 15)$$

$$= 2.31 \times (3.77 - 1.18)$$

$$= 6.00$$

The large GUS value indicates that picloram is prone to leaching. This is a consequence of two properties: it has little affinity for soil organic matter (small $K_{OM}$) and is also chemically stable in the soil (large $t_{1/2}^{soil}$).

In contrast, the organochlorine heptachlor has a very large $K_{OM}$ value and, in spite of its chemical stability, it has a GUS index indicating only limited tendency to leach into groundwater.

$$GUS = \log(109) \times (3.77 - \log 7600)$$
$$= -0.23$$

Another compound that is relatively resistant to leaching is carbaryl, a carbamate pesticide.

$$GUS = \log(19) \times (3.77 - \log 250)$$
$$= 1.75$$

In this case its resistance to leaching has more to do with the fact that it readily degrades. The undecomposed material is itself moderately mobile. It is important to restate that the mobility measured here is that associated with leaching in water. Many organochlorine compounds and others move long distances through the air as described above.

---

There are several comprehensive listings of pesticides that include extensive documentation of their properties. Additional Resources 4 is one of these collections that is available as a searchable online database. In the pesticide information profiles given in this collection, the term 'partition coefficient' refers to partitioning between octanol and water and is defined as log $K_{OW}$, often also called log $P$. The term 'adsorption coefficient' refers to the $K_{OC}$ value, which you will recall from Chapter 14 is related to the $K_{OM}$ value by $K_{OC} = 1.7 \times K_{OM}$.

Many other attempts have been made to model the physical, chemical, and biological behaviour of biocides in the field. The task is challenging, given the broad range of compounds and especially because each environmental situation is complex and unique. Of all the variables, perhaps the most unpredictable one is the microbiological activity, which, in turn, depends on the specific ecological situation. When depth is included as a variable in developing models for persistence and mobility, the changes in the physical, chemical, and microbiological properties of soil add a new dimension to the complexity. A further element required in refined models takes account of the fact that a single relation does not necessarily define retention. In some situations, rapid and nearly reversible retention is observed, presumably due to surface adsorption. In the same system, slower, irreversible processes can also take place and this is attributed to *intraparticle diffusion*. Intraparticle diffusion implies that, over time, the biocide migrates into the interior of the solid phase through micropores or within the solid matrix itself. Material that has moved away from the surface is unable to exchange readily with the surrounding solution. This inhibits the mobility of the solute and has implications for both chemical clean up and bioremediation of contaminated soils.

In spite of the limitations, indices and models are important in making qualitative and semi-quantitative predictions about the behaviour of particular biocides in specific soil environments.

### Literature link  The fate of pesticides in soil

A review of issues that determine the extent to which a pesticide will migrate into water is the subject of a paper entitled *The mobility and degradation of pesticides in soils and the pollution of ground water resources* by M. Arias-Estevez, E. Lopez-Periago, E. Martinez-Carballo, J-C. Mejuto and L. Garcia-Rio (*Agric. Ecos. and Environ.*, **123** (2008) 247. The paper does not break any new ground but sets out the subjects that need further investigation in order to understand the behaviour of pesticides in soil and water. The information is essential as background for making rational decisions about authorization and application of particular pesticides in specific situations. Estimates have been made showing that in some cases less than 0.1% of pesticide that is applied to a crop reaches the target organism. The rest contaminates plants, soil, water and air and frequently moves from one site to another. As a result, measurable amounts of one or more pesticide have been found in over half the water samples analysed in the United States. All pesticides found in groundwater and much that is found in surface water have passed through the soil. For this reason, it is important to understand the dynamics of soil–water interactions and the effects of the soil matrix on degradation processes.

A slightly modified version of the authors' diagram illustrates the components and relationships that must be examined in order to determine best practices required for environmental safety.

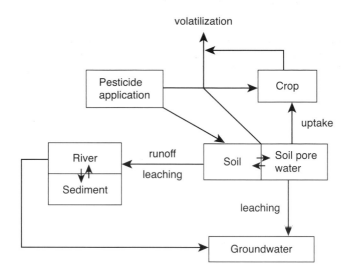

Degradation depends on many factors including the nature of the pesticide itself. In addition, the microbiological, chemical and physical features of the soil environment all have a strong influence on decomposition rates. For example, several cases are described where sorption onto organic matter or other soil constituents inhibits the degradation of the pesticide. But sorption itself is not simple; in many cases it can best be described as a two-step process, the first being rapid retention of the pesticide on the surface of the soil particle, and this is followed by an aging process in which the pesticide becomes much more strongly bound and, at the same time, less available to be degraded. Several physical and chemical explanations of the aging process have been given, including migration into deep pores within the soil and irreversible covalent binding to sites on the OM.

The differential sorption / binding processes also have an effect on mobility and movement into the surface and groundwater. For this reason, caution is required in relating potential mobility to simple laboratory experiments such as determining $K_{OW}$ values or measuring adsorption isotherms. In situations where measurements of availability are made on freshly treated soils, overestimates can be made. On the other hand, in the field it is not unusual for water to follow preferential flow pathways in which it does not contact the 'average' distribution of pesticide; in such cases, underestimates of the possibility of ground water contamination are likely.

In doing risk assessments on the behaviour of pesticides, leachability indices are useful but must always be employed carefully along with a thoughtful examination of case-specific factors that can affect stability and mobility.

---

**Main point 20.4** Leachability is a function of both stability and mobility. Stable, mobile compounds are potentially leachable, while readily degradable biocides and / or those that have a large affinity for the solid phase are less likely to move far from their point of application.

---

## ADDITIONAL RESOURCES

1. Krieger, R., *Handbook of pesticide toxicology*, Academic Press, San Diego, 2009.
2. MacKay, D., W.Y. Shiu, K-C. Ma, and S.C. Lee, *Handbook of physical-chemical properties and environmental fate for organic chemicals: Vol 4, Nitrogen and sulphur-containing compounds and pesticides*, 2nd edn, CRC Press, Boca Raton, 2006.

3. Larson, R.A. and E.J. Weber, *Reaction mechanisms in environmental organic chemistry*, Lewis Publishers, Boca Raton, Florida; 1994.

4. Extoxnet, *Pesticide Information Profiles (PIPS)*, http://extoxnet.orst.edu/pips/ghindex.html, accessed October, 2009.

5. Resource Futures International, *Persistent Organic Pollutants and the Stockholm Convention: A Resource Guide*, the World Bank and CIDA, http://siteresources.worldbank.org/INTPOPS/214574–1115813449181/20486510/PersistentOrganicPollutantsAResourceGuide2001.pdf, accessed October 2009.

## PROBLEMS

1. Consider the structure of the pesticide metolachlor and indicate structural features that would be involved in binding with organic matter in the soil.

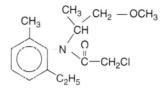

metolachlor

2. Compare the chemical forces with respect to the retention of the pesticides dieldrin and malathion by soil organic and mineral phases.

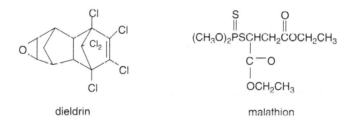

dieldrin                    malathion

3. A rainstorm causes the water to penetrate 5 cm into a soil containing 3.6% OM. Qualitatively predict the relative extent of downward movement of the herbicides aldicarb and trifluralin if they have been applied to the soil just before the rain begins.

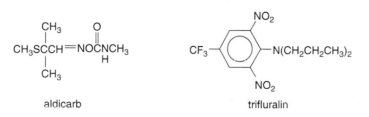

aldicarb                    trifluralin

4. The presence of a halogen atom at an odd-numbered carbon, with respect to a primary substituent site on a benzene ring, reduces the stability of the compound in comparison to a situation where the halogen is on an even-numbered carbon. Explain.

5. The half-life ($t_{1/2}$) for the hydrolytic degradation of the carbamate insecticide carbaryl is reported to be 31 days at 6°C and 11 days at 22°C. Calculate the activation energy for the reaction. Write an equation for the hydrolysis of this pesticide.

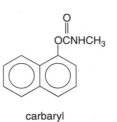

carbaryl

6. At 22°C, the organophosphorus compound diazinon has a degradation $t_{1/2}$ of 80 days in river water and 52 days in the same water after filtration. In contrast, the triazine cyprazine has corresponding $t_{1/2}$ values of 190 days and 254 days. Suggest reasons why filtration increases $t_{1/2}$ in one case and decreases it in the other.

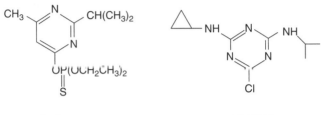

diazinon            cyprazine

7. Predict chemical degradation processes that might occur for the herbicide metribuzin.

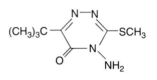

metribuzin

8. The values of log $K_{OM}$ for the insecticide lindane and the herbicide 2,4-D are 2.87 and 1.54, respectively. For each pesticide, suppose rain water dissolves 12 ppb from the plants and surface soil and then carries it downward. If the soil has 2.3% organic matter, calculate the equilibrium concentration for each pesticide in the soil.

9. Using Additional Resources 4, look up the structures and log $K_{OW}$ values of the 'Dirty dozen'. Describe the general correlation that exists between log $K_{OW}$ and the number of chlorines in the individual molecules.

10. Calculate the amount of toluene (log $K_{OW}$ = 2.69) that would bioconcentrate in a 1.0 kg fish using the following data and approximations. The lake in which the fish lives is contaminated with toluene at a level of 10 μg L$^{-1}$, due to having a gasoline station and marina in close proximity for many years. Equilibrium has been attained and the fish proximate[7] composition is 9% fat, 4% bone and other non-absorbing material, 7% air, and 80% aqueous. Assume that the fat is equivalent to octanol. How might this result compare to a biomagnification value for this same fish?

---

[7] Composition described in terms of classes of substances present is generally referred to as proximate composition, because the classes or groups, for example fats or minerals, are those first arrived at in the process of analysis; in proximate analysis the groups are measured as such, rather than as individual proteins or specific minerals. Food and Agriculture Organization of the United Nations, FAO Corporate Document Repository http://www.fao.org/wairdocs/tan/x5957e/x5957e01.htm, accessed November 2009.

Repeat this hypothetical calculation with one of the 'Dirty dozen' persistent organic pollutants having a log $K_{OW}$ = 5.0 using a concentration of 10 ng L$^{-1}$ (note much lower concentration). How does this result compare to the result obtained for toluene?

11. The equation (20.52) for the groundwater ubiquity score (GUS) has been modified from that given in the reference to use $K_{OM}$ values in place of $K_{OC}$ values. Show that the two forms of the equation are the same.

12. The organophosphorus pesticide phosmet is used as an insecticide for treating many orchard crops. Consult the Extoxnet website (Additional Resource 4), and using the representative half-life in soil, calculate the GUS score. What is the significance of this score and what are the limitations in its interpretation?

13. A triazine pesticide is present in groundwater. In this particular environment, the $K_d$ value for the pesticide is 6.5 mL g$^{-1}$. The soil has a particle density of 2.7 g mL$^{-1}$ and a porosity of 44%. The lateral flow rate of the groundwater is 0.75 m d$^{-1}$. Estimate how far the pesticide will move in one year.

# Chapter 21
# The future Earth

We began our survey of environmental chemistry by summarizing the beginning and early stages of the Earth's natural history. The Earth was formed when gases on the periphery of the solar nebula cooled and contracted to form planets. The newly formed planet was a solid sphere with a core made up of iron alloys and a crust of metal oxides and silicates. During the Earth's five billion year history, major changes occurred within and above the crust. Early on, water was formed and eventually came to cover two-thirds of the Earth's surface at an average depth of over 3 km in the great oceans. Changes in the atmosphere also took place. Most significantly, free oxygen gas was released from its combination with other elements in the crust. The presence of atmospheric oxygen was essential as a support for some living organisms, while, at the same time, other organisms maintained and regulated its supply. And so the world as we know it took shape and became a place where microorganisms, plants, and animals of all kinds could thrive and support one another. In fact, all the inorganic and living components of the Earth, through interactions and cycles, have been acting together to make our planet what it is. We have studied some of the major interactions and cycles.

Human life is short in the context of this more than five billion years of geological time, and we see the Earth's composition and processes in the light of what we observe at this point in its long history. It is easy to think that the natural world has always been the same. But, obviously, this has not been true in the past, and changes in our surroundings will continue to occur. Volcanoes erupt, spewing gases into the atmosphere and altering the global climate. Fresh mineral material is spread over the continents and under the oceans. The continents continue to shift, causing cracking and upwelling of the crust. The highlands are eroded by rainfall; physical and chemical weathering of the eroded material alters the shape and composition of the individual particles. Sediment fills the lakes, bringing in nutrients so that they eutrophy and become wetlands.

All these events are occurring today and will continue to occur. The Earth adjusts to these changes. Species evolve or become extinct in response to their changing surroundings. The ability of living and inorganic components to modify the environment and to adapt to evolving conditions leads some people to consider the Earth itself as an individual organism.[1]

In this book we have tried to emphasize that environmental chemistry begins with a knowledge of the natural processes that have taken place and are on-going now: the processes within the atmosphere, the hydrosphere, the geosphere, and the biosphere and the processes that move across boundaries, linking all parts of the environment together.

---

[1] J.E. Lovelock, *Gaia: a new look at life on Earth*, Oxford University Press, Oxford; 1979, 1987.

## Humans on Earth

Humans (*Homo sapiens sapiens*) have been inhabitants of the Earth for the past 200 000 years. We live here and, by being alive, we are participants in many of the natural processes and cycles. Our lungs take in oxygen and expel carbon dioxide. Our bodies require biomass and nutrients to survive and they release wastes that are then recycled through water, soil, plants, and other animals. But we do more than just breathe and eat. Each of the civilizations that humans have created over millennia has had a broader impact on its surroundings. At this stage in human history, that impact is very great indeed.

There are two reasons for this—the number of human beings and the *environmental footprint* that each human creates and leaves behind as a legacy for the future world. Both factors are important. In some parts of the world—in particular much of south and east Asia, Africa, South and Central America—it is the number of people that is most significant. In other regions—North America and Australia, for example—it is the ecological footprint of each person that is more important. We saw (Table 8.7) that, in 2008, annual *per capita* energy use in the United States and Canada was 307 GJ, while in Africa corresponding usage was 14 GJ. If we think of energy as the common currency of physical development, these figures imply that the average North American has a far greater impact on his / her surroundings than the average African. This statement is broadly true and it is certainly the case that many developments in industry and agriculture leave a large imprint on the global environment. Gases are released from factories and vehicles; liquid and solid wastes are discharged into water and on land—sometimes partially treated, sometimes untreated. Every year, new synthetic compounds are developed and, inadvertently or deliberately, some are released into the air or water or on to the soil where they interact with on-going natural processes. All these impacts put together are the source of the growing ecological footprint associated with modern humans at the beginning of the twenty-first century. Multiply this by the growing human population—five times greater than it was even 100 years ago—and it becomes clear that the cumulative human influences on the environment is massive.

## The atmosphere

The chemistry of the atmosphere has been changing constantly since the beginning of Earth's history. Just prior to the last glacial period, which began about 150 000 years ago, the atmospheric carbon dioxide mixing ratio as measured in Antarctic ice cores rose from about 200 to 290 ppmv, over a 20 000 year stretch of time. Recently, there has been an increase of similar magnitude—from about 275 to 387 ppmv—but the present jump has taken place over a mere 150 years. Simultaneously, the past century has been a period during which there were substantial increases in atmospheric mixing ratios of other trace gases like methane and nitrous oxide. To add to this, new human-produced gases—the chlorofluorocarbons, halons, sulphur hexafluoride, and many others—have been developed and released. While present in only trace quantities, the new compounds are able to interact with the 'natural' components and with solar radiation in the troposphere and stratosphere. The effects of trace gases on the ozone cycle in the stratosphere are well documented and large increases in the flux of solar UV-B radiation reaching parts of the Earth have been clearly measured. Details about how growth in greenhouse-gas mixing ratios will affect the climate are debatable, but the fact that there has been unprecedented rapid growth cannot be questioned. This is evidence of the significant footprint that humans are contributing to the global environment. Whatever the impact may be, it is one that will affect the entire world, not just the areas where the chemicals are released. That is the nature of atmospheric environmental problems. They know no boundaries as we have seen in additional examples from the literature on acid precipitation and on Arctic pollution.

## Water

Contaminated water also does not keep within well-defined boundaries, although its mobility may be somewhat less than that of air. The global water resource experiences the footprint of human activities and the impact is also energy related in many cases. We have become familiar with the periodic occurrence of oceanic oil spills from supertankers in various places around the world. Perhaps it is surprising that the well-publicized major accidents account for only about 10 to 15% of petroleum that is discharged into the oceans. A larger proportion comes from bilge pumping (an illegal practice), land-based industrial sources, and run-off from roads and parking lots in urban areas. There are also inputs into the oceans from the atmosphere.

Oil slicks are the most obvious evidence of petroleum contamination. Natural processes of volatilization, dissolution, and abiotic and biotic degradation operate to make the slick disappear, especially its low molar mass components. The more resistant compounds are emulsified via oceanic turbulence and some of this emulsion eventually forms tar balls. The tar balls have small surface areas and are relatively inert so they degrade only slowly and can be carried great distances before sinking into the sediment or washing up on shore. Over long periods of time they break down into smaller particles that can be resuspended, degraded, or taken up by organisms in the water or sediment column. Some of the compounds, in particular aromatic and polyaromatic hydrocarbons, have toxic and carcinogenic properties and interfere with various metabolic reactions in many marine species. All these processes take place throughout the oceans, often far from the site of the original spill.

A study[2] in the Black Sea describes the types and amounts of petroleum residues found in water and sediment and creates a picture that is typical of situations around the world. The western part of the Black Sea, shared by Turkey, Bulgaria, Romania, and the Ukraine, receives petroleum-related contaminants from discharges released by industries along the Danube, Dnieper, and Dniester Rivers. Added to these are direct inputs from marine transport, crude-oil spills, and the atmosphere. High concentrations of aliphatic and aromatic (including polyaromatic) hydrocarbons have been found in the river estuaries, but substantial amounts are also found offshore in both the sediment and dissolved phases, usually associated with combustion-derived particulate material.

## The solid Earth

The solid Earth has undergone major changes over the planet's five billion year history. We discussed the processes by which soils, the basis of agriculture and forestry, have formed (and are still forming) from massive rock and other materials. It is usually only over long periods of time that productive soils form from geologically derived materials. And, of course, there has always been a balance between soil formation and loss due to wind and water. In some cases, the transported materials are deposited elsewhere on land, replenishing the agricultural resource in productive areas like the lower reaches of the Yellow River in northern China or the Delta of the Nile in Egypt. Eventually, some soil is delivered to the oceans where it is permanently lost to agriculture. Sometimes, the human footprint intersects these natural processes in a dramatic way, as when there is deforestation of headlands leading to enhanced erosion. In other cases, great dams are built, changing not only the water flow and topography but also the chemical processes related to eutrophication and rate of flushing of solid and dissolved species in the river.

Disposal sites for solid waste material now occupy thousands of hectares of land. In some cases, these areas are carefully monitored but, from time to time, accidents occur, bringing into focus the mark that human activities leave on the solid Earth.

In southern Spain, 45 km north-west of Seville, a deposit of copper–zinc–lead ore abundant in pyrite has been mined off and on since the Roman period.[3] In recent years, a profitable mining operation (the Los Frailes mine) for the three base metals has been established and a tailings dam was built with a capacity of about 70 million t to deal with the waste rock. The tailings slurry

[2] Maldonado, C., J.M. Bayona, and L. Bodineau, Sources, distribution and water column processes of aliphatic and polycyclic aromatic hydrocarbons in the northwestern Black Sea water, *Environ. Sci. Technol.*, **33** (1999), 2693–702.
[3] Sassoon, M., Los Frailes aftermath, *Mining Environ. Management*, July 1998, pp. 8–12.

has a pH of 2 to 4 and, after settling, the liquid is pumped into a treatment facility to recover the metals before the wastewater is discharged into the Rio Agrio. This river empties into the Rio Guadiamar further downstream.

Below the tailings dam, the river flows through a rich farming area that produces citrus fruit, peaches, sunflower, wheat, maize, cotton, and olives, before reaching marsh lands that surround a National Park. The marshes are breeding grounds for more than 250 species of migratory birds and the area is protected to preserve the unique fauna and flora of the area.

On 26 April 1998, the tailings dam burst sending a 2 m wall of slurry coursing down the rivers. The amount of slurry released is uncertain but may have been as much as 5 million m³, containing up to 2 million t of solids. Approximately 2000 ha of agricultural land were completely inundated with slurry and a further 2000 ha were affected less severely. Eighty per cent of the tailings settled out along the first 13 km of the river bed and its surroundings.

Initially, farmers sustained great losses due to the burial and destruction of standing crops. What will be the long-term effects? At least in part, natural processes—the same ones that are responsible for formation of soils—have already been able to 'neutralize' some of the impact from the sudden release of the heavy slurry. The pH of the river returned to near normal within 10 days. However, zinc and lead levels remained high for some time and it is not certain whether production of food crops can resume in the near future. Humans were called upon to assist and 3 million m³ of tailings have now been removed from the river banks.

## Planetary boundaries

The catalogue of environmental issues is a long one and out of this list we have chosen only a few examples showing how human activities have imposed physical and chemical changes on our surroundings. Some problems are acute and obviously require immediate attention; others are more subtle and may begin to become visibly serious only after a very long time.

Given the extensive catalogue of problems, an important question is 'Are there environmental boundary conditions that cannot be crossed without upsetting the Earth's stability in such a way that human life becomes impossible'? We are not talking here about saving the Earth itself. The Earth will continue to change and exist, but what are the conditions that will enable it to exist in a state that can sustain life and support the human population? This question has been pondered and addressed by scientists and other futurists and one theory[4] speaks about 'planetary boundaries' that enclose a safe space within which civilizations can thrive. In a way, the theory prioritizes environmental issues into nine integrative and interlinked boundary categories:

| Category | Measurable indicator parameters |
|---|---|
| • Climate change | carbon dioxide levels and radiative forcing |
| • Rate of biodiversity loss | rate of extinction of species |
| • Nitrogen cycle | nitrogen removal rate from the atmosphere |
| • Phosphorus cycle | amount of phosphorus flowing into oceans |
| • Stratospheric ozone depletion | stratospheric ozone mixing ratios |
| • Ocean acidification | saturation state of $CaCO_3$ in surface water |
| • Global fresh-water use | consumption rates for human requirements |
| • Change in land use | land converted to crop production |
| • Atmospheric aerosol loading | concentrations of atmospheric particulates |
| • Chemical pollution | emission rates of POPs and other pollutants |

[4] Rockström, J. and 28 others, A safe operating space for humanity, *Nature* **461** (2009), 472.

It is obvious that, except perhaps for the biodiversity category, each one of these critical subjects has a central chemistry component. The authors of the safe operating space theory describe boundary conditions in terms of concentration limits or consumption rates above which they believe it is possible that the Earth will experience irreversible and in some cases abrupt environmental change—change that will seriously impair the ability of our surroundings to support human development. Ominously, the authors also suggest that in at least the first three of these categories we have already overstepped the boundaries and urgent remedial action is required.

## Dealing with complex environmental challenges

A follow-up question then becomes—how should we respond to these perceived and actual human impacts? Discussions centre around two different approaches. One is to look for the scientific / technical solution for each issue either in response to damage already observed or in anticipation of problems that might occur. Better control devices to minimize or eliminate emissions, materials that degrade photolytically or biologically, integrated pest management that includes crop rotation, biological control agents, and only selective use of pesticides—all are examples of using improved science to minimize impacts. There is also considerable current interest in *green chemistry* for industrial manufacturing processes.[5] This term refers to the redesigning, where possible, of manufacturing technologies in such a way that they employ more benign chemicals—water rather than organic compounds as solvent, for example—so that the by-products of manufacturing are themselves relatively harmless materials. Each of these approaches requires creative thinking and can markedly improve the environment by minimizing the direct impact we make.

*Green chemistry* is at the frontier of chemical science and attempts to reduce the environmental impact of the harmful chemical enterprise by developing a technology base that is inherently non-toxic to living things and the environment.

Topics of green chemistry include:

- the use of sustainable resources;
- the use of biotechnology alternatives to chemistry-based solutions methodologies;
- the design of new 'greener' and safer chemicals and materials;
- the development of environmentally improved synthetic routes and methods to important products;
- the application of innovative technology to established industrial procedures;
- tools for measuring environmental impact;
- chemical aspects of renewable energy.

An example[6] of a new process that uses the principles of green chemistry is the synthesis of ibuprofen, a very common over-the-counter pain killer that has been available since the 1960s. The original patent for its production included a six-step synthesis (the Boots' synthesis) where only 40% of the raw materials used in production ended up as part of the drug. Most of the other 60% of material ended up as part of the waste streams from the various steps of the synthesis.

In the mid-1980s, when patent protection was no longer operative, other companies were free to try new technologies and synthetic routes for producing ibuprofen. BHC was one such

---

[5] Anastas, P. and T.C. Williamson, *Green chemistry*, Oxford University Press, Oxford; 1998. The following webpage is an excellent beginning point for searching and linking to many sites related to green chemistry: http://www.epa.gov/greenchemistry/, accessed July 2010.

[6] Further detailed information about this *green chemistry* approach to producing ibuprofen can be found at http://www.chemsoc.org/networks/learnnet/green/ibuprofen/index.htm, accessed July 2010.

company, and they were able to shorten the synthesis from six steps to three and this improved the overall efficiency of raw material consumption to 77%. Also, the catalysts used (HF, Raney nickel, and Pd) are all recoverable in this synthesis, while the $AlCl_3$ used in the Boots' synthesis was not recoverable. Considering the vast quantity of ibuprofen produced annually, an improvement of 37% in efficiency significantly reduces the amount of waste generated from this one process.

But modification of technology cannot be the sole approach. For one thing, almost all technology involves some input of energy. For example, one attractive approach for reducing harmful smog-producing emissions in urban areas is to design vehicles that run on hydrogen–oxygen fuel cells. The energy-generating reaction that powers the vehicle is

$$2H_2\,(g) + O_2\,(g) \rightarrow 2H_2O\,(g)\ \Delta G = -457\,kJ \qquad (21.1)$$

Indeed, as combustion products, such fuel-cell engines emit mostly water vapour (but also small amounts of hydrogen peroxide and nitric oxide, if air is used as the oxygen source) and would go a long way toward reducing the load of volatile hydrocarbons in the atmosphere of a large city.

It is necessary, however, to produce hydrogen for the fuel cell, and this is often accomplished by reversing the energy-producing reaction. One method involves electrolysis of water and the required energy is frequently provided by a fossil fuel-fired electric power station operating at around 30% efficiency. Other methods for hydrogen production make use of steam-reforming or plasma-arc processes for which both the raw material and the energy source required for production are some form of fossil fuel. Using natural gas for these purposes, hydrogen synthesis can be described in the following two steps:

$$CH_4 + H_2O \xrightarrow{\text{heat}} 3H_2 + CO \qquad (21.2)$$

$$CO + H_2O \xrightarrow{\text{heat}} H_2 + CO_2 \qquad (21.3)$$

Note that the sum of these two reactions is exactly the reverse of the very first reaction presented in this book (reaction 1.1), which described the route by which water was generated on the Earth's surface, very early in the history of our planet.

In the production of the clean hydrogen fuel by either means, carbon dioxide is an important by-product. There is an environmental advantage associated with centralizing the production of hydrogen at a power station in that excellent emission control practices can be followed, and whatever emissions are released can be dispersed at a high altitude remote from heavily populated areas. But, inevitably, control devices at power stations require material inputs and will produce their own waste products requiring disposal. As always, carbon dioxide is almost unavoidably, a released product.

---

**Fermi question**

About 650 billion m³ of hydrogen are produced annually by processes such as those described here and as a by-product of petroleum production. What fraction of the Earth's current total energy requirements could be satisfied by this production?

---

Aside from scientific / technical solutions, the other mechanism that humans must consider for minimizing our environmental footprint is to reduce substantially the activities that impact the environment. This 'solution' must be especially directed toward the 25% of people, mostly in wealthy countries, who consume 75% of the world's resources. It involves consumption of less materials, reduced use of transportation (particularly the personal motorized vehicle), reduced dependence on non-renewable energy for providing a comfortable living environment, and many other ways of curtailing resource depletion. It is clear that the huge imbalance in energy use between, for example, North America and Africa cannot be explained away and justified on the basis of climate and size of country.

Clearly, in these closing words, we have been moving beyond the field of environmental chemistry and beyond the purpose of this book. And so, we return to our stated goal, which was to provide a description of the chemical basis for understanding our surroundings, the global environment. Whether one is dealing with ideas of human health, with design of devices and processes for controlling emissions, or setting policy, the importance of understanding the underlying nature of an issue cannot be overemphasized.

# Appendices

**Appendix A.1** Properties of the Earth.

| | |
|---|---|
| Mass of the Earth | $5.98 \times 10^{24}$ kg |
| Mass of the atmosphere | $5.27 \times 10^{18}$ kg |
| Mass of the oceans | $1.37 \times 10^{21}$ kg |
| Mass of fresh water (surface) | $1.27 \times 10^{17}$ kg |
| Mass of pore and groundwater | $9.5 \times 10^{18}$ kg |
| Mass of ice | $2.9 \times 10^{19}$ kg |
| Mass of atmospheric water | $1.3 \times 10^{16}$ kg |
| Mass of living OM (dry wt, carbon) | $8 \times 10^{14}$ kg |
| Mass of dead OM (dry wt, carbon) | $3.5 \times 10^{15}$ kg |
| Average radius of the Earth | 6378.2 km |
| Total area of Earth's surface | $5.10 \times 10^{14}$ m² |
| Area of continents | $1.48 \times 10^{14}$ m² |
| Continental area covered by ice | $1.72 \times 10^{13}$ m² |
| Volume of oceans | $1.35 \times 10^{18}$ m³ |

**Appendix A.2** Properties of air and water.

| Property | Value |
|---|---|
| Air (dry, $P^{\circ}$) | |
| Average molar mass (troposphere) / Dalton | 28.96 |
| Density / kg m$^{-3}$ | |
| $\quad$ 0°C | 1.293 |
| $\quad$ 20°C | 1.205 |
| Viscosity / g m$^{-1}$ s$^{-1}$ | |
| $\quad$ 0°C | $1.7 \times 10^{-2}$ |
| $\quad$ 20°C | $1.9 \times 10^{-2}$ |
| **Water** | |
| Molar mass / Dalton | 18.015 |
| Density / kg m$^{-3}$ | |
| $\quad$ 0°C | 999.87 |
| $\quad$ 20°C | 998.23 |
| Viscosity / g m$^{-1}$ s$^{-1}$ | |
| $\quad$ 0°C | 1.79 |
| $\quad$ 20°C | 1.00 |

**Appendix A.3** Area, biomass, and productivity of ecosystem types[a].

| Type | Area / $10^{12}$ m$^2$ | Mean plant biomass / kg C m$^{-2}$ | Productivity / kg C m$^{-2}$ y$^{-1}$ |
|---|---|---|---|
| Tropical forest | 24.5 | 18.8 | 0.83 |
| Temperate forest | 12.0 | 14.6 | 0.56 |
| Boreal forest | 12.0 | 9.0 | 0.36 |
| Woodland and shrubland | 8.0 | 2.7 | 0.27 |
| Savannah | 15.0 | 1.8 | 0.32 |
| Grassland | 9.0 | 0.7 | 0.23 |
| Tundra and alpine meadow | 8.0 | 0.3 | 0.065 |
| Desert scrub | 18.0 | 0.3 | 0.032 |
| Rock, ice, sand | 24.0 | 0.01 | 0.015 |
| Cultivated land | 14.0 | 0.5 | 0.29 |
| Swamp and marsh | 2.0 | 6.8 | 1.13 |
| Lake and stream | 2.5 | 0.01 | 0.23 |
| Open ocean | 332.0 | 0.0014 | 0.057 |
| Upwelling zones | 0.4 | 0.01 | 0.23 |
| Continental shelf | 26.6 | 0.005 | 0.16 |
| Algal bed and reef | 0.6 | 0.9 | 0.90 |
| Estuaries | 1.4 | 0.45 | 0.81 |

[a] Source: Harte, J. *Consider a spherical cow*, University Science Books, Mill Valley, California; 1988.

**Appendix B.1** The elements.

| Atomic number | Element | Symbol | Atomic mass / Dalton | Abundance[a] Crustal / mg kg$^{-1}$ | Oceanic[b] / mol L$^{-1}$ | Freshwater / mol L$^{-1}$ | Atmosphere / ppmv |
|---|---|---|---|---|---|---|---|
| 1 | Hydrogen | H | 1.00794 | 1520 | | | 0.53 |
| 2 | Helium | He | 4.002602 | 0.008 | $1 \times 10^{-9}$ | | 5.2 |
| 3 | Lithium | Li | 6.941 | 20 | $2.5 \times 10^{-5}$ | $1.7 \times 10^{-6}$ | |
| 4 | Beryllium | Be | 9.01218 | 2.6 | $7 \times 10^{-12}$ | | |
| 5 | Boron | B | 10.81 | 10 | $4.2 \times 10^{-4}$ | $1.7 \times 10^{-6}$ | |
| 6 | Carbon | C | 12.011 | 480 | $2.0 \times 10^{-3}$ | | 387 |
| 7 | Nitrogen | N | 14.00674 | 25 | | | 780800 |
| 8 | Oxygen | O | 15.9994 | 466000 | | | 209500 |
| 9 | Fluorine | F | 18.99840 | 950 | $7.0 \times 10^{-5}$ | $5.3 \times 10^{-6}$ | |
| 10 | Neon | Ne | 20.1797 | $7 \times 10^{-5}$ | $1 \times 10^{-7}$ | | 18 |
| 11 | Sodium | Na | 22.98977 | 28300 | 0.481 | $2.2 \times 10^{-4}$ | |
| 12 | Magnesium | Mg | 24.3050 | 20900 | $5.5 \times 10^{-2}$ | $1.6 \times 10^{-4}$ | |
| 13 | Aluminium | Al | 26.98154 | 81300 | $3.5 \times 10^{-8}$ | $1.9 \times 10^{-6}$ | |
| 14 | Silicon | Si | 28.0855 | 272000 | $1 \times 10^{-6}$ | $1.9 \times 10^{-4}$ | |
| 15 | Phosphorus | P | 30.97376 | 1000 | $5 \times 10^{-8}$ | $1.3 \times 10^{-6}$ | |
| 16 | Sulphur | S | 32.066 | 260 | $2.9 \times 10^{-2}$ | | |
| 17 | Chlorine | Cl | 35.4527 | 130 | 0.561 | | |
| 18 | Argon | Ar | 39.948 | 1.2 | $1.2 \times 10^{-5}$ | | 9300 |
| 19 | Potassium | K | 39.0983 | 25900 | $1.01 \times 10^{-2}$ | $3.4 \times 10^{-5}$ | |
| 20 | Calcium | Ca | 40.078 | 36300 | $1.06 \times 10^{-2}$ | $3.6 \times 10^{-4}$ | |
| 21 | Scandium | Sc | 44.95591 | 16 | $1 \times 10^{-11}$ | $8.9 \times 10^{-11}$ | |
| 22 | Titanium | Ti | 47.88 | 5600 | $1 \times 10^{-8}$ | $2.1 \times 10^{-7}$ | |
| 23 | Vanadium | V | 50.9415 | 160 | $3 \times 10^{-8}$ | $2.0 \times 10^{-8}$ | |

(*Continued*)

**Appendix B.1** (*Continued*)

| Atomic number | Element | Symbol | Atomic mass / Dalton | Abundance[a] | | | |
|---|---|---|---|---|---|---|---|
| | | | | Crustal / mg kg$^{-1}$ | Oceanic[b] / mol L$^{-1}$ | Freshwater / mol L$^{-1}$ | Atmosphere / ppmv |
| 24 | Chromium | Cr | 51.9961 | ~100 | $3 \times 10^{-9}$ | $1.9 \times 10^{-8}$ | |
| 25 | Manganese | Mn | 54.93805 | 950 | $2 \times 10^{-9}$ | $1.5 \times 10^{-7}$ | |
| 26 | Iron | Fe | 55.847 | 50000 | $1.7 \times 10^{-9}$ | $7.2 \times 10^{-7}$ | |
| 27 | Cobalt | Co | 58.93320 | 20 | $1 \times 10^{-10}$ | $3.4 \times 10^{-9}$ | |
| 28 | Nickel | Ni | 58.70 | 80 | $2 \times 10^{-9}$ | $3.8 \times 10^{-8}$ | |
| 29 | Copper | Cu | 63.546 | 50 | $1.3 \times 10^{-9}$ | $1.6 \times 10^{-7}$ | |
| 30 | Zinc | Zn | 65.39 | 75 | $8 \times 10^{-10}$ | $4.6 \times 10^{-7}$ | |
| 31 | Gallium | Ga | 69.723 | 18 | $4 \times 10^{-13}$ | $1.3 \times 10^{-9}$ | |
| 32 | Germanium | Ge | 72.61 | 1.8 | $3 \times 10^{-12}$ | | |
| 33 | Arsenic | As | 74.9216 | 1.5 | $2.0 \times 10^{-8}$ | $2.3 \times 10^{-8}$ | |
| 34 | Selenium | Se | 78.96 | 0.05 | $4 \times 10^{-12}$ | $2.5 \times 10^{-9}$ | |
| 35 | Bromine | Br | 79.904 | 0.37 | $8.6 \times 10^{-4}$ | $2.5 \times 10^{-7}$ | |
| 36 | Krypton | Kr | 83.80 | $1 \times 10^{-5}$ | $1 \times 10^{-10}$ | | 1.14 |
| 37 | Rubidium | Rb | 85.4678 | 90 | $1.5 \times 10^{-6}$ | $1.8 \times 10^{-8}$ | |
| 38 | Strontium | Sr | 87.62 | 370 | $9.4 \times 10^{-5}$ | $6.9 \times 10^{-7}$ | |
| 39 | Yttrium | Y | 88.90585 | 30 | $1 \times 10^{-10}$ | $7.9 \times 10^{-9}$ | |
| 40 | Zirconium | Zr | 91.224 | 190 | $1 \times 10^{-10}$ | | |
| 41 | Niobium | Nb | 92.90638 | 20 | $1 \times 10^{-12}$ | | |
| 42 | Molybdenum | Mo | 95.94 | 1.5 | $1.1 \times 10^{-7}$ | $5.2 \times 10^{-9}$ | |
| 43 | Technetium[c] | Tc | 98.9062 | | | | |
| 44 | Ruthenium | Ru | 101.07 | ~$1 \times 10^{-3}$ | | | |
| 45 | Rhodium | Rh | 102.90550 | ~$2 \times 10^{-4}$ | | | |
| 46 | Palladium | Pd | 106.42 | ~$6 \times 10^{-4}$ | $2 \times 10^{-13}$ | | |
| 47 | Silver | Ag | 107.8682 | 0.07 | $1 \times 10^{-12}$ | $2.8 \times 10^{-9}$ | |

| | | | | | | |
|---|---|---|---|---|---|---|
| 48 | Cadmium | Cd | 112.411 | 0.11 | $1 \times 10^{-11}$ | 0.086 |
| 49 | Indium | In | 114.82 | 0.049 | $9 \times 10^{-13}$ | |
| 50 | Tin | Sn | 118.710 | 2.2 | $2 \times 10^{-11}$ | |
| 51 | Antimony | Sb | 121.75 | 0.2 | $3 \times 10^{-9}$ | $8.2 \times 10^{-9}$ |
| 52 | Tellurium | Te | 127.60 | $5 \times 10^{-3}$ | $1.5 \times 10^{-12}$ | |
| 53 | Iodine | I | 126.90447 | 0.14 | $3.7 \times 10^{-7}$ | |
| 54 | Xenon | Xe | 131.29 | $2 \times 10^{-6}$ | $8 \times 10^{-13}$ | $5 \times 10^{-8}$ |
| 55 | Caesium | Cs | 132.9054 | 3 | $2.3 \times 10^{-9}$ | $2.6 \times 10^{-10}$ |
| 56 | Barium | Ba | 137.327 | 500 | $3.5 \times 10^{-8}$ | $4.4 \times 10^{-7}$ |
| 57 | Lanthanum | La | 138.9055 | 32 | $1.6 \times 10^{-11}$ | $3.6 \times 10^{-10}$ |
| 58 | Cerium | Ce | 140.115 | 68 | $4 \times 10^{-11}$ | $5.7 \times 10^{-10}$ |
| 59 | Praseodymium | Pr | 140.90765 | 9.5 | $3 \times 10^{-12}$ | $5.0 \times 10^{-11}$ |
| 60 | Neodymium | Nd | 144.24 | 38 | $1.3 \times 10^{-11}$ | $2.8 \times 10^{-10}$ |
| 61 | Promethium[c] | Pm | (145) | | | |
| 62 | Samarium | Sm | 150.36 | 7.9 | $3 \times 10^{-12}$ | $5.3 \times 10^{-11}$ |
| 63 | Europium | Eu | 151.965 | 2.1 | $7 \times 10^{-13}$ | $6.6 \times 10^{-12}$ |
| 64 | Gadolinium | Gd | 157.25 | 7.7 | $4 \times 10^{-12}$ | $5.1 \times 10^{-11}$ |
| 65 | Terbium | Tb | 158.92534 | 1.1 | $6 \times 10^{-13}$ | $6.3 \times 10^{-12}$ |
| 66 | Dysprosium | Dy | 162.50 | 6 | $5 \times 10^{-12}$ | $3.0 \times 10^{-10}$ |
| 67 | Holmium | Ho | 164.93032 | 1.4 | $1 \times 10^{-12}$ | $6.1 \times 10^{-12}$ |
| 68 | Erbium | Er | 167.26 | 3.8 | $4 \times 10^{-12}$ | $2.4 \times 10^{-11}$ |
| 69 | Thulium | Tm | 168.93421 | 0.48 | $6 \times 10^{-13}$ | $5.9 \times 10^{-12}$ |
| 70 | Ytterbium | Yb | 173.04 | 3.3 | $3 \times 10^{-12}$ | $2.3 \times 10^{-11}$ |
| 71 | Lutetium | Lu | 174.97 | 0.51 | $6 \times 10^{-13}$ | $5.7 \times 10^{-12}$ |
| 72 | Hafnium | Hf | 178.49 | 5.3 | $4 \times 10^{-11}$ | |
| 73 | Tantalum | Ta | 180.948 | 2 | $1 \times 10^{-11}$ | |
| 74 | Tungsten | W | 183.85 | 1 | $5.1 \times 10^{-8}$ | $1.6 \times 10^{-10}$ |
| 75 | Rhenium | Re | 186.21 | $4 \times 10^{-4}$ | $2 \times 10^{-11}$ | |

(Continued)

**Appendix B.1** (*Continued*)

| Atomic number | Element | Symbol | Atomic mass / Dalton | Abundance[a] | | | |
|---|---|---|---|---|---|---|---|
| | | | | Crustal / mg kg⁻¹ | Oceanic[b] / mol L⁻¹ | Freshwater / mol L⁻¹ | Atmosphere / ppmv |
| 76 | Osmium | Os | 190.2 | $\sim 1 \times 10^{-4}$ | | | |
| 77 | Iridium | Ir | 192.2 | $\sim 3 \times 10^{-6}$ | | | |
| 78 | Platinum | Pt | 195.08 | $\sim 1 \times 10^{-3}$ | $6 \times 10^{-13}$ | | |
| 79 | Gold | Au | 196.9665 | $1.1 \times 10^{-3}$ | $5 \times 10^{-11}$ | $2.0 \times 10^{-11}$ | |
| 80 | Mercury | Hg | 200.59 | 0.05 | $2 \times 10^{-12}$ | $3.5 \times 10^{-10}$ | |
| 81 | Thallium | Tl | 204.38 | 0.6 | $7 \times 10^{-11}$ | | |
| 82 | Lead | Pb | 207.2 | 14 | $1 \times 10^{-11}$ | $4.8 \times 10^{-9}$ | |
| 83 | Bismuth | Bi | 208.980 | 0.048 | $2 \times 10^{-13}$ | | |
| 84 | Polonium | Po | (209) | | | | |
| 85 | Astatine | At | (210) | | | | |
| 86 | Radon | Rn | (222) | | | | |
| 87 | Francium | Fr | (223) | | | | |
| 88 | Radium | Ra | 226.025 | $6 \times 10^{-7}$ | $9 \times 10^{-17}$ | | |
| 89 | Actinium | Ac | 227.028 | | | | |
| 90 | Thorium | Th | 232.038 | 12 | $4 \times 10^{-1}$ | | |
| 91 | Protactinium | Pa | 231.036 | | $9 \times 10^{-15}$ | | |
| 92 | Uranium | U | 238.03 | 2.4 | $1.4 \times 10^{-8}$ | $1 \times 10^{-9}$ | |
| 93 | Neptunium | Np | 237.048 | | | | |
| 94 | Plutonium | Pu | (244) | | | | |
| 95 | Americium | Am | (243) | | | | |

[a] Source. Most of the values for terrestrial and oceanic concentrations are from Emsley, J., *The elements*, 2nd edn, Clarendon Press, Oxford; 1991. Fresh-water values are reported as mean concentrations in river water in Libes, S.M., *An introduction to marine biogeochemistry*, John Wiley and Sons Inc., New York; 1992.

[b] Ocean concentrations are averages; in some cases values refer to the surface water.

[c] These are radioactive elements and are not normally found naturally except for trace quantities in uranium deposits.

**Appendix B.2** Thermochemical properties of selected elements and compound[a].

| | $\Delta H_f^\circ$ / kJ mol$^{-1}$ | $\Delta G_f^\circ$ / kJ mol$^{-1}$ | $S^\circ$ / J mol$^{-1}$ K$^{-1}$ |
|---|---|---|---|
| Aluminium | | | |
| Al (s) | 0 | 0 | +28.33 |
| Al$^{3+}$ (aq) | $-531$ | $-485$ | $-321.7$[b] |
| Al$_2$O$_3$ (s, corundum) | $-1675.7$ | $-1582.3$ | +50.92 |
| Al$_2$O$_3$ •3H$_2$O (s, gibbsite) | $-2586.67$ | 2310.41 | +136.90 |
| Calcium | | | |
| Ca (s) | 0 | 0 | +41.42 |
| Ca$^{2+}$ (aq) | 542.83 | $-553.58$ | $-53.1$[b] |
| CaO (s) | 635.09 | $-604.05$ | +39.75 |
| CaCO$_3$ (s, calcite) | $-1206.92$ | $-1128.84$ | +92.6 |
| Carbon (including some common organic compounds) | | | |
| C (s, graphite) | 0 | 0 | +5.740 |
| C (g) | +716.68 | +671.29 | +158.99 |
| CO (g) | $-110.53$ | $-137.15$ | +197.57 |
| CO$_2$ (g) | $-393.51$ | $-394.36$ | +213.63 |
| CO$_2$ (aq) | $-413.80$ | $-385.98$ | +117.6[b] |
| H$_2$CO$_3$ (aq) | $-699.65$ | $-623.08$ | +187.4[b] |
| HCO$_3^-$ (aq) | $-691.99$ | 586.84 | +91.2[b] |
| CO$_3^{2-}$ (aq) | $-677.14$ | $-527.86$ | $-56.9$[b] |
| CCl$_4$ (l) | $-135.44$ | $-65.28$ | +216.40 |
| CS$_2$ (l) | +89.70 | +65.27 | +151.34 |
| HCN (aq) | +107.1 | +119.7 | |
| CN$^-$ (aq) | +150.6 | +172.4 | +94.1[b] |
| CH$_4$ (g, methane) | $-74.81$ | $-50.75$ | +186.16 |
| CH$_3$ (g, methyl radical) | +145.69 | +147.92 | +194.2[b] |
| C$_2$H$_6$ (g, ethane) | $-84.68$ | $-32.92$ | +229.49 |
| C$_2$H$_4$ (g, ethene) | +52.26 | +68.08 | +219.45 |
| C$_2$H$_2$ (g, ethyne) | 226.73 | +209.17 | +200.83 |
| C$_3$H$_8$ (g, propane) | $-104.5$ | $-23.4$ | +269.9 |
| C$_4$H$_{10}$ (g, n-butane) | $-126.5$ | $-17.15$ | +310.1 |
| C$_5$H$_{12}$ (g, n-pentane) | $-146.5$ | $-8.37$ | +348.9 |
| C$_8$H$_{18}$ (l, n-octane) | $-249.95$ | $-6.71$ | +361.21 |
| C$_8$H$_{18}$ (g, n-octane) | $-208.45$ | $-16.72$ | +466.84 |
| C$_6$H$_6$ (l, benzene) | +49.0 | +124.7 | +172 |
| C$_6$H$_6$ (g, benzene) | +82.9 | +129.7 | +269.2 |
| C$_{10}$H$_8$ (s, naphthalene) | +78.53 | | |
| CH$_3$OH (l, methanol) | $-238.66$ | $-166.35$ | +126.8 |
| C$_2$H$_5$OH (l, ethanol) | $-277.69$ | $-174.89$ | +160.7 |
| C$_6$H$_5$OH (l, phenol) | $-165.0$ | $-50.9$ | +146.0[b] |

(*Continued*)

**Appendix B.2** (*Continued*)

| | $\Delta H_f^\circ$ / kJ mol$^{-1}$ | $\Delta G_f^\circ$ / kJ mol$^{-1}$ | $S^\circ$ / J mol$^{-1}$K$^{-1}$ |
|---|---|---|---|
| HCOOH (l, formic acid) | −424.72 | −361.42 | +128.95 |
| CH$_3$COOH (aq, acetic acid) | −485.76 | −396.56 | +159.8 |
| CH$_3$COO$^-$ (aq, acetate) | −486.01 | −369.39 | +86.6[b] |
| HCHO (g, formaldehyde) | −108.57 | −102.55 | +218.66 |
| CH$_3$CHO (l, acetaldehyde) | −192.30 | −128.20 | +160.2 |
| CH$_3$CHO (g, acetaldehyde) | −166.19 | −128.91 | +250.2 |
| CH$_3$COCH$_3$ (l, acetone) | −248.1 | −155.4 | +200.4[b] |
| C$_6$H$_{12}$O$_6$ (s, β-D-glucose) | −1268 | −910 | +212[b] |
| Chlorine | | | |
| Cl$_2$ (g) | 0 | 0 | +222.96 |
| Cl (g) | +121.68 | +105.70 | +165.09 |
| Cl$^-$ (aq) | −167.16 | −131.24 | +56.5[b] |
| HCl (g) | −92.31 | −95.30 | +186.80 |
| HCl (aq) | −167.16 | −131.23 | +56.5[b] |
| Hydrogen | | | |
| H$_2$ (g) | 0 | 0 | +130.58 |
| H (g) | +217.97 | +203.26 | +114.60 |
| H$_2$O (l) | −285.83 | −237.18 | +69.91 |
| H$_2$O (g) | −241.82 | −228.59 | +188.72 |
| H$_2$O$_2$ (l) | −187.78 | −120.42 | +109.6 |
| H$_3$O$^+$(aq) | −285.83 | −237.18 | |
| Iron | | | |
| Fe (s) | 0 | 0 | +27.28 |
| Fe$^{2+}$ (aq) | −89.1 | −78.9 | −137.7[b] |
| Fe$^{3+}$ (aq) | −48.5 | −4.7 | −315.9[b] |
| Fe$_2$O$_3$ (s, haematite) | −824.2 | −742.2 | +87.40 |
| Fe$_3$O$_4$ (s, magnetite) | −1118.4 | −1015.4 | +146.4 |
| Nitrogen | | | |
| N$_2$ (g) | 0 | 0 | +191.50 |
| N (g) | +472.70 | +455.58 | +153.19 |
| NO (g) | +90.25 | +86.55 | +210.65 |
| N$_2$O (g) | +82.05 | +104.20 | +219.74 |
| NO$_2$ (g) | +33.18 | +51.29 | +239.95 |
| N$_2$O$_5$ (g) | +11.3 | +115.0 | +355.7 |
| HNO$_3$ (aq) | −207.36 | −111.25 | +146.4[b] |
| NH$_3$ (g) | −46.11 | −16.42 | +192.34 |
| NH$_3$ (aq) | −80.29 | −26.57 | +111.3[b] |
| NH$_4^+$ (aq) | −132.51 | −79.31 | +113.4[b] |
| NH$_2$CONH$_2$ (s, urea) | −333.51 | −197.44 | +104.60 |

**Appendix B.2** (*Continued*)

| | $\Delta H_f^\circ$ / kJ mol$^{-1}$ | $\Delta G_f^\circ$ / kJ mol$^{-1}$ | $S^\circ$ / J mol$^{-1}$K$^{-1}$ |
|---|---|---|---|
| Oxygen | | | |
| $O_2$ (g) | 0 | 0 | +205.03 |
| O (g) | +249.17 | +231.75 | +160.95 |
| $O_3$ (g) | +142.7 | +163.2 | +238.82 |
| OH$^-$ (aq) | −229.99 | −157.28 | −10.75[b] |
| Sulphur | | | |
| S (s, rhombic) | 0 | 0 | +31.80 |
| $SO_2$ (g) | −296.83 | −300.19 | +248.11 |
| $SO_3$ (g) | −395.72 | −371.08 | +256.65 |
| $HSO_4^-$ (aq) | −887.34 | −755.99 | +131.8[b] |
| $SO_4^{2-}$ (aq) | −909.27 | −744.60 | +20.1[b] |
| $H_2S$ (g) | −20.63 | −33.59 | +205.68 |
| $H_2S$ (aq) | −39.7 | −27.86 | +121[b] |
| HS$^-$ (aq) | −17.6 | +12.08 | +62.08[b] |
| $SF_6$ (g) | −1209 | −1105.4 | +291.71 |

[a] Source: The National Bureau of Standards tables of chemical thermodynamic properties. *J. Phys. Chem. Ref. Data*, **11** Supplement 2 (1982) as reported in Breck, W.G., R.J.C. Brown, and J.D. McCowan, *Thermochemical tables to accompany chemistry for science and engineering*, McGraw-Hill Ryerson Ltd., Toronto; 1989. These values have been calculated using $P = P^\circ = 101\,325$ Pa.

[b] Some additional values were obtained from Atkins, P., Physical chemistry, 6th edn, W.H. Freeman and Co., New York; 1998 and are calculated with $P = 1$ bar $= 100\,000$ Pa

**Appendix B.3** Mean bond enthalpies $\Delta H$ / kJ mol$^{-1}$ at 298 K[a].

|     | H   | C     | N     | O     | F   | Cl  | Br  | I   | S   | P   | Si  |
| --- | --- | ----- | ----- | ----- | --- | --- | --- | --- | --- | --- | --- |
| H   | 436 |       |       |       |     |     |     |     |     |     |     |
| C   | 412 | 348 s |       |       |     |     |     |     |     |     |     |
|     |     | 612 d |       |       |     |     |     |     |     |     |     |
|     |     | 838 t |       |       |     |     |     |     |     |     |     |
|     |     | 518 a |       |       |     |     |     |     |     |     |     |
| N   | 388 | 305 s | 163 s |       |     |     |     |     |     |     |     |
|     |     | 613 d | 409 d |       |     |     |     |     |     |     |     |
|     |     | 890 t | 946 t |       |     |     |     |     |     |     |     |
| O   | 463 | 360 s | 157   | 146 s |     |     |     |     |     |     |     |
|     |     | 743 d |       | 497 d |     |     |     |     |     |     |     |
| F   | 565 | 484   | 270   | 185   | 155 |     |     |     |     |     |     |
| Cl  | 431 | 338   | 200   | 203   | 254 | 242 |     |     |     |     |     |
| Br  | 366 | 276   |       |       |     | 219 | 193 |     |     |     |     |
| I   | 299 | 238   |       |       |     | 210 | 178 | 151 |     |     |     |
| S   | 338 | 259   |       |       | 496 | 250 | 212 |     | 264 |     |     |
| P   | 322 |       |       |       |     |     |     |     |     | 201 |     |
| Si  | 318 |       | 374   | 466   |     |     |     |     |     |     | 226 |

[a] s, Single bond; d, double bond; t, triple bond; a, aromatic bond. Source: Atkins, P., *Physical chemistry*, 6th edn, W.H. Freeman and Co., New York; 1998.

**Appendix B.4** Dissociation constants for acids and bases in aqueous solution at 25°C$^a$.

| Acid | Protonated species | $K_a$ | $pK_a$ | Base | Deprotonated species | $K_b$ | $pK_b$ |
|---|---|---|---|---|---|---|---|
| Acetic acid | $CH_3COOH$ | $1.8 \times 10^{-5}$ | 4.75 | Acetate | $CH_3COO^-$ | $5.6 \times 10^{-10}$ | 9.25 |
| Aluminium (III) | $Al(H_2O)_6^{3+}$ | $7.2 \times 10^{-6}$ | 5.14 | Hydroxyaluminium (III) | $Al(OH)^{2+}$ | $1.4 \times 10^{-9}$ | 8.86 |
| Ammonium | $NH_4^+$ | $5.6 \times 10^{-10}$ | 9.25 | Ammonia | $NH_3$ | $1.8 \times 10^{-5}$ | 4.75 |
| Arsenic acid | $H_3AsO_4$ | $5.8 \times 10^{-3}$ | 2.24 | Dihydrogen arsenate | $H_2AsO_4^-$ | $1.7 \times 10^{-12}$ | 11.76 |
| Dihydrogen arsenate | $H_2AsO_4^-$ | $1.10 \times 10^{-7}$ | 6.96 | Hydrogen arsenate | $HAsO_4^{2-}$ | $9.1 \times 10^{-8}$ | 7.04 |
| Hydrogen arsenate | $HAsO_4^{2-}$ | $3.2 \times 10^{-12}$ | 11.50 | Arsenate | $AsO_4^{3-}$ | $3.1 \times 10^{-3}$ | 2.50 |
| Arsenious acid | $As(OH)_3$ | $5.1 \times 10^{-10}$ | 9.29 | Dihydrogen arsenite | $H_2AsO_3^-$ | $2.0 \times 10^{-5}$ | 4.71 |
| Boric acid | $B(OH)_3$ | $7.2 \times 10^{-10}$ | 9.14 | Borate | $B(OH)_4^-$ | $1.4 \times 10^{-5}$ | 5.86 |
| Carbon dioxide$^b$ | $CO_2$ | $4.5 \times 10^{-7}$ | 6.35 | Hydrogen carbonate | $HCO_3^-$ | $2.2 \times 10^{-3}$ | 7.65 |
| Hydrogen carbonate | $HCO_3^-$ | $4.7 \times 10^{-11}$ | 10.33 | Carbonate | $CO_3^{2-}$ | $2.1 \times 10^{-4}$ | 3.67 |
| Formic acid | $HCOOH$ | $1.8 \times 10^{-4}$ | 3.75 | Formate | $HCOO^-$ | $5.6 \times 10^{-11}$ | 10.25 |
| Hydrofluoric acid | $HF$ | $3.5 \times 10^{-4}$ | 3.46 | Fluoride | $F^-$ | $2.9 \times 10^{-11}$ | 10.54 |
| Hydrogen cyanide | $HCN$ | $4.9 \times 10^{-10}$ | 9.31 | Cyanide | $CN^-$ | $2.0 \times 10^{-5}$ | 4.69 |
| Hydrogen sulphate | $HSO_4^-$ | $1.0 \times 10^{-2}$ | 2.00 | Sulphate | $SO_4^{2-}$ | $1.0 \times 10^{-12}$ | 12.00 |
| Hydrogen sulphide | $H_2S$ | $1.0 \times 10^{-7}$ | 7.00 | Hydrogen sulphide ion | $HS^-$ | $1.0 \times 10^{-7}$ | 7.00 |
| Hydrogen sulphide ion | $HS^-$ | $1.1 \times 10^{-12}$ | 11.96 | Sulphide | $S^{2-}$ | $9.1 \times 10^{-3}$ | 2.04 |
| Hypochlorous acid | $HClO$ | $3.0 \times 10^{-8}$ | 7.52 | Hypochlorite | $ClO^-$ | $3.3 \times 10^{-7}$ | 6.48 |
| Iron (III) | $Fe(H_2O)_6^{3+}$ | $6.3 \times 10^{-3}$ | 2.19 | Hydroxyiron (III) | $FeOH^+$ | $1.6 \times 10^{-12}$ | 11.80 |
| Methylammonium | $CH_3NH_3^+$ | $2.2 \times 10^{-11}$ | 10.66 | Methylamine | $CH_3NH_2$ | $4.5 \times 10^{-4}$ | 3.34 |
| Nitrous Acid | $HNO_2$ | | | Nitrite | $NO_2^-$ | | |
| Phenol | $C_6H_5OH$ | $1.3 \times 10^{-10}$ | 9.89 | Phenate | $C_6H_5O^-$ | $7.7 \times 10^{-5}$ | 4.11 |
| Phosphoric acid | $H_3PO_4$ | $7.1 \times 10^{-3}$ | 2.15 | Dihydrogen phosphate | $H_2PO_4^-$ | $1.4 \times 10^{-12}$ | 11.85 |
| Dihydrogen phosphate | $H_2PO_4^-$ | $6.3 \times 10^{-8}$ | 7.20 | Hydrogen phosphate | $HPO_4^{2-}$ | $1.6 \times 10^{-7}$ | 6.80 |

(*Continued*)

**Appendix B.4** (*Continued*)

| Acid | Protonated species | $K_a$ | $pK_a$ | Base | Deprotonated species | $K_b$ | $pK_b$ |
|---|---|---|---|---|---|---|---|
| Hydrogen phosphate | $HPO_4^{2-}$ | $4.2 \times 10^{-13}$ | 12.38 | Phosphate | $PO_4^{3-}$ | $2.4 \times 10^{-2}$ | 1.62 |
| Silicic acid | $Si(OH)_4$ | $2.2 \times 10^{-10}$ | 9.66 | Trihydrogen silicate | $H_3SiO_4^-$ | $4.6 \times 10^{-5}$ | 4.34 |
| Sulphurous acid | $H_2SO_3$ | $1.72 \times 10^{-2}$ | 1.76 | Hydrogen sulphite | $HSO_3^-$ | $5.81 \times 10^{-13}$ | 12.24 |
| Hydrogen sulphite | $HSO_3^-$ | $6.43 \times 10^{-8}$ | 7.19 | Sulphite | $SO_3^{2-}$ | $1.56 \times 10^{-7}$ | 6.81 |

[a] Source: Most values are taken from Atkins, P., *Physical chemistry*, 6th edn, W.H. Freeman and Co., New York; 1998.
[b] Note the relation between aqueous carbon dioxide and carbonic acid (p. 256).

**Appendix B.5**  Standard redox potentials in aqueous solutions[a].

| Reduction half-reaction[b] | $E^o$ / V | $pE^o$ | $pE^o$ (w)[c] |
|---|---|---|---|
| $O_3 + 2H^+ + 2e^- \rightarrow O_2 + H_2O$ | +2.075 | +35.1 | +28.1 |
| $H_2O_2 + 2H^+ + 2e^- \rightarrow 2H_2O$ | +1.763 | +29.8 | +22.8 |
| $MnO_4^- + 4H_3O^+ + 3e^- \rightarrow MnO_2 + 6H_2O$ | +1.692 | +28.6 | +19.3 |
| $2HClO + 2H^+ + 2e^- \rightarrow Cl_2 + 2H_2O$ | +1.630 | +27.6 | +20.6 |
| $Cl_2 + 2e^- \rightarrow 2Cl^-$ | +1.396 | +23.6 | +23.0 |
| $Cr_2O_7^{2-} + 14H^+ + 6e^- \rightarrow 2Cr^{3+} + 7H_2O$ | +1.36 | +23.0 | +6.67 |
| $2NO_3^- + 12H^+ + 10e^- \rightarrow N_2 + 6H_2O$ | +1.25 | +21.1 | +12.7 |
| $O_3 + H_2O + 2e^- \rightarrow O_2 + 2OH^-$ | +1.24 | +21.0 | +28.0 |
| $MnO_2 + 4H^+ + 2e^- \rightarrow Mn^{2+} + 2H_2O$ | +1.230 | +20.8 | +6.80 |
| $O_2 + 4H^+ + 4e^- \rightarrow 2H_2O$ | +1.229 | +20.8 | +13.8 |
| $NO_3^- + 4H^+ + 3e^- \rightarrow NO + 2H_2O$ | +0.955 | +16.1 | +6.77 |
| $NO_3^- + 3H^+ + 2e^- \rightarrow HNO_2 + H_2O$ | +0.940 | +15.9 | −5.11 |
| $ClO^- + H_2O + 2e^- \rightarrow Cl^- + 2OH^-$ | +0.89 | +15.0 | +22.0 |
| $NO_3^- + 10H^+ + 8e^- \rightarrow NH_4^+ + 3H_2O$ | +0.882 | +14.9 | +6.15 |
| $NO_3^- + 2H^+ + 2e^- \rightarrow NO_2^- + H_2O$ | +0.837 | +14.2 | +7.15 |
| $Fe^{3+} + e^- \rightarrow Fe^{2+}$ | +0.771 | +13.0 | +13.0 |
| $CH_2O + 4H + 4e^- \rightarrow CH_4 + H_2O$ | +0.411 | +6.94 | −0.06 |
| $O_2 + 2H_2O + 4e^- \rightarrow 4OH^-$ | +0.40 | +6.76 | +13.8 |
| $SO_4^{2-} + 8H^+ + 6e^- \rightarrow S + 4H_2O$ | +0.353 | +5.96 | −3.37 |
| $SO_4^{2-} + 9H^+ + 8e^- \rightarrow HS^- + 4H_2O$ | +0.248 | +4.20 | −3.75 |
| $CO_2 + 8H^+ + 8e^- \rightarrow CH_4 + 2H_2O$ | +0.170 | +2.87 | −4.13 |
| $2H^+ + 2e^- \rightarrow H_2$ | 0.00 | 0.00 | −7.00 |
| $CO_2 + 4H^+ + 4e^- \rightarrow \{CH_2O\} + H_2O$ | −0.071 | −1.20 | −8.20 |
| $S + 2e^- \rightarrow S^{2-}$ | −0.50 | −8.45 | −8.45 |
| $2H_2O + 2e^- \rightarrow H_2 + 2OH^-$ | −0.828 | −14.0 | −7.00 |

[a] Note: $E^o \div 0.0591 = pE^o$. For further detailed information refer to Chapter 10, eqn 10.44. Sources: Harris, D.C., *Quantitative chemical analysis*, 4th edn, W.H. Freeman and Co., New York; 1995 and Stumm, W. and J.J. Morgan, *Aquatic chemistry*, John Wiley and Sons, New York; 1981.
[b] For simplicity all equations use $H^+$ as a substitute for the hydronium ion $H_3O^+$.
[c] An explanation of $pE^o$ (w) is provided in Chapter 15.

**Appendix C.1**  Fundamental constants.

| | | |
|---|---|---|
| Avogadro's number | $N_A$ | $6.0221367 \times 10^{23}\,mol^{-1}$ |
| Boltzmann's constant | $k$ | $1.38066 \times 10^{-23}\,J\,K^{-1}$ |
| Faraday's constant | $F$ | $9.6485309 \times 10^4\,C\,mol^{-1}$ |
| Gas constant | $R$ | $8.314510\,J\,K^{-1}\,mol^{-1}$ $0.082057\,L\,atm\,K^{-1}\,mol^{-1}$ |
| Planck's constant | $h$ | $6.6260755 \times 10^{-34}\,J\,s$ |
| Speed of light (vacuum) | $c$ | $2.99792458 \times 10^8\,m\,s^{-1}$ |
| Standard pressure | $P^o$ | $1.01325 \times 10^5\,Pa$ |

**Appendix C.2**  SI prefixes and fundamental geometric relations.

| | | |
|---|---|---|
| atto | a | $10^{-18}$ |
| femto | f | $10^{-15}$ |
| pico | p | $10^{-12}$ |
| nano | n | $10^{-9}$ |
| micro | μ | $10^{-6}$ |
| milli | m | $10^{-3}$ |
| centi | c | $10^{-2}$ |
| deci | d | $10^{-1}$ |
| hecto | h | $10^{2}$ |
| kilo | k | $10^{3}$ |
| mega | M | $10^{6}$ |
| giga | G | $10^{9}$ |
| tera | T | $10^{12}$ |
| peta | P | $10^{15}$ |
| exa | E | $10^{18}$ |

Circumference of a circle $= 2\pi r = \pi d$
Surface area of a circle $= \pi r^2$
Surface area of a sphere $= 4\pi r^2$
Volume of a sphere $= \dfrac{4}{3}\pi r^3$

# Appendix D.1  **Introductory instructions for *PHREEQC* for *Windows***

The following paragraphs briefly describe how *PHREEQC for Windows,* a computer program employed for calculating chemical equilibrium systems can be used. *PHREEQC for Windows* is available free, by download, from the US Geological Survey's website.[1] Experience with computer programming, in particular the Basic computer language, will help.

Once you have acquired and properly installed the software, you are now ready to run the program. Start the program, and a blank *Input* page will appear on the screen. You may begin by either i) typing new text (code) using the Basic language (some help is provided on the right side of the screen) or ii) recommended for first time users, by loading an existing *example* file as described next.

In the upper left corner, select 'file' and then 'open'. Navigate to your own subdirectory or the subdirectory *Examples* that comes with the program. This subdirectory contains a number of example files (*.phrq files) that can be used to learn how the program works. Select one of the files and then 'open'. The *Input* screen will now contain the 'code of instructions' for the specified calculation. For example, file *ex2* first provides the title for the calculation, in this case '*The temperature dependence of solubility of gypsum and anhydrite*', and then the rest of code follows.

---

[1] http://wwwbrr.cr.usgs.gov/projects/GWC_coupled/phreeqc/, accessed September 2009.

The tab *Database* contains information from the database file Phreeqc.dat. This Default database is quite extensive, but can also be modified for specific applications. The tabs *Grid* and *Chart* are unused at this time.

You are now ready to begin the calculation. Select 'Calculations' from the main menu and choose one of the following options, as appropriate.

| | |
|---|---|
| Start | – to start the calculation |
| Debug | – to check your Input file for any errors |
| Files | – Input, Output, and Database - choose the location of where to save Output files, and where to obtain the Input and Database files |
| Gram formula weight | – Molecular weight calculator, provide the formula, it returns the formula weight in g mol$^{-1}$ |

It would be good to first check the file locations to ensure they are being directed to and from the proper subdirectories. Then select 'Debug' to ensure there are no errors in the program code and finally, when ready, select 'Start'.

When the calculation is finished, a new window will be open on the computer screen, showing the locations of the Input, Output, and Database files, along with the time the calculation took to complete. Select 'Done' and a new *Output* tab (with your results) will open. In this example, *Output* file is rather extensive and its content should be carefully investigated. Also, the *Grid* tab also contains the saturation index data that is plotted. The graph, shown under the *Chart* tab is, *Saturation Index—plotted against—Increasing Temperature*, for both gypsum and anhydrite.

Many other examples are provided that demonstrate more completely some of the ability of this program.

# Index

Page numbers in *italics* refer to illustrations and tables.